FLAME RETARDANT POLYMER NANOCOMPOSITES

BICENTENNIAL 1807 WILEY 2007 BICENTENNIAL

THE WILEY BICENTENNIAL–KNOWLEDGE FOR GENERATIONS

Each generation has its unique needs and aspirations. When Charles Wiley first opened his small printing shop in lower Manhattan in 1807, it was a generation of boundless potential searching for an identity. And we were there, helping to define a new American literary tradition. Over half a century later, in the midst of the Second Industrial Revolution, it was a generation focused on building the future. Once again, we were there, supplying the critical scientific, technical, and engineering knowledge that helped frame the world. Throughout the 20th Century, and into the new millennium, nations began to reach out beyond their own borders and a new international community was born. Wiley was there, expanding its operations around the world to enable a global exchange of ideas, opinions, and know-how.

For 200 years, Wiley has been an integral part of each generation's journey, enabling the flow of information and understanding necessary to meet their needs and fulfill their aspirations. Today, bold new technologies are changing the way we live and learn. Wiley will be there, providing you the must-have knowledge you need to imagine new worlds, new possibilities, and new opportunities.

Generations come and go, but you can always count on Wiley to provide you the knowledge you need, when and where you need it!

WILLIAM J. PESCE
PRESIDENT AND CHIEF EXECUTIVE OFFICER

PETER BOOTH WILEY
CHAIRMAN OF THE BOARD

FLAME RETARDANT POLYMER NANOCOMPOSITES

Edited by

Alexander B. Morgan
University of Dayton Research Institute
Dayton, Ohio

Charles A. Wilkie
Marquette University Department of Chemistry
Milwaukee, Wisconsin

BICENTENNIAL 1807 WILEY 2007 BICENTENNIAL

WILEY-INTERSCIENCE

A JOHN WILEY & SONS, INC., PUBLICATION

Published by John Wiley & Sons, Inc., Hoboken, New Jersey.
Published simultaneously in Canada.

Library of Congress Cataloging-in-Publication Data:

Morgan, Alexander B.
Flame retardant polymer nanocomposites / Alexander B. Morgan, Charles A. Wilkie.
p. cm.
Includes index.
ISBN 978-0-471-73426-0 (cloth)
1. Fire resistant polymers. 2. Nanostructured materials. 3. Polymeric composites. I. Wilkie; C. A. (Charles A.) II. Title.
TH1074.5.M67 2007
628.9′223—dc22

2006024023

10 9 8 7 6 5 4 3 2 1

CONTENTS

CONTRIBUTORS

Günter Beyer, Kabelwerk Eupen AG, Malmedyer Strasse 9, B-4700 Eupen, Belgium

Serge Bourbigot, Laboratoire Procédés d'Élaboration des Revêtements Fonctionnels, LSPES UMR/CNRS 8008, École Nationale Supérieure de Chimie de Lille, F-59652 Villeneuve d'Ascq Cedex, France

Sophie Duquesne, Laboratoire Procédés d'Élaboration des Revêtements Fonctionnels, LSPES UMR/CNRS 8008, École Nationale Supérieure de Chimie de Lille, F-59652 Villeneuve d'Ascq Cedex, France

Jeffrey W. Gilman, National Institute of Standards and Technology, Gaithersburg, MD 20899-8665

M. J. Heidecker, Materials Science and Engineering Department, Pennsylvania State University, University Park, PA 16802

A. Richard Horrocks, Fire Materials Laboratory, Centre for Materials Research and Innovation, University of Bolton, BL3 5AB Bolton, UK

Yuan Hu, State Key Lab of Fire Science, University of Science and Technology of China, Hefei, 230026 Anhui, China

Baljinder K. Kandola, Fire Materials Laboratory, Centre for Materials Research and Innovation, University of Bolton, BL3 5AB Bolton, UK

Takashi Kashiwagi, Fire Research Division, National Institute of Standards and Technology, Gaithersburg, MD 20878-8665

Sergei V. Levchik, Supresta U.S. LLC, 430 Saw Mill River Road, Ardsley, NY 10502

E. Manias, Materials Science and Engineering Department, Pennsylvania State University, University Park, PA 16802

Alexander B. Morgan, University of Dayton Research Institute, Nonmetallic Materials Division, 300 College Park, Dayton, OH 45429

H. Nakajima, Materials Science and Engineering Department, Pennsylvania State University, University Park, PA 16802

Marc R. Nyden, National Institute of Standards and Technology, Gaithersburg, MD 20899-8665

G. Polizos, Materials Science and Engineering Department, Pennsylvania State University, University Park, PA 16802

Bernhard Schartel, Federal Institute for Materials Research and Testing, BAM Unter den Eichen 87, 12205 Berlin, Germany

Lei Song, State Key Lab of Fire Science, University of Science and Technology of China, Hefei, 230026 Anhui, China

Stanislav I. Stoliarov, SRA International, Egg Harbor Township, NJ 08234

Charles A. Wilkie, Marquette University, Department of Chemistry, Milwaukee, WI 53201

Mauro Zammarano, Building and Fire Research Laboratory, National Institute of Standards and Technology, Gaithersburg, MD 20899-8665; NIST Guest Researcher from CimTecLab, Area Science Park, 34012 Trieste, Italy

PREFACE

Since the early 1990s, the subject of polymer nanocomposites has expanded greatly, to its current status as a major field of polymer materials research. It is now realized that polymer nanocomposites, as a class of materials, were in use long before this field of research was officially named in the early 1990s. Indeed, work published as early as 1961, and patents going back to the 1940s, have shown that layered silicates (or clays) can be combined with polymers in low amounts to produce new materials with greatly improved properties. However, it was the work in the 1990s that properly identified these clay-containing materials as polymer nanocomposites and kindled today's interest in these materials. One could argue that polymer nanocomposites are just part of the nanotechnology boom, but there is more to it. The fundamental understanding of how two dissimilar materials interface at the nanometer scale has tremendous implications for performance and properties at the macro scale. Therefore, the study of polymer nanocomposites is not just about capturing the buzz from nanotechnology; it is about understanding structure–property relationships and interfacial science at the molecular and macromolecular scale.

With the recent understanding that the addition of clays or other nanoparticles to a polymer forms a polymer nanocomposite, these materials have been investigated for many potential applications. One of the first well-publicized commercial uses was in polyamide-6 [poly(hexamethylamide) or nylon-6] for automotive applications developed by Toyota. Specifically, the improved heat distortion temperature of the nanocomposite allowed it to be used as part of the engine, resulting in a weight savings in a car. Additional early applications for nanocomposite technology have included improved gas barrier properties (beverage and food packaging), electrical conductivity for electromagnetic applications, and improved mechanical strength and toughness for engineering use. Flammability applications for polymer clay nanocomposites were discovered a little later, and only recently has the material found its way into commercial use. Polymer nanocomposites for flammability applications are attractive because the formation of a nanocomposite not only improves the fire properties but can also improve other properties (e.g., mechanical properties), and it has the potential to bring true multifunctionality to materials.

Multifunctionality has the great potential to simplify materials science and engineering by having one material do the work of several. For example, a plastic case for an electronic device can have several requirements. It will require particular mechanical properties (e.g., modulus, impact strength), thermal properties (not melt or sag under normal use conditions), flammability properties (meet regulations depending on the fire risk scenario), and electromagnetic properties (frequency shielding). Also, cost, density, color, and recyclability will need to be considered if it is a commercial product. With such a long list of requirements, it can be very difficult to find one material that can meet all needs. For example, polycarbonate can be used to achieve the desired mechanical and thermal properties, and with the right additives, flammability, density, and color can be obtained as well. For cost-effectiveness, polycarbonate is usually mixed with acrylonitrile–butadiene–styrene terpolymer in consumer electronics. Another feature not often obtained in the casing for electronic devices is electromagnetic shielding. To obtain this shielding requirement, such as in the use of a laptop case, special paints are used, and not surprisingly, this solution increases cost, limits color choices, and can make recycling difficult. An acceptable combination of materials can be difficult to find during typical research and development operations, and frequently, the choice made by the engineer is a compromise that can lead to other problems. If just one material could meet all requirements, fabrication of parts and goods would become easier, costs might decrease, and innovation could be enabled. The class of materials that has the greatest chance of obtaining true multifunctionality is that of polymer nanocomposites.

Polymer nanocomposites have shown great improvements over traditional composites in mechanical, thermal, gas barrier, conductivity, flammability, electromagnetic shielding, and other properties, and this has spawned a huge amount of research. There are already several key references and books that look at the polymer nanocomposite field as a whole, and even focus on particular areas, but no book to date has focused on the improvements in materials flammability. As indicated previously, it has only recently been understood that the nanocomposite structure is responsible for the improvements in material properties, especially flammability, and so only now is there enough research to warrant a book focused on polymer nanocomposite flammability. Significant changes in fire safety regulations and perceptions of existing flame retardant additives have served as catalysts for increased emphasis on polymeric material flammability reduction. This increased emphasis demands not only lowered flammability but also improvements in environmental impact for the final flame-retarded part, as well as maintaining the difficult balance of properties discussed previously.

Since a polymeric material can reduce flammability and improve mechanical and thermal properties and possibly other properties as well, there is a great deal of promise that polymeric nanocomposites will not just meet this need for flammability reduction, but also exceed it, thus providing fire safety and improved properties for a wide range of consumer goods.

This book focuses on polymer nanocomposites for flammability applications and includes supporting information important to this subject. The information is divided into sections for specific topic searching, and the book is divided into three parts to help those new to the fields of materials flammability research and polymer nanocomposites: theory and fundamentals, specific flame retardant systems, and current applications and future work.

On the subject of theory and fundamentals, there are five chapters: flammability fundamentals, nanocomposite fundamentals, the impact of nanocomposite formation on flammability, modeling of thermal degradation by fire, and the flammability of specific polymers.

The chapters on specific flame retardant systems are meant to serve as detailed sources of information, allowing the reader to gather essential facts on very specific flame retardant and polymer systems. Since flame retardant solutions can vary greatly depending on polymer chemistry and intended application or regulatory test, it can be difficult to organize all available knowledge on flame retardant nanocomposites. This information is organized by flame retardant classes; within each classification there is extensive discussion of the various combinations of nanocomposites with flame retardants as solutions. The chapters are devoted to the combination of nanocomposite formation with intumescent systems, mineral additives, and halogen- and phosphorus-based fire retardants. The last chapter dealing with specific flame retardant systems focuses on thermoset flame retardant nanocomposites. This chapter is separated from the others because thermosets are prepared much differently than thermoplastics and behave quite differently under fire conditions.

The final chapters of the book are designed to show the newest advances in the field as well as to show practical uses for polymer nanocomposites in flammability application and to provide insight into the future direction of the field. Since the field of polymer nanocomposite research is rather new, new results are published regularly, including work with new types of nano-dimensional materials. The majority of work in this book refers specifically to polymer layered-silicate (clay) nanocomposites, but as shown in Chapter 10, results with carbon nanotubes, nanofibers, and colloidal inorganic particles that have shown reductions in flammability are reviewed. Chapter 11 focuses on the use of polymer nanocomposites for specific applications and their successes and pitfalls to date. In the last chapter what is known today is summarized and where the field is heading is indicated. This chapter can perhaps be viewed as a forward-looking statement concerning the types of work that must be carried out in the future. Some of the fundamental unknowns behind this technology are addressed in detail, showing the researcher ways to proceed for nanocomposite solutions to flammability issues.

We appreciate all of the efforts that the chapter authors have made to provide an up-to-date account of activities regarding the use of nanocomposites in flame retardancy. We trust that the book will be useful and that it will advance worldwide knowledge on this topic. We would like to thank Don Klosterman and Lynn Bowman of UDRI for their assistance in obtaining references on nano-reinforced composites and nanoparticle health and safety, respectively, and Dr. Anteneh Worku of the Dow Chemical Company for his assistance in obtaining references and reviewing. Finally, we thank our wives, Julie Ann Morgan and Nancy Wilkie, for their tireless support.

Dayton, Ohio
Milwaukee, Wisconsin
16 November 2006

ALEXANDER B. MORGAN
CHARLES A. WILKIE

ACRONYMS

POLYMERS

ABS	Acrylonitrile–butadiene–styrene copolymer
EVA	Poly(ethylene-co-vinyl acetate)
DGEBA	Diglycidyl ether of bisphenol A
HDPE	High-density polyethylene
LDPE	Low-density polyethylene
PA6	Polyamide-6
PA66	Polyamide-6,6
PA12	Polyamide-12
PAN	Polyacrylonitrile
PBT	Poly(butylene terephthalate)
PC	Polycarbonate
PCL	Polycaprolactone
PDMS	Poly(dimethyl siloxane)
PE	Polyethylene
PE-g-MA	Polyethylene-graft-maleic anhydride
PET	Poly(ethylene terephthalate)
PLA	Poly(lactic acid)
PMMA	Poly(methyl methacrylate)
POM	Poly(oxymethylene)
PP	Polypropylene
PP-g-MA	Polypropylene-graft-maleic anhydride
PS	Polystyrene
PTFE	Poly(tetrafluoroethylene)
PU	Polyurethane

PVC	Poly(vinyl chloride)
SAN	Styrene–acrylonitrile copolymer
SBS	Styrene–butadiene–styrene copolymer
TPU	Thermoplastic polyurethane

FLAME RETARDANTS

AO	Antimony oxide
APP	Ammonium polyphosphate
ATH	Aluminum hydroxide (also known as alumina trihydrate)
BFR	Bromine-containing flame retardant
CPW	Chlorinated paraffin wax
DB	Decabromodiphenyl oxide
DOPO	9,10-dihydro-9-oxa-10-phosphaphenanthrene-10-oxide
MCA	Melamine cyanurate
MH	Magnesium hydroxide
MPP	Melamine polyphosphate
NFR	Nitrogen-containing flame retardant
PER	Pentaerythritol
PFR	Phosphorus-containing flame retardant
RDP	Resorcinol diphosphate
TCP	Tricresylphosphate
TPP	Triphenylphosphate
TXP	Trixylylphosphate

CONE CALORIMETER/FLAMMABILITY MEASUREMENTS

FIGRA	Fire growth rate
HRR/RHR	Heat release rate/rate of heat release
LOI	Limiting oxygen index
MLR	Mass loss rate
SEA	Specific extinction area
THR/THE	Total heat release/total heat evolved
t_{ign}/TTI/t_{ig}	Time to ignition
UL-94	Underwriter's Laboratory Test #94
VSP	Volume of smoke production

NANOCOMPOSITE ANALYSIS TECHNIQUES

AFM	Atomic force microscopy
CP-MAS-NMR	Cross-polarization–magic angle spinning–nuclear magnetic resonance
DMA	Dynamic mechanical analysis
DSC	Differential scanning calorimetry
DTA	Derivative of TGA curve

NMR	Nuclear magnetic resonance
SEM	Scanning electron microscopy
TEM	Transmission electron microscopy
TGA	Thermogravimetric analysis
XRD	X-ray diffraction

NANOPARTICLES/NANOCOMPOSITE TERMINOLOGY

CNF/VGNCF	Carbon nanofiber/Vapor grown carbon nanofiber
CNT	Carbon nanotubes
FSM	Fluorinated synthetic mica
GO	Graphite oxide
LDH	Layered double hydroxide
MMT	Montmorillonite
MWNT/MWCNT	Multiwall carbon nanotubes
o-MMT/OMMT	Organically modified montmorillonite
PLS/PLSN	Polymer layered-silicate/Polymer-layered silicate nanocomposite
POSS	Polyhedral oligomeric silasesquioxanes
SWNT/SWCNT	Single-wall carbon nanotubes

1

INTRODUCTION TO FLAME RETARDANCY AND POLYMER FLAMMABILITY

SERGEI V. LEVCHIK
Supresta U.S. LLC, Ardsley, New York

1.1 INTRODUCTION

Together with numerous advantages that synthetic polymeric materials provide to society in everyday life, there is one obvious disadvantage related to the high flammability of many synthetic polymers. Polymers are used in manufacturing not only bulk parts but also films, fibers, coatings, and foams, and these thin objects are even more combustible than molded parts.

Fire hazard is a combination of factors, including ignitability, ease of extinction, flammability of the volatile products generated, amount of heat released on burning, rate of heat release, flame spread, smoke obscuration, and smoke toxicity, as well as the fire scenario.[1–3] Fire fatalities are usually reported as resulting from the lethal atmosphere generated by fires. Carbon monoxide concentrations measured in real fires can reach up to 7500 ppm,[4] which would probably result in a loss of consciousness in 4 minutes.[3] Other components of acute toxicity found in real fires play a secondary role: Hydrogen cyanide was measured at levels between 5 and 75 ppm, and for irritants such as hydrogen chloride and acrolein, 1 to 280 and 0.3 to 15 ppm were found, respectively.[4]

A recent statistical study covering almost 5000 fatalities showed that the vast majority of fire deaths are attributable to carbon monoxide poisoning, which results in lethality at concentrations much lower than believed previously.[5] Moreover, the same study showed that blood carbon monoxide loadings in fire

Flame Retardant Polymer Nanocomposites, edited by Alexander B. Morgan and Charles A. Wilkie

victims did not change significantly with the advent of synthetic polymers. Carbon monoxide yields (but not concentrations) in big fires are almost independent of the chemical composition of the material burning.[6] There is evidence suggesting[7] that there may be longer-term effects from exposure to fire atmospheres that are currently not completely understood.

According to fire statistics, more than 12 million fires break out every year in the United States, Europe, Russia, and China, killing some 166,000 people and injuring several hundreds of thousands. Although calculating the direct worldwide losses and costs of fire is difficult, $500 million is an estimate based on some national data.[8] Despite the increased use of synthetic polymers, U.S. residential fire deaths have declined steadily over the years, from about 6000 in 1977 to about 3500 in 1993, even though the population has increased.[9] Although fire problems are less severe now, U.S. fire casualties are still higher than in most developed nations.[10] The decrease in the rate of casualties is a result of many factors, including better design of appliances, electronic equipment, cars, heating equipment, houses, and so on, and ending with changes in the habits of people, such as a drop in the smoking population. The role of flame retardant polymeric materials is also a very important contributor.

In 1988, the National Bureau of Standards [now the National Institute of Standards and Technology (NIST)] ran room combustion tests comparing flame retardant with non–flame retardant plastics used in printed wiring boards, television set and business machine enclosures, cables, and upholstered furniture.[11] The results showed that flame retardant materials allow more than a 15-fold longer escape time, 75% less heat release, significantly less smoke, and a lower concentration of toxic gases. Fire retardants decrease toxicity in fires. The effect is due to a decrease in the amount of burning material.[1]

Statistical analysis shows that the fire fatality rate in the UK is much lower than that in the United States for fires where upholstered furniture is the item first ignited. The decrease in fire fatalities per capita in the UK was very rapid during the first decade following passage of UK fire safety regulations on upholstery, and is continuing. The U.S. fire fatality rate for the same types of fires has been decreasing much more slowly.[12] The Consumer Products Safety Council (CPSC) in the United States is in the final stage of introducing federal standards for upholstered furniture and mattresses, which should increase fire safety in homes in the United States and bring them into line with the UK.

In 1998, the fire safety of television sets and computer monitors manufactured in various countries was studied by a group of flame retardant experts associated with the European Chemical Industry Council. Various ignition sources were utilized, from simulation of a household candle to a trash basket full of paper. The results showed that TV sets purchased in Germany and the Nordic countries ignited easily, even with the smallest ignition source. Normally, these sets did not contain any flame retardant, in order to pass "green" labeling, or contained minimal amounts of flame retardant, to meet the European IEC 60065 test. In contrast, TV sets purchased in the United States or Japan, which were designed to meet UL-1410 or UL-1950 (analogous to IEC 60950) tests, were self-extinguishing even after exposure to a more severe ignition source.

It is clear that flame retardants are an important part of polymer formulations for applications in which polymers have a significant chance of being exposed to an ignition source (electrical and electronic goods), where polymers are easy ignitable (upholstered furniture), or where fast spread of a fire may cause serious problems (associated with building materials and transportation) when evacuating people. This chapter provides a short introduction to the principles of polymer combustion and a short overview of the mechanisms of action of the major classes of commercial flame retardants. Although intended to be especially useful for people new to these topics, experts may also find some new information.

1.2 POLYMER COMBUSTION AND TESTING

In many respects the combustion of polymers is similar to the combustion of many other solid materials; however, the tendency of polymers to spread flame away from a fire source is critical because many polymers melt and tend to produce flammable drips or flow. Therefore, it is always important to test the combustability of polymeric products under conditions close to those of the final applications or even in assembly with other materials. For example, flame spread can be measured in both the vertical and horizontal positions, but for almost all plastic materials the vertical test is more severe than the horizontal.[11]

1.2.1 Laboratory Flammability Tests

Flammability of polymers is assessed primarily through ignitability, flame spread, and heat release. Depending on the application of the polymeric material, one or more of these flammability criteria should be measured in appropriate flammability tests. Numerous flammability tests are known and are performed either on representative samples or on an assembled product. Tests can be small, intermediate, or full scale. Although similar trends in the rating of materials can be found based on small- and large-scale tests, in general there is no direct correlation between these tests.

International and national standards have been developed based on various flammability tests, and they are reviewed elsewhere.[13] Some relatively simple and inexpensive laboratory tests have found broad application. These tests are used primarily in industrial laboratories for screening of materials during product development or quality control, or in the academic community for studies of polymer flammability. In this chapter we describe some of the commonly used laboratory test methods.

Underwriters' Laboratories UL-94 test is designed to assess the "flammability of plastic materials for parts in devices and appliances." The test measures ignitability and flame spread of polymeric materials exposed to a small flame. It is accepted for standardization in many countries and also internationally. Five classifications are included in this test, but we introduce only the V-0, V-1, and V-2 classifications, because they are cited most often in the flame retardant literature. To assess this classification, a bar shape specimen of plastic 120 × 13 mm is positioned vertically and held from the top. Depending on the intended use of the plastic, bars may be 3.2, 1.6, or 0.8 mm thick. Thinner specimens are usually more flammable. Some surgical cotton is placed 300 mm below the specimen to detect combustible drips that will ignite the cotton. A Bunsen burner flame (ca. 19 mm high; calibrated) is applied to the specimen twice (10 s each). After each application the time of self-sustained combustion is recorded. A second application of the flame follows immediately after self-extinguishment of the specimen in the first application. A V-0 classification is given to material that is extinguished in less than 10 s after any flame application. The mean combustion time for the five specimens tested (10 flame applications) should not exceed 5 s, and no combustible drips can be observed. A V-1 classification is received by a sample with maximum combustion time < 50 s and mean combustion time for five specimens < 25 s. No combustible drips should be observed. The sample is classified V-2 if it satisfies the combustion time criteria of V-1, but flammable drips igniting the cotton are allowed.

Another test commonly used in laboratory practice is the limiting oxygen index (LOI) test. This method has been included in some national and international standards (e.g., ASTM D2863 and ISO 4589). The specimen size and shape is not strictly specified in the LOI test, but bars of about 100 × 65 × 3 mm are generally used when testing rigid plastics. The specimen is positioned vertically in a glass chimney and is held from the bottom. The chimney is purged continuously with a mixture of nitrogen and oxygen. The flame of a Bunsen burner is applied to the top of the specimen until the entire surface is ignited. If the specimen did not ignite after 30 s, the concentration of oxygen is increased. Ideally, the specimen should show stable candlelike combustion. If the specimen continues burning more than 3 min after removal of the ignition source or if more than 5 cm of the length of the sample is consumed, a new specimen should be installed and tested at a lower oxygen concentration. The LOI value is the limiting concentration of oxygen at which the sample tested self-extinguishes in less than 3 min with less than 5 cm of the material consumed. The LOI test does not represent a real fire scenario, but it is good as a screening tool because it gives a numerical value instead of a discrete classification (e.g., V-0, V-1, V-2).

The cone calorimeter test is a bench-scale (medium-sized) test developed at NIST[14] which quickly gained popularity in the academic community as well as for standardization purposes (e.g., ISO 5660-1, ASTM E-1354). It is also used as a tool for fire protection engineering because it allows prediction of large-scale test results. A cone calorimeter measures consumption of oxygen from a burning sample 100 × 100 mm in area and up to 50 mm thick. The heat release is calculated from the oxygen consumption data. The specimen is exposed to a constant heat flux from a conical-shaped irradiation source, which serves to simulate a variety of fire scenarios. The combustion is initiated by a small sparking igniter, which ignites gases evolved from the heated specimen. In addition to the heat release rate, the cone calorimeter apparatus can monitor time to ignition, weight loss of a sample during combustion, rate of smoke generation, carbon monoxide, carbon dioxide, and optionally, some corrosive gases, such as HCl and HBr.

1.2.2 Polymer Combustion

All polymer fires start with an ignition event, where a source of heat comes into contact with a fuel generated by the heating of the polymer. This event initiates a flow of flammable degradation products, which react with oxygen from the air to produce a flame and heat. Some of the heat is transferred back to the surface of the fuel, maintaining the flow of flammable volatile degradation products.[3] Low ignitability of the polymers is the first line of defense against fire. Although all organic polymers do ignite, the higher the temperature that a material has to reach before it ignites, the safer it is. For most materials, the ignition temperature is in the range 275 to 475°C. Ignitability is assessed via time to ignition or minimum heat input for ignition. Fire performance improves if either of these increases.[1,15]

Ignitability depends to a large extent on how quickly the surface can be raised to the ignition temperature. Special consideration has to be given to polymers that melt before thermal decomposition. Usually, at a low heat exposure, melting precedes ignition and the polymer can flow or drip, removing heat from the surface. This phenomenon is beneficial for flame retardancy of uncharrable polymers. On the other hand, at a higher heat exposure, ignition may occur before the surface is heated to sufficient depth for the melted material to flow, and such polymers may ignite relatively easy.

Polymeric foamed materials are very specific in terms of ignitability and flame spread. It has been shown that differences in the surface area of foamed polymers and cell size have a larger effect on flammability than do density or differences in chemical structure.[16] The chemical structure, of course, may dictate the surface area or porosity in the formation of foam. For example, flexible polyurethane foams can be ignited by a smoldering cigarette. A textile material normally used to enclose the foam, as is common in upholstered furniture and mattresses, actually helps ignition if suitable flame retardant–treated textiles are not used.

The possibility of extinguishing a polymer flame depends on the mechanism of thermal decomposition of the polymer. Whereas ignition of a polymer correlates primarily with the initial temperature of decomposition, steady combustion is related to the tendency of the polymer to yield a char, which is produced at the expense of combustible volatile fragments. Therefore, the dependence of steady combustion on the amount of char seems to be simple, and in an early study it was established that the oxygen index shows a very good correlation with the char yield.[17] In reality, char also serves as a physical barrier for heat flux from the flame to the polymer surface, as well as a diffusion barrier for gas transport to the flame.[18] Therefore, the contribution of the char can be more significant than is expected from a simple reduction in combustible gases.

Four general mechanisms are important for thermal decomposition of polymers: (1) random chain scission, in which the polymer backbone is randomly split into smaller fragments; (2) chain-end scission, in which the polymer depolymerizes from the chain ends; (3) elimination of pendant groups without breaking of the backbone; and (4) cross-linking.[19] Only a few polymers decompose predominantly through one mechanism; in many cases a combination of two or more mechanisms is in effect. For example, polyethylene and polypropylene tend primarily to decompose via random chain scission, which in the case of polyethylene is also accompanied by some cross-linking. Poly(methyl methacrylate) and polystyrene tend to depolymerize, poly(vinyl chloride) primarily undergoes elimination of pendant groups (dehydrochlorination), and polyacrylonitrile cross-links. In terms of flammability, random scission and depolymerization polymers are usually more flammable than polymers that cross-link or remove pendant groups. Cross-linking[20] leads to precursors of char and as a result, to lower flammability. Elimination of pendant groups results in double bonds, which can also give cross-links or lead to aromatization.

In general, polymers with aromatic or heterocyclic groups in the main chain are less combustible than polymers with an aliphatic backbone.[21] Polymers with short flexible linkages between aromatic rings tend to cross-link and char. These polymers are thermally stable and show relatively good flame retardancy. For example, bisphenol A–based polycarbonate, phenol formaldehyde resins, and polyimides are self-extinguishing and show either a V-2 or V-1 rating in the UL-94 test. On the other hand, polymers with relatively long flexible (aliphatic) linkages are still relatively combustible despite aromatics in the backbone. Examples of these polymers are poly(ethylene terephthalate), poly(butylene terephthalate), polyurethanes, and bisphenol A–based epoxy resin.

Charring of polymers proceeds through various stages: (1) cross-linking, (2) aromatization, (3) fusion of aromatics, and (4) graphitization.[18] The ability of a polymer to perform in one or several of these stages leading to char formation depends primarily on the polymer structure. However, this performance can be improved significantly by the use of flame retardants, which are discussed later in the chapter. Although many polymers tend to cross-link at early stages of thermal decomposition, this does not necessarily result in char formation. Char is formed only if the cross-linked polymer contains aromatic fragments and/or conjugated double bonds and is prone to aromatization during thermal decomposition.[20]

Fused aromatic rings in the char tend to assemble into small stacks, which are precursors of graphite. These pregraphitic domains are embedded in the amorphous char. This type of char, called *turbostratic char*, is usually formed at 600 to 900°C, temperatures typically found on the surface of burning polymers. Char that contains more pregraphitic domains is more stable to thermal oxidation and therefore less likely to burn away and expose the polymer surface to the heat of the flame. On the other hand, highly graphitized chars are rigid and may have cracks, which do not retard diffusion of combustible materials to the flame. The best-performing char would be amorphous uncracked char with a requisite pregraphitic domain content.

1.3 FLAME RETARDANCY

1.3.1 General Flame Retardant Mechanisms

Although flame retardants may differ from one another in terms of chemical structure, certain general mechanisms of action are applicable to various classes of flame retardants. The first line of separation normally distinguishes gas-phase-active and condensed-phase-active flame retardants. Gas-phase-active flame retardants act primarily through scavenging free radicals responsible for the branching of radical chain reactions in the flame. This is the chemical mechanism of action in the gas phase. Other flame retardants generate large amounts of noncombustible gases, which dilute flammable gases, sometimes dissociate endothermically, and decrease the temperature by absorbing heat. This slows combustion and may eventually result in extinguishment of the flame. This is the physical mechanism of action in the gas phase.

Condensed-phase mechanisms of action are more numerous than the gas-phase mechanisms. Charring, discussed briefly above, is the most common condensed-phase mode of action. Again, charring could be promoted either by chemical interaction of the flame retardant and the polymer or by physical retention of the polymer in the condensed phase. Charring could also be promoted by catalysis or oxidative dehydrogenation.

Some flame retardants show almost exclusively a physical mode of action. Examples are aluminum hydroxide and magnesium hydroxide. On the other hand, there is no single flame retardant that will operate exclusively through a chemical mode of action. Chemical mechanisms are always accompanied by one or several physical mechanisms, most commonly endothermic dissociation or dilution of fuel. Combinations of several mechanisms can often be synergistic.

1.3.2 Specific Flame Retardant Mechanisms

1.3.2.1 Halogen-Containing Flame Retardants Halogen-containing flame retardants represent the most diversified class of retardants.[22] To be effective, halogen-containing flame retardants need to release halogen in the form of radical or halogen halide at the same temperature range or below the temperature of decomposition of the polymer.[23,24] Theoretically, four classes of chemical compounds can be used as halogenated flame retardants: those containing fluorine, chlorine, bromine, or iodine. Fluorinated organics are normally more stable than any other polymers and do not release fluorine radicals or hydrogen fluoride. Nevertheless, there are a few examples of the commercial use of fluorinated flame retardants operating differently from all other halogenated flame retardants, and they will be discussed later. By contrast, iodinated organics have very low thermal stability and cannot be processed with most commercial polymers. In addition, fluorine and iodine are more expensive than chlorine or bromine, which also limits development of flame retardants based on these two halogens.

Chlorinated aromatic products are relatively stable and therefore not very efficient, but chlorinated aliphatic and cycloaliphatic flame retardants are well known. The chlorine content in some chlorinated paraffins can reach 70%, and some improved grades can be used in polyolefins and in high-impact polystyrene (HIPS).[25] A broad range of brominated flame retardants are commercially available. Brominated flame retardants help maintain a good balance of physical properties, such as good impact and tensile strength and a high heat distortion temperature. These flame retardants are generally suitable for many plastics; however, their principal use is in engineering plastics and epoxy resins.[26,27] In this case the emphasis is on aromatic products. Although aliphatic brominated flame retardants are often more efficient than aromatics, their use has been limited to certain polymers.[28] For similar structures there is usually a correlation between degree of bromination and thermal stability. Fully brominated aromatics have low volatility and are used in engineering resins with a relatively high processing temperature. Polymeric and oligomeric brominated aromatic flame retardants are also widely used. In addition to good thermal stability, they show better physical properties. One of the main disadvantages of many brominated aromatic flame retardants is their low resistance to ultraviolet (UV) light; however, there are specially designed commercial flame retardants that show good UV stability.

Figure 1.1 compares the flame retardant efficiency of aliphatic brominated flame retardant and aromatic brominated flame retardant. Because the thermal decomposition of the aliphatic flame retardant starts at temperatures below the thermal decomposition of polypropylene, it shows very good performance in polypropylene. In contrast, because the aromatic brominated fire retardant is significantly more stable, optimum debromination is not achieved at the temperature of decomposition of polypropylene, and this flame retardant shows inferior performance.

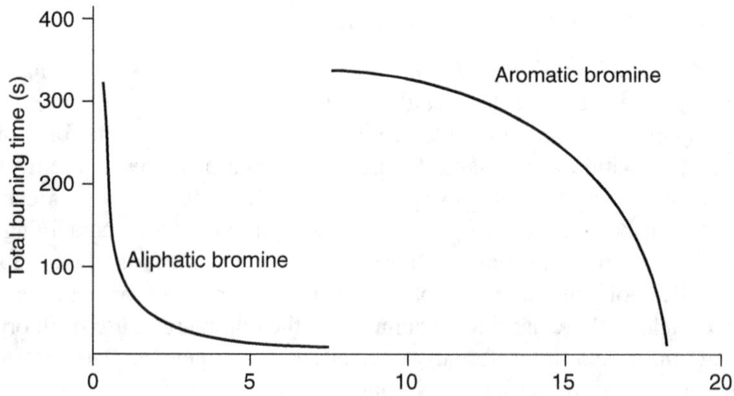

FIGURE 1.1 Dependence of total flaming time of polypropylene measured in a UL-94 test on bromine content for an aliphatic brominated flame retardant and an aromatic brominated flame retardant. (From Ref. 23, copyright © 2001, Routledge/Taylor & Francis Group, with permission.)

It is generally accepted that the main mechanism of flame retardant action of halogenated flame retardants is in the gas phase, and it is primarily the chemical mode of action. The reaction begins with the abstraction of halogen radical from the flame retardant. This halogen immediately abstracts hydrogen from either the flame retardant additive or the polymer. An example of such a sequence of reactions, with the participation of bromine and an aliphatic polymer, is

$$\text{R–Br} \longrightarrow \dot{\text{R}} + \dot{\text{Br}} \tag{1.1}$$

$$\dot{\text{Br}} + \text{CH}_2\text{–CH}_2 \longrightarrow \dot{\text{C}}\text{H–CH}_2 + \text{HBr} \tag{1.2}$$

$$\dot{\text{C}}\text{H–CH}_2 \xrightarrow{-\dot{\text{H}}} \text{CH=CH} \tag{1.3}$$

In the absence of a synergist, hydrogen halides volatilize and enter the flame. Hydrogen halides will quickly react with hydrogen or hydroxyl radicals and regenerate the halogen. Examples of such reactions with HBr are shown below in reactions (1.4) and (1.5). Further bromine radicals will react with hydrocarbons in the gas phase and regenerate HBr as shown in reaction (1.6), with the process repeating until bromine leaves the flame.

Atomic hydrogen and hydroxyl radicals are very important for sustaining combustion. The hydrogen radical is responsible for the chain-branching free-radical reactions in the flame [reaction (1.7)], whereas the hydroxyl radical is responsible for the oxidation of CO to CO_2 [reaction (1.8)], which is a highly exothermic reaction and is responsible for the larger part of the heat generation in the flame.

$$\text{HBr} + \dot{\text{H}} \longrightarrow \text{H}_2 + \dot{\text{Br}} \tag{1.4}$$

$$\text{HBr} + \dot{\text{O}}\text{H} \longrightarrow \text{H}_2\text{O} + \dot{\text{Br}} \tag{1.5}$$

$$\dot{\text{Br}} + \text{R}'\text{H} \longrightarrow \text{HBr} + \dot{\text{R}} \tag{1.6}$$

$$\dot{\text{H}} + \text{O}_2 \longrightarrow \dot{\text{O}}\text{H} + \text{O} \tag{1.7}$$

$$\text{O}\dot{\text{H}} + \text{CO} \longrightarrow \text{CO}_2 + \dot{\text{H}} \tag{1.8}$$

In some other reactions, the more reactive radicals ($H^\bullet$, $OH^\bullet$, $CH_3{}^\bullet$) are replaced by the less active $Br^\bullet$ radicals.[29] If $Br^\bullet$ meets $H^\bullet$ in the presence of a neutral molecule (third body), HBr is regenerated. It has been found by spectroscopy that the introduction of halogen-containing inhibitors into the flame clearly reduces the concentration of $H^\bullet$, $OH^\bullet$, and $HCO^\bullet$ radicals, whereas there is an increase in the content of the diradicals $C_2{}^{\bullet\bullet}$ and soot. As the concentration of inhibitor is increased, the flame temperature decreases. Small additions of halogen inhibitors (on the order of a few mol%) can reduce the rate of flame propagation up to 10-fold and have a marked effect on the ignition limits. On the other hand, halogens accelerate the formation of soot in the flame.

It is well established[21] that Sb_2O_3 is synergistic with halogen-containing flame retardants because it facilitates delivery of halogen atoms in the gas phase and prolongs residence of the halogens in the flame zone so that more "hot" radicals can be scavenged. Antimony trioxide reacts with HCl or HBr in the condensed phase, forming $SbCl_3$ or $SbBr_3$, respectively, both of which are relatively volatile. $SbCl_3$ boils at 223°C and $SbBr_3$ boils at 288°C. Halogenation of Sb_2O_3 may or may not proceed through a number of intermediate oxyhalides $Sb_nO_mX_p$, some of which can go into the gas phase as well.[30] It is also well established that Sb_2O_3 catalyses dehalogenation of the flame retardant,[31] so halogens can be moved into the flame at a lower temperature. In the flame, antimony trihalide (e.g., $SbBr_3$) is reduced step by step to metallic antimony [reaction (1.9)], which could be further oxidized by the oxygen [reaction (1.10)] or hydroxyl radical [reaction (1.11)].

$$SbBr_3 + \dot{H} \xrightarrow{-HBr} SbBr_2 + \dot{H} \xrightarrow{-HBr} SbBr + \dot{H} \xrightarrow{-HBr} Sb \tag{1.9}$$

$$Sb + O \longrightarrow SbO \tag{1.10}$$

$$Sb + \dot{O}H \longrightarrow SbOH \tag{1.11}$$

Interference with the antimony–halogen reaction will affect the flame retardancy of the polymer.[32] For example, metal cations from color pigments or an inert filler such as calcium carbonate or talc may lead to the formation of stable metal halides, rendering the halogen unavailable for reaction with antimony oxide. The result is that neither the halogen nor the antimony is transported into the vapor zone. Silicones have also been shown to interfere with the flame retardant action of halogenated flame retardants.

It is also believed that the large heat capacity of hydrogen halides and their dilution of the flame results in a decrease in the mass concentration of combustible gases and the temperature of the flame.[33] The physical effect of halogen halides is comparable to that of inert gases, CO_2, and water. There is no contradiction between the radical trap theory and the physical theory; apparently, they complement each other. The contribution of each mechanism depends on the temperature of decomposition of the flame retardant additive and the polymer.

As mentioned earlier, the halogen radicals evolved from the flame retardant in the condensed phase abstract the hydrogen from the polymer and produce unsaturation [reactions (1.2) and (1.3)]. The double bonds are known to be precursors of char formation through either cross-linking or aromatization.[18] If hydrogen is abstracted from the aromatic ring, this ring has a chance to couple with another ring and start forming polyaromatic structures, which are precursors of graphitic domains in the char. This char formation is an important condensed-phase contribution of halogen-based flame retardants,[34] which is often overlooked.

There is another condensed-phase mode of action that is specific for aliphatic bromine, and it is the opposite of char formation. Bromine radicals generated thermally at low temperature in the polymer melt can cause chain scission at tertiary C atoms.[35,36] Examples of polymers where this mechanism is operational are polystyrene (foams) and polypropylene (preferably thin parts, films, or fibers). The decreased molecular weight causes fast dripping of the hot polymer, which cools the flame and eventually extinguishes it:

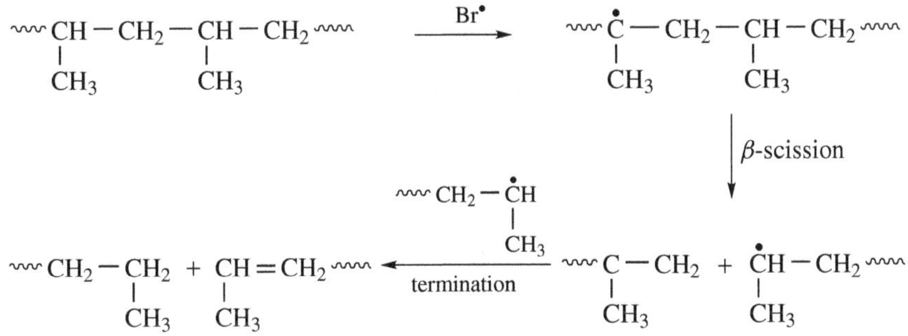

(1.12)

Poly(tetrafluoroethylene) (PTFE) is used at a very low level (0.01 to 0.5 wt%) in combination with other flame retardants to suppress flaming drips. The flame retardant action of PTFE is not related to any chemical reaction of fluorine or halogen fluoride. During polymer processing at 200 to 300°C, PTFE particles soften, the shear force of extrusion elongates the particles up to 500%, and microfibrils are formed. Upon combustion the microfibrils shrink back when the polymer melts and a network that prevents dripping is formed. This flame retardant action of PTFE is a physical phenomenon.

Potassium perfluorobutanesulfonate is added to polycarbonate at a low loading of 0.05 to 0.2 wt% which allows preservation of transparency and clarity of the polymer. Even at such a low concentration, perfluorobutanesulfonate provides a V-0 rating to the polycarbonate.[37] In this case the sulfonate group is primarily responsible for the flame retardant effect of the product, whereas the perfluorobutane group, due to its strong electron-withdrawing effect, increases the acidity of the sulfonate group. No contribution of hydrogen fluoride is known for potassium perfluorobutanesulfonate.

1.3.2.2 Phosphorus-Based Flame Retardants Phosphorus-based flame retardants are the second most widely used class of flame retardants. Recent efforts in the development of new flame retardants have shifted strongly toward phosphorus and other halogen-free systems. Among phosphorus-based flame retardants, one should distinguish (1) elemental red phosphorus, (2) inorganic phosphates, (3) numerous organic phosphorus-based products, and (4) chlororganophosphates. Although many phosphorus flame retardants exhibit general modes of action, there are specifics for each class mentioned above.

It is generally accepted that phosphorus flame retardants are significantly more effective in oxygen- or nitrogen-containing polymers, which could be either heterochain polymers or polymers with these elements in pendant groups. Effective phosphorus flame retardants are more specific than halogen-based products to certain polymers. This relates to the condensed-phase mechanism of action, where the phosphorus flame retardant reacts with the polymer and is involved in its charring.[21]

The flame retardancy of cellulose has been studied in great detail, which gave good insight for understanding the interaction of phosphorus flame retardants with polymers containing hydroxyl groups.[38] Phosphorus flame retardants, in the form of either acids derived from decomposition of ammonium phosphate salts or of phosphate esters, react (esterify or transesterify) with the hydroxyl groups of the cellulose.[39] Upon further heating, phosphorylated cellulose undergoes thermal decomposition and a significant amount of char is formed at the expense of combustible volatile products that would be produced by virgin cellulose. Some nitrogen-containing compounds, such as urea, dicyandiamide, and melamine, will accelerate phosphorylation of cellulose through formation of a phosphorus–nitrogen intermediate, and thus synergize the flame retardant action of phosphorus.[40] Phosphorus–nitrogen synergism is not a general phenomenon but depends on the structure of the phosphorus and nitrogen flame retardants as well as the polymer structure.

Similar to cellulose, phosphate esters can transesterify other polymers. For example, polycarbonates can undergo rearrangement during thermal decomposition, where phenolic OH groups are formed which then become the target for attack by aromatic phosphate esters[41] [reaction (1.13)]. Thus, phosphorus is grafted on the polymer chain. Char will be formed upon thermal decomposition of this grafted polymer. Similar phosphorylation chemistry was found for polyphenylene ether (PPE; a component of a PPE/HIPS blend), which also tends to rearrange upon heating and form phenolic OH groups.[42]

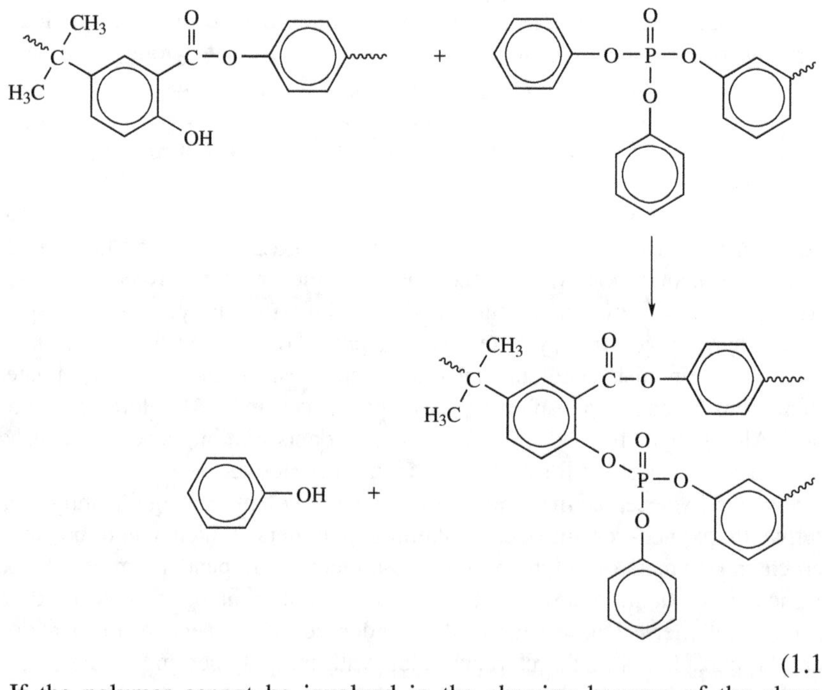

(1.13)

If the polymer cannot be involved in the charring because of the absence of reactive groups, a highly charring coadditive is used in combination with the phosphorus flame retardant. The coadditive is usually a polyol, which can undergo phosphorylation similar to that of cellulose; pentaerythritol is a typical example of such a polyol. Melamine can be used in conjunction with this system as well. These combinations of flame retardants are called *intumescent systems* because they form a viscous swollen char on the surface of the burning polymer. The char impedes the heat flux to the polymer surface and retards diffusion of volatile pyrolysis products to the flame. This mechanism of action is mostly physical because the polymer itself is not necessarily involved in the charring process, but its volatilization is retarded significantly. Intumescent systems for various polymers have been reviewed by Bourbigot et al.[43,44] The chemistry of formation of the intumescent chars was described thoroughly by Camino et al.[45,46]

Phosphorus flame retardants can remain in the solid phase and promote charring or volatilize into the gas phase, where they act as potent scavengers of $H^\bullet$ or $OH^\bullet$ radicals. Volatile phosphorus compounds are among the most effective inhibitors of combustion. A recent study showed[47] that phosphorus at the same molar concentration is, on average, five times more effective than bromine and 10 times more effective than chlorine. The mechanism of radical scavenging by phosphorus was suggested by Hastie and Bonnell.[48] The most abundant phosphorus radicals in the flame are $HPO_2^\bullet$, $PO^\bullet$, $PO_2^\bullet$, and $HPO^\bullet$, in decreasing order of significance. Some examples of radical scavenging with participation of $HPO_2^\bullet$ and $PO^\bullet$ radicals are shown in reactions (1.14) to (1.18). A third body is required in the reactions involving $PO^\bullet$ radicals.

$$HP\dot{O}_2 + \dot{H} \longrightarrow PO + H_2O \quad (1.14)$$

$$HP\dot{O}_2 + \dot{H} \longrightarrow PO_2 + H_2 \quad (1.15)$$

$$HP\dot{O}_2 + O\dot{H} \longrightarrow PO_2 + H_2O \quad (1.16)$$

$$P\dot{O} + \dot{H} + M \longrightarrow HPO + M \quad (1.17)$$

$$P\dot{O} + O\dot{H} + M \longrightarrow HPO_2 + M \quad (1.18)$$

If conditions are right, phosphorus-based molecules can volatilize and are oxidized, producing active radicals in the flame. On the other hand, phosphorus flame retardants tend to react with the polymer or to oxidize to phosphoric acid in the condensed phase. This favors mostly condensed-phase mechanisms. It is challenging to design a phosphorus-based flame retardant that will volatilize into the flame at relatively low temperatures but will not be lost during polymer processing.

Red phosphorus is the most concentrated source of phosphorus for flame retardancy. In fact, it is very effective in some polymers, such as thermoplastic polyesters or polyamides, where self-extinguishing UL-94 V-0 performance can be achieved at loadings of less than 10 wt%. Despite the apparent chemical simplicity of this additive, its mechanism of action is not completely understood. Most researchers agree[49,50] that in oxygen- or nitrogen-containing polymers, red phosphorus reacts with the polymer and induces char formation. Although there is a belief that red phosphorus is oxidized and hydrolyzed by water before it reacts with the polymer,[51,52] there is also strong evidence that red phosphorus can react directly with polyesters or polyamides in an inert atmosphere[50,53] and in the absence of moisture.[54] There is also some evidence in favor of a free-radical mechanism of interaction between red phosphorus and polyamide-6.[50] Red phosphorus shows relatively weak flame retardant effects in hydrocarbon polymers (e.g., polyolefins or polystyrene). It is believed[55] that in these polymers red phosphorus depolymerizes to white phosphorus, P_4, which volatilizes and provides gas-phase action.

Chloroalkyl phosphates, [e.g., tri(1,3-dichloroisopropyl) phosphate or tri(2-chloroisopropyl) phosphate or dichloroneopentyl tetrakis(2-chloroethyl) diphosphate] are used primarily in polyurethane foams. It would be logical to assume that chlorine and phosphorus both contribute to the flame retardant efficiency; however, this will depend on the configuration of the test (e.g., upward versus downward or horizontal combustion). The chloroalkyl phosphates are relatively volatile and tend to evaporate when heated with a flame. In downward combustion, the additive, in addition to evaporation, has a chance to react with the polymer, which provides a tarlike residue on the top of the foam,[56] whereas in upward combustion the additive quickly evaporates, yielding a high concentration of nonflammable flame retardant gases which extinguish the flame.[57]

1.3.2.3 Melamine Flame Retardants Melamine is a unique product with 67 wt% nitrogen in the molecule and fairly high thermal stability. Melamine also forms thermally stable salts with strong acids. Melamine itself, melamine cyanurate, melamine phosphate, melamine pyrophosphate, and melamine polyphosphate are commercially available for various flame retardant applications. The mechanism of flame retardant action of melamine is different from the mechanism of melamine salts or may be part of the mechanism of action of the salts. In addition, melamine phosphates have specific advantages because of the presence of phosphorus in the molecule.

Melamine is most commonly used in flexible polyurethane foams in combination with chloroalkyl phosphates and in intumescent coatings in combination with ammonium polyphosphate and pentaerythritol. Nevertheless, there is extensive patent literature on the use of melamine in thermoplastics and elastomers, which was reviewed by Weil and Choudhary.[58] The review gives good insight into the mechanism of flame retardant action of melamine. It is known that melamine does not melt, but sublimes at about 350°C (actual volatilization starts at a lower temperature). Upon sublimation, significant energy is absorbed, which decreases the surface temperature of the polymer. This is especially important for polyurethane foams having very low thermal inertia. In a hot flame, melamine may decompose further, with creation of cyanamid, which is also a very endothermic process.[58,59]

Upon heating, to whatever extent it does not sublime away, melamine can undergo progressive condensation with evolution of ammonia and formation of thermally stable condensates, known as *melam*, *melem*, and *melon*.[60] This reaction competes with melamine volatilization and is more pronounced if the volatilization of melamine is impeded: for example, by trapping in the charred polymer.

Formation of the residue is considered to be a condensed-phase contribution of melamine, whereas ammonia evolution dilutes the flame with noncombustible gases.

Upon heating, melamine-based salts dissociate, and re-formed melamine volatilizes in a manner similar to pure melamine. However, in the case of melamine salts, a larger portion of melamine undergoes progressive condensation than does pure melamine[61]; therefore, the condensed-phase contribution of the salts is larger. If the anion contains phosphorus, the phosphoric acid released will phosphorylate many polymers and produce a flame retardant effect similar to that of other typical phosphorus-based additives (see above). Melamine condensates and phosphoric acid react further at temperatures above 600°C, where triazine rings are opened and cross-linked. A $(PON)_x$ type of structure known as phosphorus oxynitride is formed.[62] Phosphorus oxynitride is very thermally stable and in some polymers can contribute to condensed-phase mechanisms.[63]

Melamine cyanurate is used primarily in unfilled polyamides.[64] Upon thermal decomposition, melamine is partially volatilized, whereas cyanuric acid catalyzes chain scission of polyamides. This leads to a decrease in melt viscosity and enhanced melt flow and dripping, which removes heat from the polymer and the polymer is extinguished.[65,66] The vaporizing melamine probably prevents drips from flaming. The fire retardant effect of melamine cyanurate deteriorates significantly in glass-filled polyamides because glass fibers prevent free melt flow.[67]

1.3.2.4 Inorganic Hydroxides Flame Retardants Inorganic hydroxides or mixed hydroxide–inorganic salts that can release water upon heating above 200°C can be used as flame retardants in many types of polymers. The two most commonly used products are aluminum hydroxide (ATH) and magnesium hydroxide (MH). In fact, ATH is, by weight, the largest commercially manufactured flame retardant, its main use being in wire and cable insulation and other elastomeric products, synthetic marble and synthetic onyx, latex for carpet back-coatings, phenolics, epoxies, and unsaturated polyesters.[68] An extensive review of manufacturing, properties, and uses of ATH and other inorganics has been published by Horn.[69] Mechanisms of fire retardant action of the mineral fillers and their effect on polymer properties have been reviewed by Hornsby and Rothon.[70]

ATH begins to release water at about 220°C with an endotherm of 1.17 kJ/g, whereas MH starts releasing water at about 330°C with an endotherm of 1.356 kJ/g. Thermogravimetric and differential scanning calorimetry curves obtained on heating of ATH and MH are shown in Figures 1.2 and 1.3, respectively. There is little doubt that the main mechanism of fire retardant action of these hydroxides is heat absorption and dilution of the flame with water vapors. Another mechanism could be the catalytic effect of anhydrous alumina, which will help acid-catalyzed dehydration of some polymers and as a result can enhance charring.[38] Since both anhydrous alumina and magnesia are white highly refractory powders, they provide heat insulation by reflecting heat when they accumulate on a surface.

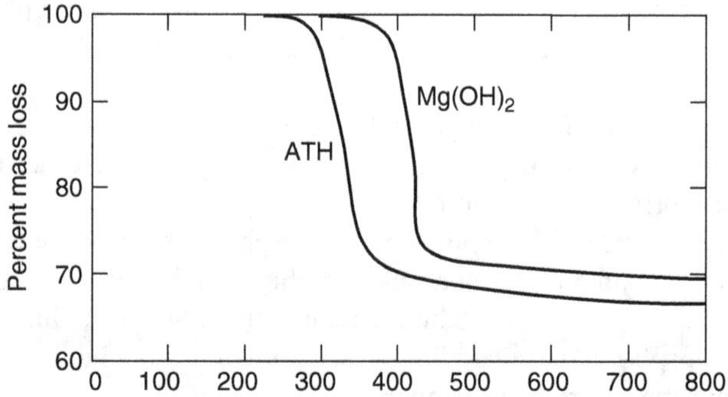

FIGURE 1.2 Thermogravimetry of ATH and $Mg(OH)_2$. (From Ref. 69, copyright © 2000, Routledge/Taylor & Francis Group, with permission.)

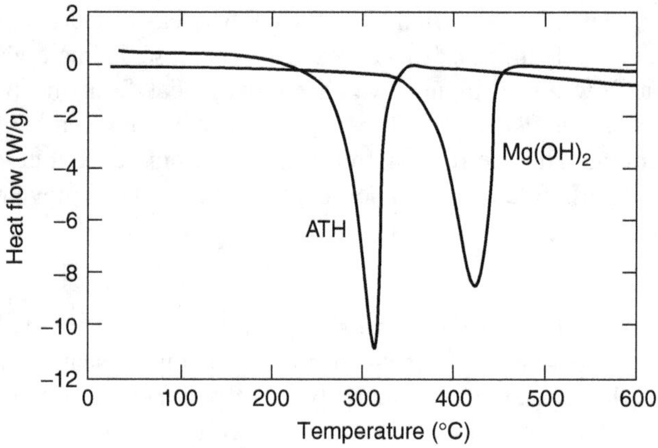

FIGURE 1.3 Differential scanning calorimetry of ATH and $Mg(OH)_2$. (From Ref. 69, copyright © 2000, Routledge/Taylor & Francis Group, with permission.)

ATH and MH are used primarily in wire and cables in poly(vinyl chloride) (PVC), polyethylene, and various elastomers. There is also some limited application of MH in polyamide-6. To pass flame retardancy tests, 35 to 65 wt% of metal hydroxide is required. Decreasing the loading of metal hydroxides will result in a significant gain in physical properties, especially low-temperature flexibility; therefore, combinations with red phosphorus, silicones,[69] boron compounds, nanoclays[71] (treated montmorillonites), and charring agents have been explored.[72] Surface treatment of metal hydroxides also helps to improve physical properties and sometimes improves flame retardancy, due to better dispersion.

1.3.2.5 Borate Flame Retardants Water-soluble borates such as sodium borate (borax) and boric acid have long been used to flame-retard cellulosic materials (e.g., paper boards, wood, and some technical textiles). On the other hand, water-insoluble and more thermally stable zinc borates have found use in thermoplastics. The mechanism of fire retardant action of these two types of borates is quite different.

It is believed that soluble borates can esterify the OH groups of cellulose and promote char formation similar to that of phosphorus. For example, a comparison of the performance of ammonium pentaborate, which decomposes and releases boric acid, and ammonium polyphosphate, which releases polyphosphoric acid, showed some similarity.[73] Borates and boric acid also release some water, which provides a heat sink. Sodium borate and boric acid or anhydride or their mixtures are low-melting solids. Their viscous glassy melts can cause intumescence by evolved decomposition gases, mostly water, or they can just cover the surface of the pyrolyzing polymer or char, healing cracks and providing a barrier to heat and decomposition products.

Several grades of zinc borates are commercially available, which release different amounts of water. Although in formulas for borates, water is often shown as a water of hydration, in fact, borates are rather complex hydroxide salts.[74] Upon heating and polymer combustion, zinc borates dehydrate endothermically, and vaporized water absorbs heat and dilutes oxygen and gaseous flammable components.[75] For example, zinc borate $2ZnO \cdot 3B_2O_3 \cdot 3.5H_2O$, known as Firebrake ZB (U.S. Borax), loses about 13.5 wt% water at 290 to 450°C and absorbs 503 J/g. Thermogravimetric curves of thermal decomposition of various borates are shown in Figure 1.4. Zinc borates are often used in halogen-containing systems and most often in PVC. In PVC, zinc borates significantly increase the amount of char formed during combustion. Zinc borates react with hydrogen chloride released from the thermal decomposition of PVC. Then zinc chloride catalyzes dehydrohalogenation and promotes cross-linking. This leads to an increase

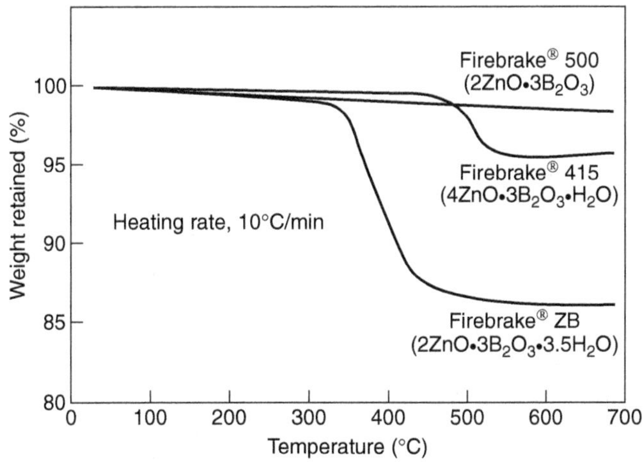

FIGURE 1.4 Thermogravimetry of commercial zinc borates. (From Ref. [76], copyright © 2001, American Chemical Society, with permission.)

in char yield, and, even more important, a significant decrease in smoke formation. At sufficiently high temperatures, zinc borate can melt to produce a glassy layer, but this usually does not happen in small flames. Instead, zinc borate sinters and helps improve the insulating properties of the char and inhibits afterglow combustion.

Zinc borate can also change the oxidative decomposition pathway of halogen-free polymers. It is not completely clear if this is happening because of an inhibition effect of boron oxides toward the oxidation of hydrocarbons[75] or the oxidation of graphite structures in the char,[38] or is due purely to the formation of a protective sintered layer. In combination with ATH, zinc borate creates a porous ceramiclike residue, which has much better insulative properties than those of pure anhydrous alumina. It was shown[77] that zinc borate accelerates dehydration of magnesium hydroxide and creates a ceramiclike structure with dehydrated MgO.

1.3.2.6 Silicon Flame Retardants Under the heading here we include any chemical compound containing Si. For a long time silicons were considered as useful coadditives in flame retardant systems, but recent developments, especially with polycarbonates have again drawn significant attention to silicon. The flame retardancy of silicons has been reviewed by Kashiwagi and Gilman.[78]

Talc is a naturally occurring magnesium silicate which is finding broad application as a filler in polyolefins. Apparently, it provides a moderate flame retardant effect, but because talc is inexpensive, it is used as a partial substitute for more expensive flame retardants. Fumed silica is used as a filler in epoxy resins for the encapsulation of electronic devices at a relatively high loading, up to 80 to 90 wt%. Because of the relatively small amount of combustible resin, this composition can be flame retarded by the addition of a very small amount of a conventional flame retardant. It is not clear if the silica contributes to the flame retardancy by any mechanism other than heat dispersion. Nanodispersed clay, which is one of the main topics of this book, is an aluminosilicate. The mechanism of its flame retardant action is discussed in other chapters of the book.

Octaphenylcyclotetrasiloxane in combination with potassium of sulfonated diphenylsulfone is used commercially in polycarbonate, where clarity of the polymer is important. Recently, some specific branched methylphenylsiloxanes were found particularly effective in polycarbonate (PC) and in PC/acrylonitrile–butadrene–styrene (ABS) blends with a low (ABS) content.[79,80] It is believed that due to the inclusion of aromatic groups in the siloxane, it becomes significantly more soluble and more easily dispersed in PC than straight polydimethylsiloxane. It was shown that these siloxanes tend to migrate from the inside of the PC resin to the surface during combustion and accumulate quickly on the surface. Such movement resulted from differences in viscosity and solubility between the siloxane and the PC at high temperatures. The branched methylphenylsiloxanes showed a higher thermal stability than that of linear dimethylsiloxanes and a greater tendency to induce charring. In contrast, Nishihara et al.[81] showed that linear polysiloxanes are more advantageous flame retardants in PC than are branched polysiloxanes because of higher mobility in the molten plastic.

1.3.2.7 Synergism The concept of synergism is very often used in the optimization of flame retardant formulations; however, synergism is sometimes misinterpreted. By definition, *synergism* means enhanced performance of the mixture of two or more components compared to the simple additive performance of the components at the same concentration. Synergism in flame retardancy, and the general concept of synergism, have been reviewed extensively by Weil.[40,82]

The two mostly common examples of synergism, halogens with antimony and phosphorus with nitrogen, were discussed earlier. Apart from Sb_2O_3, halogen-containing flame retardants are synergistic with other metal oxides, including Bi_2O_3, SnO_2, MoO_3, Fe_2O_3, and ZnO. In some formulations these metal oxides can substitute for Sb_2O_3 partially or completely. Zinc borates or zinc sulfide can be used in the same role of partial substitution of Sb_2O_3. In many instances these metal oxides also provide additional advantages of smoke suppression.

A very sharp synergistic effect between ammonium polyphosphate (APP) and some inorganic minerals, salts, and oxides in a narrow concentration range was discovered independently by Levchik et al.[83–86] and by Lewin et al.[87–89] Later, similar effects were noted in systems containing ammonium polyphosphate and zeolite.[90,91] Although different speculative mechanisms of catalysis of charring (e.g., by zeolites), or thermal oxidative promotion of charring by manganese dioxide were proposed, these mechanisms probably play a minor role. The principal mechanism appears to be interaction of polyphosphoric acid formed during thermal decomposition of APP and metal-containing compounds. Since only divalent and higher-valency metals show this effect, it is reasonable to assume that metal cations help to cross-link polyphosphoric acid and increase its viscosity. This, in turn, helps to create a more thermally insulative char structure. If the mineral compound is added in large quantities, solid crystalline phosphates are formed, and this results in cracking of the char and the loss of insulating properties. This also explains why this synergistic effect is observed in a very narrow concentration range.

Because of increased attention to halogen-free systems in recent years, there has also been a significant effort to enhance the fire retardant performance of aluminum hydroxide (ATH) and magnesium hydroxide (MH), because these additives are used at very high loading levels. It is interesting that just a simple combination of ATH and MH can be synergistic.[40] This probably relates to the extension of the temperature interval for the elimination of water. Combinations of MH and zinc borate were found to be synergistic in poly(ethylene-co-vinyl acetate), (EVA) according to a cone calorimeter study.[77] It was found that zinc borate catalyzes dehydration of MH. In addition, zinc borate helps to sinter particles of MgO together, which, in turn, leads to better retention of combustible polymer in the condensed phase and eventual charring. Further addition to MH and zinc borate to nanoclay and low-melting glass allowed achievement of a V-0 rating in the UL-94 test.[72]

1.3.3 Criteria for Selection of Flame Retardants

Criteria for the selection of flame retardants are usually based on:

- The efficiency of a particular type of flame retardant in a particular polymer system
- The processing conditions of the polymer
- Compatibility and the ability to preserve valuable physical properties
- The cost–performance trade-off

As mentioned above, halogenated flame retardants are more universal than phosphorus-based flame retardants because the halogenated retardants are effective primarily in the flame zone, which is chemically similar for many polymers. However, other criteria listed above require halogenated flame retardants to be tailored to specific polymers. For example, aliphatic halogenated flame retardants are used primarily in thermoset resins or in expandable polystyrene, because of their limited thermal stability. Flame retardants that are soluble in polystyrene are not good for HIPS, because solubility results in plasticization and a dramatic decrease in the heat distortion temperature. On the other hand, partially soluble additives (e.g., decabromodiphenyl oxide) are very suitable for HIPS because they help keep an acceptable heat distortion temperature and good impact properties. Although ABS is chemically similar to HIPS, additives that are soluble in polystyrene (e.g., tetrabromobisphenol A or brominated epoxy oligomers) are preferable. Because ABS has a higher rubber content than HIPS, the use of insoluble additives is detrimental for polymer toughness.

Phosphorus-based flame retardants are usually more suitable for engineering plastics that undergo charring than for commodity polymers. In some plastics, such as PC–ABS or poly(phenylene oxide)–HIPS blends, phosphorus-based flame retardants are more effective then halogenated flame retardants. Antimony trioxide, which is a part of halogen-containing formulations, is a Lewis acid and may destabilize some condensation polymers. Furthermore, the impact properties of engineering polymers may suffer due to the presence of powdery antimony trioxide.

Inorganic hydroxides are used at very high loading levels. Only certain polymers, e.g., polyolefins, can tolerate such high loading without a significant loss of physical properties. Furthermore, relatively low thermal stability, especially of ATH, significantly limits the use of inorganic hydroxides. Other polymeric systems in which ATH is used are PVC, unsaturated polyesters, and latex backcoatings of polyamide or polyester carpets.

1.3.4 Highly Dispersed Flame Retardants

Flame retardants of very small particle size were always of great interest. As mentioned earlier, fumed silica, which apparently has some flame retardant action, is widely used in epoxy formulations for encapsulation of electronic elements.

Another example is use of 0.1- to 2.0-μm size Sb_2O_3, which helps not only with good flame retardancy but also with good pigmentation of PVC.[92] Colloidal-size (0.03 μm) antimony pentoxide, which has a much lower refractive index than Sb_2O_3, can be used in transparent PVC applications.[93] In transparent polycarbonate applications, very small amounts (in the range of 0.02 wt%) of halogenated sulfonate salts, also of submicron particle size,[94] are used. Very fine particle metal oxides can also be used in the flame retardancy of polycarbonate[95], however, apparently this use did not find commercial application. A significant amount of melamine of micrometer and submicrometer particle size which is dispersable in the polyol is used in the flame retardancy of polyurethane foams.[96]

It was always thought that flame retardants of submicrometer particle size would have an essential advantage over flame retardants of regular particle size (micrometer and above) in terms of efficiency. Practice proved that this is true only to a certain extent and depends very much on the type of flame retardant and the flame retardant test used. For example, some phosphate esters and brominated flame retardants are soluble in a polymer matrix. Obviously, it is impossible to achieve better than this distribution for any solid flame retardant, and it is known that these soluble flame retardants do not show extraordinary efficiency compared to their solid counterparts dispersed in the polymers. There is a large class of flame retardants that will melt before they start interacting with the polymer and provide a flame retardant effect. It is clear that little can be achieved by using very fine particles of such flame retardants. A similar comment applies to flame retardants that decompose and totally disintegrate before interacting with the polymer.

A number of publications have shown the advantages of using highly dispersed ATH[97,98] or MH.[99] The average size of the particles of these specially prepared hydroxides is in the range 100 to 300 nm and the authors qualify them as nanofillers. Usually, no or very little advantage is seen with these nanoscale hydroxides in terms of the LOI and UL-94 tests, but some advantages are observed in cone calorimetry. In another study, an attempt was made to flame-retard poly(methyl methacrylate) with fumed silica.[97] Even at relatively high loadings of the silica, only marginal improvement in LOI values was observed. A decrease in the heat release rate measured in cone calorimetry is the commonly seen advantage of nanoscale particles, including nanoclays (discussed in detail in other chapters). Although many mechanistic studies on flame retardancy of nanocomposites are in progress, there is an often accepted point of view that because of their small size, nanoparticles can sinter and create a ceramic–carbon coke on the surface of a polymer which insulates it from heat. Because the flames are small in the LOI and UL-94 tests, they do not provide enough heat for sintering, and that effect of nanoparticles is not seen.

There is another physical mode of action of micro- or nanoscale particles, often overlooked, which is related to the change in rheology of the polymer melt. Even a few percent loading can decrease melt flow significantly. This change in melt viscosity does not itself make it possible to pass the flame retardant test, but in combination with other flame retardants it can be an important tool for improving performance. For example, formulations passing the UL-94 test with a V-2 rating can be upgraded to V-1 or even V-0 with the addition of < 1 wt% of a nanofiller. The effect in the LOI test could be negative or positive. If melt flow contributes to high LOI numbers and will be suppressed by the presence of a nanofiller, the LOI value may actually decrease. This is just an example of a controversy that often appears in the literature and sometimes leads to erroneous conclusions.

1.4 CONCLUSIONS AND FUTURE OUTLOOK

Without a doubt, many chemical substances of high toxicity are released in large-scale fires.[100] But if the occupants have a chance to escape the fire, these toxicants become irrelevant. If the occupants do not escape, they will be victims regardless of the relative toxicity of the flame retardants. The ability of flame retardants to delay fire spread must be recognized as being more important than the relative toxicities of these chemicals when they are decomposed in a fire situation. A study[101] using the life-cycle assessment model, incorporating the emissions from fires, has been applied to furniture. It was found that the largest environmental impact comes from non–flame retardant furniture because of the extensive evolution of polyaromatic hydrocarbons (PAHs), which are strong carcinogens. On the other hand, the environmental impact due to the evolution of chlorinated or brominated dioxins is much less significant. Numerous studies confirm that there is no significant difference in the toxicity of combustion gases from flame-retarded and non-flame-retarded materials.[19] The difference is in the concentration. Since flame retardant materials burn more slowly and often self-extinguish, they generate less toxic gases. Therefore, a true environmental benefit can be achieved only if there are fewer and smaller fires.

The most publicized issue in flame retardants nowadays is the potential replacement of some brominated flame retardants with nonhalogenated flame retardants due to environmental concerns with some halogenated materials.[102] There is also a belief, especially in Europe and the Far East, that halogen-containing flame retardant can evolve small amount of dioxins or dibenzofurans when heated and that plastics containing these flame retardants are therefore not suitable for recycling or incineration. Because of lack of alternative flame retardants, the use of halogen-containing flame retardants has been restricted in Europe and Japan. This has led some manufacturers to eliminate voluntarily the use of flame retardants: mostly because of "environmental" reasons and but because of cost saving as well. Thus, the drive to reduce cost and be more competitive while having a "green" image led to badly compromised fire safety. Considerable loss of life occurred from small ignition sources causing severe burning of non-flame-retarded TV sets. The European regulation regarding electrical and electronic device waste disposal, which requires separate treatment of halogen-containing parts, is another driver for the use of nonhalogen flame retardants or the complete avoidance of flame retardants.

It is clear that there is a great demand for environmentally friendly (usually construed to mean halogen-free) and easily recyclable flame retardant thermoplastics. However, this requirement is in conflict with another environmental requirement, biodegradation. Normally, thermally and hydrolytically stable products, which are required for multiple recycling, tend to be persistent in nature. Therefore, for the future design of flame retardants it is important to make a distinction between one-time short-period-use products, which are biodegradable, and long-term stable products, which are subject to recycling. However, even very thermally and hydrolytically stable flame retardants should eventually be destroyed, either thermally or chemically, under controlled conditions. Newly developed flame retardants should comply with these requirements.

REFERENCES

1. Hirschler, M. Fire performance of organic polymers, thermal decomposition, and chemical composition, in: G.L. Nelson and C.A. Wilkie, Eds. *Fire and Polymers: Materials and Solutions for Hazard Prevention.* ACS Symposium Series, Vol. 797. American Chemical Society, Washington, DC, 2001, pp. 293–306.
2. Purser, D.A. Toxic product yields and hazard assessment for fully enclosed design fires. *Polym. Int.* **2000**, 49, 1232–1255.
3. Irvine, D.J.; McCluskey, J.A.; Robinson, I.M. Fire hazards and some common polymers. *Polym. Degrad. Stab.* **2000**, 67, 383–396.
4. Troitzsch, J.H. Fire gas toxicity and pollutants in fires: the role of flame retardants, in: *Flame Retardants 2000.* Interscience Communications, London, 2000, pp. 177–184.
5. Hirschler, M.M. Fire retardance, smoke toxicity and fire hazard, in: *Flame Retardants '94.* Interscience Communications, London, 1994, pp. 225–238.
6. Hirschler, M.M. Fire hazard and smoke toxicity: post-flashover fire issues or Incapacitation via Irritancy, in: *Flame Retardants 2000.* Interscience Communications, London, 2000, pp. 193–204.
7. Stevenson, G.C. Countervailing risks and benefits in the use of flame retardants, in: *Flame Retardants 2000.* Interscience Communications, London, 2000, pp. 131–145.
8. Manor-Orit, G.-P. Flame retardants and the environment. *Spec. Chem.*, **2005**, 25(7), 36(4).
9. Green, J. 25 years of flame retarding plastics, in: *Proceedings of the 7th Conference on Recent Advances in Flame Retardancy of Polymeric Materials*, Stamford, CT, 1996.
10. Nelson, G.L. The changing nature of fire retardancy in polymers, in: A.F. Grand and C.A. Wilkie, Eds., *Fire Retardancy of Polymeric Materials.* Marcel Dekker, New York, 2000, pp. 1–26.
11. Babrauskas, V.; Harris, R.H., Jr.; Gann, R.G.; Levin, B.C.; Lee, B.T.; Peacock, R.D.; Paabo, M.; Twilley, W.; Yoklavich, M.F.; Clark, H.M. *Fire Hazard Comparison of Fire-Retarded and Non-Fire-Retarded Products.* NBS Special Publication 749. National Bureau of Standards, Gaithersburg, MD, 1988.
12. Hirschler, M.M. Residential upholstered furniture in the United States and fire hazard, in: *Proceedings of the 15th Conference on Recent Advances in Flame Retardancy of Polymeric Materials*, Stamford, CT, 2004.

13. Troitzsch, J., Ed. *Plastics Flammability Handbook: Principles, Regulations, Testing, and Approval*, 3rd ed. Carl Hanser Verlag, Munich, Germany, 2004.
14. Babrauskas, V. Burning rates, in: *SFPE Handbook of Fire Protection Engineering*, 2nd ed. National Fire Protection Association, Quincy, MA, 1996, pp. 3–53.
15. Babrauskas, V. *Ignition Handbook*. Fire Science Publishers, Issaquah, WA, 2003.
16. Williams, M.K.; Nelson, G.L.; Brenner, J.R.; Weiser, E.S.; Clair, T.L. Cell surface area and foam flammability, in: *Proceedings of the 12th Conference on Recent Advances in Flame Retardancy of Polymeric Materials*, Stamford, CT, 2001.
17. Van Klevelen, D.W. Some basic aspects of flame resistance of polymeric materials. *Polymer,* **1975**, 16, 615–620.
18. Levchik, S.; Wilkie, C.A. Char formation, in: A.F. Grand and C.A. Wilkie, Eds., *Fire Retardancy of Polymeric Materials*. Marcel Dekker, New York, 2000, pp. 171–215.
19. Hirschler, M.M. Chemical aspects of thermal decomposition, in: A.F. Grand and C.A. Wilkie, Eds., *Fire Retardancy of Polymeric Materials*. Marcel Dekker, New York, 2000, pp. 27–79.
20. Wilkie, C.A.; Levchik S.V.; Levchik G.F. Is there a correlation between crosslinking and thermal stability? in: S. Al-Malaika, A. Golovoy, and C.A. Wilkie, Eds., *Specialty Polymer Additives: Principles and Application*. Blackwell Science, Oxford, England, 2001, pp. 359–374.
21. Aseeva, R.M.; Zaikov, G.E. *Combustion of Polymer Materials*. Carl Hanser Verlag, Munich, Germany, 1986, p. 149.
22. Georlette, P.; Simons, J.; Costa, L. Halogen-containing fire-retardant compounds, in: A.F. Grand and C.A. Wilkie, Eds., *Fire Retardancy of Polymeric Materials*. Marcel Dekker, New York, 2000, pp. 245–284.
23. Georlette, P. Applications of halogen flame retardants, in: A.R. Horrocks and D. Price, Eds., *Fire Retardant Materials*. Woodhead Publishing, Cambridge, England, 2001, pp. 264–292.
24. Murphy, J. Flame retardants: trends and new developments. *Plast. Add. Compound.* **2001**, Apr., pp. 16–20.
25. Stevenson, D.; Lee, V.; Stein, D.; Shah, T. Flame retardant formulations for HIPS and polyolefins using chlorinated paraffins, in: *Proceedings of the Spring FRCA Conference*, San Antonio, TX, 2002, pp. 79–90.
26. Bie, F. The crucial question in fire protection. *Eng. Plast.* **2002**, 92(2), 27–29.
27. Litzenburger, A. Criteria for, and examples of optimal choice of flame retardants. *Polym. Polym. Compos.* **2000**, 8, 581–592.
28. Finberg, I.; Bar Yaakov, Y.; Georlette, P.; Squires, G.; Geran, T. Fire retardant efficiency and properties of aliphatic bromine compounds in styrenic copolymers, in: *Proceedings of the 12th Conference on Recent Advances in Flame Retardancy of Polymeric Materials*, Stamford, CT, 2001.
29. Boryniec, S.; Przygocki, W. Polymer combustion processes, 3: Flame retardants for polymeric materials. *Prog. Rubber Plast. Technol.* **2001**, 17, 127–148.
30. Pitts, J.J.; Scott, P.H.; Powell, D.G. Thermal decomposition of antimony oxychloride and mode in flame retardancy. *J. Cell. Plast.* **1970**, 6, 35–37.

31. Starnes, W.H., Jr. Kang, Y.M.; Payne, L.B. Reductive dechlorination of a cycloaliphatic fire retardant by antimony trioxide and nylon 6,6: implications for the synergism of antimony and chlorine, in: G.L. Nelson and C.A. Wilkie, Eds., *Fire and Polymers: Materials and Solutions for Hazard Prevention.* ACS Symposium Series, Vol. 797. American Clinical Society, Washington, DC, 2001, pp. 228–239.
32. Green, J. Mechanisms for flame retardancy and smoke suppression: a review. *J. Fire Sci.* **1996**, 14, 426–442.
33. Lewin, M. Physical and chemical mechanisms of flame retarding of polymers, in: M. Le Bras, G. Camino, S. Bourbigot, and R. Delobel, Eds., *Fire Retardancy of Polymers: The Use of Intumescence.* Royal Society of Chemistry, London, 1998, pp. 3–32.
34. Pearce, E.M.;. Shalaby, S.W.; Barker, R.H. Retarding combustion of polyamides, in: M. Lewin, S.M. Atlas, and E.M. Pearce, Eds., *Flame-Retardant Polymeric Materials*, Vol. 1. Plenum Press, New York, 1975, pp. 239–290.
35. Kaspersma, J.H. FR mechanism aspects of bromine and phosphorus compounds, in: *Proceedings of the 13th Conference on Recent Advances in Flame Retardancy of Polymeric Materials*, Stamford, CT, 2002.
36. Prins, A.-M.; Doumer, C.; Kaspersma, J. Glow wire and V-2 performance of brominated flame retardants in polypropylene, in: *Flame Retardants 2000.* Interscience Communications, London, 2000, pp. 77–85.
37. Tanaka, A.; Kanai, T. Thermal decomposition behavior and flame retardancy of polycarbonate containing organic metal salts: effect of salt composition. *J. Appl. Polym. Sci.* **2004**, 94, 2131–2139.
38. Lewin, M.; Weil, E.D. Mechanisms and modes of action in flame retardancy of polymers, in: A.R. Horrocks and D. Price, Eds., *Fire Retardant Materials.* Woodhead Publishing, Cambridge, England, 2001, pp. 31–68.
39. Kandola, B.K.; Horrocks, A.R.; Price, D.; Coleman, G.V. Flame retardant treatments of cellulose and their influence on cellulose pyrolysis. *J. Macromol. Sci. Rev. Macromol. Chem. Phys.* **1996**, C36, 721–794.
40. Weil, E.D. Synergists, adjuvants and antagonists in flame retardant systems, in: A.F. Grand and C.A. Wilkie, Eds., *Fire Retardancy of Polymeric Materials.* Marcel Dekker, New York, 1999, pp. 115–145.
41. Murashko, E.A.; Levchik, G.F.; Levchik, S.V.; Bright, D.A.; Dashevsky, S. Fire retardant action of resorcinol bis(diphenyl phosphate) in PC–ABS blend, II: Reactions in the condensed phase. *J. Appl. Polym. Sci.* **1999**, 71, 1863–1872.
42. Murashko, E.A.; Levchik, G.F.; Levchik, S.V.; Bright, D.A.; Dashevsky, S. Fire retardant action of resorcinol bis(diphenyl phosphate) in a PPO/HIPS blend. *J. Fire Sci.* **1998**, 16, 233–249.
43. Bourbigot, S.; Le Bras, M.; Duquesne, S.; Rochery, M. Recent advances for intumescent polymers. *Macromol. Mater. Eng.* **2004**, 289, 499–511.
44. Bourbigot, S.; Le Bras, M. Flame retardants, in: J. Troitzsch, Ed., *Plastics Flammability Handbook.* Carl Hanser Verlag, Munich, Germany, 2004, pp. 133–157.
45. Camino, G.; Delobel, R. Intumescence, in: A.F. Grand and C.A. Wilkie, Eds., *Fire Retardancy of Polymeric Materials.* Marcel Dekker, New York, 2000, pp. 217–243.
46. Camino, G.; Lomakin, S. Intumescent materials, in: A.R. Horrocks and D. Price, Eds., *Fire Retardant Materials.* Woodhead Publishing, Cambridge, England, 2001, pp. 318–336.

47. Babushok, V.; Tsang, W. Inhibitor rankings for alkane combustion. *Combust. Flame*, **2000**, 124, 488–506.

48. Hastie, J.W.; Bonnell, D.W. *Molecular Basis of Inhibited Combustion Systems*. NBS Research Report NBSIR 80–2169. National Bureau of Standards, Gaithersburg, MD, 1980.

49. Alfonso, G.C.; Costa, G.; Pasolini, M.; Russo, S.; Ballistreri, A.; Montaudo, G.; Puglisi, C. Flame-resistant polycaproamide by anionic polymerization of ε-caprolactam in the presence of suitable flame-retardant agents. *J. Appl. Polym. Sci.* **1986**, 31, 1373–1382.

50. Levchik, G.F.; Levchik, S.V.; Camino, G.; Weil, E.D. Fire retardant action of red phosphorus in nylon 6, in: M. Le Bras, G. Camino, S. Bourbigot, and R. Delobel, Eds., *Fire Retardancy of Polymers: The Use of Intumescence*. Royal Society of Chemistry London, 1998, pp. 304–315.

51. Ballistreri, A.; Foti, S.; Montaudo G.; Scamporino, E.; Arnesano, A.; Calgari, S. Thermal decomposition of flame retardant acrylonitrile polymers, 2: Effect of red phosphorus. *Makromol. Chem.* **1981**, 182, 1301–1306.

52. Ballistreri, A.; Montaudo, G.; Puglisi, C.; Scamporino, E.; Vitalini, D.; Calgari, S. Mechanism of flame retardant action of red phosphorus in polyacrylonitrile. *J. Polym. Sci. Polym. Chem.* **1983**, 21, 679–689.

53. Kuper, G.; Hormes, J.; Sommer, K. In situ x-ray absorption spectroscopy at the K-edge of red phosphorus in polyamide 6,6 during a thermo-oxidative degradation. *Makromol. Chem. Phys.* **1994**, 195, 1741–1753.

54. Levchik, G.F.; Vorobyova, S.A.; Gorbarenko, V.V.; Levchik, S.V.; Weil, E.D. Some mechanistic aspects of the fire retardant action of red phosphorus in aliphatic nylons. *J. Fire Sci.* **2000**, 18, 172–182.

55. Braun, U.; Schartel, B. Fire retardancy mechanisms of red phosphorus in thermoplastics, in: *Proceedings of the Additives 2003 Conference*, San Francisco, CA, 2003.

56. Ravey, M.; Keidar, I.; Weil, E.D.; Pearce, E.M. Flexible polyurethane foam, II: Fire retardation by tris(1,3-dichloro-2-propyl) phosphate, A: Examination of the vapor phase (the flame). *J. Appl. Polym. Sci.* **1998**, 68, 217–229.

57. Ravey, M.; Weil, E.D.; Keidar, I.; Pearce, E.M. Flexible polyurethane foam, II: Fire retardation by tris(1,3-dichloro-2-propyl) phosphate, B: Examination of the condensed phase (the pyrolysis zone). *J. Appl. Polym. Sci.* **1998**, 68, 231–254.

58. Weil, E.D.; Choudhary, V. Flame-retarding plastics and elastomers with melamine. *J. Fire Sci.* **1995**, 13, 104–126.

59. Weil, E.D.; Zhu, W. Some practical and theoretical aspects of melamine as a flame retardant, in: G.L. Gordon, Ed., *Fire and Polymers II: Materials and Tests for Hazard Prevention*. ACS Symposium Series, Vol. 599. American Chemical Society, Washington, DC, 1994, pp. 199–220.

60. Bann, B.; Miller, S.A. Melamine and derivatives of melamine. *Chem. Rev.* **1958**, 58, 131–172.

61. Costa, L.; Camino, G.; Luda di Cortemiglia, M.P. Mechanism of thermal degradation of fire-retardant melamine salts, in: G.L. Nelson, Ed., *Fire and Polymers: Hazard Identification and Prevention*. ACS Symposium Series, Vol. 425. American Chemical Society, Washington, DC, 1990, pp. 211–238.

62. Weil, E.D. Melamine phosphate flame retardants. *Plast. Compound.* **1994**, May–June, pp. 31–39.

63. Levchik, S.V.; Levchik, G.F.; Balabanovich, A.I.; Weil, E.D.; Klatt, M. Phosphorus oxynitride, a thermally stable fire retardant additive for polyamide 6 and polybutylene terephthalate. *Angew. Makromol. Chem.* **1999**, 264, 48–55.

64. Kaprinidis, N.; Zingg, J. Overview of flame retardant compositions UV stable flame retardant systems and antimony free flame retardant products for polyolefins: halogen free melamine based flame retardants for polyamides, in: *Proceedings of the Spring FRCA Conference*, New Orleans, LA, 2003, pp. 168–175.

65. Levchik, S.V.; Balabanovich, A.I.; Levchik, G.F.; Costa, L. Effect of melamine and its salts on combustion and thermal decomposition of polyamide 6. *Fire Mater.* **1997**, 21, 75–83.

66. Endtner, J.M. Development of new halogen-free flame retardant engineering plastics by application of automated optical investigation methods, in: *Proceedings of the 16th Conference on Recent Advances in Flame Retardancy of Polymeric Materials*, Stamford, CT, 2005.

67. Casu, A.; Camino, G.; De Giorgi, M.; Flath, D.; Morone, V.; Zenoni, R. Fire-retardant mechanistic aspects of melamine cyanurate in polyamide copolymer. *Polym. Degrad. Stab.* **1997**, 58, 297–302.

68. Fink, U. The market situation, in: J. Troitzsch, Ed., *Plastics Flammability Handbook*. Carl Hanser Verlag, Munich, Germany, 2004, pp. 8–32.

69. Horn, W.E., Jr. Inorganic hydroxides and hydroxycarbonates: their function and use as flame retardant additives, in: A.F. Grand and C.A. Wilkie, Eds., *Fire Retardancy of Polymeric Materials*. Marcel Dekker, New York, 2000, pp. 285–352.

70. Hornsby, P.R.; Rothon, R.N. Fire retardant fillers for polymers, in: M. Le Bras, C.A. Wilkie, S. Bourbigot, S. Duquesne, and C. Jama, Eds., *Fire Retardancy of Polymers: New Applications of Mineral Fillers*. Royal Society of Chemistry, London, 2005, pp. 19–41.

71. Beyer, G. Nanocomposites: a new concept for flame retardant polymers: *Polym. News* **2001**, 26, 370–378.

72. Weil, E.D.; Lewin, M.; Rao, D. A search for an interactive flame retardant system for ethylene–vinyl acetate, in: *Proceedings of the 15th Conference on Recent Advances in Flame Retardancy of Polymeric Materials*, Stamford, CT, 2004.

73. Levchik, G.F.; Levchik, S.V.; Selevich, A.F.; Lesnikovich, A.I. The effect of ammonium pentaborate on combustion and thermal decomposition of polyamide 6. *Vesti Akad. Nauk. Belarusi Ser. Khim. Nauk*, **1995**, 3, 34–39.

74. Shen, K.K.; Griffin, T.S. Zinc borate as a flame retardant, smoke suppressant, and afterglow suppressant in polymers, in: G.L. Nelson, Eds., *Fire and Materials: Hazards Identification and Prevention*. ACS Symposium Series, Vol. 425. American Chemical Society, Washington, DC, 1990, pp. 157–177.

75. Yang, Y.; Shi, X.; Zhao, R. Flame retardancy behavior of zinc borate. *J. Fire Sci.* **1999**, 17, 355–361.

76. Shen, K.K. Zinc borates: 30 years of successful development as multifunctional fire retardants, in: G.L. Nelson and C.A. Wilkie, Eds., *Fire and Polymers: Materials and Solutions for Hazard Prevention*. ACS Symposium Series, Vol. 797. American Chemical Society, Washington, DC, 2001 pp. 228–239.

77. Bourbigot, S.; Carpentier, F.; Le Bras, M.; Fernandez, C. Mode of action of zinc borates in flame-retardant EVA–metal hydroxide systems, in: S. Al-Malaika,

A. Golovoy, and C.A. Wilkie, Eds., *Specialty Polymer Additives: Principles and Applications*. Blackwell Science, Oxford, England, 2001, pp. 271–292.

78. Kashiwagi, T.; Gilman, J.W. Silicon-based flame retardants, in: S. Al-Malaika, A. Golovoy, and C.A. Wilkie, Eds., *Specialty Polymer Additives: Principles and Applications*. Blackwell Science, Oxford, England, 2001, pp. 353–389.
79. Iji, M.; Serizawa, S. Silicone derivatives as new flame retardants for aromatic thermoplastics used in electronic devices. *Polym. Adv. Technol.* **1998**, 9, 543–600.
80. Iji, M.; Serizawa, S. New silicone flame retardant for polycarbonate and its derivatives, in: S. Al-Malaika, A. Golovoy, and C.A. Wilkie, Eds., *Specialty Polymer Additives. Principles and Applications*. Blackwell Science, Oxford, England, 2001, pp. 293–302.
81. Nishihara, H.; Suda, Y.; Sakuma, T. Halogen- and phosphorus-free flame retardant PC plastic with excellent moldability and recyclability. *J. Fire Sci.* **2003**, 21, 451–464.
82. Weil, E. Additivity, synergism and antagonism in flame retardancy, in: W.C. Kuryla and A.J. Papa, Eds., *Flame Retardancy of Polymeric Materials*, Vol. 3. Marcel Dekker, New York, 1975, pp. 185–243.
83. Levchik, S.V; Levchik, G.F.; Selevich, A.F.; Camino, G.; Costa, L. Effect of talc on nylon 6 fire retarded with ammonium polyphosphate, in: *Proceedings of the 2nd Beijing International Symposium/Exhibition on Flame Retardants*, Beijing, China, 1993, pp. 197–202.
84. Levchik, S.V.; Levchik, G.F.; Camino, G.; Costa, L. Mechanism of action of phosphorus-based flame retardants in nylon 6, II: Ammonium polyphosphate/talc. *J. Fire Sci.* **1995**, 13, 43–58.
85. Levchik, S.V.; Levchik, G.F.; Camino, G.; Costa, L.; Lesnikovich, A.I. Mechanism of action of phosphorus-based flame retardants in nylon 6, III: Ammonium polyphosphate/manganese dioxide. *Fire Mater.* **1996**, 20, 183–190.
86. Levchik, G.F.; Levchik, S.V.; Lesnikovich, A.I. Mechanisms of action in flame retardant reinforced nylon 6. *Polym. Degrad. Stab.* **1996**, 54, 361–363.
87. Endo, M.; Lewin, M. Flame retardancy of polypropylene by phosphorus-based additives, in: *Proceedings of the 4th Conference on Recent Advances in Flame Retardancy of Polymeric Materials*, Stamford, CT, 1993.
88. Lewin, M.; Endo, M. Intumescent systems for flame retarding of polypropylene, in: G.L. Nelson, Ed., *Fire and Polymers, II: Materials and Tests for Hazard Prevention*. ACS Symposium Series, Vol. 599. American Chemical Society, Washington, DC, 1995, pp. 91–108.
89. Lewin, M.; Endo, M. Catalysis of intumescent flame retardancy of polypropylene by metallic compounds. *Polym. Adv. Technol.* **2003**, 14, 3–11.
90. Bourbigot, S.; Le Bras, M. Synergy in intumescence: overview of the use of zeolites, in: M. Le Bras, G. Camino, S. Bourbigot, and R. Delobel, Eds., *Fire Retardancy of Polymers: The Use of Intumescence*. Royal Society of Chemistry, London, 1998, pp. 222–235.
91. Le Bras, M.; Bourbigot, S. Fire retarded intumescent thermoplastic formulations, synergy and synergistic agents: review, in: M. Le Bras, G. Camino, S. Bourbigot, and R. Delobel, Eds., *Fire Retardancy of Polymers: The Use of Intumescence*. Royal Society of Chemistry, London, 1998, pp. 64–75.

92. Morley, J.C. Flame retardants and smoke suppressants, in: E.J. Wickson, Ed., *Handbook of Polyvinyl Chloride Formulating*. Wiley, New York, 1993, pp. 551–577.
93. Myszak, E.A., Jr. Use of submicron inorganic flame retardants in polymeric systems, in: *Proceedings of the 4th Conference on Recent Advances in Flame Retardancy of Polymeric Materials*, Stamford, CT, 1993.
94. Innes, J.; Innes, A. Char forming flame retardants for polycarbonate, presented at the Fall Meeting of the American Fire Safety Counsel, Las Vegas, NV, 2004.
95. Pan, W.-H. (General Electric). U.S. Patent 5274017, 1993.
96. Stern, G.; Horacek, H. Nitrogen-containing compounds: a forgotten, newly reestablished group of halogen-free flame retardants, in: *Proceedings of the 3rd Conference on Recent Advances in Flame Retardancy of Polymeric Materials*, Stamford, CT, 1992.
97. Myszak, E.D., Jr.; Sobus, M.T. New generation of inorganic colloids for flame retardancy and UV stabilization of polymers, in: *Proceedings of the 7th Conference on Recent Advances in Flame Retardancy of Polymeric Materials*, Stamford, CT, 1996.
98. Okoshi, M.; Nishizawa, H. Flame retardancy of nanocomposites. *Fire Mater.* **2004**, 28, 423–429.
99. Zhang, Q.; Tian, M.; Wu, Y.; Lin, G.; Zhang, L. Effect of particle size on the properties of $Mg(OH)_2$-filled rubber composites. *J. Appl. Polym. Sci.* **2004**, 94, 2341–2346.
100. Smith, T.H.F. An overview of the toxicological aspects of flame retardant chemicals, in: *Proceedings of the 2nd Conference on Recent Advances in Flame Retardancy of Polymeric Materials*, Stamford, CT, 1991.
101. Landry, S.; Tange, L.; Blomqvist, P.; Rosell, L.; Simonson, M.; Anderson, P. Fire-LCA model: furniture including polyurethane and textile as case study, in: *Proceedings of API EXPO2004*, Las Vegas, NV, 2004, pp. 133–143.
102. Markarin, J. Flame retardants: higher performance and wider product choice. *Plast. Add. Compound.* **2003**, Nov.–Dec. pp. 32–36.

2

FUNDAMENTALS OF POLYMER NANOCOMPOSITE TECHNOLOGY

E. MANIAS, G. POLIZOS, H. NAKAJIMA, AND M. J. HEIDECKER
Pennsylvania State University, University Park, Pennsylvania

2.1 INTRODUCTION

The term *nanocomposite* is widely employed to describe an extremely broad range of materials, where one of the components has a dimension on the submicron scale. A better and far more restrictive definition would require that a true nanocomposite be a *fundamentally new material* (hybrid) in which the nanometer-scale component or structure gives rise to *intrinsically new properties*, which are not present in the respective macroscopic composites or the pure components. The latter definition necessitates that the nanostructure has dimensions smaller than a characteristic scale that underlies a physical property of the material. For example, for the electronic properties of a conductor or semiconductor, this scale would relate to the de Broglie wavelength of the electron (ranging from a few nanometers for a metal to hundreds of nanometers for a semiconductor), for the mechanical properties of a polymer it would relate to the size of the polymer coil or crystal (again ranging from a few nanometers to hundreds of nanometers), and for the thermodynamic properties of a polymer glass it would relate to the cooperativity length (a few nanometers).

In this chapter we restrict our discussion even further, focusing on one subclass of polymer–inorganic nanocomposites, where the polymers are typically thermoplastics and the inorganic component is a high aspect ratio nanoscale filler. Particular emphasis will be given to principles that apply to pseudo-two-dimensional layered inorganic fillers (such as 2:1 aluminosilicates,[1–9] from where

Flame Retardant Polymer Nanocomposites, edited by Alexander B. Morgan and Charles A. Wilkie

most of our examples will be drawn, and layered double hydroxides[10]), and to a smaller extent to pseudo-one-dimensional fillers (such as carbon nanotubes[11]). In these systems, concurrent improvements across multiple properties are typically achieved—with simultaneous enhancement of the mechanical, thermal, and thermomechanical response—in addition to new properties—such as improved barrier, flammability, and biodegradability behaviors—compared to the unfilled polymer. Consequently, the resulting nanocomposite material is better described by the term *hybrid* (denoting large-scale changes in multiple material characters) rather than polymer *composite* (a term traditionally associated with an incremental improvement in one or two key properties[12–14]).

For these nanocomposite systems, the fundamentally new properties typically originate from the change in the polymer nature in the vicinity of the filler, such as polymers adsorbed on filler surfaces or confined in between fillers, and as such, they depend strongly on the effective surface area of the fillers (i.e., the surface area of a single filler when completely dispersed or the surface area of the typical filler cluster). Thus, good dispersions of fillers would result in a true nanocomposite at rather low filler loadings, close to the percolation threshold of these high aspect ratio fillers (cf. below 3 vol% for typical layered silicates[15] or 1 vol% for single-walled nanotubes[16]). On the other hand, in the absence of dispersion, neither the nanometer-scale geometry of these fillers nor their ultrahigh surface area is exploited, and the resulting composite falls into the class of conventional composites despite the nanometer size of the individual inorganic fillers.

In the case of nanometer-thin layered inorganic fillers, it has long been known that polymers can effectively disperse clay minerals when the minerals are appropriately modified.[1,2] The field has recently gained considerable momentum, due mainly to two major findings that pioneered the revival of these materials: First was the report of a nylon-6/montmorillonite (MMT) material from Unitika and Toyota researchers,[17,18] where very moderate inorganic loadings resulted in concurrent and remarkable enhancements of thermal and mechanical properties. Second, Giannelis et al. found that it is possible to melt-mix polymers with clays without the use of organic solvents.[19] Since then, the high promise for industrial applications has motivated vigorous research, which revealed concurrent dramatic enhancements in polymers by the dispersion of various nanometer-thin inorganic layered fillers.[10,20–23] Where the property enhancements originate from the nanocomposite structure, these improvements are generally applicable across a wide range of polymers.[6,10]

In contrast, carbon nanotubes were discovered much more recently, first observed by Iijima,[24] and since then they have been the focus of considerable research activity. This pseudo-one-dimensional form of carbon has remarkable physical and mechanical properties, such as structure-tunable electronic properties, ultrahigh thermal conductivity, and unmatched mechanical properties (e.g., stiffness, strength, and resilience). These characteristics, combined with recent advances enabling high-volume production of multi- and single-walled nanotubes, offer tremendous opportunities for the development of ultrahigh-performance nanotube-reinforced nanocomposite materials.[11]

At this point we should also mention that this chapter is not intended to provide an extensive review of the polymer nanocomposites field. The reader interested in such reviews can refer to a number of related books,[1–5] numerous compilations of relevant symposia and conference proceedings, and recent review articles.[6–8,10,11] This chapter is, rather, an attempt toward a brief eclectic overview of topics highlighting the fundamentals that underlie the materials discussed in the remainder of the book.

2.2 FUNDAMENTALS OF POLYMER NANOCOMPOSITES

2.2.1 Thermodynamics of Nanoscale Filler Dispersion

As for polymer blends, the thermodynamics of mixing for polymers and nanofillers can be described through a balance of entropic and enthalpic factors, which determines whether a pristine or organically modified filler will be dispersed in a polymer.[25–27] Especially for nanoparticles, favorable thermodynamics of mixing are essential since these ultrasmall particles are held together with very high apparent attractive forces [cf. eq. (2.3)] when immersed in liquid or polymeric media, and purely mechanical methods of mixing are not expected to be effective. Moreover, given the extensive amount of surface area that imposes entropic penalties for adsorbed, physisorbed, or intercalated macromolecules, the dispersion of nanofillers necessitates sufficiently favorable enthalpic contributions to overcome the entropic penalties.

For example, following the interfacial tension formalization of van Oss–Chaudhury–Good,[28] we consider two flat filler [e.g., layered silicate (s)] layers separated by an organic layer [e.g., alkyl surfactant film (a) or an intercalated polymer film]. In this case, successive layers are held together with an adhesive energy:

$$\Delta F_{sas} = -2\gamma_{sa} = -2\left(\sqrt{\gamma_s^{\mathrm{LW}}} - \sqrt{\gamma_a^{\mathrm{LW}}}\right)^2 - 4\left(\sqrt{\gamma_s^{+}} - \sqrt{\gamma_a^{+}}\right)\left(\sqrt{\gamma_s^{-}} - \sqrt{\gamma_a^{-}}\right) \tag{2.1}$$

when assuming additivity of apolar [Lifschitz–van der Waals (LW)] and polar [electron donor–acceptor, or Lewis acid–base (AB)] interaction terms,[28] and using standard geometric combination rules:

$$\gamma_{ij} = \gamma_{ij}^{\mathrm{LW}} + \gamma_{ij}^{\mathrm{AB}} \quad \text{with} \begin{cases} \gamma_{ij}^{\mathrm{LW}} \simeq \left(\sqrt{\gamma_i^{\mathrm{LW}}} - \sqrt{\gamma_j^{\mathrm{LW}}}\right)^2 \\ \gamma_{ij}^{\mathrm{AB}} \simeq 2\left(\sqrt{\gamma_i^{+}} - \sqrt{\gamma_j^{+}}\right)\left(\sqrt{\gamma_i^{-}} - \sqrt{\gamma_j^{-}}\right) \end{cases} \tag{2.2}$$

The i and j subscripts correspond to the various system components (layered silicate s, alkyl surfactant film a, and polymer p) and the LW and AB superscripts to the nature of interactions (apolar LW and polar AB). These relations can be converted into the Hamaker constant formalization by setting $\gamma_i^{\mathrm{LW}} = A_i/24\pi l_o^2$ with $l_o = 1.58$ Å. In the case of 2:1 aluminosilicates organically modified by alkyl surfactants, the attractive interaction energy of eq. (2.1) would correspond (Figure 2.1) to an adhesive *pressure* between the parallel flat surfaces of

$$P = \frac{A}{6\pi d^3} = \frac{-12\pi l_o^2 \Delta F_{sas}}{6\pi d^3} \tag{2.3}$$

where d is the thickness of the organic interlayer film. Given that typical alkyl surfactant modifications—butyl to dioctadecyl—correspond to a surfactant layer thickness of 0.5 to 1 nm, the corresponding adhesive pressure between successive silicate layers is at least[32] 10^5 to 10^4 bar (cf. Figure 2.1). Thus, favorable enthalpic interactions are absolutely necessary for filler dispersion and nanocomposite formation.

Focusing on polymer nanocomposites based on organically modified layered silicates, Vaia et al. published a tractable approach to calculate the entropic and enthalpic contributions to the free energy of mixing[25] and have used this to predict miscibility of polystyrene with alkylammonium–modified silicates[26] (montmorillonite and fluorohectorite). According to this model, the entropic contributions are unfavorable and rather small: Specifically, the conformational entropy penalty of polymer confinement is compensated by an increase in conformational freedom of the tethered surfactants upon dispersion for gallery increases of up to 0.7 nm, and adopts small unfavorable values for larger gallery increases (see Figure 4 of Vaia et al.[25]). Consequently, small per-monomer favorable enthalpic interactions

Material	γ^{LW}	γ^+	γ^-
Water[29]	21.8	25.5	25.5
Montmorillonite[26]	66	0.7	36
Alkane[29] (C_{12}-C_{18})	26	0	0
Polypropylene[29]	26	0	0
Polyethylene[29]	33	0	0
Polystyrene[29]	42	0	1.1
PMMA[29]	40.6	0	12
Carbon nanotube*[, 30]	18.4	12	12
PET[31]	43.5	0.01	6.8
Nylon 6, 6[29]	36.4	0.02	21.6

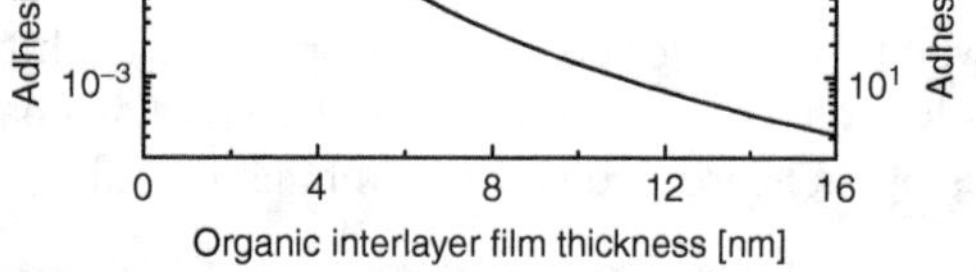

FIGURE 2.1 (*Left*) Various surface tension components γ (mJ/m^2) for materials discussed in the text (*from $\gamma^{\mathrm{AB}} \simeq 24$ mJ/m^2 assuming that $\gamma^+/\gamma^- = 1$). (*Right*) Adhesive pressure versus interlayer thickness as predicted by eq. (2.3) for two flat montmorillonite surfaces separated by apolar organic films (e.g., an olefin). For small film thicknesses (<2.5 to 3 nm) this continuum approach is not valid; rather, the adhesive pressure has discontinuous stable maxima[32] (much higher than the dashed line) which correspond to integer numbers of monomer layers.

can drive dispersion of these nanofillers in the polymer and promote the formation of a nanocomposite. These favorable enthalpic interactions are an *excess enthalpy*, akin to the χ parameter definition in Flory–Huggins theory; for silicate (s) modified by a surfactant (a) and a polymer (p), this excess enthalpic interaction per area can be approximated[25] by $\Delta H \sim \epsilon_{ps} + \epsilon_{pa} - (\epsilon_{aa} + \epsilon_{as})$, where ϵ_{ij} is a measurement of the pairwise interaction between components i and j [which can be quantified through pairwise atomic interaction parameters, cohesive energy densities, solubility parameters, or interfacial tension (Hamaker constants) formulations[29,32]]. For most polymers and surfactants $\epsilon_{pa} - \epsilon_{aa} \ll \epsilon_{ps} - \epsilon_{as}$, and to a first approximation for polymer/surfactant-modified inorganic nanocomposites, *favorable enthalpy for mixing is achieved when the polymer–inorganic interactions are more favorable than the surfactant–inorganic interactions.*

Following our prior nomenclature, dispersion would dictate a negative interaction energy change (upon mixing), which corresponds to a positive interfacial tension difference ($\gamma_{as} - \gamma_{ps}$). For an apolar ($\gamma_a^{\pm} \simeq 0$) alkyl surfactant (e.g., dodecane to nonadecane,[29] $\gamma_a^{\mathrm{LW}} \simeq 26$ mJ/m^2) used to organically modify a typical silicate (e.g., montmorillonite, with[26] $\gamma_s^{\mathrm{LW}} \simeq 66$ mJ/m^2, $\gamma_s^+ \simeq 0.7$ mJ/m^2, and $\gamma_s^- \simeq 36$ mJ/m^2), miscibility would be achieved with any polymer for which

$$\gamma_{\text{excess}}^{\text{total}} = (\sqrt{\gamma_p^{\mathrm{LW}}} - \sqrt{66})^2 + 2(\sqrt{\gamma_p^+} - \sqrt{0.7})(\sqrt{\gamma_p^-} - \sqrt{36}) - 9.1 \text{ mJ/m}^2 < 0 \tag{2.4}$$

This is satisfied for most polymers (e.g., Table XIII-5 in van Oss' book[29]) except perfluorinated polymers and most of the polyolefins (polypropylene, polyisobutylene, etc). Miscibility is also promoted for all apolar polymers ($\gamma_p^{\pm} \simeq 0$) with 26 mJ/m^2 $< \gamma_p^{\mathrm{LW}} <$ 125 mJ/m^2, and for polar polymers with $\gamma_{ps}^{\mathrm{LW}} \simeq 26$ mJ/m^2 and $\gamma_{ps}^{\mathrm{AB}} < 0$ (i.e., as Vaia[26] states, $\gamma_p^+ > 0.7$ mJ/m^2 and $\gamma_p^- < 36$ mJ/m^2, or $\gamma_p^+ < 0.7$ mJ/m^2 and $\gamma_p^- > 36$ mJ/m^2). Thus, *for most polymers* the commonly used organic modification by *alkyl-cationic surfactants is adequate to create sufficient excess enthalpy and promote nanocomposite formation* with montmorillonite.

In a different approach,[27] a longer macromolecular "surfactant" that would increase the layer separation to 5 to 10 nm necessitates much smaller favorable enthalpic contributions since the adhesive pressure to be overcome is about a thousand times smaller. This last theoretical prediction has been verified for polypropylene (PP)[33] in the absence of excess enthalpic interactions (i.e., $\gamma_{\mathrm{PP}}^{\mathrm{LW}} = 26$ mJ/m^2 $\simeq \gamma_a^{\mathrm{LW}}$ and $\gamma_{\mathrm{PP}}^{\pm} = 0$, and eq. (2.4) yields $\gamma_{\text{excess}}^{\text{total}} \simeq 0$), which in turn implies that for short surfactants the entropic penalties from the physisorbed PP will hinder spontaneous miscibility, whereas the entropic gains from longer surfactants would promote miscibility.[27] At this point, we would like to make three more comments:

1. It should be obvious that free energy calculations cannot be done on a per molecule basis, but rather, the free energy of the system or the free energy per volume must be calculated. Thus, certain parameters that were omitted herein [such as the monomeric volumes of polymer and surfactant and the

grafting density of the surfactant on the filler—in the case of silicates this would be proportional to the cation exchange capacity (CEC)] must also enter the calculations.[26] The arguments above [e.g., eq. (2.4)] can be used when there is a substantial fraction of *both* polymer and surfactant in contact with the filler surface: for example, in the case of 2:1 aluminosilicates $0.65 < \text{CEC} < 1.7$ meq/g (or equivalently, surfactant grafting densities of one surfactant per $2 < A < 0.8\ \text{nm}^2$), and still provide only approximate values or criteria. A more detailed discussion is provided elsewhere.[26]

2. In the case of polypropylene (PP), the approach described above yields a zero excess enthalpic interaction for an alkyl-modified silicate [since[29] $\gamma_{\text{PP}}^{\text{LW}} = 25.7\ \text{mJ/m}^2 \simeq \gamma_a^{\text{LW}}$ and $\gamma_{\text{PP}}^{\pm} \simeq 0$, eq. (2.4) yields $\gamma_{\text{excess}}^{\text{total}} \simeq 0$], which implies that the entropic factors, albeit small in magnitude, will hinder spontaneous miscibility.
3. Under the approximations and assumptions mentioned above and without considering any entropic contributions, the interfacial (adhesive) energy per area of a polymer and a silicate is given by[28]

$$\Delta F_{ps}^{\text{total}} = \Delta F_{ps}^{\text{LW}} + \Delta F_{ps}^{\text{AB}} = (\gamma_{ps}^{\text{LW}} - \gamma_p^{\text{LW}} - \gamma_s^{\text{LW}}) + (\gamma_{ps}^{\text{AB}} - \gamma_p^{\text{AB}} - \gamma_s^{\text{AB}}) \tag{2.5}$$

Substituting γ_{ps}^{LW} and γ_{ps}^{AB} from eq. (2.2) yields

$$\Delta F_{ps}^{\text{total}} = -2\sqrt{\gamma_p^{\text{LW}}\gamma_s^{\text{LW}}} - 2\left(\sqrt{\gamma_p^{+}\gamma_s^{-}} + \sqrt{\gamma_p^{-}\gamma_s^{+}}\right) \tag{2.6}$$

which for a strictly apolar polymer becomes

$$\Delta F_{ps}^{\text{total}} = -2\sqrt{\gamma_p^{\text{LW}}\gamma_s^{\text{LW}}}$$

2.2.2 Synthetic Routes for Nanocomposite Formation

For traditional composite materials, high performance requires, in a first approach, homogeneous and thermodynamically stable dispersion of the fillers in the polymer matrix. To this end, the two major hurdles to be overcome are (1) deaggregation of the filler assemblies (clusters of fillers often containing tens, hundreds, or even millions of filler particles, associated with very strong interparticle forces[32]), and (2) achieving sufficiently strong polymer–filler interfaces, required for good mechanical coupling between the matrix and the filler. Both these requirements are also necessary in polymer-based nanocomposites, and depending on the nanofiller, there exist additional hurdles that need to be overcome toward nanocomposite formation. Examples of such challenges include entropic effects of polymers in nanoscopic confinements between two-dimensional fillers, as discussed in Section 2.2.1; deaggregation of intertwined one-dimensional filler clusters, as in carbon nanotube bundles or ropes; and overcoming the much faster kinetics (compared with colloidal micrometer-sized fillers) of nanofiller reaggregation.

As for the thermodynamic consideration in Section 2.2.1, we attempt to highlight these challenges by describing in some detail the most common synthetic routes for nanocomposite formation employed for polymer/layered-inorganic hybrids. Most examples are drawn from layered-silicate fillers, but the conclusions are general across most nanofillers, and one should be able to envision similar strategies for nanocomposite formation based on other types of nanofillers.

2.2.2.1 Solution-Aided Dispersion and Brute-Force Melt Processing In most cases, polymer–inorganic systems that do not possess favorable thermodynamics for nanocomposite formation can be "trapped" in dispersed—even exfoliated—structures through solvent casting, sonication, or high-shear-rate/high-temperature extrusion. Such trapped structures are usually easy to achieve* but in most cases are neither thermodynamically stable nor amenable to further processing: for example, in Figure 2.2, x-ray diffraction (XRD) of precipitated PP–MMT hybrids from a co-suspension of polypropylene and o-MMT (organically modified montmorillonite) in trichlorobenzene (similar structures can be obtained from aggressive melt processing, such as high-shear-rate extrusion,[34–37] or dynamic packing injection molding[38]).

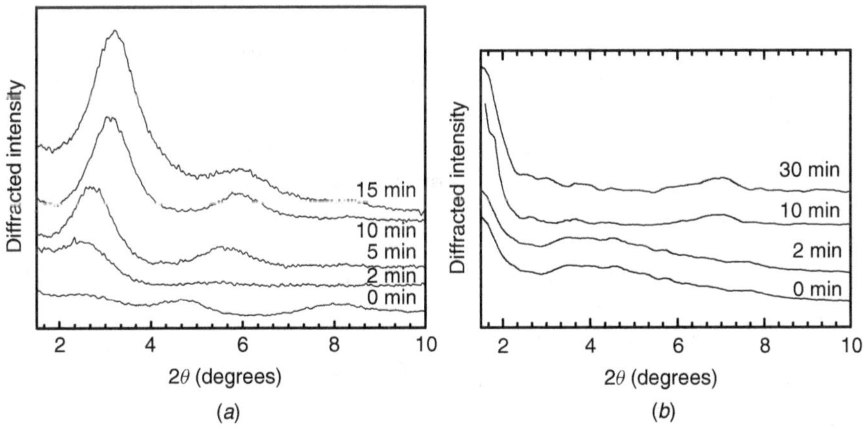

FIGURE 2.2 Structure evolution–stability of (*a*) neat-PP/2C18–MMT and (*b*) PP-g-MA/2C18-MMT nanocomposites that were initially (0 min) trapped apart. XRD studies of compression-molded samples are shown. The neat-PP/2C18–MMT very fast collapses to intercalated–immiscible tactoids, whereas for the MA–functionalized PP, the trapped dispersed structure is maintained even under prolonged high-temperature processing. This suggests that the MA groups have sufficiently strong interactions with the MMT to prevent the polymer from sliding away from the inorganic layers. (Adapted from Ref. 49, copyright © 2001, American Chemical Society, with permission.)

*Easy or successful brute-force melt processing in the case of layered inorganic fillers is obviously limited to fillers of relatively small lateral size, given the very strong adhesive forces per area between such particles (Figure 2.1).

However, upon subsequent processing by compression molding (at 180°C, 15 tons) of these hybrids, the polymer melts and the trapped hybrid structure relaxes toward a thermodynamically favorable state. If the o-MMT dispersion is not thermodynamically favorable, the layers will collapse in low *d*-spacing parallel stacks (e.g., neat-PP/dimethyldioctadecylammonium–MMT; Figure 2.2*a*) during high-temperature processing, leading to a conventionally filled "macro" composite. However, when there exists favorable free energy for o-MMT/polymer mixing, the exfoliated filler structures are retained [e.g., polypropylene containing maleic anhydride (MA) functional groups and dimethyldioctadecylammonium MMT; Figure 2.2*b*]. Typically, this approach can yield stable dispersions only for polymers with strong specific interactions with MMT [e.g., polymers that hydrogen-bond to the silicates, such as poly(vinyl alcohol),[39] polyurethanes,[40,41] and polyamide-6[42–44]]. It is striking that only 0.5 mol% of MA can have the same effect in PP. As expected, mechanical shear markedly reduces the time necessary for structure relaxation, and the structure of Figure 2.2*b* is recovered after 8 min of mixing (extrusion at 180°C). In concert, even after very moderate mixing (1 to 3 min at 180°C) trapped systems of neat-PP/2C18–MMT result in an immiscible or intercalated structure with a wide XRD reflection, extending from 1.8 to 2.7 nm in *d*-spacing. Along the same lines, when sonication is employed in polymer–nanofiller co-suspensions, instead of aggressive melt processing of the polymer with the nanofiller, similar trends can be observed and a well-dispersed structure can be stable when favorable interactions are present, as for example in polystyrene/imidazolium–montmorillonite systems.[45] The sonication approach is, in general, a highly successful route for polymer nanocomposites based on carbon nanotubes, since the sonication can effectively disperse the nanotube bundles in solvent and subsequently in polymer matrix, and is commonly employed despite criticisms that sonication may cause tube breakdown.

This approach is qualitatively similar to the *swelling agent approach*, as for example, by Wolf et al.[46] In such approaches an alkylammonium-exchanged montmorillonite is intercalated by an organic swelling agent such as ethylene glycol, naphtha, or heptane (all with boiling points below the processing or extrusion temperature).[46] Subsequently, the swollen organo-modified clay is compounded with PP in a twin-screw extruder at 250°C. At this processing temperature, the swelling agent evaporates, leading to the formation of a nanocomposite that is XRD silent. In principle, this is the same as the *solution intercalation approach*, where a solvent is employed to mix the o-MMT with the polymer, and a mostly exfoliated structure is trapped upon evaporation of the solvent. For fillers that cannot be surface-modified by grafted surfactants (such as graphite), the swelling agent approach is probably the most effective route for achieving filler dispersions.

In all the cases above and in the absence of polymer cross-linking or favorable thermodynamics to retain the dispersion achieved (by solvent, mechanical shear/vibration, swelling agent, etc.), the fillers will reaggregate upon further processing, and all the high-performance character due to nanoscale filler dispersion will be lost.

2.2.2.2 Static Melt Intercalation This method involves the mechanical mixing of a polymer with an appropriately modified filler and subsequent annealing above the softening temperature of the polymer.[19] This approach provides the best route to test[26] with sensitivity the thermodynamic arguments detailed above and to yield well-defined systems for fundamental studies. However, due to the quiescent processing conditions (absence of external shear), which eliminate any mechanical contribution for the dispersion of fillers, and to the very slow intercalation–exfoliation kinetics,[47,48] such methods are typically very slow, thus having very limited applicability in industry.

We mention only one example for this method, polypropylene (PP) in organically modified montmorillonite (o-MMT), so as to elucidate how the thermodynamics of mixing can be tested.[33,49] The challenge with PP is to design systems where the polymer–MMT interactions are more favorable than the surfactant–MMT interactions. As mentioned above, for an alkyl surfactant used as the organic modification in o-MMT, there is no excess enthalpy for mixing with PP ($\gamma_{\text{excess}}^{\text{total}} \simeq 0$), or in other words, the polymer–MMT interactions are equal to the surfactant–filler interactions. In agreement with the thermodynamic arguments presented above, minute amounts (0.5 to 1 mol%) of randomly incorporated polar or polarizable ($\gamma^{\text{AB}} \neq 0$) functional groups, such as methylstyrene, hydroxyl, and maleic anhydride, can promote PP/o-MMT miscibility[49] under static melt intercalation. Also, small blocks (1 to 5 mol%) of poly(methyl methacrylate) (PMMA) added to PP were shown to be sufficient to drive miscibility (in this case, $\gamma_{\text{PMMA}}^{\text{AB}} = 0$, but $\gamma_{\text{PMMA}}^{\text{LW}} \simeq 40 > \gamma_{a}^{\text{LW}} \simeq 26$ mJ/m^2), since the favorable thermodynamics for the PMMA can overcome the purely entropic barrier for the PP intercalation. Even in the extreme case, where the miscible block becomes as short as a single group, miscibility can still be achieved[33] when this group possesses sufficiently strong interactions for the filler (as, e.g., an ammonium group[33]). On the other hand, if mixing is to be promoted for nonfunctionalized PP, a surfactant must be chosen with poorer interactions with the filler than the olefinic polymer (i.e., $\gamma^{\text{AB}} = 0$ and $\gamma^{\text{LW}} < \gamma_{\text{PP}}^{\text{LW}} \simeq 26$ mJ/m^2); such surfactants are, for instance, fluorinated or semifluorinated alkyls ($\gamma_{\text{FE}}^{\text{LW}} \simeq 18$ mJ/m^2). This strategy has also been proven experimentally.[49]

2.2.2.3 Melt Processing This is a very frequently used approach[6,8,16] in which the polymer and the (usually organically modified) filler are incorporated together in a traditional polymer processing method, most commonly, extrusion or kneading, and less frequently, injection molding. In concert with the principles of static melt intercalation, favorable thermodynamics for mixing are introduced by the design of functionalities on the polymer and by the choice of the organic modification for the fillers. In addition to any thermodynamic contributions, mechanical shear provides a kinetic driving force for further dispersion of the fillers in the polymer matrix and accelerates substantially the kinetics of filler dispersion.

The latter effect is particularly important for polymers that possess very high attractions for the filler surfaces and can be kinetically arrested under static melt intercalation.[48] In many cases, end users of polymer nanocomposites are hesitant to incorporate nanofillers directly (in the form of ultrafine powders) in their current processing practices, and the *concentrate* or *masterbatch* two-step approach is preferred. In this case, first a polymer nanocomposite (concentrate) is formulated at relatively high filler loadings of about 25 wt%, which can be processed and palletized to look like a normal polymer resin. This concentrate is subsequently diluted (i.e., let down) to the desired filler loading by pure polymer resin (cf. below).

2.2.2.4 Masterbatch Approaches Beyond any industrial reservations for incorporating nanoparticles directly into the final stages of processing, there also exist in some cases scientifically justified reasons to follow the concentrate or masterbatch approach. For example, in the first studies aiming to develop PP/o-MMT materials,[34–37,46,50] polypropylene oligomers modified with either maleic anhydride (MA) or hydroxyl groups (OH) were first mixed with octadecylammonium-exchanged montmorillonite, creating a masterbatch at high filler loadings which was subsequently blended with neat PP, usually assisted by strong mechanical shear in an extruder or mixer. In this way, the MA–polypropylene disperses the o-MMT, given the favorable thermodynamics, and in the second step PP and PP-g-MA are effectively at theta conditions, and the extrusion is promoting mixing due only to entropic reasons (cf. morphologies of miscible polymer blends). Although at first glance, this approach may seem similar to the one denoted above as "brute force," in the masterbatch case there do exist favorable thermodynamics for mixing, which not only result in more effective dispersions, but also stabilize the dispersed nanocomposite structure. However, the structure and properties of the resulting hybrid materials still depend strongly on the processing conditions, and in the case of PP, for example, they range from very moderate dispersions and property improvements[34,36,37,46,50] to good dispersions and better-performing hybrids.[35] Obviously, an MA–polypropylene pretreatment with very low maleic anhydride content does not promote nanocomposite formation,[36] and very high maleic anhydride content makes the masterbatch so robust that MMT does not mix further with neat PP.[34,37] Furthermore, the PP-g-MA can have marked effects on PP crystallization and, consequently, cause the mechanical properties to deteriorate, especially when the PP-g-MA is of substantially lower molecular weight or isotacticity than the PP matrix, or contains high levels of branching. Therefore, it is frequently necessary to develop several variants of a masterbatch [based on functional polymers with varied characteristics: e.g., in the case of PP with various molecular weights of PP-g-MA or in the case of polyethylene (PE) with various polymer microstructures (LDPE, LLDPE, HDPE, etc.)], depending on the specific characteristics of the polymer matrix for which they are intended.

2.2.2.5 In Situ Polymerization Schemes One of the cornerstone studies, and probably the single most important study in pioneering the revival of the polymer/

layered-silicate nanocomposites field, was the work by the Toyota group in which they polymerized polyamide-6 in the presence of, and end-tethered on, the surfaces of montmorillonite layers.[17,18] Since then, the strategy of in situ polymerization of a monomer in a co-suspension with inorganic filler has been employed successfully for a variety of polymers, with and without end tethering the macromolecules on the filler surfaces, and through various polymerization reactions, for a variety of polymers and fillers (detailed examples are discussed in a review article[8]). In most cases, nanocomposites formed by in situ polymerization result in structures that are kinetically trapped (cf. the solution approach above) in a well-dispersed structure. In general, these structures do possess favorable thermodynamics to retain the filler dispersion upon subsequent processing (such as compression or injection molding of the hybrid after polymerization), since this method requires that the monomer initially disperses the inorganic particles sufficiently. However, if more polymer is added in the subsequent processing step (cf. an attempt to use the in situ polymerized hybrid as a masterbatch), in most cases there occurs a loss of the exfoliated structure achieved during the in situ polymerization step, and typically a less dispersed structure is obtained. For example, attempts to dilute the polyamide-6/montmorillonite nanocomposite[17] with pure polyamide-6 or an in situ poly(ϵ-caprolactone)/montmorillonite hybrid[51,52] with pure poly(ϵ-caprolactone) result in collapse of the mostly exfoliated in situ structure. Typically, this well-dispersed in situ structure becomes intercalated upon addition of the homopolymer, where the inorganic fillers adopt a parallel stacking with a polymer bilayer (an intercalated layer about two monomers thin) in the interlayer gallery.

2.2.2.6 Extension to Other Fillers These ideas can be extended to other high aspect ratio fillers when taking their idiomophies into account. The ideas can be transferred almost as stated above to other two-dimensional and pseudo-two-dimensional layered fillers [e.g., layered double hydroxide (LDH)[10] or graphite] when addressing their differences from layered aluminosilicates; for example, LDH would require anionic surfactants, whereas graphite is not amenable to grafted modifications and an intercalated swelling agent is needed (cf. the masterbatch or solution approaches above).

For one-dimensional nanofillers, however, there are important differences that may necessitate different choices for nanocomposite formation. For example, in the case of carbon nanotubes, polymer–matrix nanocomposites can be fabricated using almost all of the schemes discussed above, but the effectiveness and importance of these schemes are very different from those of polymer/layered-inorganic nanocomposites. Dispersion of nanotubes is hindered not only by their high affinity for one another, but also by their ability to intertwine with one another, forming bundles or ropes. These often-large agglomerations are typically formed during synthesis of the nanotubes (especially for single-walled carbon nanotubes), which need to be well unbundled before attempting dispersion in a polymer matrix. At the same time, the reactive bonding of surfactants on the nanotube surfaces, although possible via multiple chemistries,[53,54] most often causes deterioration of their remarkable physical properties (more so for single-walled than for multiwalled carbon nanotubes), in particular their thermal and electron conductivities, as well as their stiffness and strength.*

After these thoughts, and following our earlier discussion of polymer–silicate nanocomposites, it seems obvious that the nanocomposite formation schemes that depend on favorable thermodynamics (e.g., melt blending) or brute-force mechanical mixing are of limited use here, whereas the solution mixing and the in situ polymerization schemes should be much more effective.[11,56] In fact, the most common approach for polymer–nanotube composite formation involves first unbundling the nanotube aggregates in solvent (most often aided by sonication and physisorbed surfactants, and centrifugal separation) and subsequent solution-aided dispersion in a polymer matrix. These solution-aided dispersions can effectively trap the nanotubes in a well-dispersed morphology after solvent evaporation (see, e.g., Refs. 57 and 58). Alternatively, instead of employing a physisorbed surfactant and two steps of solution dispersion, nanocomposites can be formed in a one-step solution process (much like their polymer–silicate counterparts, by co-dissolving the host polymers and nanotubes in a common solvent), as, for example, with poly(vinyl alcohol)[59] and polystyrene.[16] For the same reason (i.e., employing 'solvent' to unbundle the nanotubes), in situ polymerization has also proven to be an effective method for producing well-dispersed nanocomposites. A characteristic example of this approach is the polymerization of PMMA in the presence of solution-dispersed nanotubes, leading to high-molecular-weight polymers and very good nanocomposite morphologies (see, e.g., Refs. 60–62).

Finally, unlike polymer/layered-silicate nanocomposites, melt processing is far less common for nanotube-reinforced nanocomposites. Melt processing relies on mechanical shear and thermodynamics to unbundle the nanotubes and disperse them further in a polymer matrix. Since neither of these two processes is expected to be very effective for ordinary polymers and nanotubes, typically the nanocomposites produced in this fashion have significant filler aggregation and comparably poor performance [e.g., high-density polyethylene, polypropylene, and polyamide-6/acrylonitrile–butadiene–styrene (ABS) have been melt-processed with nanotubes[63–65]]. Since direct melt processing is inherently ineffective in dispersing nanotubes into polymers, melt processing will probably remain limited in practice except for those systems for which polymer–nanotube masterbatches can be developed at reasonable cost and with good nanotube dispersion.

2.2.3 Dispersion Characterization: Common Techniques and Limitations

Due to its ease of use and availability, simple Bragg-reflection powder x-ray diffraction is most commonly used to probe nanocomposite structure, especially for polymer/layered-inorganic filler hybrids where the d_{001} basal reflection is

*This does not automatically imply that the respective nanocomposites are also characterized by deteriorated properties. For example, where good dispersions and/or covalent bonding occurs between the polymer matrix and the functionalized nanotubes, the nanocomposites can have very good mechanical property enhancements.[55]

indicative of filler–filler separation. However, the XRD can only detect the distance of periodically stacked layers; disordered (bunched together but not parallel stacked) or exfoliated layers are not detected, and large d-spacings (higher than 50 nm) are sometimes not detectable by powder XRD. In general, for medium (ca. 1 μm) lateral size platelets, such as those in natural clays, even with favorable thermodynamics for nanocomposite formation, the structure is characterized by the coexistence of exfoliated, intercalated, and disordered layers. Thus, a silent XRD may hide a large number of disordered tactoids, whereas an XRD with an intercalated peak does not reveal the extent of exfoliation. In both cases, the nanocomposite properties are commonly affected *dramatically* by structures that are not manifested in the XRD, and thus *XRD can be highly misleading when employed as a single tool for quantifying nanocomposite structure or even filler dispersion.* Although detailed quantitative analysis of such XRD data[66] in the low 2θ range, coupled with careful sample preparation and use of model reference samples, can yield substantially more information about the nanocomposite structure,[66] powder XRD is insufficient to capture and characterize the nanocomposite structure. Furthermore, when polymer–inorganic nanocomposites are based on fillers that are not two-dimensional in geometry (and thus do not have basal spacings, as for example carbon nanotubes and spherical or ellipsoidal nanoparticles), XRD is completely incapable of even a first-order qualitative determination of dispersion or structure.

Small-angle x-ray scattering (SAXS) is probably the most informative widely available technique to characterize nanocomposite structure. The main hurdle with this method is converting the information collected in the k-space quantitatively into parameters that describe the real space morphology of the hybrids. As an example, for polymer/layered-inorganic fillers, simple[67] and more realistic[68] models of discoid scatters in organic matrices have been proposed that can be used to interpret scattering data into real space parameters for such nanocomposites. In a simple approach,[67] after relatively simple analysis of the scattering data, average descriptors of the structure can be obtained which are of some value for quantifying the hybrid structure. A more complete description of structure necessitates much more careful design and implementation of scattering studies and more tedious analysis.[68] Even where models for specific structures have been developed and methods for an experimental approach and analysis have been outlined, as for example in the case of layered inorganic nanoparticles,[68] the amount of work involved to implement such approaches in real polymer nanocomposite systems has proven to be a barrier to the widespread use of SAXS as a common morphology characterization practice.

Transmission electron microscopy (TEM) is also widely employed, in its simplest bright-field mode, as a tool for direct visualization of the nanocomposite structure of polymer nanocomposites. This is possible because there exists sufficient contrast for the transmitted electrons between the polymer matrix and most fillers (inorganic particles, carbon in nanotubes or graphite, and almost all oxides) without polymer staining. In the extreme case, high-resolution TEM[69] can even provide a qualitative picture of the inorganic filler crystal structure, or can be combined with point electron diffraction to interrogate crystal structures in specific filler or polymer regions. Although TEM does not suffer from the same shortcomings as XRD, since it can visualize nanoscale fillers directly without the need for parallel stacking, it does have other limitations: First, it is very painstaking to obtain quantitative information about any of the characteristic parameters that describe the nanocomposite morphology. Such information can only be derived from image analyses of many and independent TEM images, so as to ensemble typical structures in the composite with some statistical importance. Second, since TEM is essentially a projection method, it is difficult to characterize structures normal to the large surface area of fillers; for example, almost all TEMs published for polymer/layered-silicate nanocomposites show images with the silicates positioned on the image edge-on, since layers parallel or oblique to the sample surface project as extended dark areas in a TEM image. Despite these limitations, we believe that informative TEMs should, at a minimum, complement XRD or other morphology studies, even if only to capture the hierarchical structures of the hybrid qualitatively at various length scales. Probably the additional information provided by TEMs is crucial when accompanying featureless XRD structures such as silent (no basal reflections) polymer/layered-nanofiller nanocomposites (which in most cases are wrongfully interpreted as exfoliated structures), polymer–nanotube hybrids, and polymer–nanoparticulate composites.

Finally, morphological information can also be obtained *indirectly* from methods that reflect the composite morphology into other macroscopic properties. Within the focus of this book, two examples of such methods can be mentioned: rheological measurements and cone calorimetry flammability methods. Both methods can sensitively detect well-dispersed nanofillers in a polymer matrix and can distinguish them from the respective conventional composites based on the same polymer and fillers but without nanometer-scale dispersion of the latter. We shall not provide further details on this; we just point the interested reader discussions of the cone calorimetry approach in subsequent chapters, and to a few representative references for the rheology.[7,16,70]

In summary, in lieu of providing a recipe for the characterization of nanocomposite morphology, we illustrate the limitations of the foregoing characterization techniques through an example of the plethora of parameters needed to describe the morphology of a polymer/layered-silicate nanocomposite (Figure 2.3). Even in this case, which can actually be interrogated by XRD characterization, only the distribution of basal (layer–layer distances within parallel stacked clusters) spacings can be obtained by XRD. SAXS can, in addition, provide some additional parameters,[67] such as mean number of layers per stack and "projected" lateral dimension of layers, while through more realistic models and analysis,[68] only approximate values can be obtained for the rest of the important parameters (Figure 2.3). In almost all cases, a representative set of TEM structure observations should also be obtained (in addition to diffraction or scattering characterization), which should provide a qualitative description of structure, although there may be shortcomings in quantifying the various morphological parameters

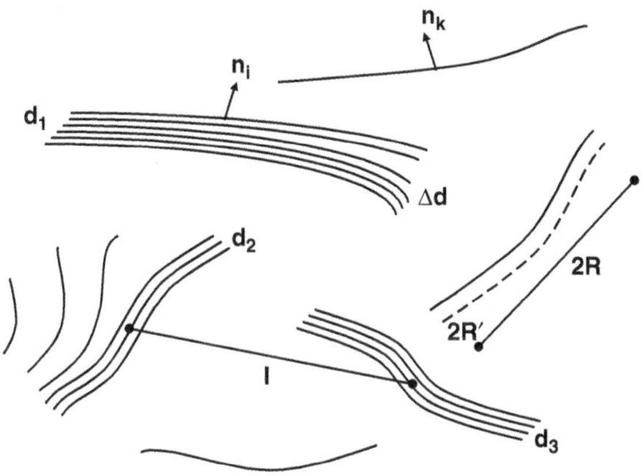

FIGURE 2.3 Relevant parameters needed to describe the morphology of a polymer/layered-silicate nanocomposite. Layer parameters: layer thickness (H), lateral contour size ($2R'$), and corresponding projected lateral size ($2R$). Layer stack parameters: distribution of layer–layer distances within a stack (d_{001}: d_1, d_2, d_3), distortion in d (Δd), and mean number of layers per stack (N). Distribution of stacks parameters: mean particle–particle distance between center of mass of stacks (l), relative particle–particle orientation [$\psi(\vec{n}_i, \vec{n}_k)$], and fraction of layer stacks consisting only of individual layers (χ). (Adapted from Ref. 68.)

(due to the local-only observation of morphologies, even by numerous TEM images).

2.3 EFFECTS OF NANOFILLERS ON MATERIAL PROPERTIES

2.3.1 Effects on Polymer Crystallization

2.3.1.1 Polymer-Specific Effects It is expected that the incorporation of nanoparticles in a semicrystalline polymer matrix would substantially affect the crystallization behavior of the polymer. Depending on polymer–filler interactions, three types of behavior can develop.

(a) Development of New Crystal Structures Where strong specific interactions exist between a filler and a polymer, a new crystal structure can develop in the vicinity of the filler, which is often not the same as the crystal structure of the unfilled polymer under normal crystallization conditions. The best example of such behavior is the case of polyamide-6/montmorillonite nanocomposites, in which the γ-crystal phase of polyamide is promoted next to the fillers.[42–44] This behavior originates from the strong hydrogen bonding of the amide groups with the silicate (SiO_x) surfaces, and is, for the same reason, also observed in poly(vinyl alcohol)/MMT nanocomposites.[39,71] A less frequent case, where new crystal structures are promoted by nanoscale fillers, also exists

for polymers that develop nonbulk crystal phases when the polymer chains are aligned parallel to the filler's solid surfaces; two examples of such nanocomposites are polyvinyldene fluoride (PVDF)[72] and syndiotactic polystyrene (sPS).[73] In all cases where the inorganic surfaces promote growth of a different crystal phase, the nanocomposite mechanical and thermal properties can be enhanced through this mechanism when the surface-nucleated crystalline phase has better mechanical and thermal characteristics than those of the bulk crystal phase. Fillers with a large surface area maximize these filler-induced enhancements of the material properties; a dramatic manifestation of such a response is found in polyamide-6/montmorillonite nanocomposites.

(b) Polymer Amorphized by Filler In very few cases, such as poly(ethylene oxide) (PEO)/Na^+–montmorillonite nanocomposites, the polymer–Na^+ interactions are favorable to mixing but not conducive to crystallinity.[74] Specifically, the crystallization of PEO nanocomposites based on alkali-cation bearing fillers is found to be inhibited, exhibiting a decrease in spherulite growth rate and crystallization temperature. Although the overall crystallization rate increases with silicate loading as a result of the extra nucleation sites that occur in the bulk PEO matrix (i.e., far from the silicate surfaces), PEO is highly amorphized near the montmorillonite surfaces. This behavior is attributed to the specific way that PEO interacts with Na^+ montmorillonite, where strong coordination of PEO to the surface Na^+ cations promotes noncrystalline (ether crown) PEO conformations.

(c) Heterogeneous Nucleation by Fillers For the vast majority of polymers, the effect of nanofillers on polymer crystallization relates only to crystal nucleation by the fillers (which typically increases proportionally to the number of individual filler clusters) and to changes in the kinetics of crystallization (which are typically characterized by a two- to fourfold decrease in the linear growth rate of crystallization). In these cases, and for filler loading below ca. 10 wt%, the equilibrium melting temperature (T_m^0) is not affected by the nanocomposite formation. For example, as shown in Figure 2.4, the T_m^0 of PP-g-MA, PET, and PEO nanocomposites and the respective bulk polymers were estimated based on Hoffman–Weeks plots, and it is shown that moderate (below 10 wt%) MMT addition does not change the T_m^0 value [T_m^0(PP-g-MA) = 183.8°C, T_m^0(PET) = 260.1°C, T_m^0(PEO) = 69.7°C]. These results are consistent with the literature reported earlier[8] and make it possible to compare the crystallization kinetics of neat polymers and their nanocomposites at the same isothermal crystallization temperature. To further elucidate the effect of MMT on the crystallization kinetics of these polymers, isothermal crystallization measurements can be carried out with differential scanning calorimetry (DSC) and complemented by direct imaging of the crystallites [cross-polarization optical microscopy and atomic force microscopy (AFM)] for systems crystallized under the same conditions (Figure 2.4). Initially (crystals grow in three dimensions and the crystallites have not yet impinged), crystallization kinetics can be expressed as

$$V_f^c = \frac{4}{3}\pi \rho_n G_R^3 t^3 \tag{2.7}$$

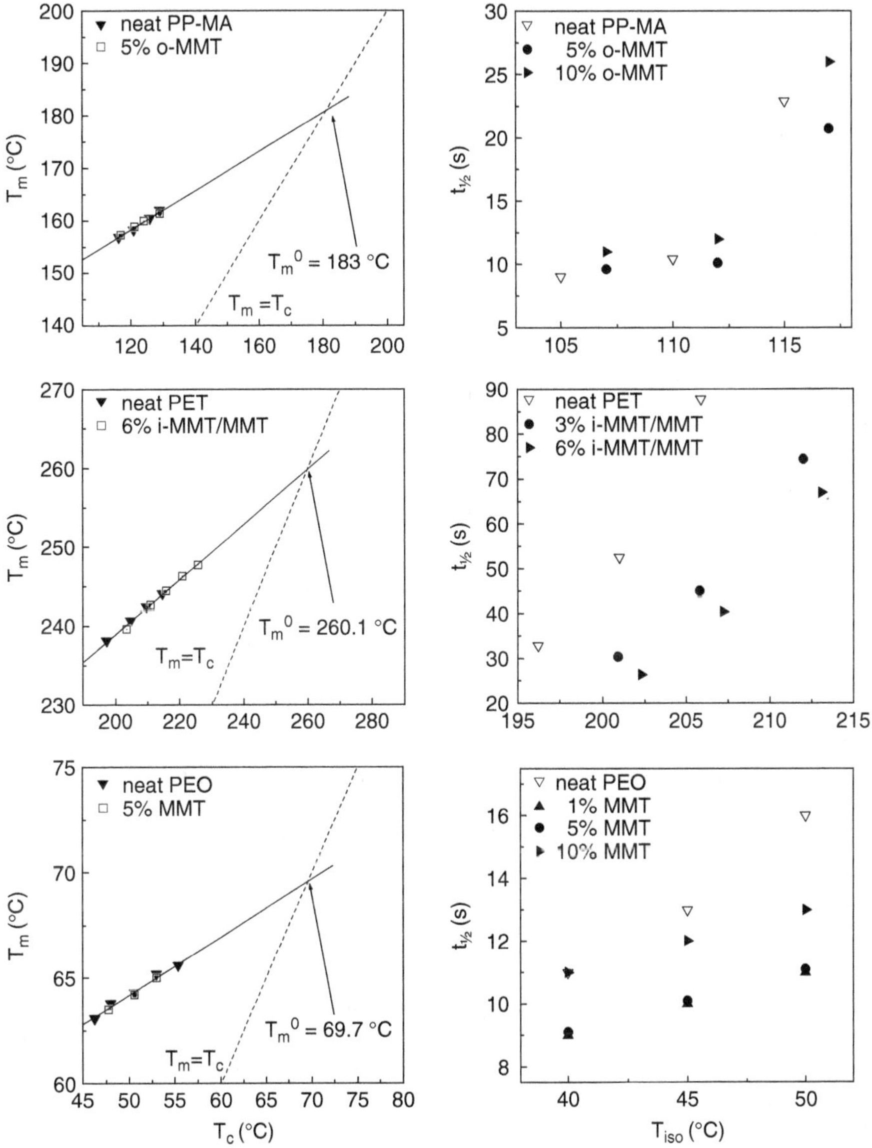

FIGURE 2.4 (*Left*) Hoffman–Weeks plots of neat polymers and their nanocomposites; the T_M^0 of the polymers is not affected by the nanocomposite formation. (*Right*) Half-time of crystallization for the same neat polymers and their nanocomposites; the overall crystallization rate is reduced for PET and PEO upon incorporation of an inorganic nanofiller, and is not affected for PP-g-MA. When accounting for changes in the nuclei density with filler incorporation, the linear growth rates G_R are slowed down in all systems shown. For both panels: PP-g-MA (*top*), PET (*middle*), PEO (*bottom*).

where V_f^c is the total crystal volume (crystallinity), ρ_n the nuclei density, G_R the linear crystal growth rate, and t the crystallization time. When V_f^c is 0.5, the corresponding crystallization time t is defined as the half-time of crystallization ($t_{1/2}$) and denotes the time necessary to reach 50% of the total enthalpy of crystallization under isothermal differential scanning calorimetry conditions (Figure 2.4). When the nuclei density, ρ_n, is measured by cross-polarized optical microscopy and/or AFM, the crystal linear growth rate can be estimated. The half-times of crystallization for neat polymers and their nanocomposites is shown in Figure 2.4 for various isothermal crystallization temperatures (T_{iso}). As expected, the *overall* crystallization rate increases with clay or filler addition, as denoted by the decrease in $t_{1/2}$ upon the addition of nanofillers (this effect is rather small for PP-g-MA). However, accounting for the nuclei density increase in the nanocomposites (for PP-g-MA, the ρ_n increased ca. six- to eightfold at 5 to 10 wt% o-MMT content; for PET, more than 500-fold at 3 to 6 wt% o-MMT; and for PEO, 20- to 50-fold for 5 to 10 wt% MMT), linear crystal growth is slowed down due to the introduction of clay across all systems. Despite the qualitative differences between PEO, PP, and PET crystallization when reinforced by MMT, and despite the quantitative differences in $t_{1/2}$, when the increase in nuclei density is accounted for, all systems show a G_R decrease of 0.25 to 0.5 upon MMT addition (for PP the nanocomposite G_R value is 0.5 of the bulk polymer value, for PET it is 0.25, and for PEO it is 0.33 of the respective bulk polymers). This agreement between such different systems strongly indicates that the geometric constraints associated with the dispersion of MMT fillers is determining the effect (decrease) on the linear crystal growth rate in these systems rather than the polymer–MMT interactions. In the latter case, one would expect a qualitatively different effect in PET and PP compared to PEO, and also substantial quantitative differences between PET and PP. All these effects manifest themselves in differential scanning calorimetry studies, especially when the behavior of the neat (i.e., unfilled) polymer is compared against that of the respective nanocomposite (Figure 2.5).

2.3.1.2 General Effects Across Polymers Despite the variety of the nanofiller effects on polymer crystallinity, which originate from the various polymer–filler interactions, there also exist important common effects on the crystallinity due to the nanocomposite structure. The most important of these general effects is probably a general reduction in the size of the polymer crystallites upon nanocomposite formation. For example, in Figure 2.6 we compare the spherulites observed for unfilled polymers and their respective 3 wt% montmorillonite nanocomposites. Independent of how the fillers affect the nucleation and/or kinetics of crystallization, there is in all cases a substantial decrease in the spherulitic size. This behavior originates from the discontinuity of space caused by the inorganic fillers, which forces spherulites to have sizes comparable with the filler–filler separation, independent of the bulk polymer spherulite size. This effect is also independent of whether crystallization in the nanocomposite is nucleated homogeneously (PEO) or heterogeneously (PP, sPS), and of whether the fillers hinder crystallization

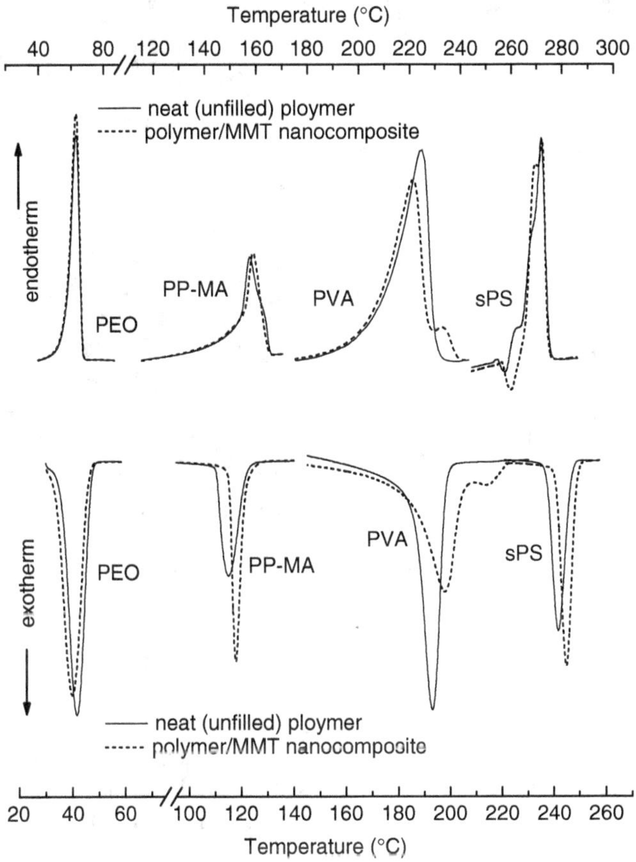

FIGURE 2.5 Differential scanning calorimetry comparison of unfilled polymers and their respective nanocomposites with montmorillonite layered silicates. (*Top*) Heating DSC scans: The crystalline melting point is markedly unaffected by the addition of fillers since the polymer crystal structure (e.g., the crystal unit cell) is not affected by the filler. A notable exception are those polymers where a new crystal structure is promoted near the filler surface, such as PVA, syndiotactic-PS, and polyamide (not shown here). (*Bottom*) Cooling DSC scans: The crystallization point is strongly affected by the fillers, bearing traces of heterogeneous nucleation (PP, sPS, PVA), crystallization of new crystal structures (PVA), or hindering of crystallization near a filler (PEO).

(PEO), promote new crystal structures (sPS), or simply act as heterogeneous nucleating agents (PP).

2.3.1.3 Effects of One-Dimensional Nanofillers Like layered-inorganic fillers, carbon nanotubes influence polymer crystallization when incorporated as filler in the polymer matrix; however, these effects do not have as wide a variety as the layered silicates discussed above. In the vast majority of reports, carbon nanotubes act simply as heterogeneous nucleating agents in crystallizable polymer

FIGURE 2.6 Comparison of cross-polarized optical microscopy pictures of unfilled polymers (*top*) and their respective nanocomposites (*bottom*) with 3 wt% of montmorillonite fillers; PEO (*left*), PP-g-MA (*middle*), and sPS (*right*).

systems.[11,56,64,75–78] For example, polypropylene crystallization in the presence of nanotubes shows increased crystallization temperature and rate of crystallization with the introduction of varying concentrations of nanotubes,[79,80] with no change in the crystalline structure or the melting point. Furthermore, the PP crystallite size decreases in the presence of nanotubes,[64,75,76] in agreement with the general behavior observed in polymer/layered-inorganic nanocomposites.

When nanotubes interact strongly with the host polymer, as for example with conjugated and ferroelectric polymers, polymer crystallization is altered, developing higher-order structures and increased degrees of crystallinity.[79,80] However, the crystallization effects discussed above for layered silicates due to polymer coordination with alkali cations (for PEO) and due to extensive hydrogen bonding (for amides) are, as expected, absent in nanotube-reinforced nanocomposites. Namely, PEO does not have any amorphous regions near nanotubes, and it follows bulklike crystallization, with the overall percent crystallinity, crystallization point, and melting point remaining unaffected, even at loadings of 7 wt% nanotubes.[81] Similarly, polyamide-6 and polyamide-12 matrices reinforced with nanotubes exhibit crystallization similar to that of the unfilled polymer.[78,82]

Finally, the one-dimensional geometry of nanotubes provides exciting opportunities for controlled nucleation and growth of single crystals along individual fibers, as for example with polyamide-6,6 and polyethylene crystallized from solution, allowing for control of crystallite periodicity and molecular-level architecture.[83] This unique capability of nanotubes can conceivably lead to special types of "functionalization" of individual nanotubes, which can be exploited to improve interactions (see our discussion of better interfacial coupling) and control dispersion in selected polymer matrices.

2.3.2 Effects on Mechanical Properties

Most polymer–clay nanocomposite studies report tensile properties as a function of MMT content (ϕ_{MMT}). As a typical example, in Figure 2.7 we compare tensile moduli from various studies of neat PP/o-MMT and MA-functionalized–PP/o-MMT nanocomposites. The characteristic behavior for polymer/layered-inorganic nanocomposite materials[6,8] is observed: Namely, there is a sharp increase in Young's modulus for very small inorganic loadings ($\phi_{\text{o-MMT}}$ <4 wt%), followed by a much slower increase beyond $\phi_{\text{o-MMT}} \simeq 5$ wt%. With increasing ϕ_{MMT}, the yield stress does not change markedly compared to the neat-polymer value, and there is only a small decrease in the maximum strain at break. PP systems filled conventionally (i.e., no nanometer-level dispersion) by the same fillers (e.g., 2C18–MMT) do not exhibit as large increases in their tensile modulus (Figure 2.7*a*).

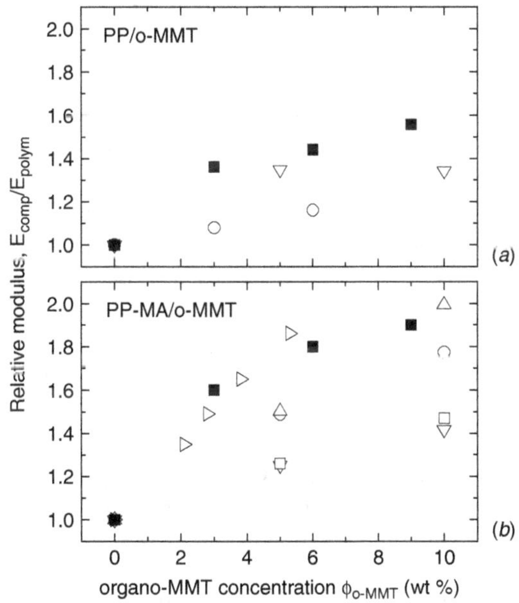

FIGURE 2.7 Tensile moduli (relative to bulk value) for various PP–MMT nanocomposites. (*a*) neat-PP hybrids: with f-MMT (■[49]), C18–MMT (▽[35]), and 2C18–MMT (○[49]). In the absence of favorable thermodynamics, the dispersion and thus the mechanical properties are a strong function of the processing conditions. (*b*) PP-g-MA/2C18–MMT melt-processed nanocomposite (■[49]) and PP hybrids formed via various PP-g-MA pretreated o-MMT master batches: C18–MMT (▷[34]) and C18–MMT (○, △[35]). Given the well-defined thermodynamics of mixing, there is a small variation of dispersion and mechanical properties across different systems and various research groups. Slight changes in the thermodynamics [e.g. when a different surfactant is employed C8–MMT (▽, □[35])] result in moduli changes. (Adapted from Ref. 49, copyright © 2001, American Chemical Society, with permission.)

This mechanical reinforcement is expected and not too exciting at first glance, especially considering that the montmorillonite filler platelets have a very high intrinsic stiffness (tensile modulus of 140 to 180 GPa). However, there are some points that can be made: The tensile results obtained from thermodynamically stable hybrids are not affected by processing conditions (since the nanocomposite structure remains the same), whereas in the absence of favorable PP/o-MMT thermodynamics, the structure and tensile properties vary strongly with the processing conditions (Figure 2.7*b*). Similar improvements in mechanical properties can also be achieved by other layered particulate fillers; however, much higher filler loadings are required (e.g., by loading 30 to 60 wt% of talc or mica [14]), since such particles are not well dispersed and the *effective* filler surface area is orders of magnitude smaller. Finally, for PP/o-MMT, the relative improvement in the moduli compared to the unfilled polymer is rather small (barely reaching 60% for PP and 100% for PP-g-MA), whereas in other systems, such as elastomers or polyethylene, improvements of 400 to 1200% in the Young's modulus can be achieved by the same o-MMT fillers. The origin of this behavior is traced to two effects:

1. The relatively poor interaction of polyolefins with o-MMT [cf. eq. (2.6); interfacial adhesion energy of ~83 mJ/m^2; see also Sec. 11.2 in Israelachvili[32]]. As the polymer–inorganic adhesion is improved (e.g., when MA functional groups are added to the polymer), the stresses are much more effectively transferred from the polymer matrix to the inorganic filler, and thus a higher increase in Young's modulus is achieved (Figure 2.7*b*).
2. The relatively high modulus of the original polymer (for the PP reported, 0.6 to 1.3 GPa). The latter effect becomes clearer when this behavior is contrasted with nanocomposites formed by the same filler in a "softer" matrix, such as elastomers or PEs that have tensile moduli in the range 0.1 to 0.3 GPa.

As further evidence of the last two points, we also show the tensile moduli of polyamide–MMT systems (Figure 2.8*a*), where substantial improvements in mechanical properties can be achieved, despite the relatively high stiffness of the polymer matrix, due to the very effective stress transfer from the polymer to the filler, mediated by strong hydrogen bonding. In the case of polyamide-6/MMT nanocomposites, independent of the original polyamide-6 matrix characteristics and of whether the hybrids were formed by in situ polymerization or melt blending,[17,18,42–44,84] there seems to be considerable agreement on the enhancement achieved in the tensile modulus that spans research groups, methods, and materials. We postulate that due to the strong interfacial adhesion (i.e., every amide group of the polymer can hydrogen bond to the silicate surface), the interfacial strength and maximum interfacial shear stress are dictated by the polymer–MMT interactions and overwhelm all other parameters that relate to processing and dispersion, polymer matrix characteristics, and/or stiffness of the filler.

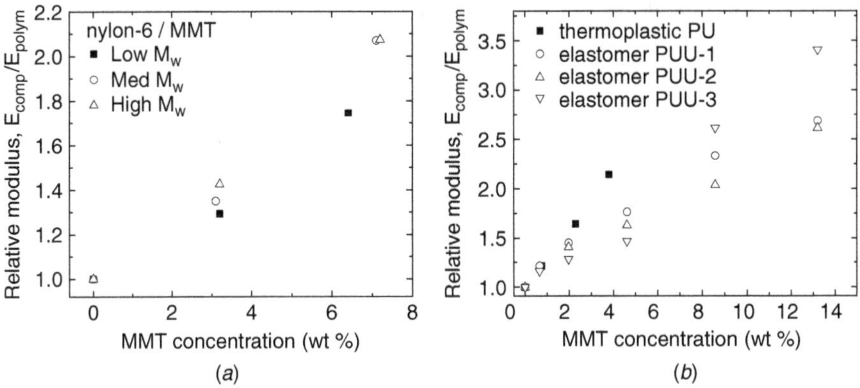

FIGURE 2.8 Tensile moduli (relative to bulk value) for various nanocomposites: (*a*) polyamide-6/MMT nanocomposites,[84] with low-, medium-, and high- molecular weight polyamide-6 matrix, as an example of high improvement in mechanical properties due to effective stress transfer from polymer to filler; (*b*) polyurethane and polyurethane copolymers/MMT nanocomposites, as an example of high improvement in mechanical properties due to a "soft" original polymer (■[85], ▽[40], ○ △[41].)

2.3.2.1 Theoretical Insights in to Mechanical Properties Even from the very brief discussion above, it becomes obvious that a priori prediction of the mechanical properties of polymer–inorganic nanocomposites is rather involved, and to date the design of such nanocomposites is based mostly on Edisonian approaches. Theoretical models developed for the prediction of mechanical properties of conventional composites, such as the Halpin–Tsai[86] and Mori–Tanaka[87] models, fail in their "straightforward" application to nanocomposite systems. There are numerous physical phenomena that need to be included in such models so as to better describe the mechanical behavior of polymer–matrix nanocomposite materials. Again drawing examples from polymer/layered-silicate nanocomposites, recent theoretical models have been developed that attempt to capture the behavior of these materials by accounting for the high aspect ratio of the fillers: for example, an effort[88] that modifies the Halpin–Tsai model to account for the buckling of filler platelets, incomplete dispersion, and nonbiaxial in-plane filler orientation; despite its additional complexity and improvements, this modified Halpin–Tsai model does not seem to be highly successful in predicting the mechanical properties of polymer/layered-silicate nanocomposites for a wide range of polymer matrices.[88] The main shortcoming in the previous approach is attributed[89] to the insufficient modeling of a "constrained region" of polymeric material surrounding the nanoscopic filler; this interfacial polymer is expected to differ in properties and morphology from the bulk polymer matrix, as has been observed experimentally. However, accounting for such a constrained region, as for example by introducing appropriate modifications[89,90] in the Mori–Tanaka model, still has a limited predictive power when applied across various polymer matrices and necessitates adjustment of the model's parameters for each nanocomposite system.[90] Even in the most focused approach, when a mechanical model is developed to describe a single polymer–inorganic nanocomposite[91]—while accounting for the imperfect interfacial coupling and the effective aspect ratio and filler volume fraction due to varied dispersion with filler loading—such a model necessitates calculation of an interfacial strength parameter (in this case an interfacial shear stress, which was calculated[91] to be 2 to 8 MPa for a poly(dimethyl siloxane) (PDMS)/MMT system).

Despite any shortcomings and approximations, these theoretical endeavors offer valuable insights in important design parameters for the mechanical performance of polymer nanocomposites. Specifically:

- Mechanical properties are determined by the *effective filler aspect ratio* and *effective filler volume fraction* when incomplete dispersion is accounted for[88,91] (rather than on the absolute filler loading and the aspect ratio of the individual fillers).
- Filler-specific mechanisms of deformation and fracture can have a considerable contribution to the mechanical properties of the nanocomposite.[88]
- The correct enumeration of the interfacial strength is crucial for correct estimation of the composite's mechanical properties,[91,90] and its small value compared to the modulus of the filler can dramatically limit a filler's reinforcing effectiveness.

In particular for the last item, interfacial strength at the polymer–filler interface can be experimentally measured directly in very few cases; for example, carbon nanotubes have been pulled out from a polymer [poly(ethylene–butene)] matrix by AFM, yielding interfacial strengths [92,93] of 10 to 90 MPa, depending on the nanotube radius. These experimental interfacial strength values correlate well with interfacial forces calculations, [30] such as those described earlier [eq. (2.6)]. Thus, one may expect that the same approach used for predicting miscibility of polymers and layered fillers may be helpful in estimating the polymer–filler interfacial strength. Given the continuum character and the assumptions behind such calculations, and the very approximate numbers available for the surface tension components of the materials involved, this approach can only provide a first-order estimation of the interfacial strength for polymer and various nanofillers. Albeit its uncertainty, this theoretical value of the polymer–filler interfacial strength may be an important design element for the mechanical properties of nanocomposites, especially since it is very difficult to envision approaches able to determine this interfacial strength experimentally.

Some examples of the application of eq. (2.6) for polymer/layered-inorganic nanocomposites could be:

- For polypropylene–montmorillonite interfaces, ignoring all necessary functionalizations for PP would yield an interfacial adhesive energy of

$\sim$83 mJ/m^2, corresponding to an interfacial strength of $\sim$10 MPa (cf. 3 to 7 MPa from tensile measurements[49]).

- For PDMS–montmorillonite, the same approach yields an interfacial energy of $\sim$91 mJ/m^2 or an interfacial strength of $\sim$11 MPa (cf. 2 to 8 MPa from theoretical models[91]).
- For polyamide–montmorillonite nanocomposites, and ignoring all crystalline-phase changes that may be caused by silicate fillers,[42–44] an adhesive interfacial adhesion of $\sim$107 mJ/m^2, corresponding to an interfacial strength of $\sim$14 MPa.
- For carbon nanotube–polypropylene yields an interfacial energy of $\sim$49 mJ/m^2 (cf. 47 mJ/m^2 from AFM experiments [92]) or an interfacial strength* of $\sim$6.2 MPa (cf. 20 to 40 MPa from multiwalled nanotubes,[93] and 2 MPa from computer simulations[94]).

To the extent that they are valid, the observations above bear significant implications for the possibilities of mechanical property improvements via nanocomposite formation. Specifically:

1. Given the nature of a polymer (i.e., γ^{LW} and $\gamma^{\perp}$), the maximum mechanical reinforcement by a completely dispersed nanofiller will be limited by the polymer–filler interfacial strength. For example, in the case of PE and PP ($\gamma^{LW} \simeq 26$ mJ/m^2 and $\gamma^{\pm} = 0$) and layered silicates, there would be a common limit of about 2 to 4 GPa for the maximum tensile modulus that can be achieved through nanocomposite formation. This is in agreement with experimental studies for these systems, which show a similar absolute value for the maximum tensile modulus obtained by PE and PP [albeit reflected in much bigger *relative* improvements of 400 to 1200% for the softer LDPE, compared to 60 to 100% for the stiffer *i*-PP (Figure 2.7)].
2. The addition of a small number of functional groups (e.g. addition of maleic anhydride groups in PP) would increase the interfacial adhesion only moderately, and would similarly increase the tensile moduli only moderately (Figure 2.7*b*).
3. The addition of large numbers of strongly interacting (with the filler) groups along the chain, such as hydrogen-bonding groups densely across the polymer backbone, would result in larger *relative* improvements in mechanical properties (Figure 2.8), but still below the upper limits set by the interfacial adhesions calculated. (The use of polyamide-6 as an example in this case is questionable, given the promotion of the γ-phase crystal for the

*Equation (2.6) is independent of geometry; however when estimating interfacial strength, the filler geometry (i.e., contact geometry) must be considered (see, e.g., Sec. 11.1 in Israelachvili [32] or Sec. VI.1 in Van Oss[29]). The value provided for the nanotube–polyethylene here (6.2 MPa) is based on the interaction of two semi-infinite flat surfaces. Calculation for a cylinder in contact with a semi-infinite flat surface yields an interfacial strength of 4.6 MPa, whereas the interaction between a cylinder emerged in a polymer should be somewhere between these two values.

nanocomposites[42–44]; however, the favorable comparison of polyamide-6 behavior with the behavior of urethane–urea systems may a posteriori justify this choice.)

4. Finally, although chemical bonding of the polymer to the filler may seem the ultimate route to reinforce the polymer–filler interface, if such covalent bonds are not introduced densely across the length of the polymer, they will result in only a limited interfacial reinforcement and a respectively moderate improvement in the mechanical properties. This has been shown in cross-linked systems with reactive (via the cross-linking groups) dispersion of silicate layered fillers.[95]

2.3.3 Effects on Barrier Properties

The permeability of small penetrant molecules through an organic matrix is determined by the solubility and diffusivity of the small molecule in the matrix as well as by the mean-square displacement (total path length traveled) divided by the sample thickness. In principle, the addition of a filler in the polymer matrix is expected to affect the solubility and diffusivity of a penetrant molecule, especially in the vicinity of the filler (i.e., in the filler–polymer interfacial region and at least one polymer R_g away from the filler surface). Also, it is expected that fillers will affect the path tortuosity (mean-square displacement of penetrant versus film thickness) directly, when penetrants are forced to travel around impermeable fillers, and indirectly, when fillers induce polymer chain alignment or alignment and modification of polymer crystallites.*

Theoretical approaches on the barrier properties of nanocomposites treat fillers as impermeable nonoverlapping particles and assume no permeability changes in the polymer matrix.[97–100] Effectively, this means that the permeability of the composite will be smaller than the permeability of the matrix (unfilled polymer) by a factor equal to path tortuosity in the composite (simply assuming that the penetrant path cannot cross any filler particles). This path tortuosity was calculated by Nielsen[97] for completely aligned filler particles (all fillers have their larger surface parallel to the film surfaces, but there is no order in the filler center of mass), and its contribution to the composite permeability was derived to be

$$\frac{P_{\text{comp}}}{P_{\text{poly}}} = \frac{1-\phi}{1+a\phi} \tag{2.8}$$

with a being the filler aspect ratio (for square fillers of length/width L and thickness W, a = L/2W) and ϕ the volume fraction of the filler. Bharadwaj[100]

*The first mechanism, associated with chain alignment and the related diffusive anisotropy of a small molecule within aligned chains, has a relatively weak effect on permeability,[96] whereas the second mechanism, associated with crystallite alignment and changes in the crystal morphologies, causes rather strong changes in permeability and is commonly employed in strain-hardened semicrystalline polymers for barrier applications.

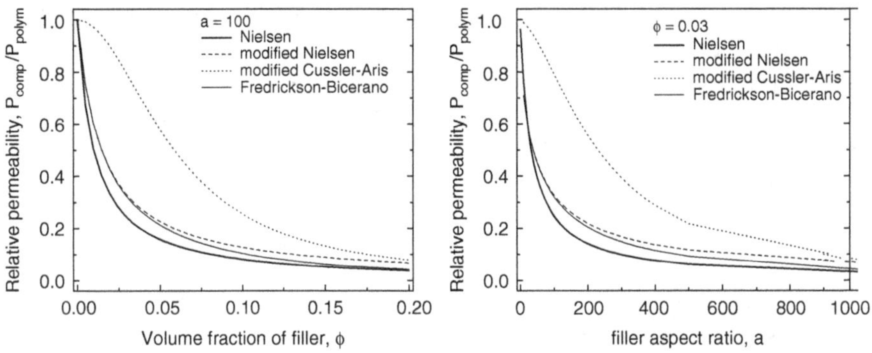

FIGURE 2.9 Comparison of theoretical models quantifying the effect of path tortuosity on the permeability of a composite: Nielsen model[97] [eq. (2.8)], Friedrickson–Bicerano [eq. (2.10)], modified Nielsen [eq. (2.9)], and Cussler–Aris [eq. (2.11)].

has modified this equation to account for nonaligned fillers by introducing an order parameter S for the filler orientation:

$$\frac{P_{\text{comp}}}{P_{\text{poly}}} = \frac{1-\phi}{1+a\phi\frac{2}{3}(S+\frac{1}{2})} \qquad \text{with} \quad S = \frac{1}{2}\langle 3\cos^2\theta - 1\rangle = \begin{cases} 1 & \parallel \text{surface} \\ 0 & \text{random} \\ -\frac{1}{2} & \perp \text{surface} \end{cases} \tag{2.9}$$

which reduces to Nielsen's equation for perfectly aligned fillers ($S = 1$). In a more detailed approach, Friedrickson and Bicerano[99] derived the same path tortuosity effects for circular fillers (radius R and thickness $2W$) and an aspect ratio $a = R/2W$:

$$\frac{P_{\text{comp}}}{P_{\text{poly}}} = \frac{1}{4}\left(\frac{1}{1+a\phi\beta_1} + \frac{1}{1+a\phi\beta_2}\right)^2 \qquad \text{with} \begin{cases} \beta_1 = (\pi/\ln a)(2-\sqrt{2})/4 \\ \beta_2 = (\pi/\ln a)(2+\sqrt{2})/4 \end{cases} \tag{2.10}$$

which can cover a wider ϕ range, from dilute to semidilute, than the modified Nielsen and modified Cussler–Aris relations (as presented in the same work,[99] modified to address circular fillers):

$$\frac{P_{\text{comp}}}{P_{\text{poly}}} = \frac{1}{1+a\phi\pi/\ln a}\begin{pmatrix}\text{modified} \\ \text{Nielsen}\end{pmatrix}, \qquad \frac{1}{1+[a\phi\pi/(4\ln a)]^2}\begin{pmatrix}\text{modified} \\ \text{Cussler–Aris}\end{pmatrix} \tag{2.11}$$

Nevertheless, eq. (2.10) generally gives results similar to those using the Nielsen approach [eq. (2.8)], when a geometric correction of $\sqrt{\pi}/2$ is applied to the filler aspect ratio (i.e., comparing equal area fillers, square for Nielsen and circular for Friedrickson–Bicerano). A comparison of the theoretical models is illustrated in Figure 2.9. Given that all models (except the Cussler–Aris) give similar behavior for the range of parameters relevant to polymer/layered-inorganic nanocomposites ($10 < a < 1000$ and $\phi \leq 15$ vol%), we henceforth use the much simpler Nielsen model, including the addition of the orientation term [eqs. (2.8) and (2.9)]. According to this model, the obvious expectations can be quantified: higher aspect ratio fillers provide substantial lower permeabilities for a given filler volume fraction (Figure 2.10*a*), and aligned fillers are more effective barriers for a given aspect ratio and filler loading (Figure 2.10*c*). Additionally, some not-so-obvious conclusions can also be drawn:

- Beyond the filler aspect ratio, the composite permeability is also controlled by the filler volume fraction and/or by filler alignment [e.g., eq. (2.9)]: Thus, low aspect ratio fillers can be as effective as higher aspect ratio fillers, although at slightly higher loadings. For example, for aligned fillers (Figure 2.10*a*), a completely exfoliated montmorillonite (a = 500) at $\phi \simeq$ 2 vol% has comparable permeability with a partially exfoliated montmorillonite (a = 200) at $\phi \simeq$ 3%, or a mostly intercalated montmorillonite (a = 100) at $\phi \simeq$ 5%. This observation has important implications in designing a barrier nanocomposite: For the same example, instead of completely exfoliating a given filler, a task that is usually difficult to achieve, the same filler in a partially exfoliated or mostly intercalated morphology could achieve the same barrier performance at slightly higher filler loadings.
- Perfectly aligned fillers result in similar permeabilities with randomly oriented fillers of higher aspect ratio and/or at higher loading. For example (Figure 2.10*b*), for an a = 300 filler, perfect alignment at ϕ = 1.5% results in the same barrier performance as that of the same filler when randomly oriented at $\phi \simeq$ 4.3%; and for an a = 500 filler, perfect alignment at 1.5 vol% is comparable in permeability with a 4.5 vol% composite with random filler orientation. Along the same lines, a perfectly aligned filler nanocomposite with a = 300 at ϕ = 1% has similar barrier performance as a = 500 at 2%. This observation also provides important guidance on how to avoid the difficult task of perfectly aligning the fillers parallel to the film surface (Figure 2.10*c*).
- The effect of filler orientation on permeability decreases in importance with a higher filler aspect ratio (Figure 2.10*c*). For example, permeability improvement for a = 1000 is only 5% better with perfect alignment (S = 1) than with a random orientation (S = 0), and for a = 500 this difference is about 10%.

Additionally, the favorable comparison of these theoretical predictions with experimental data (Figure 2.10*d*) gives some credibility to the conclusions above. In Figure 2.10*d* we plot experimental water vapor permeabilities of various solvent cast nanocomposite films. The experimental behavior follows closely the theoretical trend and is enclosed between the response of exfoliated systems (especially for low filler loadings) and that of intercalated systems (for moderate and higher loadings). This reflects the same effective filler aspect ratio

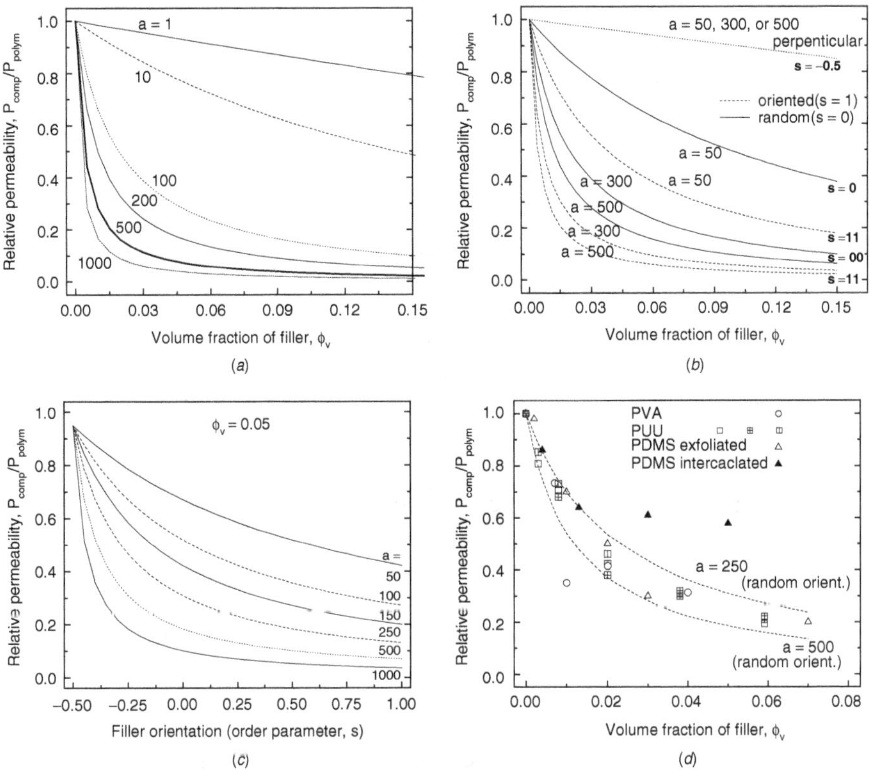

FIGURE 2.10 Theoretical predictions based on path tortuosity [eq. (2.9)], as a function of (*a*) filler aspect ratio $a = 1$ to 1000; (*b*) filler aspect ratio and alignment ($S = 1$: perfect smectic alignment—dashed lines; $S = 0$: random orientation—solid lines); (*c*) filler aspect ratio for a constant volume fraction $\phi_V = 5\%$. (*d*) Comparison of the same theoretical predictions (parameters as indicated) with experimental values for water vapor permeabilities in various polymer–montmorillonite nanocomposites. (From Refs. 39–41.)

discussed before in relation to mechanical properties. This agreement persists for all systems that have good filler dispersion (as achieved by solvent casting) and disappears for the same polymer and filler when dispersion is poor (cf. PDMS/dimethyldialkyl–montmorillonite with intercalated versus exfoliated composite structures). Finally, this agreement is rather independent of polymer and filler hydrophillicity, ranging from very hydrophillic poly(vinyl alcohol) reinforced by Na^+-montmorillonite, to moderate poly(urethane-co-ureas) to rather hydrophobic poly(dimethyl siloxane) and dialkyl-modified montmorillonite. This agreement is also independent of polymer crystallinity: ranging from semicrystalline poly(vinyl alcohol) with filler-induced crystallinity effects, to segmented semicrystalline poly(urethane-co-ureas), to amorphous poly(dimethyl siloxane). Thus, it seems that the path tortuosity effects may in fact overwhelm other important parameters (such as permeant solubility changes* and polymer crystallinity effects) when it comes to predicting permeability changes upon nanocomposite formation.

2.4 FUTURE OUTLOOK

Nanocomposites, in the sense of hybrid materials with novel properties beyond the realm of unfilled polymers and conventional composites, bear high promise for enabling new uses and applications of polymer materials. In the simplest approach, they can expand the window of applications of a given polymer, and in the best case they can enable the use of polymer–matrix composites in applications where metal or ceramic materials are currently used. One of the first untapped challenges in the field is to go beyond the simple dispersion of fillers and move toward the development of methods to create well-defined three-dimensional morphologies of nanofillers: morphologies that contain highly aligned fillers, house-of-cards structures, edge-connected (starlike) formations, and alternating two- and one-dimensional fillers.

The highest benefit of the hybrid character of nanocomposites comes from overcoming the property trade-offs associated with conventional composites: For example, nanocomposites can improve stiffness without sacrificing toughness, can enhance barrier properties without sacrificing transparency, can bestow flame retardancy without sacrificing mechanical properties, and can enhance mechanical performance and biodegradability simultaneously. When such behaviors are enhanced synergistically with effects from other additives or fillers, they can effectively push the envelope of the current state of the art. Such approaches will develop particularly exciting systems where synergistic combinations of multiple nano- and macrofillers are properly combined in a multifiller composite material.

Although it currently engages an overwhelming number of research groups, the field desperately needs well-designed scientifically-based studies to explore the fundamentals of these materials. Since barriers to entering the field are really low (no need for special equipment or expensive materials; studies can be published even when reproducing results from previous works or making minor incremental advances), the temptation is high to simply mix polymers with off-the-shelf nanofillers and just report x-ray diffraction and mechanical measurements. The real potential of these materials will remain untapped, however, until the nanoscale mechanisms responsible for macroscopic properties are unveiled and are further exploited to make radically new materials. New horizons need to be explored, especially outside the "comfort zone" of traditional polymer or materials scientists. If one considers the numerous examples of biological organic and inorganic nanostructures with unparalleled performances and combinations

*We should point out that barrier or permeability properties relate to the rate of permeant molecule diffusion through a polymer or nanocomposite material and cannot be extended to make predictions for ultimate water uptake or more general solvent uptake. If water or solvent uptake is of interest, an independent experiment is required, and for this property, the changes in solubility upon nanocomposite formation are the determining factor.

of properties that transcend any synthetic material, one can only start to imagine the limitless possibilities of this field.

Acknowledgments

Experimental data, theoretical arguments, conclusions, and opinions presented in this chapter have been developed through our research in this field over the last few years. These efforts were supported financially by the Pennsylvania State University, the National Science Foundation, the Office of Naval Research, the ACS/Petroleum Research Fund, the National Research Council, and various industrial projects, including Projects from Sumitomo Chemical, Air Products, Coca-Cola, and Bayer MaterialScience.

REFERENCES

1. Theng, B.K.G. *Formation and Properties of Clay–Polymer Complexes*. Elsevier, Amsterdam, The Netherlands, 1979.
2. Theng, B.K.G. *Chemistry of Clay–Organic Reactions*. Wiley, New York, 1974.
3. Pinnavaia, T.J.; Beall, G.W., Eds. *Polymer–Clay Nanocomposites*. Wiley, Chichester, West Sussex, England, 2000.
4. Utracki, L.A. *Clay-Containing Polymeric Nanocomposites*. Rapra Technology, Shawbury, Shrewsbury, England, 2004.
5. Mai, Y.; Yu, Z., Eds. *Polymer Nanocomposites*. Woodhead Publishing, Cambridge, England, 2006.
6. Alexandre, M.; Dubois, P. Polymer-layered silicate nanocomposites: preparation, properties and uses of a new class of materials. *Mater Sci. Eng. R Rep.* **2000**, 28, 1–63.
7. Giannelis, E.P.; Krishnamoorti, R.K.; Manias, E. Polymer–silicate nanocomposites: model systems for confined polymers and polymer brushes. *Adv. Polym. Sci.*, **1998**, 138, 107–148.
8. Ray, S.S.; Okamoto, M. Polymer/layered silicate nanocomposites: a review from preparation to processing. *Prog. Polym. Sci.* **2003**, 28, 1539–1641.
9. LeBaron, P.C.; Wang, Z.; Pinnavaia, T.J. Polymer-layered silicate nanocomposites: an overview. *Appl. Clay Sci.* **1999**, 15, 11–29.
10. Leroux, F.; Besse, J.P. Polymer interleaved layered double hydroxide: a new emerging class of nanocomposites. *Chem. Mater.* **2001**, 13, 3507–3515.
11. Thostenson, E.T.; Ren, Z.F.; Chou, T.W. Advances in the science and technology of carbon nanotubes and their composites: a review. *Compos. Sci. Technol.* **2001**, 61, 1899–1912.
12. Solomon, D.H.; Hawthorne, D.G. *Chemistry of Pigments and Fillers*. R.E. Krieger, Malabar, FL, 1991.
13. Al-Malaika, S.; Golovoy, A.; Wilkie, C.A., Eds. *Chemistry and Technology of Polymer Additives*. Blackwell Science, Oxford, England, 1999.
14. Karian, H.G., Ed. *Handbook of Polypropylene and Polypropylene Composites*. Marcel Dekker, New York, 1999.
15. Lu, C.; Mai, Y.-M. Influence of aspect ratio on barrier properties of polymer–clay nanocomposites. *Phys. Rev. Lett.* **2005**, 95, 088303.

16. Mitchell, C.A.; Bahr, J.L.; Arepalli, S.; Tour, J.M.; Krishnamoorti, R. Dispersion of functionalized carbon nanotubes in polystyrene. *Macromolecules* **2002**, 35, 8825–8830.

17. Kojima, Y.; Usuki, A.; Kawasumi, M.; Okada, A.; Fukushima, Y.; Kurauchi, T.T.; Kamigaito, O. Synthesis and mechanical properties of nylon-6/clay hybrid. *J. Mater Res.* **1993**, 8, 1179–1184, 1185–1189.

18. Kojima, Y.; Usuki, A.; Kawasumi, M.; Okada, A.; Kurauchi, T.T.; Kamigaito, O. Synthesis of nylon-6/clay hybrid by montmorillonite intercalated with ϵ-caprolactam. *J. Polym. Sci. A Polym. Chem.* **1993**, 31, 983–986.

19. Vaia R.A.; Ishii, H.; Giannelis, E.P. Synthesis and properties of 2-dimensional nanostructures by direct intercalation of polymer melts in layered silicates. *Chem. Mater.* **1993**, 5, 1694–1696.

20. Lan, T.; Kaviratna, P.D.; Pinnavaia, T.J. Mechanism of clay tactoid exfoliation in epoxy–clay nanocomposites. *Chem. Mater.* **1995**, 7, 2144–2150. Wang, M.S.; Pinnavaia, T.J. Clay polymer nanocomposites formed from acidic derivatives of montmorillonite and an epoxy-resin. *Chem. Mater.* **1994**, 6, 468–474. Pinnavaia, T.J. Intercalated clay catalysts. *Science* **1983**, 220, 365–371.

21. Giannelis, E.P.; et al. Structure and dynamics of polymer/layered silicates nanocomposites. *Chem. Mater.* **1996**, 8, 1728–1764. Polymer layered silicate nanocomposites. *Adv. Mater.* **1996**, 8, 29–35. Nanostructure and properties of polysiloxane-layered silicate nanocomposites. *J. Polym. Sci. B Polym. Phys.* **2000**, 38, 1595–1604.

22. Vaia, R.A.; Jandt, K.D.; Kramer, E.J.; Giannelis, E.P. Microstructural evolution of melt intercalated polymer–organically modified layered silicates nanocomposites. *Chem. Mater.* **1996**, 8, 2628–2635. Vaia, R.A.; Price, G.; Ruth, P.N.; Nguyen, H.T.; Lichtenhan, J. Polymer/layered silicate nanocomposites as high performance ablative materials. *Appl. Clay Sci.* **1999**, 15, 67–92.

23. Kanatzidis, M.G.; Wu, C.-G.; Marcy, H.O.; DeGroot, D.C.; Kannewurf, C.R. Conductive polymer–oxide bronze nanocomposites: intercalated polythiophene in V_2O_5 xerogels. *Chem. Mater.* **1990**, 2, 222–224. Intercalation of poly(ethylene oxide) in vanadium pentoxide xerogel. *Chem. Mater.* **1991**, 3, 992–994. Synthesis, structure, and reactions of poly(ethylene oxide)/V_2O_5 intercalative nanocomposites. *Chem. Mater.* **1996**, 8, 525–535.

24. Iijima, S. Helical microtubules of graphitic carbon. *Nature* **1991**, 354, 56–58.

25. Vaia, R.A.; Giannelis, E.P. Lattice model of polymer melt intercalation in organically-modified layered silicates. *Macromolecules* **1997**, 30, 7990–7999.

26. Vaia, R.A.; Giannelis, E.P. Polymer melt intercalation in organically-modified layered silicates: Model predictions and experiment. *Macromolecules* **1997**, 30, 8000–8009. Equation (2.2) in this chapter differs from equation (6) in Ref. 26. Private discussions with R.A. Vaia clarified that Equation (2.2) as provided here is correct, and is the same as the authors used for their calculations in Refs. 26 and 25. Equation (6) in Ref. 26 was wrongly put in press due to a typographical error.

27. Balazs, A.C.; Singh, C.; Zhulina, E. Modeling the interactions between polymers and clay surfaces through self-consistent field theory. *Macromolecules* **1998**, 31, 8370–8381.

28. van Oss, C.J.; Chaudhury, M.K.; Good, R.J. Interfacial Lifschitz–van der Waals and polar interactions in macroscopic systems. *Chem. Rev.* **1988**, 88, 927–941.

29. van Oss, C.J.; *Interfacial Forces in Aqueous Media.* Marcel Dekker, New York, 1994.

30. Nuriel, S.; Liu, L.; Barber, A.H.; Wagner, H.D. Direct measurement of multiwall nanotube surface tension. *Chem. Phys. Lett.* **2005**, 404, 263–266.

31. Wu, W.; Giese, R.F.; van Oss, C.J.; Evaluation of the Lifshitz–van der Waals/acid-base approach to determine surface tension components. *Langmuir*, **1995**, 11, 379–382.

32. Israelachvili, J. *Intermolecular and Surface Forces.* Academic Press, San Diego, CA, 1991.

33. Wang, Z.-M.; Nakajima, H.; Manias, E.; Chung, T.C. Exfoliated PP/clay nanocomposites using ammonium-terminated PP as the organic modification for montmorillonite. *Macromolecules* **2003**, 36, 8919–8922.

34. Hasegawa, N.; Kawasumi, M.; Kato, M.; Usuki, A.; Okada, A. Preparation and mechanical properties of polypropylene–clay hybrids using a maleic anhydride–modified polypropylene oligomer. *J. Appl. Polym. Sci.* **1998**, 67, 37–92.

35. Reichert, P.; Nitz, H.; Klinke, S.; Brandsch, R.; Thomann, R.; Mülhaupt, R. Polypropylene/organoclay nanocomposite formation: influence of compatibilizer functionality and organoclay modification. *Macromol. Mater. Eng.* **2000**, 275, 8–17.

36. Kato, M.; Usuki, A.; Okada, A. Synthesis of polypropylene oligomer–clay intercalation compounds. *J. Appl. Polym. Sci.* **1997**, 66, 1781–1785.

37. Kawasumi, M.; Hasegawa, N.; Kato, M.; Usuki, A.; Okada, A. Preparation and mechanical properties of polypropylene–clay hybrids. *Macromolecules* **1997**, 30, 6333–6338.

38. Wang, K., Liang, S.; Du, R.N.; Zhang, Q.; Fu, Q. The interplay of thermodynamics and shear on the dispersion of polymer nanocomposite. *Polymer* **2004**, 45, 7953–7960.

39. Strawhecker, K.; Manias, E. Structure and properties of poly(vinyl alcohol)/Na^+ montmorillonite hybrids. *Chem. Mater.* **2000**, 12, 2943–2949.

40. Xu, R.; Manias, E.; Snyder, A.J.; Runt, J. New biomedical poly(urethane urea)–layered silicate nanocomposites. *Macromolecules* **2001**, 34, 337–339.

41. Xu, R.; Manias, E.; Snyder, A.J.; Runt, J. Low permeability polyurethane nanocomposites. *J. Biomed. Mater. Res.* **2003**, 64A, 114–119.

42. Lincoln, D.M.; Vaia, R.A.; Wang, Z.G.; Hsiao, B.S.; Krishnamoorti, R. Temperature dependence of polymer crystalline morphology in nylon 6/montmorillonite nanocomposites. *Polymer* **2001**, 42, 9975–9985.

43. Lincoln, D.M.; Vaia, R.A.; Wang, Z.G.; Hsiao, B.S. Secondary structure and elevated temperature crystallite morphology of nylon-6/layered silicate nanocomposites. *Polymer* **2001**, 42, 1621–1631.

44. Lincoln, D.M.; Vaia, R.A.; Krishnamoorti, R. Isothermal crystallization of nylon-6/montmorillonite nanocomposites. *Macromolecules* **2004**, 37, 4554–4561.

45. Morgan, A.B.; Harris, J.D. Exfoliated polystyrene–clay nanocomposites synthesized by solvent blending with sonication. *Polymer* **2004**, 45, 8695–3703.

46. Wolf, D.; Fuchs, A.; Wagenknecht, U.; Kretzschmar, B.; Jehnichen, D.; Häussler, L. Nanocomposites of polyolefin clay hybrids, in: *Proceedings of Eurofiller'99, Lyon-Villeurbanne*, France, 1999, pp. 6–9.

47. Vaia, R.A.; Jandt, K.D.; Kramer, E.J.; Giannelis, E.P. Kinetics of polymer melt intercalation. *Macromolecules* **1995**, 28, 8080–8085.
48. Manias, E.; Chen, H.; Krishnamoorti, R.K.; Genzer, J.; Kramer, E.J.; Giannelis, E.P. Intercalation kinetics of poly(styrene)/silicate nanocomposites. *Macromolecules* **2000**, 33, 7955–7966.
49. Manias, E.; Touny, A.; Wu, L.; Strawhecker, K.; Lu, B.; Chung, T.C. Polypropylene/montmorillonite nanocomposite materials: a review of synthetic routes and materials properties. *Chem. Mater.* **2001**, 13, 3516–3523.
50. Oya, A.; Kurokawa, Y.; Yasuda, H. Factors controlling mechanical properties of clay mineral/polypropylene nanocomposites. *J. Mater. Sci.* **2000**, 35, 1045–1050.
51. Messersmith, P.B.; Giannelis, E.P. Polymer-layered silicate nanocomposites: in situ intercalative polymerization of ϵ-caprolactone in layered silicates. *Chem. Mater.* **1993**, 5, 1064–1066.
52. Messersmith, P.B.; Gianellis, E. Synthesis and barrier properties of poly(ϵ-caprolactone)–layered silicate nanocomposites. *J. Polym. Sci. Polym. Chem.* **1995**, 33, 1047–1057.
53. Bahr, J.L.; Tour, J.M. Covalent chemistry of single-wall carbon nanotubes. *J. Mater. Chem.* **2002**, 12, 1952–1958.
54. Sinnott, S.B. Chemical functionalization of carbon nanotubes. *J. Nanosci. Nanotechnol.* **2002**, 2, 113–123.
55. Ramanathan, T.; Liu, H.; Brinson, L.C. Functionalized SWNT/polymer nanocomposites for dramatic property enhancement. *J Polym. Sci. B Polym. Phys.* **2005**, 43; 2269–2279.
56. Xie, X.L.; Mai, Y.-W.; Zhou, X.P. Dispersion and alignment of carbon nanotubes in polymer matrix: a review. *Mater. Sci. Eng. R Rep.* **2005**, 49, 89–112.
57. Wise, K.E.; Park, C.; Siochi, E.J.; Harrison, J.S. Stable dispersion of single wall carbon nanotubes in polyimide: the role of noncovalent interactions. *Chem. Phys. Lett.* **2004**, 391, 207–211.
58. Moon, S.; Jin, F.; Lee, C.-J.; Tsutsumi, S.; Hyon, S.-H. Novel carbon nanotube/poly(*l*-lactic acid) nanocomposites; their modulus, thermal stability, and electrical conductivity. *Macromol. Symp.* **2005**, 224, 237–295.
59. Paiva, M.C.; Zhou, B.; Fernando, K.A.S.; Lin, Y.; Kennedy, J.M.; Sun, Y-P. Mechanical and morphological characterization of polymer–carbon nanocomposites from functionalized carbon nanotubes. *Carbon*, **2004**, 42, 2849–2854.
60. Putz, K.W.; Mitchell, C.A.; Krishnamoorti, R.; Green, P.F. Elastic modulus of single-walled carbon nanotube/poly(methyl methacrylate) nanocomposites. *J. Polym. Sci. B Polym. Phys.* **2004**, 42, 2286–2293.
61. Jia, Z.; Wang, Z.; Xu, C.; Liang, J.; Wei, B.; Wu, D.; and Zhu, S. Study on poly(methyl methacrylate)/carbon nanotube composites. *Mater. Sci. Eng. A* **1999**, 271, 395–400.
62. Costache, M.C.; Wang, D.; Heidecker, M.J.; Manias, E.; Wilkie, C.A. The thermal degradation of poly(methyl methacrylate) nanocomposites. *Polym. Adv. Technol.* **2006**, In press.
63. Tang, W.; Santare, M.H.; Advani, S.G. Melt processing and mechanical property characterization of multi-walled carbon nanotube/high density polyethylene (MWNT/HDPE) composite films. *Carbon* **2003**, 41; 2779–2785.

64. Manchado, M.A.L.; Valentini, L.; Biagiotti, J.; Kenny, J.M. Thermal and mechanical properties of single-walled carbon nanotubes: polypropylene composites prepared by melt processing. *Carbon* **2005**, 43, 1499–1505.
65. Meincke, O.; Kaempfer, D.; Weickmann, H.; Friedrich, C.; Vathauer, M.; Warth, H. Mechanical properties and electrical conductivity of carbon-nanotube filled polyamide-6 and its blends with acrylonitrile/butadiene/styrene. *Polymer* **2004**, 45, 739–748.
66. Vaia, R.A.; Liu, W.D. X-ray powder diffraction of polymer/layered silicate nanocomposites: model and practice. *J. Polym. Sci. B Polym. Phys.* **2002**, 40, 1590–1600.
67. Hanley, H.J.M.; Muzny, C.D.; Ho, D.L.; Glinka, C.J.; Manias, E. A SANS study of organo-clay dispersions. *Int. J. Thermophys.* **2001**, 22, 1435–1448.
68. Vaia, R.A.; Liu, W.D.; Koerner, H. Analysis of small-angle scattering of suspensions of organically modified montmorillonite: implications to phase behavior of polymer nanocomposites. *J. Polym. Sci. B Polym. Phys.* **2003**, 41, 3214–3236.
69. Drummy, L.F.; Koerner, H.; Farmer, K.; Tan, A.; Farmer, B.L.; Vaia, R.A. High-resolution electron microscopy of montmorillonite and montmorillonite/epoxy nanocomposites. *J. Phys. Chem. B.* **2005**, 109, 17868–17378.
70. Krishnamoorti, R.; Giannelis, E.P. Strain hardening in model polymer brushes under shear. *Langmuir* **2001**, 17, 1448–1452.
71. Strawhecker, K.; Manias, E. AFM studies of poly(vinyl alcohol) crystallization next to an inorganic surface. *Macromolecules* **2001**, 34, 8475–8482.
72. Giannelis, E.P. unpublished data.
73. Wang, Z.M.; Chung, T.C.; Gilman, J.W.; Manias. E. Melt-processable syndiotactic polystyrene/montmorillonite nanocomposites. *J. Polym. Sci. B Polym. Phys.* **2003**, 41, 3173–3137.
74. Strawhecker, K.; Manias, E. Crystallization behavior of poly(ethylene oxide) in the presence of Na^+ montmorillonite fillers. *Chem. Mater.* **2003**, 15, 844–849.
75. Bhattacharyya, A.R.; Sreekumar, T.V.; Liu, T.; Kumar, S.; Ericson, L.M.; Hauge, R.H.; Smalley, R.E. Crystallization and orientation studies in polypropylene/single wall carbon nanotube composite. *Polymer* **2003**, 44, 2373–2377.
76. Valentini, L.; Biagiotti, J.; Kenny, J.M.; Santucci, S. Morphological characterization of single-walled carbon nanotubes–PP composites. *Compos. Sci. Technol.* **2003**, 63, 1149–1153.
77. Grady, B.P.; Pompeo, F.; Shambaugh, R.L.; Resasco, D.E. Nucleation of polypropylene crystallization by single-walled carbon nanotubes. *J. Phys. Chem. B* **2002**, 106, 5852–5858.
78. Sandler, J.K.W.; Pegel, S.; Cadek, M.; Gojny, F.; Es, M.V.; Lohmar, J.; Blau; W.J. Schulte, K.; Windle, A.H.; Shaffer, M.S.P. A comparative study of melt spun polyamide-12 fibres reinforced with carbon nanotubes and nanofibres. *Polymer* **2004**, 45, 2001–2015.
79. Ryan, K.P.; Lipson, S.M.; Drury, A.; Cadek, M.; Ruether, M.; O'Flaherty, S.M.; Barron, V.; McCarthy, B.; Byrne, H.J.; Blau, W.J.; Coleman, J.N. Carbon-nanotube nucleated crystallinity in a conjugated polymer based composite. *Chem. Phys. Lett.* **2004**, 391, 329–333.
80. Zhang, S.; Zhang, N.; Huang, C.; Ren, K.; Zhang, Q. Microstructure and electromechanical properties of carbon nanotube/poly(vinylidene fluoride–trifluoroethylene–chlorofluoroethylene) composites. *Adv. Mater.* **2005**, 17, 1897–1901.

81. Goh, H.W.; Goh, S.H.; Xu, G.Q.; Pramoda, K.P.; Zhang, W.D. Crystallization and dynamic behavior of double-C60-end-capped poly(ethylene oxide)/multi-walled carbon nanotube composites. *Chem. Phys. Lett.*, **2003**, 379, 236–241.
82. Liu, T.; Phang, I.Y.; Shen, L.; Chow, S.Y.; Zhang, W.-D. Morphology and mechanical properties of multiwalled carbon nanotubes reinforced nylon-6 composites. *Macromolecules* **2004**, 37, 7214–7222.
83. Li, C.Y.; Li, L.; Cai, W.; Kodjie, S.L.; Tenneti, K. Nanohybrid shishkebabs: periodically functionalized nanotubes. *Adv. Mater.* **2005**, 17, 1198–1202.
84. Fornes, T.D.; Yoon, P.J.; Keskkula, H.; Paul, D.R. Nylon 6 nanocomposites: the effect of matrix molecular weight. *Polymer* **2001**, 42, 9929–9940.
85. Pattanayak, A.; Jana, S.C. Properties of bulk-polymerized thermoplastic polyurethane nanocomposites. *Polymer* **2005**, 46, 3394–3406.
86. Halpin, J.C.; Tsai, S.W. Environmental factors estimation in composite materials design. *AFML Trans.* **1967**, pp. 67–423. Halpin, J.C. *Primer on Composite Materials Analysis*, 2nd ed. Technomic Publishing, Lancaster, PA, 1992.
87. Mori, T.; Tanaka, K. Average stress in matrix and average energy of materials with misfitting inclusions. *Acta Metall.* **1973**, 21, 571–574.
88. Brune, D.A.; Bicerano, J. Micromechanics of nanocomposites: comparison of tensile and compressive elastic moduli, and prediction of effects of incomplete exfoliation and imperfect alignment on modulus. *Polymer* **2002**, 43, 369–337.
89. Wang, J.; Pyrz, R. Prediction of the overall moduli of layered-silicate reinforced nanocomposites, I: Basic theory and formulas. *Compos. Sci. Technol.* **2004**, 64, 925–934.
90. Wang, J.; Pyrz, R. Prediction of the overall moduli of layered-silicate reinforced nanocomposites, II: Analyses. *Compos. Sci. Technol.* **2004**, 64, 935–944.
91. Shia, D.; Hui, C.Y.; Burnside, S.D.; Giannelis, E.P. An interface model for the prediction of Young's modulus of layered silicate elastomer nanocomposites. *Polym. Compos.* **1998**, 19, 608–617.
92. Barber, A.H.; Cohen, S.R.; Wagner, H.D. Measurement of carbon nanotube–polymer interfacial strength. *Appl. Phys. Lett.* **2003**, 82, 4140–4142.
93. Barber, A.H.; Cohen, S.R.; Kenig, S.; Wagner, H.D. Interfacial fracture energy measurements for multi-walled carbon nanotubes pulled from a polymer matrix. *Compos. Sci. Technol.* **2004**, 64, 2283–2289.
94. Frankland, S.J.V.; Caglar, A.; Brenner, D.W.; Griebel, M. Molecular simulation of the influence of chemical cross-links on the shear strength of carbon nanotube–polymer interfaces. *J. Phys. Chem. B* **2002**, 106, 3046–3048.
95. Huh, J.Y.; Manias, E. unpublished data.
96. Chassapis, C.S.; Petrou, J.K.; Petropoulos, J.H.; Theodorou, D.N. Analysis of computed trajectories of penetrant micromolecules in a simulated polymeric material. *Macromolecules* **1996**, 29, 3615–3624.
97. Nielsen, L.E. Models for the permeability of filled polymer systems. *J. Macromol. Sci. (Chem.)* **1967**, A1, 929–942.
98. Cussler, E.L.; Hughes, S.E.; Ward, W.J.; Aris, R. Barrier membranes. *J. Membr. Sci.* **1988**, 38, 161–174.
99. Fredrickson, G.H.; Bicerano, J. Barrier properties of oriented disk composites. *J. Chem. Phys.* **1999**, 110, 2181–2188.
100. Bharadwaj, R.K. Modeling the barrier properties of polymer-layered silicate nanocomposites. *Macromolecules* **2001**, 34, 9189–9192.

3

FLAME RETARDANT MECHANISM OF POLYMER–CLAY NANOCOMPOSITES*

JEFFREY W. GILMAN

National Institute of Standards and Technology, Gaithersburg, Maryland

3.1 INTRODUCTION

For more than a decade potential environmental problems associated with organo-bromine flame retardant systems have motivated the search for non-halogen-based approaches to reduce polymer flammability. Initially, research focused on development of new phosphorus-based flame retardants, and numerous publications and patents have been issued in this area.[1–8] Similarly motivated research has also produced nonhalogen flame retardant approaches based on other elements, such as boron[9] and silicon.[10] At the same time, work on the use of additives, or fillers, with nanometer-scale primary particle sizes, produced polymer nanocomposites. These materials exhibit enhancement in a variety of physical properties at one-tenth the loading required when micrometer-size additives are used.[11]

*This work was carried out by the National Institute of Standards and Technology (NIST), an agency of the U.S. government and by statute is not subject to copyright in the United States. Certain commercial equipment, instruments, materials, or companies are identified in this paper in order to adequately specify the experimental procedure. This in no way implies endorsement or recommendations by NIST. The policy of NIST is to use metric units of measurement in all its publications, and to provide statements of uncertainty for all original measurements. In this document, however, data from organizations outside NIST are shown, which may include measurements in nonmetric units or measurements without uncertainty statements.

Flame Retardant Polymer Nanocomposites, edited by Alexander B. Morgan and Charles A. Wilkie

Activity at the intersection of these two fields has produced a new class of flame retardants based on nanocomposites.

3.1.1 Initial Discoveries

In this section we briefly review the initial development of clay nanocomposites and present initial work on their flammability properties. In the next section we review several studies aimed at determining the flame retardant mechanism of polymer–clay nanocomposites. This is not intended as an exhaustive review of all nanocomposite flammability research; instead, an attempt is made to present an objective view of the mechanism proposed for how clay nanocomposites reduce polymer flammability, based on data from NIST and other research groups.

The study of polymers combined with layered silicates at the nanoscale to form nanocomposites appears to have begun in the late 1940s with a patent application by Carter et al.[12] in 1947, and again in the early 1960s with patents from Nahin and Backlund.[13] Publication of papers from Blumstein,[14] Dekking,[15] and Freidlander[16] also appeared during this period. The later papers focused on in situ polymerization of vinyl monomers in the gallery between montmorillonite lamella. Most of this early work involved intercalated polymer–clay nanocomposites comprised of much higher loadings of clay (50% mass fraction) than are used today in nanocomposites (5% mass fraction). However, they opened the door on a new field of study, which shows no signs of diminishing.[17]

Polymer–clay nanocomposites with lower loadings (1 to 10% mass fraction) characterize the types of materials that are the focus of more recent studies, such as those disclosed in initial patents from General Motors (GM),[18] Imperial Chemical Industries (ICI),[19] and DuPont[20] in the 1970s and mid-1980s. The GM patent primarily claims the use of clays as substitutes for antimony oxides, while the ICI patent teaches the use of "delaminated vermiculite" to impart self-extinguishing and charring properties to expanded polystyrene beads. The DuPont patents also discuss the flame retardant properties of polymer–clay nanocomposites, but only as antidrip additives to formulations heavily filled with conventional flame retardants. The inventors note an increase in char formation, but they attribute this to the polyester. Kamigato and co-workers at Toyota also filed patents as early as 1981 on the in situ polymerization of styrene, isoprene, vinyl acetate, and caprolactam.[21] One claim in the patent is that clay-rich polymer composites are flame retardant, but no data are provided. A similar unsupported statement appears in a 1976 Japanese patent application from Unitika.[22] Although some of these patents show that organoclays enable self-extinguishing properties, or a V-0 rating in the UL-94 test, and that they may pass other fire tests, such as the UL-910 Steiner tunnel test, no other study of char-forming flame retardant properties of nanocomposites appeared in the literature until the mid-1990s.

Groups at NIST and Cornell both reported that nanocomposites alone, with no other flame retardant, reduced the parent polymer's flammability and enhanced char formation. Giannelis et al.[23,24] found self-extinguishing properties for polycaprolactone (PCL) and aliphatic–polyetherimide layered-silicate nanocomposites when the nanocomposites were exposed to small open-flame tests. Researchers at NIST used cone calorimetry and radiative gasification to show that polyamide-6 (PA6), polystyrene (PS), and maleated polypropylene (PP-g-MA) montmorillonite (MMT) nanocomposites had enhanced char formation and gave up to 75% lower flammability, as measured by the peak heat release rate or peak mass loss rate.[25–27] In most cases the carbonaceous char yield was limited to 2 to 5% mass fraction; consequently, the total heat release rate (THR) was not affected significantly. In addition, ignition times were either not improved or sometimes shorter. However, the unique character of this new approach to flame retardant polymeric materials was the dual benefit of reduced peak heat release rate and improved physical properties, a combination not usually found with conventional flame retardants. A significant number of papers have since been published on this topic, and some have shed light on the mechanism[28–42] by which clay nanocomposites have reduced flammability.

3.2 FLAME RETARDANT MECHANISM

3.2.1 Polystyrene Nanocomposites

In 1998 a government and industry consortium carried out a detailed study of the flame retardant mechanism of polymer–clay nanocomposites.[43] A goal of this consortium was to determine the effects of varying specific parameters on the flammability properties of polymer–clay nanocomposites and to use this information to elucidate the flame retardant mechanism. One devise used to study the flame retardant mechanism of nanocomposites was the radiative gasification apparatus. The gasification apparatus is used to study condensed-phase pyrolysis processes under firelike heat fluxes. The use of nitrogen pyrolysis allows the sample to be viewed during pyrolysis without interference by the flame normally present during combustion. A diagram of the gasification apparatus is shown in Figure 3.1. Details of the experimental methods and validation of the correlation between the mass loss rate (MLR) measured in the gasification apparatus and the heat release rate (HRR) measured in a cone calorimeter have been published.[26,44]

The initial results published indicated that a clay-reinforced carbonaceous char formed during combustion (or pyrolysis) of nanocomposites. This is particularly significant for systems whose base resin normally produces little, or no, char when burned alone (PS, PP-g-MA, PA6, and EVA [poly(ethylene-co-vinyl acetate)]). The char formation in PS clay nanocomposites is easily seen in the images from nitrogen atmosphere gasification shown in Figure 3.2.

Sonobe and co-workers reported the use of montmorillonite clay as a template to prepare pregraphitic materials from non-char-forming polymers in 1990.[45–47] It is presumably the same confinement and "coking" phenomena that give the additional carbonaceous char during burning in a variety of polymer–clay nanocomposites. It should be noted that although the peak heat release rate is reduced by 75%, the total HRR is not reduced significantly. Therefore, the carbonaceous

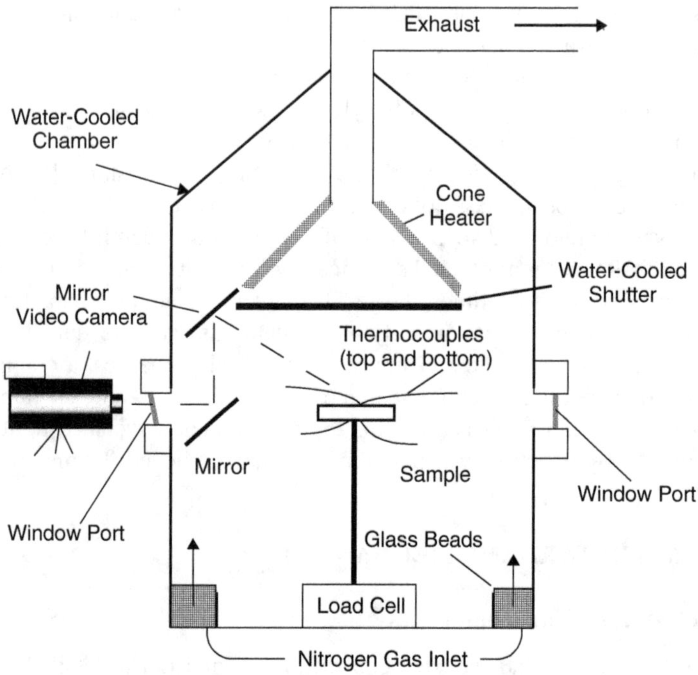

FIGURE 3.1 Radiative gasification apparatus.

char and clay combine to form an insulator that slows the heat transfer and the rate at which pyrolysis occurs.

A detailed study of the char formation process in PS clay nanocomposites was performed by Morgan et al., using the NIST radiative gasification apparatus.[48] The gasification apparatus is equipped with a water-cooled shutter which is used to interrupt the pyrolysis at specific times and consequently, allows preparation of a series of PS–clay nanocomposite samples with a range of pyrolysis histories. Mass loss rate data are gathered during the experiment. The mass loss rate behavior of a material is an excellent guide to predict the heat release rate; MLR represents the amount of fuel released as a function of time. The more mass released, the more fuel released, which in turn will lead to heat release upon combustion with oxygen. Of course, not all mass released is combustible; certain flame retardants (such as halogens) can release a substantial amount of mass that is nonflammable. In the case of the polymer–clay nanocomposites, the volatile pyrolyzed mass is assumed to be completely flammable, since no gas-phase flame retardants are present in the sample.

Figure 3.3 shows the mass loss rate data for PS, PS with micro-dispersed sodium-MMT, and PS–MMT (mass fraction 10%) nanocomposite with a mixed intercalated–delaminated structure [for a transmission electron micrograph (TEM), see Figure 3.4*a*]. The times at which samples were exposed to pyrolysis were 82, 95, 200, 400, and 1150 s. These times correspond to particular events in

FIGURE 3.2 Images of PS and PS clay nanocomposite [5% mass fraction of organoclay (MMT)] at three points during the radiative gasification of the samples. This pyrolysis, done in a nitrogen atmosphere, reveals the extensive charring of PS due to the nano-dispersed clay.

the pyrolyzing and burning behavior of the PS–polymer nanocomposite. The 82 s sample corresponds to events shortly after ignition. The 95-s sample reflects the peak MLR, which can be related to the peak heat release rate. The next sample, collected at 200 s, relates to the start of the plateau of steady burning behavior. The following sample at 400 s then corresponds to the end of this plateau of steady burning behavior, beginning to lead toward eventual extinguishment and exhaustion of mass loss for the sample. The final sample collected at 1150 s reflects the end of burning behavior and represents a fully combusted sample.

Images of the five pyrolysis samples are shown in Figures 3.5 to 3.9. These images show that initially the clay–carbonaceous char layer is thin, but as the pyrolysis times increase, the samples are comprised of a greater fraction of char until the char dominates the structure in the 400 s and 1150 s samples. Pyrolysis residues were collected from three regions on each of the five

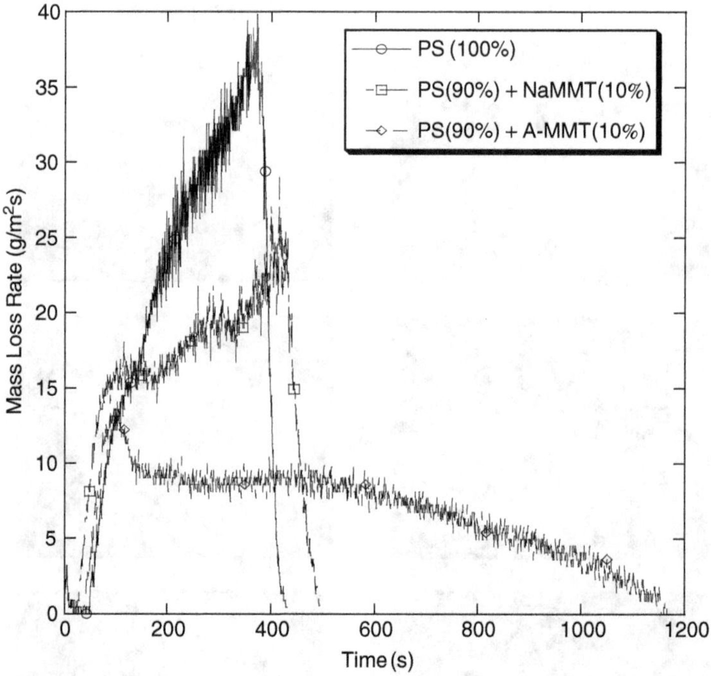

FIGURE 3.3 Mass loss rate data for PS, PS with micro-dispersed NaMMT, and a PS/mass fraction 10% MMT nanocomposite with a mixed intercalated–delaminated structure. The times at which partially pyrolized samples were exposed to pyrolysis were 82, 95, 200, 400, and 1150 s.

pyrolysis samples: surface, middle, and bottom (see Figure 3.10). These 15 samples were characterized using x-ray diffraction (XRD) and thermogravimetric analysis (TGA).

A summary plot of the XRD data [*d*-spacing (001 peak)] of sample versus pyrolysis time is shown in Figure 3.11. The XRD data reveal that the pyrolysis of the PS–clay nanocomposite yields a clay–char whose structure is independent of pyrolysis time after some minimum exposure. As the total pyrolysis time increases, clay–char is formed deeper and deeper into the sample; this is confirmed by the increasing char thickness observed in Figures 3.5 to 3.9 and with the XRD data in Figure 3.10. After this initial pyrolysis, the clay–char has a *d*-spacing of 1.3 nm. This dense spacing of MMT layers in the clay–char is confirmed in the TEM of the 1150 s residue sample, shown in Figure 3.4*b*. Similar XRD results were reported in initial studies of PS nanocomposite and PA6 nanocomposite chars[25] and in recent work by Kashiwagi et al. on PA6 nanocomposites.[30]

Morgan[48] also performed TGA characterization (shown in Figure 3.12) of the PS nanocomposite residue samples; these data revealed additional details about the characteristics of the clay–char. The mean mass fraction loss from the TGA

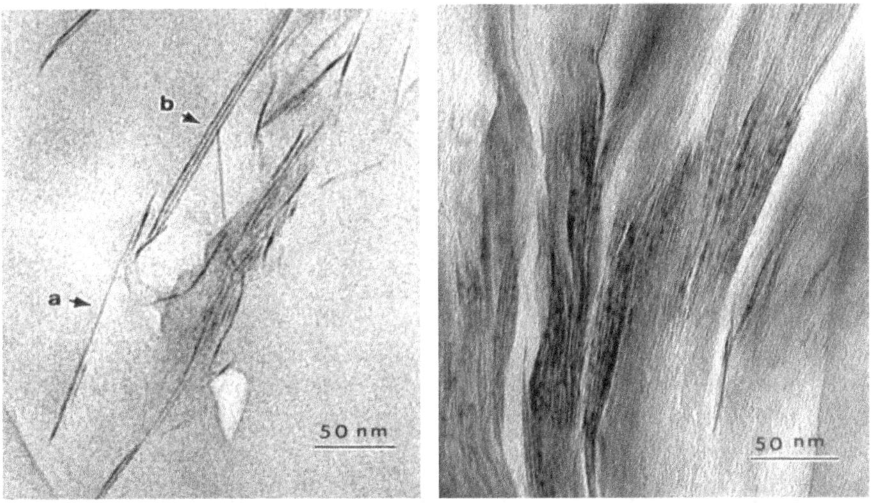

(a) (b)

FIGURE 3.4 TEM images of MMT layers in a PS/mass fraction 10% MMT nanocomposite (*a*) before and (*b*) after pyrolysis.

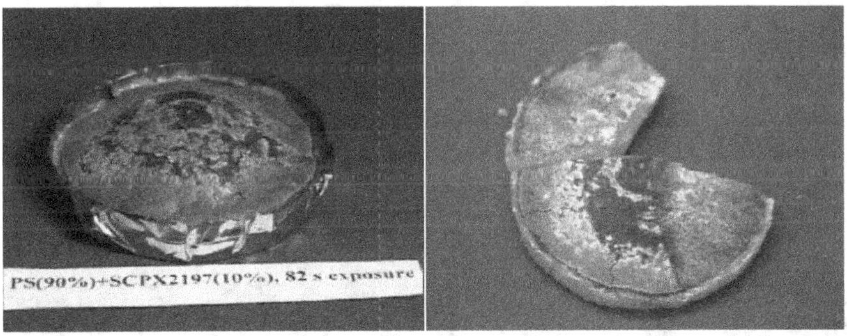

FIGURE 3.5 Image of a PS/mass fraction 10% MMT nanocomposite pyrolized for 82 s. The image on the right shows a cross section and has had the surface char partially removed. (*See insert for color representation of figure.*)

in air was $28.7 \pm 6.2\%$ (σ), for all four of the longer exposure time samples (95 s, 200 s, 400 s, and 1150 s). Assuming that the remaining material (mass fraction 71.3%) is clay and correcting for the higher density of clay relative to carbonaceous char (2.1 g/cm^3 versus 1.0 g/cm^3), a $1:1$ volume ratio of clay to carbonaceous materials was observed in the clay–char. This agrees qualitatively with the TEM of the 1150 s residue sample shown in Figure 3.4*b*.

Comparison of the TGA data in N_2 to that in air revealed other information about the clay–char (Figure 3.12). The mean mass fraction loss from the TGA in N_2, averaged over 95 s, 200 s, 400 s, and 1150 s exposures, is $17.5 \pm 6.8\%$ (σ),

FIGURE 3.6 Image of a PS/mass fraction 10% MMT nanocomposite pyrolized for 95 s. The image on the right shows a cross section and partial removal of the surface char. (*See insert for color representation of figure.*)

FIGURE 3.7 Image of a PS/mass fraction 10% MMT nanocomposite pyrolized for 200 s. The image on the right shows a cross section and partial removal of the surface char. (*See insert for color representation of figure.*)

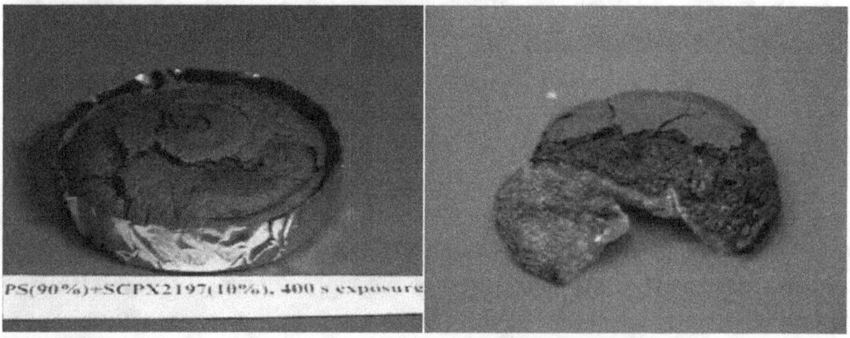

FIGURE 3.8 Image of a PS/mass fraction 10% MMT nanocomposite pyrolized for 400 s. The image on the right shows a cross section and partial removal of the surface char. (*See insert for color representation of figure.*)

FIGURE 3.9 Image of a PS/mass fraction 10% MMT nanocomposite pyrolized for 1150 s. The image on the right shows a cross section and partial removal of the surface char. (*See insert for color representation of figure.*)

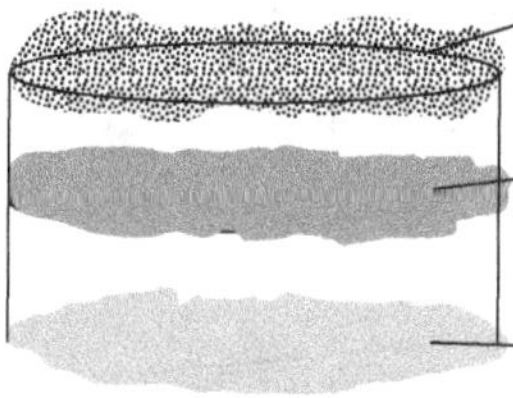

FIGURE 3.10 Three regions (surface, middle, and bottom) of a pyrolysis sample that were sampled for analysis using XRD and TGA.

considerable less than the 28.7% from the measurement in air. The difference between these mass losses suggests that there are two types of carbonaceous materials in the carbon part of the clay–char: a carbonaceous material that can be gasified by heating in N_2, and a second fraction of material which requires more aggressive conditions, that is, heating in the presence of air, which allows for oxidative degradation and complete removal of the carbonaceous material. From the TGA data the mass fraction ratio of these two types of carbon is 1.5 : 1, with the more stable form of char in the minority.

In summary, the XRD and TGA data revealed that the clay–char formed during pyrolysis of the PS–MMT nanocomposite has a layered structure with an invariable 1.3-nm *d*-spacing (Figure 3.11). The clay–char contains a mass fraction of 28% carbonaceous material, of two types, which differ primarily in their thermooxidative stability. Although this char characterization study revealed little about the mechanistic details of *how* the clay–char forms, it did result in a more complete picture of the nature of the clay-reinforced carbonaceous char.

3.2.2 Polypropylene–Clay Nanocomposites

In contrast to the results above for PS–MMT nanocomposite chars from pure maleic anhydride, grafted PP (PP-g-MA) nanocomposites produce a featureless

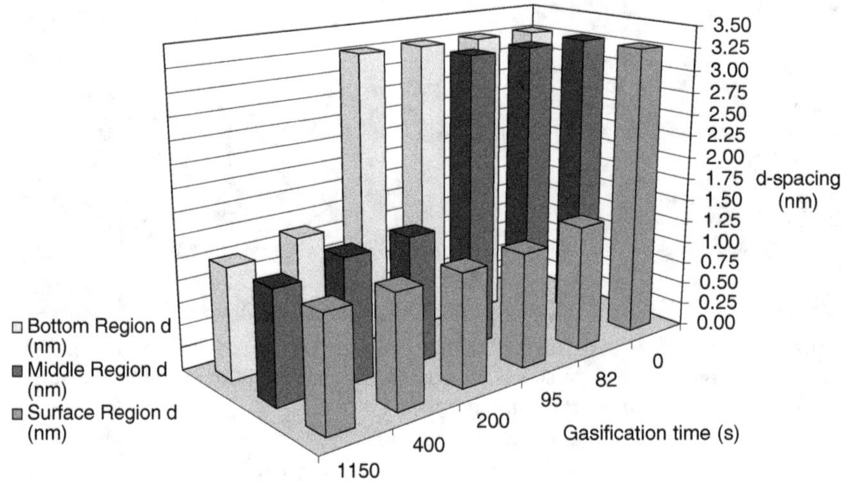

FIGURE 3.11 Clay–char d-spacings versus gasification time and sampling region, revealing a step decrease in d-spacing, from 3.27 nm (the d-spacing of the PS–AMMT nanocomposite) to 1.3 nm with increased exposure time. (*See insert for color representation of figure.*)

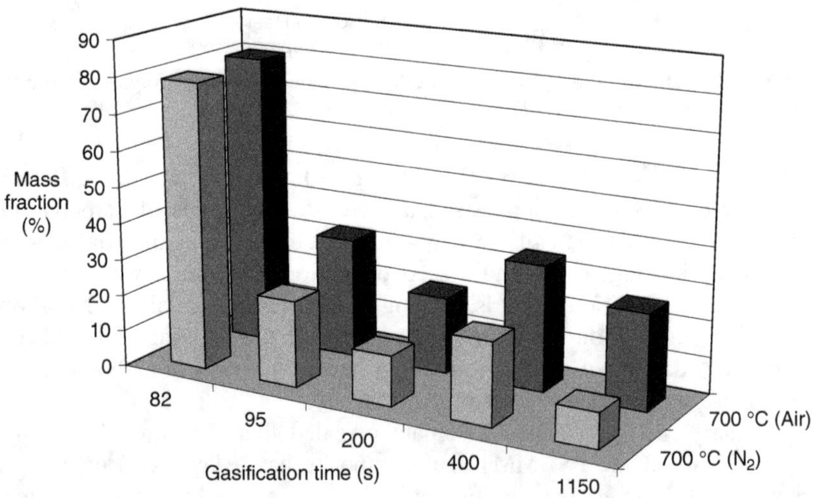

FIGURE 3.12 Mass fraction (%) loss in TGA (in N_2 and air) for surface region clay–chars at various gasification exposure times. (*See insert for color representation of figure.*)

XRD pattern; TEM (Figure 3.13) confirms the disordered structure of clay layers in the char. The charring process associated with PP–clay nanocomposites was also studied using the radiative gasification apparatus in the same way as the PS nanocomposite described previously in this chapter. This research was also part of the same NIST consortium work mentioned with the PS nanocomposite

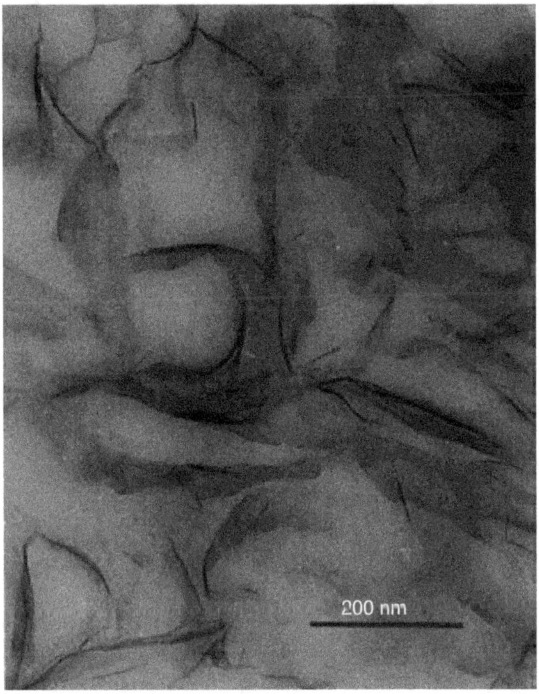

FIGURE 3.13 TEM of the combustion char of a PP-g-MA/MMT clay nanocomposite (4% mass fraction MMT), revealing the disordered structure of the clay in the residue.

FIGURE 3.14 Digital images of nitrogen gasification residues from three nanocomposites: (*a*) PP/mass fraction 5% MMT; (*b*) PP/mass fraction 15% PP-g-MA/mass fraction 2% MMT; (*c*) PP/mass fraction 15% PP-g-MA/mass fraction 5% MMT.

system. In this case the PP-g-MA was added as a compatibilizer to improve the exfoliation of the organoclay in the PP. This approach was used successfully by Toyota researchers to prepare PP–clay nanocomposites with enhanced mechanical properties.[49] The nitrogen gasification images in Figure 3.14 illustrate the effect of adding PP-g-MA (mass fraction 15%) to the PP clay nanocomposites.

The PP–MMT (mass fraction 5%) (Figure 3.14*a*) nanocomposite (without PP-g-MA) had an intercalated structure and produced no black carbonaceous char. Only the residual MMT remains after pyrolysis in nitrogen. However, both of the PP nanocomposites that contain PP-g-MA (mass fraction 15%) show extensive charring, and as the mass fraction of MMT is increased from 2% to 5%, the quality of the char improves; fewer cracks and a more continuous structure are observed. It appears that the loading level of MMT and the presence of PP-g-MA in the formulation is critical to the formation of a carbonaceous char with few cracks.

On possible hypothesis for this improvement could be due to the improvement in exfoliation brought by the PP-g-MA. However, it is more likely that the PP-g-MA plays a dual role of improving clay dispersion and forming char during polymer combustion. Data that support the theory that PP-g-MA is working directly with the clay to form the char comes from a related unpublished study of PP nanocomposite flammability by Manias. In this study, PP copolymers that contained polar comonomers[50] were combined with organoclay; this produced excellent delaminated structures without the use of PP-g-MA. However, only a weak reduction in flammability was observed for the resulting PP–clay nanocomposites: thus illustrating the important role that PP-g-MA plays, presumably as a char former, in reducing the flammability of PP–clay nanocomposites.

Additional important mechanistic distinctions associated with nanocomposite flammability have recently been identified by Kashiwagi et al. in a study of both clay and carbon nanotube (CNT) nanocomposites.[30,51] Previous work using radiative gasification has reported the aggregation of surface char on the top of polyamide-6/clay nanocomposites during pyrolysis.[26] Kashiwagi recently reinvestigated this process and proposed that the homogeneous coverage of the surface occurs by a complex dynamic process involving recession of the polymer, surfactant degradation, bubble migration, and bubble bursting.[30] In this study and a similar one on poly(methylmethacrylate) (PMMA)–CNT nanocomposites, Kashiwagi et al. proposed that a homogeneous distribution of nanoparticles in the polymer melt *during pyrolysis* must be maintained. The stability of the dispersion is believed to prevent aggregation, or phase separation, of the nanoparticles, thereby preventing formation of cracks in the residue and promoting continuous coverage of the surface by residue. Using viscoelastic measurements, they found that the production of a continuous char residue was correlated directly with gel behavior at low frequencies in the CNT nanocomposites.[52] Presumably, this behavior not only prevents phase separation of the nanoparticle (CNT, clay, etc.) but also reduces the rate of gas escape from the melt and would prevent dripping during burning of vertically mounted samples, such as in a UL-94 V test.

Data from the NIST consortium work also supports the assertion that a homogeneous residue is critical for providing the most effective heat-transfer barrier. Comparison of the mass loss rate data for three different PP/PP-g-MA/clay nanocomposites and pure PP-g-MA is shown in Figure 3.15. As the images of the residues show, little or no carbonaceous (black) material is present after gasification. The fluorinated synthetic mica (FSM) has the lowest mass loss rate; this

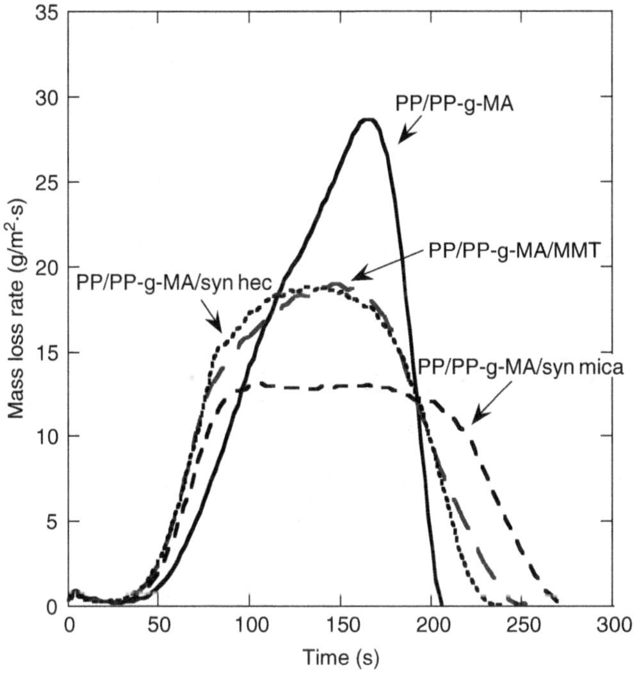

FIGURE 3.15 Effect of clay type on MLR of PP (84.6%)/PP-g-MA (7.7%)/clay (7.7%) at 50 kW/m^2 in N_2.

is attributed to the more continuous crack-free residue that formed relative to the other samples (see Figure 3.18 compared to Figures 3.16 and 3.17; see the color insert). This is similar to results from Morgan, where a synthetic clay produced a more continuous char than MMT in EVA nanocomposites when burned in a cone calorimeter.[53]

From the above it appears that in PS–and PP–clay nanocomposites, both the homogeneity of the clay residue and the presence of additional carbonaceous char are important mechanistically in providing a continuous coverage of effective heat-transfer material and thereby reducing the flammability. It should be kept in mind that since the carbonaceous char yield is low, the primary modes in which char affects the flammability are by prevention of reagglomeration of the clay particles and to provide additional insulating properties to the porous char.

Inan and co-workers' study of the flammability of PA6–clay nanocomposites provides an elegant illustration of the dominant heat transfer roll that the char plays in controlling nanocomposite flammability.[54] In these experiments PA6 nanocomposite samples were placed atop pure PA6 samples, these compression-molded "composite" samples were burned in a cone calorimeter, and the reduction in peak heat release rate for the composite sample was found to be 77% of that expected if the entire sample had been nanocomposite. Since only half of the composite contained clay, this magnitude of effect is surprising. Furthermore,

FIGURE 3.16 Gasification residue of PP/PP-g-MA/MMT with 7.7% PP-g-MA and MMT. (*See insert for color representation of figure.*)

FIGURE 3.17 Gasification residue of PP/PP-g-MA/synthetic hectorite with 7.7% PP-g-MA and hectorite. (*See insert for color representation of figure.*)

by using thermocouples implanted in the back of the composite sample, global heating rates were calculated during the cone experiments and a 30% lower heating rate was obtained for the composite sample than for pure PA6. This shows that once the nanocomposite portion of the composite sample had burned and clay-reinforced carbonaceous char was present on the top of the sample, even the underlying pure PA6 had its flammability reduced significantly by the insulation effect of the char. Inan and co-workers' study of PA6 nanocomposites also provides insight into the source of shortened ignition times sometimes observed in nanocomposites. Inan found that when no surfactant was used, and

FIGURE 3.18 Gasification residue of PP/PP-g-MA/synthetic mica with 7.7% PP-g-MA and mica. (*See insert for color representation of figure.*)

six PA nanocomposites were prepared by in situ polymerization of caprolactam, longer ignition delay times were observed relative to PA6–clay nanocomposites prepared using surfactant-treated clay and melt compounding methods.

Morgan et al. also studied the effect of surfactant on ignition delay times in PP/PP-g-MA/clay nanocomposites. In this study, careful extraction of surfactant-treated montmorillonite, which removed excess surfactant, produced 17% longer ignition delay times in the cone calorimeter relative to nanocomposite samples that contained the excess surfactant in organo-modified clay.[55]

3.2.3 Thermal Analysis of Polymer–Clay Nanocomposites

In Section 3.2.2, data were presented which show that char formation, possibly from a catalytic reaction between the polymer (PS, PA6, or a polymer compatibilizer, PP-g-MA) and the clay, is often present when low peak HRRs (or mass loss rates) are observed. However, data were also presented which show that in the absence of any substantial charring there can still be a 50 to 60% reduction in peak HRRs; if synthetic mica is used in PP/PP-g-MA nanocomposites. It appears that at least two mechanisms may be important to the function of nanocomposites: one involving char formation and a second involving the inorganic residue alone. This dual mechanism may explain why the effectiveness of clay nanocomposites varies from polymer to polymer.

In an effort to fashion a unified mechanism for clay nanocomposites, Wilkie et al. have studied the effect that clay has on the thermal degradation behavior of more than 11 different polymers. This work also attempts to correlate the thermal analysis data with flammability properties measured in a cone calorimeter.[35,56–60] In this work TGA degradation products were cryogenically trapped and analyzed using gas chromatography/mass spectroscopy (GC-MS); the thermal degradation pathways of the polymers with and without clay were investigated. Wilkie and co-workers[35,56–60] found a potentially interesting connection between the thermal degradation behavior and the fire retardancy of polymer–clay nanocomposites. Specifically, they found that if the clay produces a significant change in decomposition products, as measured by TGA/GC-MS, there is a larger flame retardancy effect. Furthermore, for polymers that decomposes via radical pathways, they find a direct relationship between the stability of the radical produced upon degradation and the magnitude of reduction in the peak heat release rate measured in a cone calorimeter. Presumably, this is a "cage effect," resulting from the slightly longer residence time that the decomposition products have in the condensed phase of the nanocomposite relative to the pure polymer. This is due to both the higher viscosity of the melt from the presence of the clay and from the cross-linking reactions. In the case of PS and EVA, the enhanced residence time, coupled with longer-lived, more stable radicals, allows secondary reactions that form char, thus the lower peak HRR. In the case of PA6, the enhanced residence time allows formation of a greater number of degradation products with molecular mass larger than that of caprolactam, the primary decomposition product of both pure PA6 and PA6 with MMT.

This work agrees with a number of studies where, in the presence of clay, enhanced cross-linking reactions or char-forming reactions have been observed: for example in the work of Gilman,[27] Kashiwagi,[30] and Bourbigot,[61] where PA6, in the presence of nano-dispersed clay, is shown to produce enhanced char yield. As discussed above, PS nanocomposites, also show this behavior. EVA nanocomposites studied by Camino and co-workers,[36] were also reported to char; this was proposed as the source of antidrip properties. However, Ferriol and co-workers'[34] results on PMMA organo-MMT nanocomposites contradict the data by Wilkie on PMMA, where Ferriol found char formation and a 50% reduction in peak HRR. Additional information that would complement the TGA/GC-MS work needs to be obtained to fully connect these small-scale experiments to bench-scale flammability performance. For example, the effects of scale (TGA : 10 mg versus cone calorimeter: 100 g) and heating rate (TGA heating rate: 20°C/min, fire heating rate: >500°C/min) need to be studied. Lyon has studied some of these issues in the development of a microcalorimetry devise and shown that a connection exists under certain conditions (i.e., when evaluating the relative flammability of pure polymers). However, the effects of scale are difficult to predict; for example, in charring systems there is always some minimum thickness of material that must be sacrificed to form the protective char; if the sample thickness is not significantly above this minimum thickness, no material will remain after the initial char forms for the char to protect, and the flame retardant effect is consequently reduced or nonexistent.

3.3 CONCLUSIONS AND FUTURE OUTLOOK

Several conclusions may be drawn from the discussion above. A definite flame retardant effect appears in the form of a reduction in the peak HRR of polymers that contain nano-dispersed clay. This is generally true for most thermoplastic polymers studied to date: PS, PA6, PP, PE, EVA, PA12, and PMMA; however it appears that different mechanisms are operative in different polymers. Presumably, this is because, depending on the polymer, the clay may change the decomposition products; it may cause cross-linking and ultimately catalyze carbonaceous char formation. In some cases, however, no char is formed and it is only the quality of the clay char that controls the flame retardant effect. Of the clays examined so far, all are effective, although synthetic clays in some cases may perform better than MMT. It appears that the maintenance of the homogeneous dispersion of the nanoplatelets in the polymer melt during pyrolysis is critical in producing a homogeneous residue. This is true whether carbonaceous char forms or if only clay makes up the residue. This reduces the rate of gas escape from the melt and prevents dripping during burning. The conditions required for this may include: meeting or exceeding the minimum clay loading (5%), a good initial dispersion, and a high-viscosity gel behavior in the melt (due to strong interfacial interactions between the clay and polymer, and the high aspect ratio of the clay) and cross-linking or carbonaceous char formation. These factors contribute to different extents in different polymer systems, and this is presumably the reason that the effectiveness varies from polymer to polymer.

Since this approach to reducing flammability reduces the peak HRR but not the THR, does not improve the ignition properties, and often may shorten the time to ignition, it is necessary to combine the nanocomposite with other flame retardant additives to pass regulatory flammability tests where ignition properties dominate. A variety of successful uses of this combined method have resulted in development of formulations that pass small-scale fire tests such as the UL-94 V[63] and pass far more aggressive tests, such as the UL-1666 riser test for wire and cable.[31,34] Morgan has recently reviewed a combination of conventional flame retardants with polymer–clay nanocomposites that yielded success in various regulatory flammability tests.[53]

Continued improvement in our ability to prepare polymer–clay nanocomposites coupled with a careful investigation of the interaction of the large number of other additives with clay nanocomposites should yield more flame retardant products that are cost-effective and fire safe. Candidate systems for studying the interactions of clays should not be limited to conventional flame retardants; it should include other nanoscale inorganic and metallic additives that can provide new catalytic pathways for cross-linking and char formation. Examples of promising such approaches include Ferriol et al.'s work with oxide nanoparticles,[34] Hu et al.'s work with iron–MMT,[64] and Tang and co-workers' interesting combination of clays with catalysts that form carbon nanotubes.[41]

REFERENCES

1. Green, J. In: *Fire Retardancy of Polymeric Materials*. Marcel Dekker, New York, 2000, p. 147.
2. Weil, E.D.; Ravey, M.; Keidar, I.; Gertner, D. Flame retardant actions of tris(1,3-dichloro-2-propyl) phosphate in flexible urethane foam. *Phosphorus Sulfur Silicon Relat. Elem.* **1996**, 110(1–4), 87.

3. Ravey, M.; Keidar, I.; Weil, E.D.; Pearce, E.M. Flexible polyurethane foam, II: Fire retardation by tris(1,3-dichloro-2-propyl) phosphate, A: Examination of the vapor phase (the flame). *J. Appl. Polym. Sci.* **1998**, 68(2), 217.
4. Ravey, M.; Pearce, E.M. Flexible polyurethane foam, III: Phosphoric acid as a flame retardant. *J. Appl. Polym. Sci.* **1999**, 74(5), 1317.
5. Price, D.; Pyrah, K.; Hull, T.R.; Milnes, G.J.; Wooley, W.D.; Ebdon, J.R.; Hunt, B.J.; Konkel, C.S. Ignition temperatures and pyrolysis of a flame-retardant methyl methacrylate copolymer containing diethyl(methacryloyloxymethyl)-phosphonate units. *Polym. Int.* **2000**, 49(10), 1164.
6. Chao, C.Y.H.; Wang, J.H. Comparison of the thermal decomposition behavior of a non-fire retarded and a fire retarded flexible polyurethane foam with phosphorus and brominated additives. *J. Fire Sci.* **2001**, 19(2), 137.
7. Zaikov, G.E.; Lomakin, S.M. Ecological issue of polymer flame retardancy. *J. Appl. Polym. Sci.* **2002**, 86(10), 2449.
8. Wei, P.; Li, H.X.; Jiang, P.K.; Yu, H.Y. An investigation on the flammability of halogen-free fire retardant PP-APP-EG systems. *J. Fire Sci.* **2004**, 22(5), 367.
9. Shen, K.K. In: *Fire and Polymers: Materials and Solutions for Hazard Prevention.* American Chemical Society, Washington, DC, 2001, p. 228.
10. Gilman, J.W.; Kashiwagi, K. In: *Fire Retardancy of Polymeric Materials.* Marcel Dekker, New York, 2000, p. 353.
11. Alexandre, M.; Dubois, P. Polymer-layered silicate nanocomposites: preparation, properties and uses of a new class of materials. *Mater. Sci. Eng. R Rep.* **2000**, 28(1–2), 1.
12. Carter, L.W.; Hendricks, J.G.; Bolley, D.S. U.S. Patent 2531396, 1950.
13. Nahin, P.G.; Backlund, P.S. U.S. Patent 3084117, **1963**.
14. Blumstein, A. Étude des polymerisations en coucheadsorbee, 1. *Bull. Soc. Chim. Fr.* **1961**, 126, 899.
15. Dekking, H.G.G. Propagation of vinyl polymers on clay surfaces, I: Preparation, structure, and decomposition of clay initiators. *J. Appl. Polym. Sci.* **1965**, 9, 1641.
16. Friedlander, H.Z.; Frink, C.R. Organized polymerization, III: Monomer intercalated in montmorillonite. *Polym. Lett.* **1964**, 2, 475.
17. Zeng, Q.H.; Yu, A.B.; Lu, G.Q.; Paul, D.R. Clay-based polymer nanocomposites: research and commercial development. *J. Nanosci. Nanotechnol.* **2005**, 5(10), 1574.
18. Mehta, R.K.S.; Weiss, P. U.S. Patent 4070315, 1976.
19. Bradbury, J.A.; Rowlands, R.; Tipping, J.W. U.S. Patent 4447491, 1984.
20. Shain, A.L. U.S. Patent 4582866, 1986.
21. Kamigaito, O.; Fukushima, Y.; Doi, H. U.S. Patent 4472538, 1984.
22. Fujiwara, S.; Skamoto, T. Sho 50 [1975]-35890 (SHO 51 [1976]-109998, 1976.
23. Giannelis, E.P. Polymer layered silicate nanocomposites. *Adv. Mater.* **1996**, 8(1), 29.
24. Giannelis, E.P. Polymer-layered silicate nanocomposites: synthesis, properties and applications. *Appl. Organomet. Chem.* **1998**, 12(10–11), 675.
25. Gilman, J.W.; Jackson, C.L.; Morgan, A.B.; Harris, R.; Manias, E.; Giannelis, E.P.; Wuthenow, M.; Hilton, D.; Phillips, S.H. Flammability properties of polymer–layered-silicate nanocomposites: polypropylene and polystyrene nanocomposites. *Chem. Mater.* **2000**, 12(7), 1866.

26. Gilman, J.; Kashiwagi, T.; Lomakin, S.; Giannelis, E.; Manias, E.; Lichtenhan, J.; Jones, P. In: *Fire Retardancy of Polymers: The Use of Intumescence*. Royal Society of Chemistry, London 1998, p. 203.

27. Gilman, J.W.; Kashiwagi, T.; Lichtenhan, J.D. Nanocomposites: a revolutionary new flame retardant approach. *SAMPE J.* **1997**, 33(4), 40.

28. Torre, L.; Lelli, G.; Kenny, J.M. Cure kinetics of epoxy/anhydride nanocomposite systems with added reactive flame retardants. *J. Appl. Polym. Sci.* **2004**, 94(4), 1676.

29. Wang, S.F.; Hu, Y.; Zong, R.W.; Tang, Y.; Chen, Z.Y.; Fan, W.C. Preparation and characterization of flame retardant ABS/montmorillonite nanocomposite. *Appl. Clay Sci.* **2004**, 25(1–2), 49.

30. Kashiwagi, T.; Harris, R.H.; Zhang, X.; Briber, R.M.; Cipriano, B.H.; Raghavan, S.R.; Awad, W.H.; Shields, J.R. Flame retardant mechanism of polyamide 6–clay nanocomposites. *Polymer* **2004**, 45(3), 881.

31. Beyer, G. Flame retardancy of nanocomposites—from research to reality: review. *Polym. Polym. Compos.* **2005**, 13(5), 529.

32. Tang, Y.; Hu, Y.; Xiao, J.F.; Wang, J.; Song, L.; Fan, W.C. PA-6 and EVA alloy/clay nanocomposites as char forming agents in poly(propylene) intumescent formulations. *Polym. Adv. Technol.* **2005**, 16(4), 338.

33. Qin, H.L.; Zhang, S.M.; Zhao, C.G.; Hu, G.J.; Yang, M.S. Flame retardant mechanism of polymer/clay nanocomposites based on polypropylene. *Polymer* **2005**, 46(19), 8386.

34. Laachachi, A.; Leroy, E.; Cochez, M.; Ferriol, M.; Cuesta, J.M.L. Use of oxide nanoparticles and organoclays to improve thermal stability and fire retardancy of poly(methyl methacrylate). *Polym. Degrad. Stab.* **2005**, 89(2), 344.

35. Jang, B.N.; Wilkie, C.A. The effect of clay on the thermal degradation of polyamide 6 in polyamide 6/clay nanocomposites. *Polymer* **2005**, 46(10), 3264.

36. Zanetti, M.; Lomakin, S.; Camino, G. Polymer layered silicate nanocomposites. *Macromol. Mater. Eng.* **2000**, 279(6), 1.

37. Yang, F.; Yngard, R.; Nelson, G.L. Flammability of polymer–clay and polymer–silica nanocomposites. *J. Fire Sci.* **2005**, 23(3), 209.

38. Quede, A.; Cardoso, J.; Le Bras, M.; Delobel, R.; Goudmand, P.; Dessaux, O.; Jama, C. Thermal stability and flammability studies of coated polymer powders using a plasma fluidized bed process. *J. Mater. Sci.* **2002**, 37(7), 1395.

39. Bourbigot, S.; Le Bras, M.; Duquesne, S.; Rochery, M. Recent advances for intumescent polymers. *Macromol. Mater. Eng.* **2004**, 289(6), 499.

40. Costache, M.C.; Jiang, D.D.; Wilkie, C.A. Thermal degradation of ethylene–vinyl acetate copolymer nanocomposites. *Polymer* **2005**, 46(18), 6947.

41. Tang, T.; Chen, X.C.; Chen, H.; Meng, X.Y.; Jiang, Z.W.; Bi, W.G. Catalyzing carbonization of polypropylene itself by supported nickel catalyst during combustion of polypropylene/clay nanocomposite for improving fire retardancy. *Chem. Mater.* **2005**, 17(11), 2799.

42. Keszei, S.; Matko, S.; Bertalan, G.; Anna, P.; Marosi, G.; Toth, A. Progress in interface modifications: from compatibilization to adaptive and smart interphases. *Eur. Polym. J.* **2005**, 41(4), 697.

43. Gilman, J.W.; Kashiwagi, T.; Morgan, A.B.; Harris, R.H.; Brassell, L.; VanLandingham, M.J. *C.L. 2005*. NISTIR 6531. National Institute of Standards and Technology, Washington, DC, 2005.
44. Zanetti, M.; Kashiwagi, T.; Falqui, L.; Camino, G. Cone calorimeter combustion and gasification studies of polymer layered silicate nanocomposites. *Chem. Mater.* **2002**, 14(2), 881.
45. Sonobe, N.; Kyotani, T.; Tomita, A. Carbonization of polyfurfuryl alcohol and polyvinyl acetate between the lamellae of montmorillonite. *Carbon* **1990**, 28(4), 483.
46. Sonobe, N.; Kyotani, T.; Tomita, A. Formation of graphite thin-film from polyfurfuryl alcohol and polyvinyl acetate carbons prepared between the lamellae of montmorillonite. *Carbon* **1991**, 29(1), 61.
47. Kyotani, T.; Yamada, H.; Sonobe, N.; Tomita, A. Heat-treatment of polyfurfuryl alcohol prepared between taeniolite lamellae. *Carbon* **1994**, 32(4), 627.
48. Morgan, A.B.; Harris, R.H.; Kashiwagi, T.; Chyall, L.J.; Gilman, J.W. Flammability of polystyrene layered silicate (clay) nanocomposites: carbonaceous char formation. *Fire Mater.* **2002**, 26(6), 247.
49. Nam, P.H.; Maiti, P.; Okamoto, M.; Kotaka, T.; Hasegawa, N.; Usuki, A. A hierarchical structure and properties of intercalated polypropylene/clay nanocomposites. *Polymer* **2001**, 42(23), 9633.
50. Manias, E.; Touny, A.; Wu, L.; Strawhecker, K.; Lu, B.; Chung, T.C. Polypropylene/montmorillonite nanocomposites: review of the synthetic routes and materials properties. *Chem. Mater.* **2001**, 13(10), 3516.
51. Kashiwagi, T.; Du, F.M.; Winey, K.I.; Groth, K.A.; Shields, J.R.; Bellayer, S.P.; Kim, H.; Douglas, J.F. Flammability properties of polymer nanocomposites with single-walled carbon nanotubes: effects of nanotube dispersion and concentration. *Polymer* **2005**, 46(2), 471.
52. Kashiwagi, T.; Du, F.M.; Douglas, J.F.; Winey, K.I.; Harris, R.H.; Shields, J.R. Nanoparticle networks reduce the flammability of polymer nanocomposites. *Nat. Mater.* **2005**, 4, 928.
53. Morgan, A.B. Flame retardant polymer layered silicate nanocomposites: a review of commercial and open literature systems. *Polym. Adv. Technol.* **2006**, 17, 206.
54. Inan, G.; Patra, P.K.; Kim, Y.K.; Warner, S.B. Flame retardancy of laponite- and montmorillonite-based nylon 6 nanocomposites: continuous nanophase and nanostructured materials. *MRS Proc.* **2004**, 788L8.46.
55. Morgan, A.B.; Harris, J.D. Effects of organoclay Soxhlet extraction on mechanical properties, flammability properties and organoclay dispersion of polypropylene nanocomposites. *Polymer* **2003**, 44(8), 2313.
56. Jang, B.N.; Costache, M.; Wilkie, C.A. The relationship between thermal degradation behavior of polymer and the fire retardancy of polymer/clay nanocomposites. *Polymer* **2005**, 46(24), 10678.
57. Jang, B.N.; Wilkie, C.A. The effects of clay on the thermal degradation behavior of poly(styrene-co-acrylonitirile). *Polymer* **2005**, 46(23), 9702.
58. Costache, M.C.; Jiang, D.D.; Wilkie, C.A. Thermal degradation of ethylene–vinyl acetate copolymer nanocomposites. *Polymer* **2005**, 46(18), 6947.
59. Jang, B.N.; Wilkie, C.A. The thermal degradation of polystyrene nanocomposite. *Polymer* **2005**, 46(9), 2933.

60. Jash, P.; Wilkie, C.A. Effects of surfactants on the thermal and fire properties of poly(methyl methacrylate)/clay nanocomposites. *Polym. Degrad. Stab.* **2005**, 88 (3), 401.
61. Bourbigot, S.; Le Bras, M.; Dabrowski, F.; Gilman, J.W.; Kashiwagi, T. PA-6 clay nanocomposite hybrid as char forming agent in intumescent formulations. *Fire Mater.* **2000**, 24(4), 201.
62. Lyon, R.E.; Walters, R.N. Pyrolysis combustion flow calorimetry. *J. Anal. Appl. Pyrol.* **2004**, 71(1), 27.
63. Gilman, J.; Kashiwagi, T. In: T.J. Pinnavia and G.W. Beall, Eds., *Polymer–Clay Nanocomposites*. Wiley, Chichester, West Sussex, England, 2000, p. 193.
64. Kong, Q.H.; Hu, Y.; Yang, L.; Fan, W.C.; Chen, Z.Y. Synthesis and properties of poly(methyl methacrylate)/clay nanocomposites using natural montmorillonite and synthetic Fe–montmorillonite by emulsion polymerization. *Polym. Compos.* **2006**, 27(1), 49.

4

MOLECULAR MECHANICS CALCULATIONS OF THE THERMODYNAMIC STABILITIES OF POLYMER–CARBON NANOTUBE COMPOSITES*

STANISLAV I. STOLIAROV
SRA International, Egg Harbor Township, New Jersey
MARC R. NYDEN
National Institute of Standards and Technology, Gaithersburg, Maryland

4.1 INTRODUCTION

It is has been demonstrated that the addition of small quantities of carbon nanotubes (CNTs) can dramatically improve the thermal and mechanical properties of polymers.[1–6] In many cases, however, this enhancement of properties is limited by the degree to which the CNTs can be dispersed uniformly within the polymer matrix. This appears to be particularly true for flame-retarding applications. For example, Kashiwagi and co-workers demonstrated that the heat release rates from burning well-dispersed nanocomposites consisting of single-walled CNTs (SWCNTs) in poly(methyl methacrylate) were significantly lower than

Flame Retardant Polymer Nanocomposites, edited by Alexander B. Morgan and Charles A. Wilkie

those from poorly dispersed samples.[7] Unfortunately, CNTs do not mix spontaneously with most polymers. Instead, they tend to form aggregates consisting of tens or even hundreds of discrete tubes that can become entangled in long rope-like structures.[6] At the source of the problem are the very properties from which the benefits of CNTs derive. Thus, CNTs reinforce the polymer matrix because they are inherently more rigid and less mobile than the polymer molecules they replace, but these attributes also limit their miscibility. Indeed, it is known that the excluded volume, which accompanies the introduction of CNTs in a polymer matrix, restricts the motion of the polymer, thereby decreasing the entropy of mixing.[8]

In recent years, some progress has been made in obtaining well-dispersed polymer–nanotube composites using a variety of approaches. These include in situ polymerization,[9] the addition of surfactants and compatibilizers,[10] polymer wrapping,[11] and functionalizing the ends and sidewalls of the tubes.[12,13] The widespread implementation of these methods, however, is limited by the absence of a quantitative understanding of the thermodynamics associated with the breakup of the nanotube bundles and accommodation of the discrete tubes within the polymer matrix.

In this chapter we introduce a simple methodology based on molecular mechanics that can be used to estimate the free energy of mixing nanotubes with polymers and apply it to predicting the thermodynamic stability of polystyrene–CNT composites as a function of nanotube radius. We anticipate that this approach can be adapted to other systems of interest by tailoring the constituent molecular models to represent the polymers, surfactants, and functional groups under consideration as part of a rational strategy to determine the best approach to the preparation of well-dispersed and stable polymer–CNT composites.

4.2 BACKGROUND AND CONTEXT

Molecular mechanics has been used extensively to investigate the structures and mechanical properties of CNTs. By employing this technique, thermodynamically stable structures for molecules and nanostructures can be obtained as minimum energy points on the multidimensional surface that represents the potential energy as a function of the atomic coordinates. In this section we review some previous studies involving molecular mechanics, which are especially relevant to our own work directed at the elucidation of factors that control the thermodynamics of polymer–CNT composites.

Typically, the potential energy of the system of interest is represented by a set of analytical functions referred to collectively as a *force field*. Associated with each of these functions are parameters that are specific for each type of atom accommodated by the force field. In many force fields, the atom types depend on the bonding environment, so there can be multiple atom types for the same element. In principle, this facilitates a more realistic description of the potential energy: For example, the force field may assign unique atom types to *sp*, sp^2, and sp^3 hybridized carbon atoms to account for the differences in electronic structure, which, in turn, affect the nature (bond length, angle, and dissociation energy) of the bonds formed by these atoms.

In our calculations, we used the PCFF force field,[14] which has the form

$$V = \sum_{ij}^{n\text{bonds}} V_{\text{bond}}(r_{ij}) + \sum_{ijk}^{n\text{angles}} V_{\text{angle}}(\theta_{ijk}) + \sum_{ijkl}^{n\text{torsions}} V_{\text{torsion}}(\phi_{ijkl}) + \sum_{ij}^{n\text{pairs}} V_{\text{nb}}(d_{ij}) \tag{4.1}$$

The component terms V_{bond}, V_{angle}, V_{torsion}, and V_{nb}, represent the variation of the potential energy with bond distance (r), bond angle (θ), torsional angle (ϕ), and the distance between nonbonded atoms (d). Although a variety of force fields have been used in the calculations described in this section, they are all based on the assumption that the variation of potential energy with geometry can be represented by simple analytical functions (Figure 4.1). The parameters (D_b, α, r_e, k_θ, θ_e, k_ϕ, n, ϕ_e) that govern the variation in potential energy with internal coordinates (r, θ, ϕ,) are usually determined by fitting the analytical expressions for V_{bond}, V_{angle}, and V_{torsion} to energies from quantum calculations. The parameters for the nonbonded interactions (ε, r^*, δ) in V_{nb} are adjusted so that the calculated densities and cohesive energies agree with experimental measurements on volatile liquids.

Some of the first molecular mechanics calculations on CNTs were reported by Robertson et al. in a paper published in 1992.[15] By computing the energies for a series of SWCNTs with various diameters ($D < 0.9$ nm) and chiralities, these authors found that the strain energy per carbon atom (relative to an unfolded, planar sheet of graphite) varied as D^{-2}. Several years later, molecular mechanics calculations by Tersoff and Ruoff demonstrated that the energy of interaction per unit surface area between adjacent nanotubes (and consequently, the cohesive energy of a CNT bundle) varied as $D^{-1/2}$.[16] This dependence was verified in

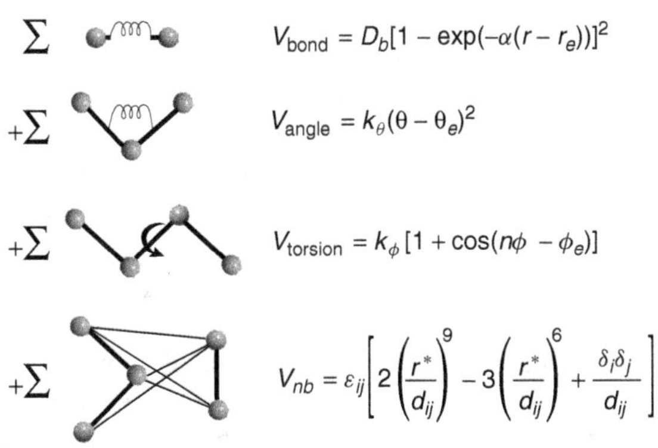

FIGURE 4.1 Component terms in the PCFF force field.

a study by Girifalco et al.[17] Tersoff and Ruoff also observed that the walls of SWCNTs in a hexagonal lattice had a tendency to flatten when their diameters exceeded about 2.5 nm.

A similar effect was reported in a paper by Gao et al.,[18] who performed molecular mechanics calculations on separated SWCNTs. They indicated that for diameters of less than about 2 nm, only the cylindrical (circular cross section) geometries are stable, whereas a collapsed structure, which is flattened at the center with circular bulges at both ends, becomes more stable when the diameter of the CNTs exceeds about 6 nm. For diameters between 2 and 6 nm, both cylindrical and flattened structures were found to be stable. The authors pointed out that the flattening of the SWCNTs was actually due to the van der Waals attraction between the opposing walls. At some point ($D \sim 6$ nm), this attractive interaction becomes sufficient to overcome the increase in (angular) strain due to the circular bulges at the ends, which must form when the tubes collapse in the center.

A study by Hertel et al. demonstrated that the walls of SWCNTs also tend to flatten when they are adsorbed on a planar substrate. They observed that the extent of this distortion is reduced for multiwalled carbon nanotubes (MWCNTs),[19] which consist of multiple concentric tubes. A straightforward extension of this observation is that MWCNTs should also tend to retain their cylindrical shapes (better than SWCNTs do) when they are bundled together in ropes.

Although most past investigations employing molecular mechanics have focused on the properties of CNTs in the absence of polymers, the interfacial characteristics of polymer–CNT composites were examined in a more recent paper by Liao and Li. [20] The authors of this study used molecular mechanics to calculate the energy required to extract a CNT from polystyrene. The value they reported (80 kJ/mol $\cdot$ nm^2) is in reasonable agreement with the average value ($\Delta E^S_{\text{np}} + \Delta E^S_{\text{pp}} = 71$ kJ/mol $\cdot$ nm^2) obtained from our own calculations (see Table 4.2).

None of the aforementioned studies focused directly on the problem of predicting the thermodynamic stability of polymer–CNT composites. This possibility was, however, explored in a recent paper by Maiti et al. [21] These authors computed Hildebrand solubility parameters for CNTs (from the cohesive energy densities of nanotube bundles) as a function of their diameters. By comparing these to accepted values of the solubility parameters for a series of polymers, they were able to make predictions about the diameters required for miscibility. It should be noted, however, that their analysis was based on the Flory–Huggins theory for regular solutions, which does not account for structural changes that can affect the nature of the interactions between solute (nanotube) and solvent (polymer) that can result in an exothermic enthalpy of mixing.[22] Our results, which are presented in the following sections of this chapter, are not limited in this way.

Since molecular mechanics calculations consist of minimizing the potential energy, without consideration of kinetic energy, they cannot account for the effects of temperature. Thus, the energies obtained from them correspond to hypothetical structures at 0 K. In principle, the temperature dependence of the properties of interest can be investigated by performing molecular dynamics simulations. In these calculations, atomic trajectories are computed by numerical integration of the classical equations of motion. The thermal motion is initiated by providing the component atoms with random momenta from a Boltzmann distribution consistent with the temperature specified. As the simulation progresses, the molecular structures are perturbed from their initial equilibrium geometries (typically, obtained from molecular mechanics and/or x-ray crystallography) in the presence of restoring forces that are computed from the gradients of the potential energy, which can be represented by a force field or even, for smaller systems, computed on the fly from electronic structure calculations.

It should be noted, however, that since molecular dynamics is based on the classical equations of motion (as opposed to the time-dependent Shrödinger equation), it does not account for quantum effects such as zero-point energy and the discrete nature of vibrational states. For example, in a molecular dynamics simulation the C–H stretching modes of a polymer are active (and therefore contribute to the heat capacity) at temperatures much lower than would be predicted by a more rigorous quantum calculation. Furthermore, although molecular dynamics offers a mechanism to incorporate kinetic energy (and temperature), the accuracy of the results are limited by the degree to which the relevant phase space is explored in the simulations. This is problematic for polymer–CNT composites because the high viscosity of the polymer matrix and the low mobility of the nanotubes ensure that the relaxation times will be long.

Currently, the computational demands dictated by the number of atoms needed to describe polymer–CNT composites restrict these simulations to nanosecond time scales, whereas the actual mixing process conducted in a laboratory requires several minutes (at a minimum). Because of this disparity, there are no assurances that the structures that evolve in the molecular dynamics simulations are representative of those that characterize the initial and final states of the actual materials. By employing molecular mechanics (together with a little intuition and some trial and error), it is possible to identify low-energy structures for the initial and final states that are at least as representative as those generated in molecular dynamics simulations. In the final analysis, however, the only assurance that either approach is capable of providing a realistic description of the mixing process is agreement between calculated properties (and/or predicted trends) and experimental measurements.

4.3 DESCRIPTION OF THE METHOD

The increase in entropy that accompanies the formation of a mixture is an important factor in determining the miscibilities of small molecules. However, we suspect that entropic effects are much less significant for mixtures involving large immobile molecules such as CNTs and polymers. This presumption is consistent with standard treatments of polymer miscibility based on lattice theory and is a consequence of the dramatic reduction in the number of lattice configurations (representing the mixture) due to the constraints on the contiguity of the component monomers (and nanotube segments).[22] The gain in entropy for mixtures of CNTs and polymers is probably even smaller than it is for polymer blends because nanotubes are inherently less flexible than polymer chains. Because of their inherent rigidity, nanotubes are less efficient in occupying the available space (in the polymer matrix) than are (more flexible) polymer chains, thereby reducing the number of possible arrangements (combinatorial entropy) in the mixture. Moreover, the small contribution that entropy changes make to the free energy of mixing should be relatively insensitive to changes in the physical (diameter and chirality) and chemical (i.e., functional groups) structures of the component nanotubes. Thus, it should be possible to predict trends in the thermodynamic stability of nanocomposites directly from their enthalpies of mixing (ΔH_{mix}). Furthermore, since the nanotubes do not occupy any more volume when they are dispersed in the polymer than they do when they are bunched together, the volume change accompanying the formation of the composite should also be very small, implying that $\Delta H_{\text{mix}} \approx \Delta E_{\text{mix}}$.

Unfortunately, an explicit calculation of the energy of mixing (ΔE_{mix}) is precluded because of the computational demands of evaluating all of the interactions between the atoms in an actual nanocomposite, which might contain nanotubes many microns in length and as many as 1000 carbon atoms in the polymer for every carbon atom in the nanotubes (i.e., a loading of approximately 0.1%). Instead, the approach we adopted makes use of localized molecular models of the polymer, nanocomposite, and the exfoliated and bundled nanotubes to estimate the relative magnitudes of the energies associated with the polymer–polymer (pp), CNT–CNT (nn), and CNT–polymer (np) interactions. The energy of mixing is then evaluated in terms of a simple path in which the nanotubes are exfoliated from a bundle and dispersed in a distorted polymer with cylindrical cavities to accommodate the nanotubes. In the laboratory, this is realized by melt blending or dissolving the polymer (to reduce the viscous forces that impede the dispersion of the nanotubes) and applying a shear force (via extrusion or sonication) to exfoliate and disperse the nanotubes. From this perspective, the energy of mixing is the difference between the energy required to exfoliate the nanotubes from a bundle and the energy needed to extract the nanotubes from the polymer matrix relative to the relaxed polymer without any nanotubes. The component processes are depicted in Figure 4.2.

Following the logic of this scheme, the energy of mixing can be evaluated from

$$\Delta E_{\text{mix}} = [\Delta E^S_{\text{nn}} - (\Delta E^S_{\text{np}} + \Delta E^S_{\text{pp}})]S \tag{4.2}$$

where

$$\Delta E^S_{\text{nn}} = \frac{\Delta E_{\text{nn}}}{S_n}, \qquad \Delta E^S_{\text{np}} = \frac{\Delta E_{\text{np}}}{S_n}, \qquad \Delta E^S_{\text{pp}} = \frac{\gamma \Delta E_{\text{pp}}}{S_n} \tag{4.3}$$

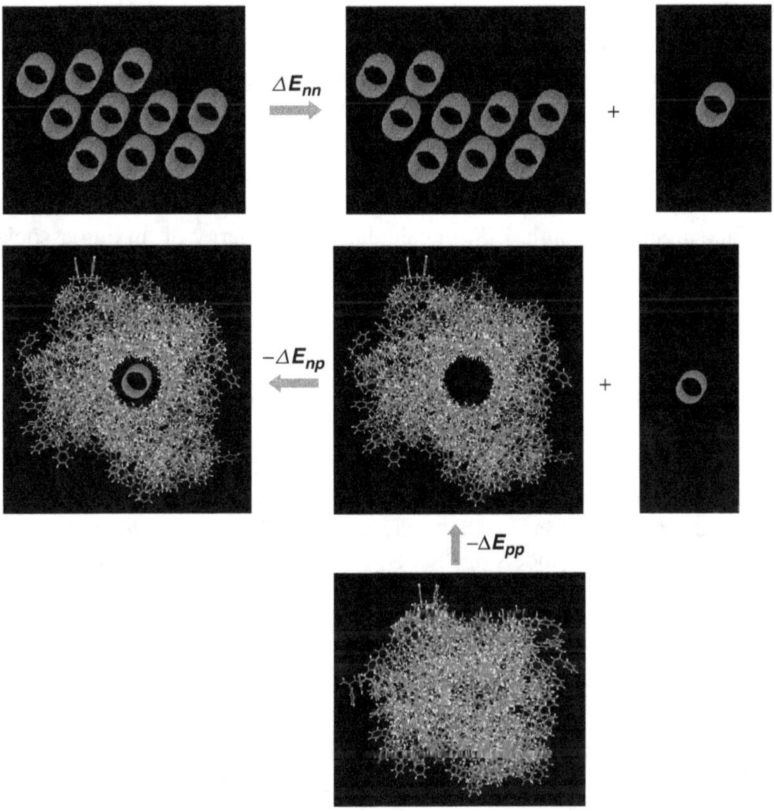

FIGURE 4.2 Processes involved in the formation of a PS–CNT composite from a nanotube bundle and polymer.

are calculated from the energy differences of the model reactions depicted in Figure 4.2, and γ is a correction factor that is discussed in Section 4.4.

In these equations, ΔE_{nn} is the energy required to exfoliate a nanotube from a bundle. The magnitude of this term reflects the strength of the interaction between nanotubes. The second term, ΔE_{np}, is the energy needed to extract a nanotube from a polymer–CNT agglomerate that represents the environment of the nanocomposite in the vicinity of the nanotube. This term accounts for the interactions between the nanotube and polymer. The last term, ΔE_{pp}, is the energy lowering that results from closing the cylindrical cavity occupied by the nanotube in the polymer–nanotube agglomerate. This results in a decrease in surface area and a corresponding decrease in energy (relative to the value obtained when there is a cavity) due to the increase in the number of attractive polymer–polymer interactions. It is determined from the energy difference between the relaxed polymer (optimized in the absence of the nanotube) and the polymer matrix in the nanocomposite (optimized in the presence of the nanotube). The origin of the negative sign with respect to the contribution of $(\Delta E_{\mathrm{np}}^{S} + \Delta E_{\mathrm{pp}}^{S})$ in eq. (4.2) is simply that in the formation of the nanocomposite, the nanotube is inserted in (rather than extracted from) the polymer matrix.

The individual terms in eq. (4.3) are normalized by dividing by the surface area of the model nanotube, S_n, to facilitate the extrapolation of the results obtained from the atomic length scales of the molecular models to the much larger dimensions that prevail in real materials (i.e., moles of atoms). Thus, as indicated in eq. (4.2), the sum of these component energies is multiplied by the total surface area of nanotubes, S, in calculating the energy of mixing associated with the formation of a real nanocomposite.

4.4 APPLICATION TO PS–CNT COMPOSITES

The method described in Section 4.3 was applied in an attempt to understand the factors that determine the thermodynamic stability of polystyrene–CNT composites. Polystyrene (PS) was chosen for the first application because it has aromatic rings which should interact favorably with the aromatic rings that comprise the nanotubes based on the premise that "like dissolves like." The calculations were performed on molecular models of PS, PS–CNT agglomerates, nanotube bundles, and separated nanotubes using a commercial software package (Material Studio)* with the PCFF force field.[14] Energy-optimized structures were determined by minimizing the energy using a cutoff of 1.5 nm for the nonbonding interactions, which were represented by a Lennard-Jones 6–9 potential (see Figure 4.1).

Molecular models of uncapped (7,0) nanotubes with radius $R = 0.28$ nm were used in the calculations. From them we constructed three models to represent the nanotube bundles. The largest one, which is depicted in Figure 4.2, consisted of ten 3.6-nm-long nanotubes arranged in a closest-packing structure. Comparable results were obtained from the smaller model, consisting of seven 3.6-nm-long nanotubes, which is shown in Figure 4.3. The third model was identical to the one in Figure 4.3 except that the nanotubes were approximately twice as long (7.3 nm). The polymer–CNT agglomerates were constructed by minimizing the energies of the intermediate structures obtained by adding successive polymer chains. The structure of one of these polymer–nanotube agglomerates is depicted in Figure 4.2. On this basis, it was determined that 12 chains were sufficient to achieve convergence of the polymer–nanotube interaction energies to the limit of infinite dilution. The lengths of the polymer chains, which were adjusted to ensure that they covered the full surface of the nanotube, were 17 and 34 monomers for agglomerates containing the 3.6- and 7.3-nm nanotubes, respectively. The density of the polymer matrix in these polymer–CNT agglomerates was about 1000 kg/m^3.

The effects of increasing the lengths of the model nanotubes from 3.6 nm to 7.3 nm can be seen in the data presented in Tables 4.1 and 4.2. These results seem to justify the intuitive notion that for a specified diameter, the energies of

*Certain commercial equipment, instruments, materials, or companies are identified in this chapter to specify the experimental procedure adequately. This in no way implies endorsement or recommendation by NIST.

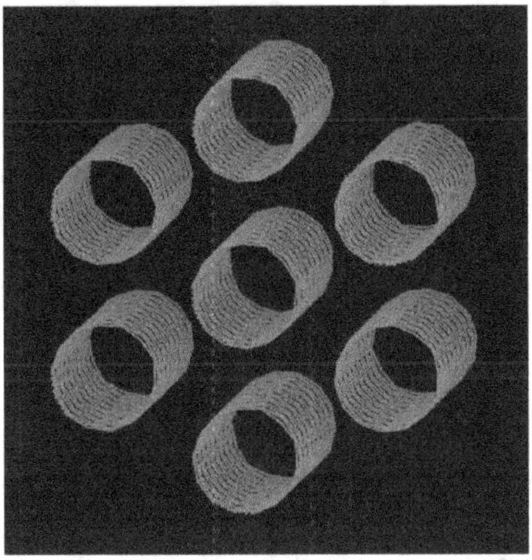

FIGURE 4.3 Smallest of the bundled nanotube models used for computing ΔE^S_{nn}.

TABLE 4.1 Models of Nanotube Bundles

Number of Nanotubes	Length of Nanotube (nm)	Surface Area of Nanotube (nm^2)	Number of Carbon Atoms/Area (nm^{-2})	b_0 (nm)	ΔE^S_{nn} (kJ/mol · nm^2)
7	3.6	6.3	39.8	0.32	159
7	7.3	12.9	39.1	0.32	163
10	3.6	6.3	39.8	0.32	162

TABLE 4.2 Models of Polymer–Nanotube Agglomerates

Length (nm)	R_{pn} (nm)	R_c (nm)	γ	ΔE^S_{np}(kJ/mol · nm^2)	ΔE^S_{pp}(kJ/mol · nm^2)
3.6	1.75	0.525	0.87	176	−95
7.3	1.90	0.525	0.88	160	−99

interaction per unit area are almost independent of the lengths of the nanotubes. Moreover, ΔE^S_{nn} appears to be independent of both the length and number of nanotubes in the bundle.

The dependence of these energies on the radius of the constituent nanotubes, however, is more complicated. Consider first the ΔE^S_{nn} term. If the number of atoms per unit surface area is independent of the radius of the nanotube (which is tantamount to assuming that the aromatic rings have the same structures), it can be shown that the energy of extraction of a CNT from a bundle (per unit surface area of the nanotube) will decrease approximately in accordance with[16,17]

$$\Delta E^S_{nn} = \frac{k'_{nn} l \sqrt{R}}{2\pi R l} = \frac{k_{nn}}{\sqrt{R}} \tag{4.4}$$

Here R is the radius and l is the length of the component nanotubes. The decrease in ΔE^S_{nn} arises from the simple fact that the average distance, $\bar{b}$, between two parallel cylindrical surfaces (with the same radius) increases from their distance of closest approach, b_0, with increasing radius (Figure 4.4). Since the atoms that comprise each of the two interacting nanotubes are on average farther apart in large-diameter nanotubes (assuming that their distance of closet approach remains the same), the attraction between them diminishes, thereby reducing the cohesive energy per unit surface area of the bundle.

A similar analysis can be applied to both ΔE^S_{np} and ΔE^S_{pp}. The latter of these two terms is proportional to the surface energy of the polymer (see the discussion above), which increases linearly with the number of atoms that are brought from the interior to the surface. The area of the nanotube cavity in the polymer matrix is $2\pi(R+d)l$, where $d = 0.25$ nm is the average distance between the polymer and the surface of the nanotube. Of course, the outer surface of the polymer must also expand to accommodate the nanotube. The correction factor,

$$\gamma = \frac{R_c}{R_c + R_{\text{pn}} - \sqrt{R_{\text{pn}}^2 - R_c^2}} \tag{4.5}$$

represents the fraction of the total increase in surface area due to the formation of the cylindrical cavity. In eq. (4.5), $R_c = R + d$ is the radius of the cylindrical cavity and R_{pn} is the radius of the polymer–nanotube agglomerate, which is also assumed to be cylindrical in shape. This correction is needed to extrapolate the results from the model calculations, where the change in the outer surface of the cylinder representing the polymer–nanotube agglomerate is significant (because the models are relatively small), to realistic dimensions, where this contribution is negligible. The values of this correction factor for the polymer–CNT agglomerates corresponding to the 3.6- and 7.3-nm-long nanotubes are listed in Table 4.2.

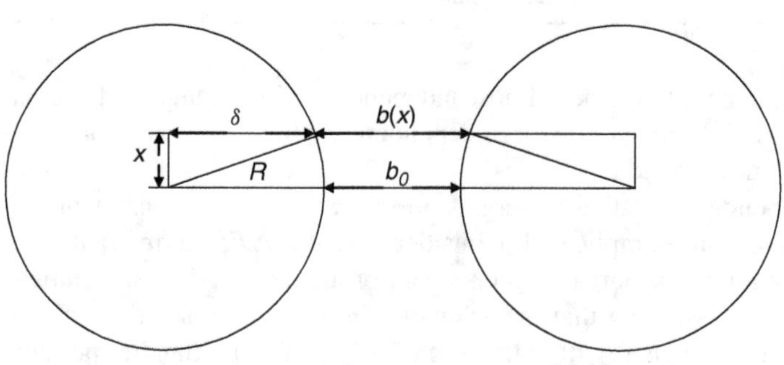

$$b(x) = b_0 + 2(R-\delta) \approx b_0 + \frac{x^2}{R}\ (R \gg x),\quad \bar{b} = \frac{1}{2R}\int_{-R/3}^{R/3} b(x)dx = \frac{b_0}{3} + \frac{R}{81}$$

FIGURE 4.4 How the distance between opposing points on two cylindrical surfaces depends on the radius.

Consider next the $\Delta E_{\mathrm{np}}^{S}$ term. From Figure 4.2, it appears that the polymer effectively encircles the nanotube. The distance between the surface of the nanotube and the polymer (d) is determined by the van der Waals radii of the interacting atoms and should be relatively independent of the diameter of the nanotube. Therefore, the nature of polymer-nanotube interaction is similar to the case of two parallel sheets, for which the energy of interaction per unit surface area is constant. This is, in fact, the limiting behavior of $\Delta E_{\mathrm{np}}^{S}$ as $R \to \infty$. However, because of their concentric arrangement, the number of interactions between atoms on the cylindrical surface of the polymer and atoms on the surface of the nanotube (per unit area of the nanotube) increases as the ratio of the surface areas. Thus, we infer that both $\Delta E_{\mathrm{np}}^{S}$ and $\Delta E_{\mathrm{pp}}^{S}$ scale the same way with nanotube radius so that

$$\Delta E_{\mathrm{np}}^{S} + \Delta E_{\mathrm{pp}}^{S} = \frac{k'_{\mathrm{np+pp}} \cdot 2\pi(R+d)l}{2\pi Rl} = k_{\mathrm{np+pp}}\left(1 + \frac{d}{R}\right) \tag{4.6}$$

The averages of the values of $\Delta E_{\mathrm{nn}}^{S}$ and $(\Delta E_{\mathrm{np}}^{S} + \Delta E_{\mathrm{pp}}^{S})$ reported in Tables 4.1 and 4.2 were extrapolated as a function of nanotube radius using eqs. (4.4) and (4.6). The results of these extrapolations are plotted in Figure 4.5. At $R = 4.5$ nm, the energy needed to exfoliate the nanotubes falls to the point where it is offset by the energy released by the attractive interactions between the nanotubes and polymer and thermodynamic neutrality of the mixing is attained. Substituting the surface area of a single nanotube for S (i.e., $S = 2\pi Rl$) in eq. (4.2), we obtain the enthalpy of mixing of 1 mol of nanotubes having the dimensions specified. Unlike the surface-normalized enthalpy, which (according to Figure 4.5) decreases monotonically with increasing nanotube radius, the enthalpy normalized by the number of nanotubes goes through a maximum at about $R = 1.5$ nm, as indicated in Figure 4.6. For the purpose of comparing thermodynamic stabilities of polymer–CNT composites, the molar (per mole of nanotubes) enthalpy may be more appropriate than the surface-normalized enthalpy. The difference in the molar enthalpy of two systems containing tubes of the same length is likely to represent the difference in the molar free energy. This is the case because (as has already been discussed) even if the change in the molar entropy due to mixing is not negligibly small, it should be close in value for nanotubes with similar flexibility. Thus, on the basis of our calculations, we predict that it should be possible to obtain stable, exfoliated nanocomposites by blending CNTs having diameters greater than about 9 nm with polystyrene. Since SWCNTs are typically much smaller than this (1.0 to 1.4 nm in diameter[23]), we conclude that thermodynamic neutrality is never attained when SWCNTs are added to PS. On the other hand, our results do indicate that it should be possible to make thermodynamically stable PS–CNT nanocomposites from MWCNTs, which have diameters ranging from about 10 nm to almost 100 nm.[24]

The conclusion that enthalpy of mixing becomes exothermic for sufficiently large diameters presumes that the relationship expressed in eq. (4.4) continues to be valid as R becomes arbitrarily large. However, as mentioned above, Tersoff and Ruoff found that the walls of SWCNTs packed in a hexagonal lattice tend

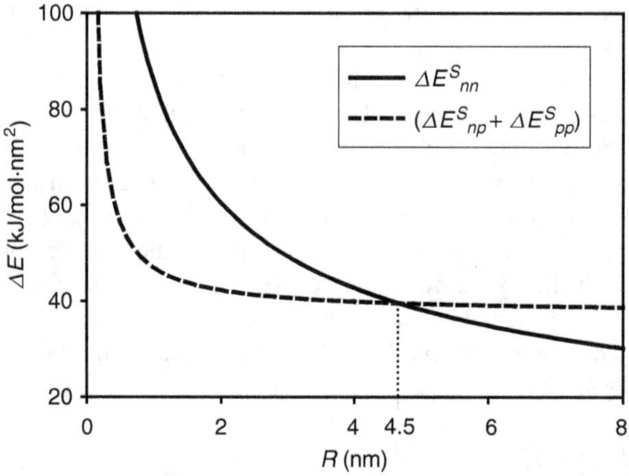

FIGURE 4.5 Variation of the nanotube–nanotube and the sum of the nanotube–polymer and polymer–polymer energies as a function of nanotube radius.

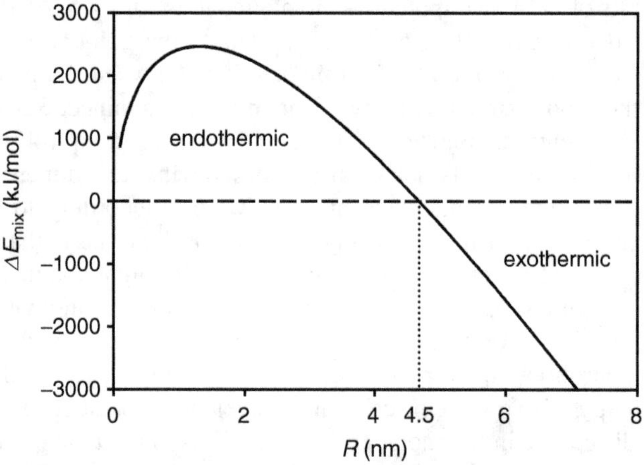

FIGURE 4.6 Variation of the energy of mixing as a function of radius for CNTs that are 10 nm long.

to flatten once their radius exceeds about 2.5 nm.[16] This is consistent with the intuitive notion that the structural integrity of a nanotube should diminish as its radius gets larger. However, MWCNTs, which are comprised of concentric tubes, are more likely to resist distortion[19] as their diameters approach the point of thermodynamic neutrality.

4.5 UNCERTAINTIES AND LIMITATIONS

There are a number of considerations that limit the accuracy and reliability of the approach outlined in this chapter. To begin with, as indicated above in the background section, we have not considered the effects of temperature on the energy of mixing. Thus, we assume that the temperature-dependent terms cancel in taking the difference between the initial (polymer + CNT bundle) and final (nanocomposite) states. To facilitate a more detailed analysis of this hypothesis, it is convenient to categorize the contributions to the derivative of the energy of mixing with respect to temperature (heat capacity) as vibrational and conformational. The latter arises from the differences in the populations of the conformational structures of the polymer (each corresponding to a minimum in the potential energy) resulting either from the interactions with nanotubes or temperature, whereas the former is associated with the vibrations of these conformers and nanotubes about their equilibrium geometries. We think that the vibrational contribution will be close to zero because the interactions between the nanotubes and polymer chains in the nanocomposite are not sufficiently different from the polymer–polymer and CNT–CNT interactions (Tables 4.1 and 4.2) to affect the nature of the vibrations (i.e., the distribution of the density of states of the vibrators) of the nanotubes and polymer chains. Thus, the contribution of vibrational energy to the heat capacity cancels because it is the same for the initial and final states.

This may not be a good assumption with respect to conformational heat capacity because there is evidence suggesting that polymers adapt different conformations (which may have different energies) in the vicinity of nanotubes.[25] Our calculations do account for this possibility because the structure of the polymer was optimized independently with and without the nanotube, but we did not investigate whether these conformations change as a function of temperature. Thus, we are assuming that the structures obtained from our molecular mechanics calculations are representative of the structures that are populated at the processing temperature of the mixture. The reasonableness of this assumption was discussed in Section 4.2.

The difficulty in obtaining representative structures of the polymer has additional ramifications that can affect the accuracy of our calculations. Thus, as indicated above, the definition of ΔE^S_{pp} requires that independent optimizations of the initial (polymer–nanotube agglomerate) and final structures (relaxed polymer after removing the nanotube) be performed. A comparison of the optimized structures before and after removing the nanotube indicated that significant structural changes to the polymer did occur. However, it is not clear whether all of these changes can be attributed to the presence of the nanotube. For example, in an effort to assess the convergence of ΔE_{mix} with respect to the number of polymer chains, we performed another set of calculations on a nanotube surrounded by 18 (17 monomer) chains depicted in Figure 4.7. From these calculations, we obtained the following results: $\Delta E^S_{\mathrm{np}} = 176$ kJ/mol · nm^2 and $\Delta E^S_{\mathrm{pp}} = -153$ kJ/mol · nm^2. Although the value for ΔE^S_{np} is in excellent agreement with the results obtained with 12 chains (Table 4.2), which implies that the limit of infinite dilution has been reached, the value for ΔE^S_{pp} is much different (the absolute value is almost 60% too large) than the values we obtained from the smaller models (Table 4.2). The source of the disparity is that in these calculations (i.e., on the models with

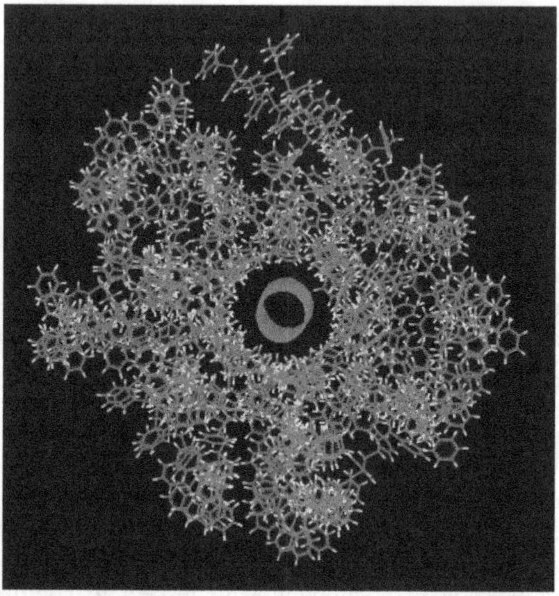

FIGURE 4.7 Nanotube (3.6 nm long) surrounded by 18 PS chains, each consisting of 17 monomers.

18 polymer chains), the structure of the model polymer optimized in the absence of the nanotube was much more compact than the polymer optimized in the presence of the nanotube. Unfortunately, it is not clear whether this change in conformation was induced by the presence of the nanotube or whether it is merely an artifact of the minimization process. In fact, we suspect that the latter explanation is more correct. That is, the polymer (in the polymer–nanotube agglomerate) adopted an artificially high-energy conformation, corresponding to a local minimum on the potential energy surface. The implication of this observation is that it is very difficult to ensure that the energy differences used in determining $\Delta E_{\mathrm{pp}}^{S}$ only reflect the effects of accommodating the nanotube, and this can result in significant errors.

In general, it is more difficult to find optimal structures for large molecules than for small molecules (because there are more degrees of freedom), so it is not surprising that this discrepancy became more apparent when we attempted to include more polymer chains in our calculations. This problem, however, is not insignificant in the values reported in Table 4.2. For example, if we renormalize the average value of $-\Delta E_{\mathrm{pp}}^{S}$ so that it represents the energy needed to increase the surface area of the polymer, we obtain 51 kJ/mol · nm^2, which overestimates the accepted value of the surface tension of PS (25 kJ/mol · nm^2)[26] by more than 100%. This disparity seems too large to be due entirely to errors in the force field (see below). Thus, we feel that the problems associated with the optimization discussed in the preceding paragraph must be responsible for a significant part of this discrepancy. On this basis, we think that the approach presented in earlier sections should be modified to provide a more accurate estimate for this contribution to the energy of mixing. Recognizing that ΔE^S_{pp} can be determined from the surface tension of the polymer, we recommend substituting a simpler calculation that does not involve reoptimization of the initial and final states of the polymer. One possibility, suggested by analogy to the graphite calculations discussed below, might be to compute the energy (per unit area) required to separate a layer from a polymer agglomerate comprised of several identical layers.

The accuracy of the force field may also be an issue. In fact, the validity of using the PCFF force field to evaluate the energies of graphitic structures has not been established. To examine this issue, we performed calculations of the cohesive energy of graphite for which there is an accepted experimental value.[27] The energy of the model of graphite depicted in Figure 4.8 was subtracted from the energy of the model obtained by moving the top sheet far enough away that the potential energy of its interaction with the remaining two sheets was effectively zero. Assuming a surface density of 39.8 atoms/nm (Table 4.1), we obtained 230 kJ/mol · nm^2, whereas the experimentally determined cohesive energy is only 160 kJ/mol · nm^2. On the basis of this comparison, we infer that the PCFF force field overestimates the interaction energies between graphite sheets by a little more than 40%.

The force field errors can be reduced by adjusting the parameters corresponding to the (polymer–polymer) nonbonding interactions such that this energy is consistent with the accepted value (assuming that the experimental measurements have been performed) for the surface tension of the polymer of interest. Similar adjustments should also be made in the parameters that determine the nanotube–nanotube interactions to ensure that the accepted value for the cohesive energy of graphite is also obtained. Since the parameters that govern the van der Waals interactions between the polymer and nanotube are typically determined from the geometric mean of the parameters for the nn and pp interactions, we would expect to see a comparable improvement in the accuracy of this contribution (np) to the enthalpy of mixing at the same time.

Unfortunately, the calculation of the diameter for which the energy of mixing nanotubes in PS becomes exothermic is extremely sensitive to the errors in the component terms discussed above. Thus, when we attempted to correct for the errors in the forcefield by scaling the calculated values for ΔE^S_{nn} and ΔE^S_{np} by 0.7

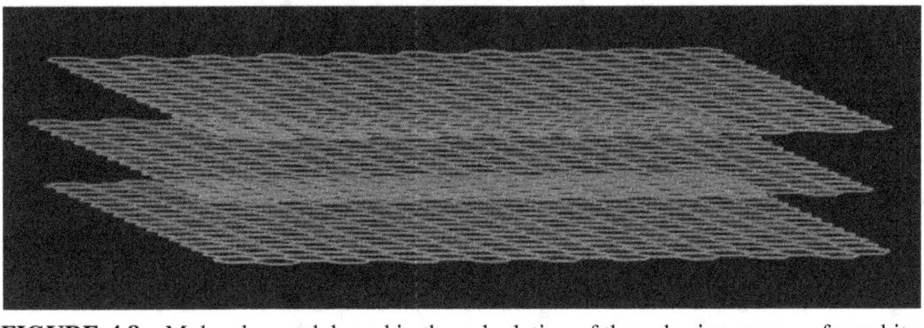

FIGURE 4.8 Molecular model used in the calculation of the cohesive energy of graphite.

(based on the error in the PCFF value for the cohesive energy of graphite) and for the (energy optimization and force field) errors reflected in ΔE^{S}_{pp} by multiplying our calculated value by 0.5, we found that exothermic neutrality occurred at a diameter of 2 nm. This is considerably different (smaller) than the value obtained directly from our calculations. For this reason we do not place much credibility on the actual value reported above. The basic observation that the surface energy of mixing should become more exothermic with increasing nanotube diameter, however, is a real effect that does provide useful guidance.

It should be kept in mind that even if we could remove all of these errors from our calculations, thermodynamic stability may actually be less important than kinetics in determining whether a nanocomposite will be stable, or even if it is possible to attain one in the first place. Thus, by employing sonication and other high-energy mixing techniques, it may be possible to obtain a nanocomposite with good dispersion even if the free energy of mixing is positive. Furthermore, even if they are thermodynamically unstable, nanocomposites can be effectively stable if the transport of nanotubes through the polymer matrix is sufficiently slow that they cannot aggregate during the service life of the material. The converse is also true. That is, it may be impossible to disperse the nanotubes even if the nanocomposite, once formed, is thermodynamically stable.

4.6 SUMMARY AND CONCLUSIONS

A method for estimating the free energy of mixing CNTs with polymers was formulated and presented in this chapter. The formation of the nanocomposite was analyzed in terms of a simple path in which the nanotubes are exfoliated from a bundle and dispersed in a distorted polymer with cylindrical cavities to accommodate the nanotubes. From this perspective, the energy of mixing is the difference between the energy required to exfoliate the nanotubes from a bundle and the energy needed to extract the nanotubes from the polymer matrix relative to the relaxed polymer without nanotubes. These energy components were evaluated by performing molecular mechanics calculations on individual localized models representing the polymer, nanotube bundles, nanotube–polymer agglomerates, and the separated nanotubes. This method was applied to polystyrene–CNT composites and the factors that determine their thermodynamic stability were identified.

To a first approximation, the interaction energies (per unit surface area of the nanotubes) were shown to be independent of the lengths but dependent on the diameters of the component nanotubes. On this basis it was determined that a thermodynamically stable nanocomposite could be obtained by mixing CNTs with diameters greater than about 9 nm in polystyrene. This may explain why it is so difficult to obtain good dispersion of SWCNTs in PS, since they rarely grow to have diameters greater than about 3 nm. On the other hand, since the diameters of MWCNTs typically exceed 10 nm, we would expect them to disperse much better than SWCNTs in polystyrene. Although the errors in our calculations are of sufficient magnitude that the precise value is questionable, it is clear from our analysis that the energy of mixing nanotubes with PS (and, by extension, all polymers) will become more exothermic as the diameters of the component nanotubes increase beyond a critical value. This suggests that in the absence of any treatment (i.e., functionalization of the nanotubes and/or the addition of compatibilizers), MWCNTs will provide better dispersion than that provided by the much smaller SWCNTs.

The approach outlined in this chapter should be applicable to other polymer–nanotube systems provided that the molecular models used in the calculations are modified to reflect their chemical natures. Although the results may not offer a definitive answer to the question of whether or not it is possible to obtain a well-dispersed nanocomposite from a given set of components, they do provide a quantitative basis for assessment of the relative stability of various compositions. We hope to be able to demonstrate this in the future by using this approach to examine the effects of the nature and degree of functionalization of SWCNTs on the thermodynamic stability of nanocomposites.

Acknowledgments

The authors acknowledge support received from the Federal Aviation Administration, from the Fire Safe Aircraft Cabin Materials Program managed by Dr. Richard Lyon, and from an intramural project (monitored by Dr. Felix Wu) sponsored by NIST's Advanced Technology Program.

REFERENCES

1. Haggermueller, R.; Gommans, H.H.; Rinzler, A.G.; Fischer, J.E.; Winey, K.I. Aligned single-wall carbon nanotubes in composites by melt processing methods. *Chem. Phys. Lett.* **2000**, 330, 219–225.
2. Ajayan, P.M.; Schadler, L.S.; Giannaris, C.; Rubio, A. Single-walled carbon nanotube–polymer composites: strength and weakness. *Adv. Mater.* **2000**, 12, 750–753.
3. Mamedov, A.A.; Kotov, N.A.; Prato, M.; Guldi, D.M.; Wicksted, J.P.; Hirsch, A. Molecular design of strong single-wall carbon nanotube/polyelectrolyte multilayer composites. *Nat. Mater.* **2002**, 1, 190–194.
4. Du, F.; Fischer, J.E.; Winey, K.I. Coagulation method for preparing single-walled carbon nanotube/poly(methyl methacrylate) composites and their modulus, electrical conductivity, and thermal stability. *J. Polym. Sci. B* **2003**, 41, 3333–3338.
5. Ramanathan, T.; Liu, H.; Brinson, L.C. Functionalized SWNT/polymer nanocomposites for dramatic property improvement. *J. Polym. Sci. B* **2005**, 43, 2269–2279.
6. Bower, C.; Kleinhammes, A.; Wu, Y.; Zhou, O. Intercalation and partial exfoliation of single-walled carbon nanotubes by nitric acid. *Chem. Phys. Lett.* **1998**, 288, 481–486.
7. Kashiwagi, T.; Du, F.; Winey, K.I.; Grotha, K.M.; Shields, J.R.; Bellayer, S.P.; Kim, H.; Douglas, J.F. Flammability properties of polymer nanocomposites with single-walled carbon nanotubes: effects of nanotube dispersion and concentration. *Polymer* **2005**, 46, 471–481.
8. Wei, C.; Srivastava, D.; Cho, K. Thermal expansion and diffusion coefficients of carbon nanotube–polymer composites. *Nano Lett.* **2002**, 2, 647–650.

9. Park, C.; Ounaies, Z.; Watson, K.A.; Crooks, R.E.; Smith, J.; Lowther, S.E.; Conell, J.W.; Siochi, E.J.; Harrison, J.S.; St. Clair, T.L. Dispersion of single-wall carbon nanotubes by in situ polymerization under sonication. *Chem. Phys. Lett.* **2002**, 364, 303–308.
10. Liu, J.; Rinzler, A.G.; Dai, H.J.; Hafner, J.H.; Bradley, R.K.; Boul, P.J.; Lu, A.; Iverson, T.; Shelimov, K.; Huffman, C.B.; Rodriguez-Macias, F.; Shon, Y.S.; Lee, T.R.; Colbert, D.T.; Smalley, R.E. Fullerene pipes. *Science* **1998**, 280, 1253–1255.
11. Zheng, M.; Jagota, A.; Semke, E.D.; Dine, B.A.; Mclean, R.S.; Lustig, S.R.; Richardson, R.E.; Tassi, N.G. DNA-assisted dispersion and separation of carbon nanotubes. *Nat. Mater.* **2003**, 2, 338–342.
12. Dyke, C.A.; Tour, J.M. Unbundled and highly functionalized carbon nanotubes from aqueous reactions. *Nano Lett.* **2003**, 3, 1215–1218.
13. Mitchell, C.A.; Bahr, J.L.; Arepalli, S.; Tour, J.M.; Krishnamoorti, K. Dispersion of functionalized carbon nanotubes in polystyrene. *Macromolecules* **2002**, 35, 8825–8830.
14. Accelrys, Inc. PCFF Version 3.1.
15. Robertson, D.H.; Brenner, D.W.; Mintmire, J.W. Energetics of nanoscale graphitic tubules. *Phys. Rev B* **1992**, 45, 12592–12595.
16. Tersoff, J.; Ruoff, R.S. Structural properties of a carbon-nanotube crystal. *Phys. Rev. Lett.* **1994**, 73, 676–679.
17. Girifalco, L.A.; Hodak, M.; Lee, R.S. Carbon nanotubes, buckyballs, ropes, and a universal graphitic potential. *Phys. Rev. B* **2000**, 62, 131104–13110.
18. Gao, G.; Cagin, T.; Goddard, W.A., III. Energetics, structure, mechanical and vibrational properties of single-walled carbon nanotubes. *Nanotechnology* **1998**, 9, 184–191.
19. Hertel, T.; Walkup, R.E.; Avouris, P. Deformation of carbon nanotubes by surface Van der Waals forces. *Phys. Rev. B* **1998**, 58, 13870–13873.
20. Liao, K.; Li, S. Interfacial characteristics of a carbon nanotube–polystyrene composite system. *Appl. Phys. Lett.* **2001**, 79, 4225–4227.
21. Maiti, A.; Wescott, J.; Kung, P. Nanotube–polymer composites: insights from Flory–Huggins theory and mesoscale simulations. *Mol. Simul.* **2005**, 31, 143–149.
22. Hildebrand, J.H.; Scott, R.L. *Regular Solutions*. Prentice-Hall; Englewood Cliffs, NJ, 1962.
23. Saito, Y.; Koyama, T.; Kawabata, K. Growth of single-layer carbon tubes assisted with iron-group metal catalysts in carbon arc. *Z. Phys. D* **1997**, 40, 421–424.
24. Ding, M.; Eitan, A.; Fisher, F.T.; Chen, X.; Dikin, D.A.; Andrew, R.; Brinson, L.C.; Schadler, L.S.; Ruoff, R.S. Direct observation of polymer sheathing in carbon nanotube–polycarbonate composites. *Nano Lett.* **2003**, 3, 1593–1597.
25. Ryan, K. Polymer crystallization as a reinforcement mechanism for polymer–carbon nanotube composites. Ph.D. dissertation, University of Dublin, Dublin, Inland, 2005.
26. Van Krevelen, D.W. *Properties of Polymers*. Elsevier, Amsterdam, The Netherlands, 1990.
27. Girifalco, L.A.; Lad, R.A. Energy of cohesion, compressibility, and the potential energy functions of the graphite system. *J. Chem. Phys.* **1956**, 25, 693–697.

5

CONSIDERATIONS REGARDING SPECIFIC IMPACTS OF THE PRINCIPAL FIRE RETARDANCY MECHANISMS IN NANOCOMPOSITES

BERNHARD SCHARTEL
Federal Institute for Materials Research and Testing, Berlin, Germany

5.1 INTRODUCTION

In the preceding chapters the use of nanocomposites to increase the fire retardancy of polymers was introduced as an interesting scientific topic, but also as a promising nanotechnology for industrial application. Polymer nanocomposites are made available with established and thus economical preparation tools such as extrusion and injection molding, but also in situ polymerization or solvent-aided methods. Indeed, their commercial use in mass products has become an accepted method of improving mechanical[1–3] and fire properties.[4,5] In comparison to established flame retardants, they are competitive due to their positive impact on mechanical properties. Furthermore, because they are eco-friendly, they are discussed as a halogen-free alternative in thermoplastics and thermosets. Therefore, it is not surprising that layered silicate polymer nanocomposites have been proposed as an up-and-coming approach to improve the fire retardancy of polymers.[6] Comparing nanocomposites with microcomposites makes clear that this technology aims beyond "simple" further miniaturization. It is based on the exploitation of new effects that arise from nanostructured materials.[7–9] Furthermore, in the preceding chapters the concept of nanocomposite was presented in detail and the principal mechanisms, such as barrier formation and changing the

Flame Retardant Polymer Nanocomposites, edited by Alexander B. Morgan and Charles A. Wilkie

viscosity, were discussed, whereas in succeeding chapters special systems and current trends are discussed.

The reason for this chapter is to address the discrepancy between the rather enthusiastic point of view proposing nanocomposites as the fire retardants of the future, and the fact that some systems—and especially, performance in some fire tests—raise a question as to whether nanoscaled inert additives should be called flame retardants at all. It becomes clear that even though the main concept and mechanisms seem established, the details of the structure–property relationship are still a subject of ongoing discussion. First of all, there is no simple property fire performance, but rather ignitability, flammability, flame spread, total heat evolved (THE), and so on, which are influenced quite differently by the various flame retardancy mechanisms reported for polymer nanocomposites. Hence, the resulting efficiency may be quite different in different fire tests and fire scenarios. Second, polymer nanocomposites can show several mechanisms, whose importance is strongly dependent on the specific interactions between the polymer matrix and the nanoscaled additive. Third, the nanoscaled morphology controls the properties and hence each of these mechanisms as well. Unfortunately, most sources on the fire behavior of polymer nanocomposites merely emphasize the well-established general concepts rather than trying to evaluate specific effects of the system being investigated. It is also quite common to communicate the advantages transparently, without annotating any limits of the concept. An aim of this chapter is to provide a better understanding of the impact on different fire test–and material-specific characteristics of the various mechanisms in polymer nanocomposites. Rather than giving a comprehensive overview, the focus is on the main general mechanisms and conclusions, illustrated by means of selected representative examples. Rather than summarizing what has happened so far, some key ideas and details are sketched systematically to clarify the potential future directions in the field.

5.2 INFLUENCE OF NANOSTRUCTURED MORPHOLOGY

5.2.1 Intercalation, Delamination, Distribution, and Exfoliation

The strong influence of morphology on fire behavior is shown by a variety of studies on layered-silicate nanocomposites.[10–12] Figure 5.1 displays the results from cone calorimeter readings on the heat release rate (HRR) taken from a study on poly(propylene–graft–maleic anhydride) (PP-g-MA, Aldrich Chemical Company, Milwaukee)[12] and on an epoxy resin (Epoxy) based on bisphenol A diglycidyl ether and 4-methyl hexahydrophthalic anhydride.[13] The thermoplastic noncharring PP-g-MA is compared to two corresponding, rather well exfoliated 5 wt% modified montmorillonite nanocomposites, both compounded using a twin-screw extruder. The quality of exfoliation differs since two different clays were used: montmorillonite modified with dimethyl dehydrogenated tallow ammonium (Cloisite 20A, Southern Clay Products, Gonzales, Texas), here called A, and montmorillonite modified with methyl tallow bis-2-hydroxyethyl ammonium (Cloisite 30B, Southern Clay Products), noted below as B. These

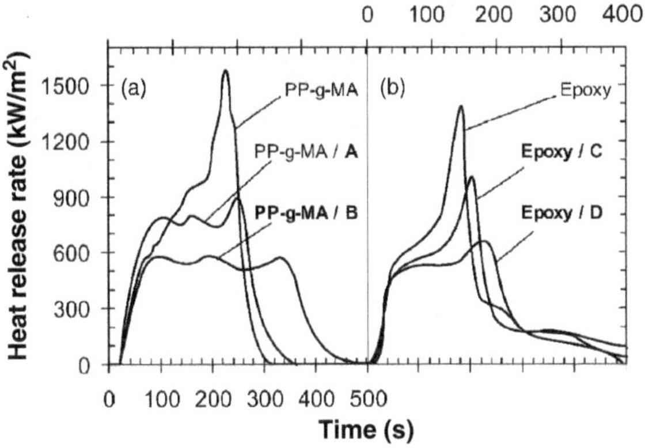

FIGURE 5.1 Heat release rate monitored in cone calorimeter experiments (irradiance = 70 kW/m^2): (*a*) PP-g-MA, PP-g-MA/5 wt% A, PP-g-MA/5 wt% B; (*b*) Epoxy, Epoxy/5 wt% C, and Epoxy/5 wt% D.

PP-g-MA/modified clay systems were chosen to illustrate the difference in nanocomposite formation due to different interactions between the modified clay used and the polymer. Of course, these interactions are material specific, since they are controlled by the chemical structure. Indeed, the PP-g-MA enables well-exfoliated structures, due to its decided polar character (Figure 5.2*a*). The epoxy resin nanocomposites were obtained by curing in the presence of 5 wt% tetraphenylphosphonium-modified montmorillonite. The only difference between the two tetraphenylphosphonium-modified montmorillonite systems (called C and D) was that they were prepared using different drying procedures. The different drying procedures resulted in different particle morphology, which can be characterized by a specific surface area (BET) of 45 m^2/g (C) and a BET of 175 m^2/g (D). The different modified clay particle morphologies resulted in a different quality of exfoliation. The epoxy resin/modified clay systems were chosen to illustrate the difference in nanocomposite formation due to preparation procedures, such as the morphology of the substances used. The in situ polymerization of epoxy resin and modified clay resulted in systems between close to and far away from a good nanocomposite (Figure 5.2*c* and *d*). The PP-g-MA and epoxy resin systems chosen, in particular PP-g-MA/5 wt% A, PP-g-MA/5 wt% B, and Epoxy/5 wt% D, represent the entire range or kind, respectively, of morphologies that are typically discussed as nanocomposites for fire retardancy, whereas Epoxy/5 wt% D marked the changeover from a nanocomposite to a microcomposite behavior.

The HRR curves for PP-g-MA and Epoxy were typical for a noncharring material, with behavior very close to that of a thermally intermediate thick regime.[14] After ignition the HRR increases to the averaged (steady-state) HRR, which can barely be seen as a shoulder, whereas the peak at the end of the experiment is the dominant characteristic. With increasing quality of nanocomposite formation, these characteristics took on increasingly plateaulike behavior for the nanocomposites. For both systems the nanocomposite showed increasing burning times, resulting in only minor changes in the THE, which indicates

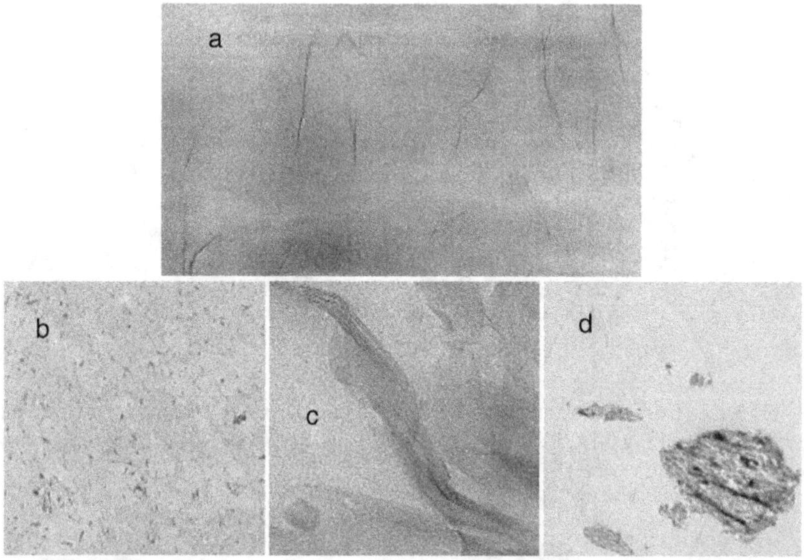

FIGURE 5.2 TEM images: (*a*) PP-g-MA/5 wt% B showing well-exfoliated (= high degree of delamination into single layers and homogeneous dispersion) clay layers; (*b*) Epoxy/5 wt% D (low resolution); (*c*) Epoxy/5 wt% D (high resolution) showing rather good exfoliated (= significant intercalation and delamination and homogeneous dispersion) clay layers; (*d*) Epoxy/5 wt% C (low resolution) showing hardly any exfoliation.

the total fire load of the specimen. The peak of HRR (PHRR) that corresponds to the fire growth was strongly reduced to values between one-third and one-half of the original values with increasing quality of exfoliation. It equaled the averaged (steady-state) HRR. Neither a significant quantitative change nor this change in principle in the HRR curves were found for corresponding microcomposites. It becomes clear that as a general result, the system-specific optimization of nanocomposite formation is crucial in terms of fire retardancy. Apart from chemical and morphological variations of the additive systems, the processing parameters and the polymeric materials used are also key factors.[15] The nanocomposite formation was reported to be influenced further by the polarity of the polymer and its molecular weight.[16] Furthermore, processing parameters such as shear rate, temperature, and resident time during a thermoplastics melt blending process control the morphology.[17] Results similar in principle to the ones shown in Figure 5.1 were obtained by different variations, all proving that the quality of the nanoscaled structure significantly determines the magnitude of the fire retardancy effect.

The quantitative assessment of nanocomposite formation is quite a challenge. The terms generally used, such as *tactoid* (= microcomposites), *intercalated*, and *exfoliated*, are rather oversimplifying models for some typical characteristics of different morphologies. The intercalation of silicate structures by polymer chains, the delamination of silicate layers, the mixing of the various compounds, and the distribution of particles or single layers in the matrix are mechanisms that are controlled kinetically, interact in a quite complex manner with each other, and are influenced by the preparation conditions beyond time and temperature, such as shear stress. Hence, nanocomposite preparation rarely ends up in a thermodynamically stable state or perfectly homogeneous samples.[18] Often, intermediate states of nanocomposite formation are reached. Partially exfoliated or strongly inhomogeneous materials were obtained. Regions with a low concentration of exfoliated silicate layers and regions with a higher concentration of intercalated and partially delaminated stacks of several silicate layers were observed at the same time. However, it should be noted that such systems are clearly closer to a perfect nanocomposite than to a microcomposite.

X-ray diffraction and transmission electron microscopy (TEM) are the methods used predominantly to characterize nanocomposite formation.[18,19] The effects used in x-ray diffraction are the disappearance of the Bragg peak due to a disturbed periodicity and the shift in the Bragg peak to changed periodicity in electron density allocation. Although interpretation of delamination and intercalation may make sense for layered-silicate nanocomposites, strictly speaking, x-ray investigation fails to monitor exfoliation (delamination + distribution) since the quality of distribution is not detected. Consequently, x-ray diffraction is particularly insufficient when rather good nanocomposites are compared with each other quantitatively.[20] TEM results are very convincing at first glance, due to their imaging character, but are accompanied by a small and often unrepresentative area illuminated. Hence, TEM investigations also show a lack in terms of quantitative evaluation. However, both systems presented in Figure 5.1, for instance, show a clear correspondence between reduction in the PHRR and increased nanocomposite formation monitored qualitatively by x-ray and TEM (Figure 5.2).[12,13,21] Promising approaches to tackling the problem of quantitative characterization of delamination and exfoliation have been reported recently: for example, the use of nuclear magnetic resonance (NMR) techniques[22,23] or rheology.[24,25]. Indeed, for the thermoplastic PP-g-MA system, a quantitative correspondence was found between the reduction in PHRR and the melt viscosity for low shear rates and temperatures, since both are controlled by the quality of nanocomposite formation.[12]

Choosing and optimizing the systems, such as the polymer matrix, nanoscaled additive, and compatibilizer, as well as the preparation of nanocomposites, are key challenges for the successful development of nanocomposites. The control and exploration of the specific properties and interactions are the main tasks that have to be tackled by further screening to develop suitable tools and a better understanding of the structure–property relationships.

5.2.2 Orientation

Silicate layers show an extreme anisotropic shape, strong interactions due to their polar–ionic character, and high stiffness compared to conventional polymers. These extraordinary properties resulted in extraordinary phase, orientation, and rheological behavior that is probably comparable discotic or sanidic liquid crystals.[26,27] Shear rates, which are common in the standard or industrial processing of thermoplastics, resulted in a strong orientation of the silicate layers in nanocomposites.[25] Since the orientational relaxation times of silicate plates can be large,[28,29] especially in comparison with polymer chains, a certain orientation order is quite often frozen in during the cooling step of injection molding or extrusion.[18] Furthermore, exfoliation processes based on intercalation and subsequent delamination resulted in the distribution of delaminated layers with a clear orientation. Delamination due to an intercalation of more and more polymer chains can be kinetically more favorable than orientational relaxation of the silicate plates. Hence, terms such as *exfoliated ordered* and *exfoliated disordered* were proposed to describe these morphologies so often obtained.[18] Furthermore, complex and anisotropic morphologies known from smectic systems were proposed to explain the rheological behavior of nanocomposites.[18,30–33] Obviously, self-organization and preparation resulted in anisotropic systems and special phases due to thermodynamics and kinetics. However, specific influence of orientation on the fire retardancy mechanisms has yet to be addressed. Perhaps the role of orientation is irrelevant, or perhaps it is a fully integrated part of the morphology and formation of nanocomposite that cannot be discussed separately.

5.2.3 Morphology During Combustion or Barrier Formation

As described above, nanoscaled morphology influences the fire performance of nanocomposites. Some sources even proposed that the reduction in PHRR is a measure for nanocomposite formation.[34] Furthermore, both intercalated and exfoliated nanocomposites were proposed for fire retardancy. Comparing both phases for various specific systems did not result in any general conclusion favoring one of the morphologies over the other.[7,8,35] It should be noted that strictly speaking, all of these ideas are oversimplifying. The morphology monitored for the intact polymer material in the solid state at room temperature does not control fire behavior directly. The interactions between silicate layers and polymer decomposition, as well as barrier formation during pyrolysis, control the fire retardancy effect. Hence, although the nanoscaled distribution is the essential starting point for fire retardancy in nanocomposites, it is not the entire story. Indeed, the morphology of the nanocomposites changes strongly due to an ablative reassembly.[8,36]

A combination of different physical and chemical mechanisms probably causes the formation of a silicate barrier layer at the sample surface, including demixing, layered silicate phase formation, charring, migration, and bubbles of decomposition products.[37,38] Even though silicate barrier formation seems to be the typical general mechanism that plays an important role for all nanocomposites, it becomes clear that the formation is a quite specific process for each system. Mechanisms such as demixing, migration, and layered silicate phase formation are controlled by the specific molecular interactions between polymer, silicate, and the organic compatibilizer. The surface energies of the various composites influence not only nanocomposite formation but also barrier formation. Furthermore, nearly all of the processes are activated thermally, so that the pyrolysis temperature influences the decomposition rates, demixing force, viscosity, diffusion, and so on.

Demixing, migration, and bubbling are influenced by the melt viscosity. Hence, barrier formation may be quite different for thermoplastics and thermosets, for instance. Some mechanisms can be ambivalent. Bubbles of decomposition products support the transport of silicate layers to the surface but are also reported to result in nonclosed surface layers since they stabilize cracks and holes. The different mechanisms take place at the same time. Some of them even influence each other and interact with the pyrolysis of the polymer. It was reported that especially, the formation of a carbon char–silicate layer resulted in surface layers, which act as a barrier for mass and heat transport.[36,39] Obviously, interactions between silicate and polymer play an important role in the formation of such an organic–inorganic layer; in other words, physical and chemical processes are important. Furthermore, the barrier formation may be quite different for noncharring and charring polymers. Recently, an essential influence of oxygen was reported for the formation of a char–silicate layer.[40] Moreover, the formation of a surface layer as closed as possible was reported to be essential for high efficiency of the barrier properties. The synergistic effect between layered silicate and carbon nanotubes was reported to be based on the formation of closer surface layers.[41] The formation of a closed surface layer is controlled by bubbling and by layered-silicate phase formation. Hence, a large set of parameters, including molecular interactions, viscosity, and mechanical stability of the surface layer, influences this characteristic. It becomes clear that the general mechanism of silicate surface layer formation is a feature whose details are quite specific to the different systems.

Obviously, the accumulation of the sample surface layer is rather complex during combustion and is not well described in detail. This is also true for the influence of increasing clay content. As a typical result, considerably reduced PHRR was observed with increasing content from 0 up to 10 wt%, but the decrease was not proportional for higher amounts (>7 wt%) of clay added (Figures 5.3 and 5.4*b*).[12] The reduction in PHRR converged to a limiting value. The enrichment in layered silicates on the surface seems to be a specific process rather than statistical precipitation or migration. For instance, increasing thickness of the residue layer was observed with increasing clay content but not the closing of cracks, resulting in incomplete prevention of the release of pyrolysis gases.[12] It is concluded that an arbitrarily high amount of clay makes no sense in terms of barrier formation, but adding clay of 5 to 7.5 wt% seems to be sufficient to prepare nanocomposites with an improved PHRR.

5.3 FIRE RETARDANCY EFFECTS AND THEIR IMPACT ON THE FIRE BEHAVIOR OF NANOCOMPOSITES

5.3.1 Inert Filler and Char Formation

Organically modified silicate clays are typically used to obtain a suitable nanocomposite formation. The content of organic compounds for a typical modified clay

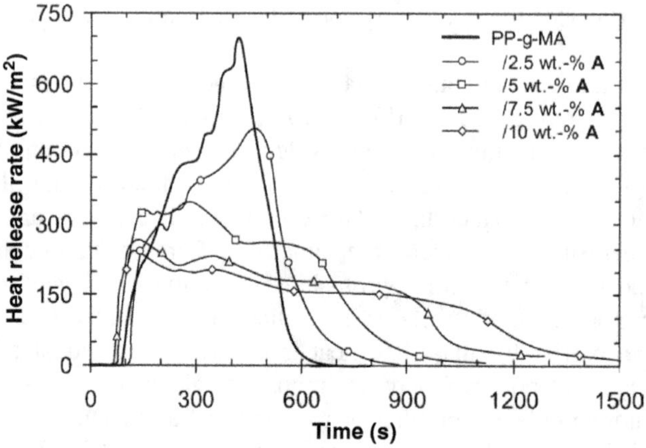

FIGURE 5.3 Heat release rate monitored in a cone calorimeter for PP-g-MA/A nanocomposites (irradiance = 30 kW/m^2), varying the filler content between 0 and 10 wt%.

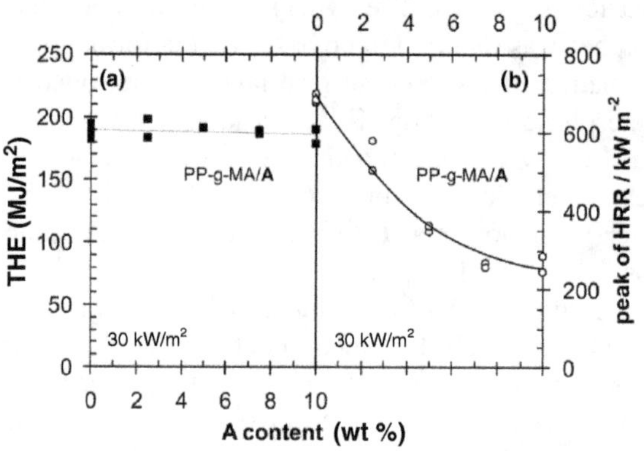

FIGURE 5.4 Cone calorimeter experiments for PP-g-MA/A nanocomposites (irradiance = 30 kW/m^2), varying the filler content between 0 and 10 wt%: (*a*) total heat evolved (squares); (*b*) PHRR rate (circles).

results in a decomposition mass loss of about 15 to 30 wt% in thermogravimetric measurements. The inorganic silicate does not decompose at temperatures relevant during the pyrolysis of polymeric materials. Hence, the modified layered silicate functioned in part as an inert filler. Indeed, hardly any residue was obtained above the inert filler content—not only in thermogravimetric experiments but also in fire experiments for noncharring polymer systems. Even if the interaction between silicate and polymer leads to a char–silicate surface layer, the additional carbonic is typically rather small (0 to 5 wt%) for noncharring polymers such as polypropylene (PP) and polystyrene (PS). Such a change in char formation is not relevant to reduce the THE. In such systems the THE reduction is of the same order of magnitude as the replacement of polymer with layered silicate. Figure 5.4*a* illustrates this inert filler effect using a layered-silicate/PP-g-MA nanocomposite as a model system for a noncharring thermoplastic. This influence was investigated on PP-g-MA/A nanocomposites by adding 0, 2.5, 5, 7.5, and 10 wt% layered silicate.[12] No significant difference between different montmorillonite nanocomposites was found for this system in terms of THE. The THE decreased linearly, corresponding to increasing replacement of the polymer. The effective heat of combustion was unchanged. A relevant gas-phase mechanism such as flame inhibition was absent. The residues corresponded to the amount of A used. No significant additional carbonic char formation was found for this system. Layered silicate acted as an inert filler. Furthermore, when amounts of only around 5 wt% were used, the flame retardancy effect remained of the same order of magnitude as the error of the data in terms of THE.

Metal hydroxides are widespread flame retardants and are also discussed as inert fillers.[42] They show endothermic decomposition into an inorganic residue accompanied by the release of water. Water is a very effective cooling agent and dilutes the fuel gases. Unlike montmorillonite, metal hydroxides provide an additional significant heat-sink mechanism. Hence, the use of compounds such as hydrotalcite was proposed instead of montmorillonite. However, metal hydroxides are typically used in amounts of 40 wt% up to 65 wt% to obtain flame-retarded polymers. The conclusion is confirmed that small amounts of layered silicates fail to make a crucial impact on fire behavior, due to their inert filler characteristic. Small amounts of additives can only improve the THE by specific interactions changing the decomposition pathway in the condensed phase or the oxidation in the gas phase. Such fire retardancy mechanisms are indicated by increased char formation or reduced effective heat of combustion. Unfortunately, most polymer nanocomposite systems are characterized by mainly physical effects, whereas chemical interactions play a minor role. The effective heat of combustion is rarely influenced significantly and the char formation increase is mostly on a rather disappointing scale between 0 and 10 wt%. Promising systems that show a crucial influence on effective heat of combustion or carbonic char formation are quite rare. Nevertheless, the formation of inorganic residue can result in an efficient barrier layer at the surface, thus influencing other important characteristics, such as the HRR (Figures 5.3 and 5.4*b*). Such barrier effects are discussed below.

5.3.2 Decomposition and Permeability

The results reported on the thermal and thermooxidative decomposition of layered silicate nanocomposites are rather contradictory and do not lead to unambiguous or consistent conclusions. The results vary from enhanced decomposition, to no significant influence, to a strong improvement depending on the source and system discussed.[21,43–46] The influence on thermal decomposition differs strongly from nanocomposite to nanocomposite. What is more, often the product release is changed rather than the primary decomposition reactions. The diffusion of the products is hindered by the decreased permeability for nanocomposites; 5 wt% layered silicate–polymer nanocomposites show a reduction in gas permeability of around 40 to 60%, even for small gas molecules such as nitrogen and oxygen.[47,48] The reduction in oxygen diffusion results in higher decomposition temperatures for thermooxidative thermogravimetric investigations with a constant heating rate.[21,46] Segmental or chain conformation changes necessary for decomposition or product release can also be reduced significantly, especially in intercalated systems or systems with strong interactions between the polymer and silicate layers. Even large effects were reported. Poly(methyl methacrylate) (PMMA) intercalated in montmorillonite showed a 40 to 50 K increase in decomposition temperature,[49] and an even higher increase of 140 K was reported for poly(dimethyl siloxane) (PDMS) intercalated in montmorillonite.[45] For both systems the restricted thermal motion of segments or decomposition products was concluded to be the principal mechanism. Adding modified clay may also change the chemical decomposition reactions. Layered silicate can act as an acid buffer, change the water content, or even catalyze chemical reactions, whereas the decomposition of the organic modifier may trigger decomposition. Recently, a series of papers was published reporting different changes in pyrolysis products for different nanocomposites.[50,51] The influence on thermal decomposition is determined by specific interactions with the nanoscaled additive during polymer decomposition. The influence on the thermal and thermooxidative decomposition is a specific characteristic for each system.

A relevant change in thermal and thermooxidative decomposition is not a general mechanism for all layered silicate nanocomposites. For all scenarios based on a stable flame zone, the fire behavior is controlled by an anaerobic pyrolysis. The changes reported for thermal and thermooxidative decomposition are of minor importance for most layered silicate–polymer nanocomposites. Product changes, such as from monomer to oligomers, some additional carbonic char (1 to 5 wt%), or decomposition temperature shifts typically around 5 to 25 K hardly cause relevant changes in effective heat of combustion or total amount of volatiles, but may influence the time to ignition. Hence, the changes in thermal and thermooxidative decomposition are ruled out as the main fire retardancy mechanism in terms of essentially decreased flame spread. Neither does it improve flammability as monitored by the UL-94 classification. For instance, PDMS nanocomposites showing a strong enhancement in thermal stability failed to achieve an UL-94 V-0 classification similar to that of PDMS.[45] The changed monomer–oligomer distribution for PS did not change significantly in the heat of combustion.[52] Only a few systems have a potential for essential fire behavior improvements due to changes in their thermal or thermooxidative decomposition. Most of the nanocomposite systems show only a negligible or small increase in carbonic char formation. An essential decrease in fuel production due to increased char formation is not found. Systems that show a crucial increase in residue are rather rare. However, such systems would be extremely promising since they would combine chemical and physical mechanisms. Remarkable effects on the chemical reactions were reported for epoxy systems. Adding layered silicates transformed a three-step decomposition into a two-step decomposition.[40,53] In principle, such effects open the potential for essential improvement of fire behavior.

5.3.3 Viscosity and Mechanical Reinforcement

Changing the decomposition pathway resulting in charring, and influencing the chemical reactions in the gas phase resulting in flame inhibition, are not the only ways to improve the fire behavior; physical mechanisms such as cooling, barrier formation, and changing the heat capacity, the thermal conductivity, or the viscosity also have a relevant influence on the fire behavior. The melt viscosity for the pyrolysis zone is not only important for barrier formation, as discussed above, but also controls the dripping behavior. Dripping behavior is crucial in many fire scenarios. Polymer nanocomposites based on anisotropic additives such as layered silicate or nanotubes showed a strong increase in viscosity for even low filler contents, especially when low shear rates were applied. The nano-dispersion resulted in structures that strongly reduce melt dripping, such as physical network structures. Indeed, the influence on the melt flow is proposed as one of the main general mechanisms in nanocomposites. Preventing dripping can be good or bad, depending on the scenario in question. High melt flows are suitable strategies to pass the glow wire test or to reach a V-2 classification in UL-94 vertical burning tests. In such cases nanocomposites may show a worse performance. For example, the thermoplastic PP-g-MA material used as an example above received a V-2 classification, but only an HB classification was achieved by the nanocomposites PP-g-MA/5 wt% A and PP-g-MA/5 wt% B.[12] Adding layered silicate prevented extinguishing through dripping in this system, since more combustible material remained in the pyrolysis zone. Nanocomposites using fibrous nanotubes in noncharring thermoplastics even tended to effects similar to wicking, as indicated by a dramatic decrease in the limiting oxygen index (LOI) for polyamide-6 (PA6)/multiwall carbon nanotube (MWNT) systems from to 26.4 to 23.7%,[54] which is similar to the known effect of glass fiber reinforcement. Such rather negative results on flammability performance were found especially for noncharring thermoplastic, for which dripping is quite common and often controls flammability. Of course, different systems and fire tests are affected very specifically by the changed melt viscosity, and this mechanism is not the only one that may influence the performance.

However, the examples given illustrate that the changed viscosity can have a crucial impact on the fire performance, in particular in case of noncharring thermoplastics. Similar obvious changes are not observed for charring materials since they may not show dripping anyway. For charring materials a decent mechanical reinforcement of the char or a changed deformation behavior were observed instead. Nanocomposites using the same MWNT systems in the charring thermoplastics PC did not result in a worsening in LOI (PC: $25 \pm 0.1\%$; PC with 2, 4, and 6 wt% MWNT $25.0 \pm 0.4\%$), and the main influence in cone calorimeter results seems to be the reduced deformation during burning.[55] Especially for intumescent systems, significant influences can be expected, since viscosity is one of the main parameters controlling formation of the multicellular structure. Recently, synergistic effects were reported for nano-distributed layered silicate in intumescent systems based on a mechanical stabilization of char.[56] Furthermore, apart from the typical flammability test, the prevention of dripping is essential in many fire scenarios and corresponding fire tests that favor the use of nanocomposite.

5.3.4 Barrier for Heat and Mass Transport

Specific aspects of barrier formation were discussed above. A silicate or silicate–char surface layer acting as a barrier for heat and mass transport is probably the main general fire retardancy mechanism of all layered-silicate nanocomposites. Most sources claim that this mechanism is responsible for the strongly improved performance in a cone calorimeter test. In particular, the strong reduction in PHRR is used to propose that layered silicates are the most promising approach for fire retardancy of polymers. However, the barrier effects and their influences on cone calorimeter results are not described in detail, so that the specific characteristics of these mechanisms are unclear.

The cone calorimeter characterizes the fire behavior of a horizontal specimen for forced flaming combustion. The PHRR corresponds to the fire growth in a fire. Figures 5.1, 5.3, and 5.4 show typical results for polymer nanocomposites in a cone calorimeter. Large reductions in PHRR up to 75% and increased burning times are found. The shape of the HHR curves changed in comparison to the polymer. Nanocomposites show a more plateaulike behavior with increasing amounts of layered silicate or improved nanocomposite formation, respectively. The HRR curve changes into a shape that is typical for char- or residue-forming materials.[14] After ignition the HRR increases to the averaged (steady-state) HRR, which also became the PHRR with increasing quality of exfoliation. The HRR subsequently decreased slightly until flame-out. Nanocomposites showed increasing burning times such that THE shows only minor reduction or no change. The total fire load of the specimen is often not influenced significantly, whereas the PHRR is reduced drastically. Without any significant chemical impact on burning behavior, such as flame inhibition or charring of the polymer, physical barrier effects became obvious as the main general fire retardancy effects of nanocomposites.[12,57] This conclusion corresponds with the inert filler behavior of most of the systems without any pronounced chemical impact on decomposition and burning behavior. However, the rather physical barrier formation need not be the entire story, but can be accompanied by chemical mechanisms taking place at the same time or can be strongly influenced chemical processes. These additional processes are not general for all nanocomposites but are clearly material specific and often the key for promising materials.

Switching the characteristics of the HRR curve of noncharring polymers to the characteristics of residue-forming polymers means that the type of PHRR is changed. The PHRR due to the increasing thermal feedback from the back side of the specimen at the end of the burning vanishes, and the averaged (steady-state) HRR reached at the beginning of the burning became dominant for nanocomposites. A vanishing of the PHRR at the end of the burning can also be reached experimentally by measuring a noncharring polymer specimen using a modified

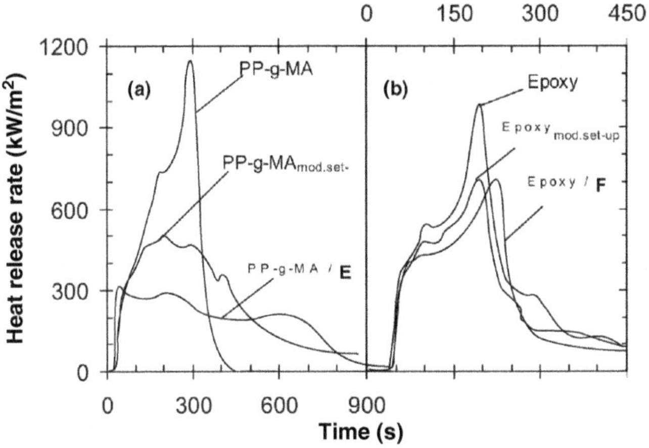

FIGURE 5.5 Heat release rate monitored in cone calorimeter experiments (irradiance = 50 kW/m^2): (*a*) PP-g-MA, PP-g-MA measured with the modified sample holder and PP-g-MA/5 wt% E; (*b*) Epoxy, Epoxy measured with the modified sample holder, and Epoxy/5 wt% F.

sample holder that reduces the thermal feedback from the back of the specimen.[58] Obviously, the barrier formation in nanocomposite at the surface and the use of a thermal conducting sample holder at the back side of the specimen are different approaches to influence the heat impact on the pyrolysis zone, but remarkably, both result in the same effect on the HRR curve. In Figure 5.5, PP-g-MA and Epoxy measured with a standard and a modified sample holder are compared with the nanocomposites PP-g-MA/5 wt% Nanomer I.28E (Nanocor, Arlington Heights, Illinois) and Epoxy/5 wt% tetraphenylphosphonium-modified montmorillonite, respectively.[13,58] Nanomer I.28E is an octadecyltrimethyl ammonium–modified montmorillonite, here called E. A spray-dried tetraphenylphosphonium-modified montmorillonite was used with a BET of 100 m^2/g (called F). It becomes clear that the change in heat transport, especially the smaller effective heat impact on the pyrolysis zone at the end of the test, is a main reason for the reduction in PHRR.

The qualitative change in the origin of the PHRR also has a specific quantitative impact. The ratio between PHRR at the end of burning and the averaged (steady-state) HRR is different for noncharring polymers. The ratio is a specific characteristic of the material under consideration. Table 5.1 summarizes the ratio of averaged (steady-state) HRR/PHRR for polymers such as PA6, PS, acrylonitrile–butadiene–styrene (ABS), poly(methyl, methacrylate) (PMMA), and so on. at low irradiation using percentages. The data are estimated from published HRR curves for an irradiation of 35 kW/m^2.[12,59,60] The averaged (steady-state) HRR was determined roughly according to the procedure proposed by Lyon.[61] The ratios between averaged (steady-state) HRR and PHRR are compared with the typical reduction in PHRR published for the corresponding nanocomposites using an irradiation of 35 kW/m^2.[35,12] The correspondence is

TABLE 5.1 Comparison of the Typical Reduction in PHRR for Nanocomposites with the Ratio of Averaged (Steady-State) HRR/PHRR Using 35 kW/m² Irradiation

Nanocomposite	Reduction of PHRR for Well-Prepared Nanocomposites (%)	Ratio Between Averaged (Steady-State) HRR and PHRR (%)
PA6	63[a]	~70[b]
PS	57[a]	~55[b]
PP-g-MA	54,[a] 46–57[c]	50–60[b]
ABS	45[a]	~45[b]
HIPS (high-impact PS)	40[a]	40–45[d]
PMMA	25[a]	~20[b]

[a] From Ref. [35].
[b] From Ref. [59].
[c] From Ref. [12].
[d] From Ref. [60].

convincing. Not only does the order and order of magnitude for the different polymers correspond, but even the specific values between 20 and 70% match each other. The reduction in PHRR is controlled not only by the specific barrier properties of the surface layer, but also by the specific fire behavior of the polymer.

The PHRR or the flame spread, respectively, is better characterized as a specific property of a certain specimen in a specific fire scenario than as a material property such as the effective heat of combustion. Hence, the PHRR and the corresponding fire retardancy effect depend on the specifics of the scenario, such as the sample holder, sample thickness, and the irradiance used in a cone calorimeter test. Materials showed decreased time to ignition, decreased burning time, and enhanced PHRR with higher irradiance, since the energy impact per unit of time was increased. It was proposed that the barrier mechanisms are also indicated by special characteristics of varying the irradiance.[62] The influence of irradiance on PHRR becomes less pronounced when it is determined by the formation of a physical barrier for heat and mass transport. Consequently, the relative flame retardancy effect of a physical barrier increased with increasing irradiance. In Figure 5.6 the PHRR values are compared for PP-g-MA and PP-g-MA/5 wt% A for 30 up to 70 kW/m² and for Epoxy and Epoxy/5 wt% D.[12,13] The PHRR was reduced up to 75% and up to 50%, respectively, at 70 kW/m², whereas the flame retardancy effect seems to vanish for lower irradiance. Hence, fire tests with low irradiance such as flammability tests (UL-94, LOI, etc.) may be influenced less by the main general mechanism, which is discussed in detail below. The influence of the sample thickness of the specimen was also reported on the fire retardancy effect of nanocomposites.[63] The fire retardancy effect vanished for thermally thin samples. This is probably due to the fact that for thermally thin specimens a different type of PHRR occurs. The PHRR becomes more and more dependent on the total heat evolved, which has not significantly changed.[64]

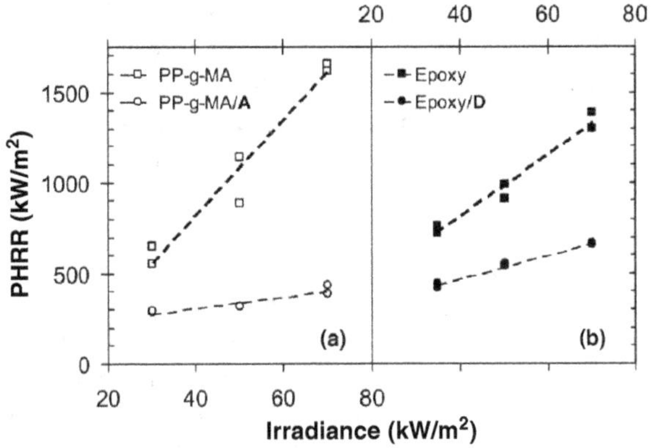

FIGURE 5.6 Peak of heat release rate plotted against irradiance: (*a*) PP-g-MA and PP-g-MA/5 wt% A; (*b*) Epoxy and Epoxy/5 wt% D.

5.4 ASSESSMENT OF FIRE RETARDANCY

5.4.1 Differentiated Analysis with Regard to Different Fire Properties

The most important fire risks are:

- Ignitability
- Flammability
- Heat release rate/flame spread
- Total heat evolved
- Fire penetration
- Smoke obscuration/smoke toxicity

The fire retardancy of real products means that one or several of these fire risks must be under control in case of a specific fire scenario. For electronic and electrical products, flammability in the glow wire test or in the UL-94 test must be ruled out. Delay or prevention of the start of a fire is the goal when an ignition source such as a glowing wire or small candlelike flame is applied. Fire tests for building products such as the new European SBI (single burning item) apparatus bring out the response of materials when they encounter a single burning item such as a wastepaper basket. The reduction or the prevention of flame spread during a developing fire is the main objective. Protection elements such as fire doors and the like are tested using a standard time–temperature curve. The fire penetration in case of a fully developed fire is the main target. Indeed, the pairs ignitability/flammability–ignition, flame spread/fire growth–developing fire, and THE or fire penetration–fully developed fire are crucial for fire testing. The three fire scenarios are different with respect to ventilation, temperature, involved length scales, and irradiation.

Most polymer nanocomposites show rather minor influences on decomposition of the polymer. Apart from a few exceptions, neither decomposition temperatures nor effective heat of combustion of the volatiles change relevantly. Consequently, the time to ignition is not improved. Neither is the initial increase in HRR typically changed for nanocomposites. These minor effects on the beginning of an HRR curve are a quite general characteristic for surface layer–forming systems. The fire retardancy effects become dominant only with preceding burning. In addition, for some nanocomposites, the time to ignition is even decreased. Earlier decomposition of the organic modifier or a changed heat absorption in nanocomposites are among the probable reasons for this observation. It must be concluded that nano-dispersed layered silicates by themselves are a rather disappointing fire retardancy approach in terms of fire property ignitability. Such a disappointing performance was also often found for the flammability of nanocomposites.[12,65,66] Typical HRR values of most nanocomposites investigated were clearly above about ~150 kW/m^2 when low irradiances such as 30 to 35 kW/m^2 were used in a cone calorimeter. The nanocomposites do not show a convincing tendency for self-extinguishing behavior. The LOI results often did not differ significantly between polymer and polymer nanocomposites, or showed only small changes.[12,65,66] In some systems, even a worsening of the LOI values was reported.[54] Corresponding results were found for UL-94 testing. It becomes clear that it is a common feature of the main general fire retardancy mechanisms in nanocomposites to provide considerable improvement in terms of fire growth and flame spread, but not in terms of preventing or delaying the onset of a fire, such as reducing the ignitability or flammability. Again, only specific systems may be really promising in this area, with additional mechanisms of flame retardancy.

Figure 5.7 illustrates the typical fire performance of layered-silicate nanocomposites for various irradiances. Corresponding results were reported for other nanocomposites.[12,65,66] The data are shown for 5 wt% phosphonium bentonite epoxy resin nanocomposites (Epoxy/E).[13] THE is plotted against the fire growth rate (FIGRA). Comparable plots were proposed to give a good graphical assessment of the fire behavior of various materials.[67,68] Advanced fire retardancy yields

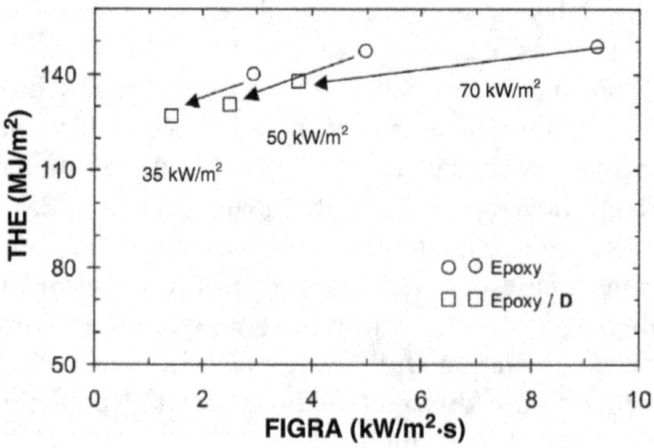

FIGURE 5.7 Total heat evolved plotted against FIGRA (peak heat release rate/time to peak heat release rate). Results are shown for an epoxy resin and epoxy resin/5 wt% D nanocomposite at various irradiances (30, 50, and 70 kW/m^2).

improvements in both THE and FIGRA. Layered silicate–polymer nanocomposites result in a remarkable decrease in fire growth, particularly for high irradiances, whereas the impact on THE remains limited. It should be noted that this limitation must not remain unchallenged. Recently, promising approaches have been discussed: combination with established flame retardants[5,69] (see also Chapters 6 to 9) or the search for systems in which the char formation of the polymer is also enhanced significantly by additional more chemical mechanisms.[36]

Polymer nanocomposites seem to have some remarkable advantages in terms of mechanical and ecological considerations. Layered silicates show significant reinforcing, whereas many common flame retardants act as plasticizers. Furthermore, they are discussed as a promising approach to halogen-free fire retardants. Their price hinders their commercialization to some extent, but does not rule it out. However, the prerequisite of a nanoscaled structure may demand advanced technology for preparation. The rather physical mechanisms proposed for nanocomposites do not significantly affect pyrolysis and combustion reactions. Nanocomposites do not significantly increase fire hazards such as CO or smoke production.[70,71]

5.4.2 Different Fire Scenarios Highlight Different Effects of Nanocomposites

Adding nanoparticles influences the fire behavior of polymers by different mechanisms. The role and importance of these mechanisms are quite different in various fire scenarios. The irradiance was observed as a major parameter controlling fire retardancy efficiency (Figure 5.7). It should be noted that the results of extrapolation to small irradiances correspond to flammability scenarios such as LOI and UL-94 tests.[61] The fire retardancy diminished with decreasing irradiance. Consequently, comprehensive cone calorimeter results indicate that no significant improvement can be expected for nanocomposites in terms of flammability tests based on the main general fire retardancy mechanisms. This conclusion corresponds to observations that most nanocomposites do not show relevantly improved self-extinguishing behavior in tests such as LOI and UL-94. The strong fire retardancy observed at high irradiance in terms of fire growth and flame spread, respectively, and the rather disappointing performance in flammability are not contradictory. The incombustible residue of nanocomposites built up a surface layer capable of reducing the HRR significantly, but rarely of extinguishing the fire. Fire scenarios highlight the barrier effect of nanocomposites when they are controlled by the fire growth and flame spread, respectively, at higher heat fluxes.

Nanocomposites rarely built up surface layers efficient enough to result in a relevant decrease in flammability, except in combination with conventional fire retardants. However, the extinction behavior in flammability tests such as LOI and UL-94 is also strongly influenced by the dripping behavior of the materials under investigation. Adding highly anisotropic nanoscaled particles has a strong influence of the melt viscosity of the polymers. They can work as antidripping agents, result in a reinforcement of char, or induce wicking. Consequently, for some systems the flammability test results highlight the change in viscosity.

Ecological considerations and investigations of fire hazards such as CO and smoke production target the inert filler characteristics of nanocomposites. The rather physical mechanisms proposed for nanocomposites are advantageous for such considerations. Nanocomposites appear to be a promising eco-friendly approach to fire retardancy in polymers.

5.5 SUMMARY AND CONCLUSIONS

This chapter illuminated the specific aspects of fire retardancy in polymer nanocomposites. In particular, it focused on the main general mechanisms. The fire behavior of nanocomposites was discussed comprehensively and assessed. Apart from the effect of nanocomposite formation and other influences, two main general mechanisms are worked out: (1) formation of a surface layer during combustion, and (2) change in melt viscosity during pyrolysis. The fire retardancy mechanisms occurring in nanocomposites have, first, a very specific impact on various fire tests, and second, show material-specific characteristics. Furthermore, the rather physical main general mechanisms may be accompanied or strongly affected by specific chemical processes in distinct systems.

The surface layer works as a barrier for pyrolysis gases and heat, whereas the change in melt viscosity influences dripping during combustion. The efficiency of these mechanisms is specific with respect to different systems and different fire tests. The influence of these mechanisms on different fire properties and for different fire scenarios is sketched in detail. High fire retardancy potential was reported for barrier layers in terms of fire growth and flame spread, respectively, under forced flaming conditions. Other important fire characteristics, such as ignitability, flammability, and THE, however, are not improved in any relevant way by most of the barrier layers observed. The strongly changed melt viscosity efficiently prevents dripping.

In the majority of cases, the physical main general fire retardancy mechanisms of nanocomposites are not sufficient to pass some of the important fire tests for polymeric materials. Hence, compounds such as organically treated layered silicates are not convincing stand-alone flame retardants. Indeed, for most systems a convincing potential is concluded only in combination with established flame retardants (Chapters 6 to 9). This may change when systems are developed that combine the physical with chemical mechanisms such as those that trigger relevant additional char formation. Different concepts have been proposed to tackle this target, such as using silicate layers as catalysts, changing the decomposition pathway, or using layered structures as microreactors.

Nanocomposite formation is a key prerequisite of efficient fire retardancy and shows a strong structure–property relationship. Major mechanisms controlling morphology are known. However, the specific optimization of the preparation for each polymer nanocomposite system persists as a major challenge in the future.

For the time being, nanocomposites are not considered to harbor potential as possible flame-retarded polymers. So far, nanocomposite formation has been used successfully as a synergist in some polymers in combination with established flame retardants. Indeed, in such systems, they have already been commercialized very successfully.[4] The possible combinations of nanocomposites with other flame retardants are countless. They are and will continue to be under consideration.[56,69,72–76] Some of them even show remarkable synergisms.

Acknowledgments

The examples used in this chapter are based on studies performed in the author's research group. They are part of M. Bartholmai's Ph.D. dissertation or were performed in cooperation with A. Hartwig and D. Pütz from IFAM in Bremen.

REFERENCES

1. Kojima, Y.; Usuki, A.; Kawasumi, M.; Okada, A.; Fukushima, Y.; Karauchi, T.; Kamigaito, O. Synthesis of nylon-6–clay hybrid by montmorillonite intercalated with ε-caprolactam. *J. Polym. Sci. A Polym. Chem.* **1993**, 31, 983–986.
2. Kojima, Y.; Usuki, A.; Kawasumi, M.; Fukushima, Y.; Okada, A.; Karauchi, T.; Kamigaito, O. Mechanical properties of nylon 6–clay hybrid. *J. Mater. Res.* **1993**, 8, 1185–1189.
3. LeBaron, P.C.; Wang, Z.; Pinnavaia, T.J. Polymer-layered silicate nanocomposites: an overview. *Appl. Clay Sci.* **1999**, 15, 11–29.
4. Schall, N.; Engelhardt, T.; Simmler-Hübenthal, H.; Beyer, G. Ger. Patent DE 199 21 472 A 1, 2000.
5. Beyer, G. Flame retardant properties of EVA-nanocomposites and improvements by combination of nanofillers with aluminum trihydrate. *Fire Mater.* **2001**, 25, 193–197.
6. Gilman, J.W.; Kashiwagi, T.; Lichtenhan, J.D. Nanocomposites: a revolutionary new flame retardant approach. *SAMPE J.* **1997**, 33, 40–46.
7. Gilman, J.W.; Kashiwagi, T.; Giannelis, E.P.; Manias, E.; Lomakin, S.; Lichtenham, J.D.; Jones, P. Nanocomposites: radiative gasification and vinyl polymer flammability, in: M. Le Bras, G. Camino, S. Bourbigot, and R. Delobel, Eds., *Fire Retardancy of Polymers: The Use of Intumescence.* Royal Society of Chemistry, London, 1998, pp. 203–221.
8. Gilman, J.W. Flammability and thermal stability studies of polymer layered-silicate (clay) nanocomposites. *Appl. Clay Sci.* **1999**, 15, 31–49.
9. Zanetti, M.; Camino, G.; Mülhaupt, R. Combustion behaviour of EVA/fluorohectorite nanocomposites. *Polym. Degrad. Stab.* **2001**, 74, 413–417.
10. Duquesne, S.; Jama, C.; Le Bras, M.; Delobel, R.; Recourt, P.; Gloaguen, J.M. Elaboration of EVA–nanoclay systems: characterization, thermal behaviour and fire performance. *Compos. Sci. Technol.* **2003**, 63, 1141–1148.
11. Bendaoudi, A.; Duquesne, S.; Jama, C.; Le Bras, M.; Delobel, R.; Recourt, P.; Gloaguen, J.-M.; Lefebvre, J.-M.; Addad, A. Effect of the processing conditions on the fire retardant and thermomechanical properties of PP–clay nanocomposites,

in: M. Le Bras, C.A. Wilkie, S. Bourbigot, S. Duquesne, and C. Jama, Eds., *Fire Retardancy of Polymers: New Applications of Mineral Fillers*. Royal Society of Chemistry, London 2005, pp. 114–125.

12. Bartholmai, M.; Schartel, B. Layered silicate polymer nanocomposites: new approach or illusion for fire retardancy? Investigations of the potentials and the tasks using a model system. *Polym. Adv. Technol.* **2004**, 15, 355–364.
13. Schartel, B.; Knoll, U.; Hartwig, A.; Pütz, D. Phosphonium-modified layered silicate epoxy resins nanocomposites and their combinations with ATH and organophosphorus fire retardants. *Polym. Adv. Technol.* **2006**, 17, 281–293.
14. Lyon, R.E. Plastics and rubber, in: C.A. Harper, Ed., *Handbook of Materials in Fire Protection*. McGraw-Hill, New York, 2004, pp. 3.1–3.51.
15. Kim, S.W.; Jo, W.H.; Lee, M.S.; Ko, M.B.; Jho, J.Y. Effects of shear on melt exfoliation of clay in preparation of nylon 6/organoclay nanocomposites. *Polym. J.* **2002**, 34, 103–111.
16. Fornes, T.D.; Yoon, P.J.; Keskkula, H.; Paul, D.R. Nylon 6 nanocomposites: the effect of matrix molecular weight. *Polymer* **2001**, 42, 9929–9940.
17. Dennis, H.R.; Hunter, D.L.; Chang, D.; Kim, S.; White, J.L.; Cho, J.W.; Paul, D.R. Effect of melt processing conditions on the extent of exfoliation in organoclay-based nanocomposites. *Polymer* **2001**, 42, 9513–9522.
18. Vaia, R.A. Structural characterization of polymer-layered silicate nanocomposites, in: T.J. Pinnavaia and G.W. Beall, Eds., *Polymer–Clay Nanocomposites*. Wiley, Chichester, West Sussex, England, 2000, pp. 229–266.
19. Alexandre, M.; Dubois, P. Polymer-layered silicate nanocomposites: preparation, properties, and uses of a new class of materials. *Mater. Sci. Eng.* **2000**, 28, 1–63.
20. Morgan, A.B.; Gilman, J.W. Characterization of polymer-layered silicate (clay) nanocomposites by transmission electron microscopy and x-ray diffraction: a comparative study. *J. Appl. Polym. Sci.* **2003**, 87, 1329–1338.
21. Tidjani, A.; Wald, O.; Pohl, M.-M.; Hentschel, M.P.; Schartel, B. Polypropylene-graft-maleic anhydride-nanocomposites, I: Characterization and thermal stability of nanocomposites produced under nitrogen and in air. *Polym. Degrad. Stab.* **2003**, 82, 133–140.
22. VanderHart, D.L.; Asano, A.; Gilman, J.W. Solid-state NMR investigation of paramagnetic nylon-6 clay nanocomposites, 2: Measurement of clay dispersion, crystal stratification, and stability of organic modifiers. *Chem. Mater.* **2001**, 13, 3796–3809.
23. VanderHart, D.L.; Asano, A.; Gilman, J.W. NMR measurements related to clay-dispersion quality and organic-modifier stability in nylon-6/clay nanocomposites. *Macromolecules* **2001**, 34, 3819–3822.
24. Wagener, R.; Reisinger, T.J.G. A rheological method to compare the degree of exfoliation of nanocomposites. *Polymer* **2003**, 44, 7513–7518.
25. Krishnamoorti, R.; Ren, J.X.; Silva, A.S. Shear response of layered silicate nanocomposites. *J. Chem. Phys.* **2001**, 114, 4968–4973.
26. Demus, D. Phase types, structures, and chemistry of liquid crystals, in: H. Stegemeyer, guest Ed., *Liquid Crystals*; in: H. Baumgärtel, E.U. Franck, and W. Grünbein, Eds., *Topics in Physical Chemistry*. Steinkopff Verlag, Darmstadt, Germany, 1994, pp. 1–50.

27. Marrucci, G. Rheology of nematic polymers, in: A. Ciferri, Ed., *Liquid Crystallinity in Polymers: Principles and Fundamental Properties*. VCH Publishers, New York, 1991, pp. 395–422.

28. Malwitz, M.M.; Butler, P.D.; Porcar, L.; Angelette, D.P.; Schmidt, G. Orientation and relaxation of polymer–clay solutions studied by rheology and small-angle neutron scattering. *J. Polym. Sci. B Polym. Phys.* **2004**, 42, 3102–3112.

29. Ren, J.X.; Casanueva, B.F.; Mitchell, C.A.; Krishnamoorti, R. Disorientation kinetics of aligned polymer layered silicate nanocomposites. *Macromolecules* **2003**, 36, 4188–4194.

30. Krishnamoorti, R.; Yurekli, K. Rheology of polymer layered silicate nanocomposites. *Curr. Opin. Colloid Interface Sci.* **2001**, 6, 464–470.

31. Schmidt, G.; Nakatani, A.I.; Han, C.C. Rheology and flow-birefringence from viscoelastic polymer–clay solutions. *Rheol. Acta* **2002**, 41, 45–54.

32. Schmidt, G.; Nakatani, A.I.; Butler, P.D.; Han, C.C. Small-angle neutron scattering from viscoelastic polymer–clay solutions. *Macromolecules* **2002**, 35, 4725–4732.

33. Okamoto, M.; Nam, P.H.; Maiti, P.; Kotaka, T.; Hasegawa, N.; Usuki, A. A house of cards structure in polypropylene/clay nanocomposites under elongational flow. *Nano Lett.* **2001**, 1, 295–298.

34. Su, S.P.; Jiang, D.D.; Wilkie, C.A. Polybutadiene-modified clay and its nanocomposites. *Polym. Degrad. Stab.* **2004**, 84, 279–288.

35. Wilkie, C.A. An introduction to the use of fillers and nanocomposites in fire retardancy, in: M. Le Bras, C.A. Wilkie, S. Bourbigot, S. Duquesne, and C. Jama, Eds., *Fire Retardancy of Polymers: New Applications of Mineral Fillers*. Royal Society of Chemistry, London, 2005, pp. 3–15.

36. Zanetti, M.; Kashiwagi, T.; Falqui, L.; Camino, G. Cone calorimeter combustion and gasification studies of polymer layered silicate nanocomposites. *Chem. Mater.* **2002**, 14, 881–887.

37. Lewin, M. Some comments on the modes of action of nanocomposites in the flame retardancy of polymers. *Fire Mater.* **2003**, 27, 1–7.

38. Kashiwagi, T.; Harris, R.H.; Zhang, X.; Briber, R.M.; Cipriano, B.H.; Raghavan, S.R.; Awad, W.H.; Shields, J.R. Flame retardant mechanism of polyamide-6 nanocomposites. *Polymer* **2004**, 45, 881–891.

39. Gilman, J.W.; Kashiwagi, T.; Nyden, M.; Brown, J.E.T.; Jackson, C.L.; Lomakin, S.; Giannelis, E.P.; Manias, E. Flammability studies of polymer layered silicate nanocomposites: polyolefin, epoxy, and vinyl ester resins, in: M.A. Malden, S. Al-Malaika, A. Golovoy, and C.A. Wilkie, Eds., *Chemistry and Technology of Polymer Additives*. Blackwell Science, London, 1999, pp. 249–265.

40. Pastore, H.O.; Frache, A.; Boccaleri, E.; Marchese, L.; Camino, G. Heat induced structure modifications in polymer–layered silicate nanocomposites. *Macromol. Mater. Eng.* **2004**, 289, 783–786.

41. Beyer, G. Filler blend of carbon nanotubes and organoclays with improved char as a new flame retardant system for polymers and cable applications. *Fire Mater.* **2005**, 29, 61–69.

42. Horn, W.E. Inorganic hydroxides and hydroxycarbonates: their function and use as flame-retardants, in: A.R. Horrocks and D. Price, Eds., *Fire Retardant Materials*. Woodhead Publishing, Cambridge, England, 2001, pp. 285–352.

43. Zanetti, M.; Camino, G.; Reichert, P.; Mülhaupt, R. Thermal behaviour of poly(propylene) layered silicate nanocomposites. *Macromol. Rapid Commun.* **2001**, 22, 176–180.

44. Bourbigot, S.; Gilman, J.W.; Wilkie, C.A. Kinetic analysis of the thermal degradation of polystyrene–montmorillonite nanocomposite. *Polym. Degrad. Stab.* **2004**, 84, 483–492.

45. Burnside, S.D.; Giannelis, E.P. Synthesis and properties of new poly(dimethylsiloxane) nanocomposites. *Chem. Mater* **1995**, 7, 1597–1600.

46. Zanetti, M.; Camino, G.; Thomann, R.; Mülhaupt, R. Synthesis and thermal behaviour of layered silicate–EVA nanocomposites. *Polymer* **2001**, 42, 4501–4507.

47. Lan, T.; Kaviratna, P.D.; Pinnavaia, T.J. On the nature of polyimide–clay hybrid composites. *Chem. Mater.* **1994**, 6, 573–577.

48. Messersmith, P.B.; Giannelis, E.P. Synthesis and barrier properties of poly(ε-caprolactone)-layered silicate nanocomposites. *J. Polym. Sci. A Polym. Chem.* **1995**, 33, 1047–1057.

49. Blumstein, A. Polymerization of adsorbed monolayers, 2: Thermal degradation of inserted polymer. *J. Polym. Sci. A Polym. Chem.* **1965**, 3, 2665–2673.

50. Jang, B.N.; Wilkie, C.A. The effect of clay on the thermal degradation of polyamide 6 in polyamide 6/clay nanocomposites. *Polymer* **2005**, 46, 3264–3274.

51. Jang, B.N.; Wilkie, C.A. The thermal degradation of polystyrene nanocomposite. *Polymer* **2005**, 46, 2933–2942.

52. Su, S.P.; Wilkie, C.A. The thermal degradation of nanocomposites that contain an oligomeric ammonium cation on the clay. *Polym. Degrad. Stab.* **2004**, 83, 347–362.

53. Hartwig, A.; Sebald, M. Preparation and properties of elastomers based on a cycloaliphatic diepoxide and poly(tetrahydrofuran). *Eur. Polym. J.* **2003**, 39, 1975–1981.

54. Schartel, B.; Pötschke, P.; Knoll, U.; Abdel-Goad, M. Fire behaviour of polyamide 6/multiwall carbon nanotube nanocomposites. *Eur. Polym. J.* **2005**, 41, 1061–1070.

55. Schartel, B.; Braun, U.; Knoll, U.; Bartholmai, M.; Goering, H.; Neubert, D.; Pötschke, P. Mechanical, thermal and fire behaviour of bisphenol A polycarbonate/multiwall carbon nanotube nanocomposites. In preparation.

56. Bourbigot, S.; Le Bras, M.; Duquesne, S.; Rochery, M. Recent advances for intumescent polymers. *Macromol. Mater. Eng.* **2004**, 289, 499–511.

57. Papazoglou, E.S. Flame retardants for plastics, in: C.A. Harper, Ed., *Handbook of Materials in Fire Protection.* McGraw-Hill, New York, 2004, pp. 4.1–4.88.

58. Schartel, B.; Bartholmai, M.; Knoll, U. Some comments on the use of cone calorimeter data. *Polym. Degrad. Stab.* **2005**, 88, 540–547.

59. Hirschler, M.M. Heat release from plastics materials, in: V. Babrauskas, and S.J. Grayson, Eds., *Heat Release in Fires*. Elsevier Science, Barking, Essex, England, 1992, pp. 375–422.

60. Braun, U.; Schartel, B. Flame retardant mechanisms of red phosphorus and magnesium hydroxide in high impact polystyrene. *Macromol. Chem. Phys.* **2004**, 205, 2185–2196.

61. Lyon, R.E. Ignition resistance of plastics, in: M. Lewin, Ed., *Recent Advances in Flame Retardancy of Polymers*, Vol. 13. Business Communications Co., Norwalk, CT, 2002, pp. 14–25.

62. Schartel, B.; Braun, U. Comprehensive fire behaviour assessment of polymeric materials based on cone calorimeter investigations. *e-Polymers* **2003**, art. no. 13. http://www.e-polymers.org/papers/schartel_010403.pdf.

63. Kashiwagi, T.; Shields, J.R.; Harris, R.H., Jr.; Awad, W.H., Jr. Flame retardant mechanism of a polymer clay nanocomposite, in: M. Lewin, Ed., *Recent Advances in Flame Retardancy of Polymers*, Vol. 14. Business Communications Co., Norwalk, CT, 2003, pp. 14–26.

64. Babrauskas, V. Heat release rates, in: P.J. DiNenno, D. Drysdale, C.L. Beyler, W.D. Walton, R.L.P. Custer, J.R. Hall, Jr., and J.M. Watts, Jr., Eds., *The SFPE Handbook of Fire Protection Engineering*, 3rd ed. National Fire Protection Association, Quincy, MA, 2002, pp. 3-1 to 3-37.

65. Hartwig, A.; Pütz, D.; Schartel, B.; Bartholmai, M.; Wendschuh-Josties, M. Combustion behaviour of epoxide based nanocomposites with ammonium and phosphonium bentonites. *Macromol. Chem. Phys.* **2003**, 204, 2247–2257.

66. Schartel, B. Fire retardancy based on polymer layered silicate nanocomposites, in: M. Okamoto, Ed., *Advances in Polymeric Nanocomposite*. CMC Publishing, Osaka, Japan, 2004; pp. 242–257.

67. Petrella, R.V. The assessment of full-scale fire hazards from cone calorimeter data. *J. Fire Sci.* **1994**, 12, 14–43.

68. Babrauskas, V. Fire test methods for evaluation of fire-retardant efficacy in polymeric materials, in: A.F. Grand and C.A. Wilkie, Eds., *Fire Retardancy of Polymeric Materials*. Marcel Dekker, New York, 2000, pp. 81–113.

69. Gilman, J.W.; Kashiwagi, T. Polymer-layered silicate nanocomposites with conventional flame retardants, in: T.J. Pinnavaia and G.W. Beall, Eds., *Polymer–Clay Nanocomposites*. Wiley, Chichester, West Sussex, England, 2000, pp. 193–206.

70. Hull, T.R.; Wills, C.L.; Artingstall, T.; Price, D.; Milnes, G.J. Mechanisms of smoke and co suppression from EVA composites, in: M. Le Bras, C.A. Wilkie, S. Bourbigot, S. Duquesne, and C. Jama, Eds., *Fire Retardancy of Polymers: New Applications of Mineral Fillers*. Royal Society of Chemistry, London, 2005, pp. 372–385.

71. Hull, T.R.; Price, D.; Liu, Y.; Wills, C.L.; Brady, J. An investigation into the decomposition and burning behaviour of ethylene–vinyl acetate copolymer nanocomposite materials. *Polym. Degrad. Stab.* **2003**, 82, 365–371.

72. Chigwada, G.; Wilkie, C.A. Synergy between conventional phosphorus fire retardants and organically-modified clays can lead to fire retardancy of styrenics. *Polym. Degrad. Stab.* **2003**, 81, 551–557.

73. Le Bras, M.; Bourbigot, S. Mineral fillers in intumescent fire retardant formulations: criteria for the choice of a natural clay filler for the ammonium polyphosphate/pentaerythritol/polypropylene system. *Fire Mater.* **1996**, 20, 39–49.

74. Zanetti, M.; Camino, G.; Canadese, D.; Morgan, A.B.; Lamelas, F.J.; Wilkie, C.A. Fire retardant halogen–antimony–clay synergism in polypropylene layered silicate nanocomposites. *Chem. Mater.* **2002**, 14, 189–193.

75. Chigwada, G.; Jash, P.; Jiang, D.D.; Wilkie, C.A. Fire retardancy of vinyl ester nanocomposites: synergy with phosphorus-based fire retardants. *Polym. Degrad. Stab.* **2003**, 81, 551–557.

76. Beyer, G. Flame retardancy of nanocomposites: from research to technical products. *J. Fire Sci.* **2005**, 23, 75–87.

6

INTUMESCENCE AND NANOCOMPOSITES: A NOVEL ROUTE FOR FLAME-RETARDING POLYMERIC MATERIALS

Serge Bourbigot and Sophie Duquesne

Laboratoire Procédés d'Élaboration des Revêtements Fonctionnels, LSPES UMR/CNRS 8008, École Nationale Supérieure de Chimie de Lille, Villeneuve d'Ascq Cedex, France

6.1 INTRODUCTION

Today, the probability of a catastrophic fire razing an entire town in peacetime is remote. Due to the plethora of mainly governmental legislation, fire protection plays an extremely important role in reducing fire risk. Nevertheless, the development of science and technology provides the availability of sophisticated products but, concurrently, increases the use of combustible materials.[1] Various methods can be used to protect materials more effectively against attack by fire. An efficient way is to use flame retardants and/or particles (micro- or nano-dispersed) incorporated directly in the materials (e.g., thermoplastics or thermosets) or in a coating covering their surface (e.g., structural steel or textiles).[2] This approach (incorporation of flame retardant either directly in the polymer or in a coating) is chosen in this work to provide low flammability to polymeric materials because it is an acceptable compromise between cost and properties and because it brings great flexibility to design materials with multifunctional properties.

Demand for designing materials with novel functionalities and also to be multifunctional is growing for many applications. They should possess unique

Flame Retardant Polymer Nanocomposites, edited by Alexander B. Morgan and Charles A. Wilkie

mechanical, thermal, thermal–mechanical, electrical, and thermal–electrical properties and low flammability, with sustainability in a variety of harsh environments for space, automotive, electronic, and infrastructure elements. This is an important challenge in the materials science and engineering industry. Due to this evolutionary change in the materials science and engineering research focus, the integration of many conventional materials, such as carbon, clay, ceramic, aluminum particles, and so on, at the nanoscale to make nanocomposites has emerged to build new blocks of revolutionary materials with superior and optimized properties.

The pioneering work of Gilman et al. has demonstrated that the presence of nano-dispersed montmorillonite clay in polymeric matrices produces a substantial improvement in fire performance.[3–5] Gilman and other groups have described this approach and developed hybrid polymeric materials, including organo-modified clays,[6–9] nanoparticles of TiO_2,[10] nanoparticles of silica,[11] layered double hydroxides (LDHs),[12,13] carbon nanotubes (CNTs),[14,15] or polyhedral silsesquioxanes (POSSs).[16–18] All these materials exhibit low flammability along with other properties, such as enhanced mechanical properties. Typically, the peak heat release rate (HRR) is decreased by 50 to 70% in a cone calorimeter experiment. However, UL-94 and the limiting oxygen index (LOI) of polymer nanocomposites are poor. As an example, the peak HRR of polyamide-6 (PA6)/clay nanocomposites is decreased by 63% compared to virgin PA6 at 35 kW/m^2, while UL-94 test fails (no rating) and the LOI is only 23 vol%.[19] Cone experiments are made in a horizontal position; thus, dripping cannot occur and accumulation of clay at the surface can play its role in forming a protective barrier. On the contrary, the low viscosity of the materials when heated leads to dripping when they are in a vertical position (LOI and UL-94) and the protective barrier flows away from the flame. The substrate is no longer protected and burns; that is why the nanocomposite approach needs to be enhanced. It is the main goal of this chapter to survey the combination of traditional flame retardants, in particular intumescent flame retardants, with nanofillers. It is expected that this approach will provide the opportunity to design fire-safe materials that meet the specifications required by legislation and show enhancements in other properties, such as mechanical properties.

The chapter is organized in four parts. In Section 6.2 we review intumescence briefly to provide a basic understanding of the mechanism of action by intumescence. This is followed by the use of zeolites as synergists in intumescent systems (Section 6.3). The reason that combining intumescence systems with zeolites provides superior performance and why clay should also be a candidate to provide synergistic effect in intumescent systems are explained. The mechanism of action is described and the discussion is focused on the role of the chemical structure of zeolite and why clay should be a crucial ingredient in intumescent formulations. Section 6.4 is an investigation of the performance of intumescent systems containing organo-modified clay. A few formulations are examined in terms of flame retardancy using LOI, UL-94, and cone calorimetry, and mechanical properties are also considered. The role of organoclay, nano-dispersed in a polymeric matrix, on intumescent fire retardancy is addressed and a mechanism of action is proposed. In Section 6.5 we investigate the potential use of nanofillers other than clay in intumescent systems. The influence of the chemical nature of these is addressed. In Section 6.6 we survey recent published works and provide a critical view regarding the use of nanofillers combined with intumescent fire retardants.

6.2 BASICS OF INTUMESCENCE

The word *intumescence* comes from the Latin *intumescere*, which means "to swell up." It is an apt description of an intumescent material, which when heated beyond a critical temperature begins to swell and then to expand. The result of this process is a foamed cellular charred layer on the surface which protects the underlying material from the action of the heat flux or flame.[20]. Intumescent flame-retarding polymers or textiles are essentially a special case of a condensed-phase mechanism.[21–25] Intumescent systems interrupt the self-sustained combustion of the polymer at its earliest stage (i.e., the thermal degradation with the evolution of gaseous fuels). The intumescence process results from a combination of charring and foaming at the surface of the burning polymer. The resulting foamed cellular charred layer, whose density decreases as a function of temperature, protects the underlying material from the action of the heat flux or of the flame. So the charred layer acts as a physical barrier that slows heat and mass transfer between the gas and condensed phases.

A typical example of an intumescent system is polypropylene (PP) containing ammonium polyphosphate (APP)/pentaerythritol (PER) (intumescent additives: ammonium polyphosphate [APP: $(NH_4PO_3)_n$, $n = 700$]/pentaerythritol (PER) = 3 : 1 (wt/wt) at 30 wt% loading) or an intumescent commercial additive (Exolit AP750, Clariant [ammonium polyphosphate with an aromatic ester of tris(2-hydroxyethyl)-isocyanurate[26] at 30 wt% loading]). Evaluation of the fire performance shows that the formulation containing AP750 performed better than that with APP–PER (Table 6.1) but that in both cases a V-0 rating was achieved in the UL-94 test.

These results are confirmed by cone calorimetry[27] (Figure 6.1). The presence of intumescent systems in PP causes strong decreases in the rate of heat release (RHR) values compared to those of the virgin polymer (the peak RHR of PP is about 1800 kW/m^2). Moreover, the RHR curve of PP–AP750 is very flat, and the RHR values reach only 80 kW/m^2 while those of PP–APP/PER reach

TABLE 6.1 LOI Values of PP-Based Intumescent Systems Compared to Virgin PP

Formulation	LOI (vol%)	UL-94 Rating (3.2 mm)
PP	18	No rating
PP–APP/PER	32	V-0
PP–AP750	38	V-0

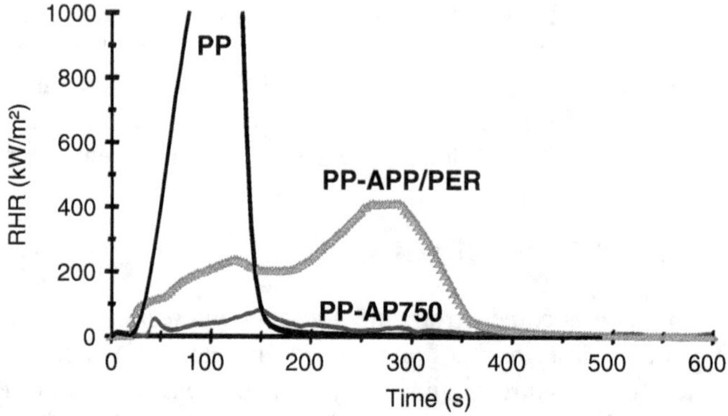

FIGURE 6.1 RHR curves of PP–APP/PER and PP–AP750 versus virgin PP (external heat flux = 50 kW/m^2). (From Ref. 27, copyright © 1997, Sage Publications, with permission.)

400 kW/m^2. It is noteworthy that the RHR curve of PP–APP/PER is typical of intumescent systems exhibiting two peaks. The first peak is assigned to the ignition and to the flame spread on the surface of the material, and then, when the RHR values become constant, to the protection by the intumescent coating. During this time period, the polymer is protected by the intumescent shield. The second peak is explained by the destruction of the intumescent structure and the formation of a carbonaceous residue.

A direct application of the intumescence phenomenon is the protection of metallic materials in the construction industry. In the case of fire, these materials lose their mechanical strength, leading to the collapse of building structures. The use of intumescent paint acts as a heat barrier to protect the material of interest. Recent work from our laboratory[28] shows that good thermal protection of metallic substrate can be achieved using intumescent coatings based on a thermoset epoxy–amine resin system into which the fire retardant agents boric acid and ammonium polyphosphate derivatives have been incorporated. The coatings were evaluated on a large scale in an industrial furnace according to the UL-1709 standard. Figure 6.2 shows the evolution of temperature as a function of time on the back side of steel plates coated with various formulations. Steel usually loses its main structural properties at around 500°C. For safety reasons and because the thermocouple is on the back side of the steel plate, 400°C was chosen as the failure temperature (horizontal line on Figure 6.2).

The time to failure of the steel plate covered with thermoset resin (curve B) is close to the time to failure of the steel plate alone (curve A). When the APP derivative is added to the thermoset resin (curve C), an improvement in performance is observed (time to failure of 11.3 min compared to 5 min for the uncoated steel). Intumescence and charring take place, but the char falls off the plate before the end of the experiment (change of slope at 610°C). Addition of boric acid (curve D) to the resin also leads to improved performance; the time of failure is increased to 18.2 min. Development of intumescence is also observed;

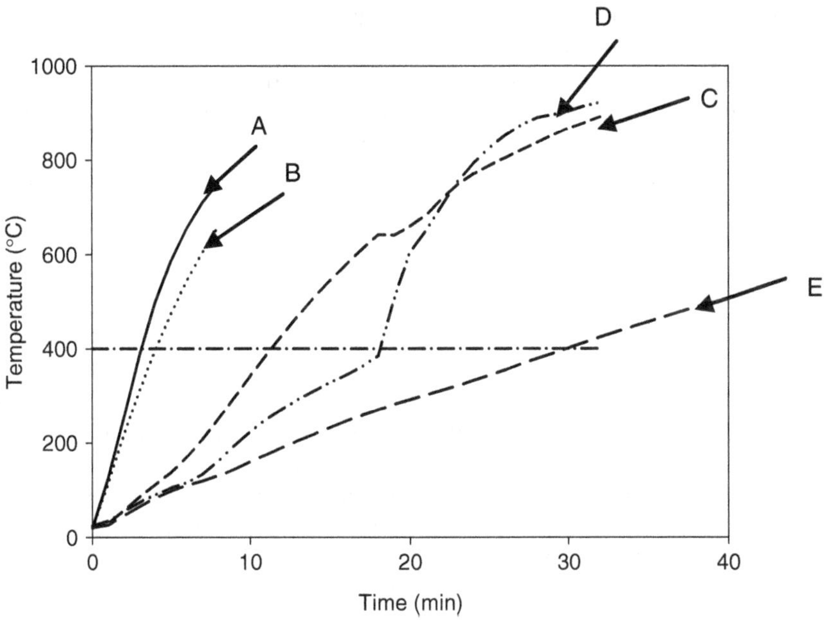

FIGURE 6.2 Evolution of temperature as a function of time on the back side of a steel plate coated with various intumescent coatings: A, virgin steel plate; B, virgin thermoset epoxy resin; C, thermoset resin containing APP derivative; D, thermoset resin containing boric acid; E, thermoset resin containing APP and boric acid. Specification of the test (OTI 95 634) was to burn a given volume of propane (0.3 kg/s) at a given heat flux (200 to 250 kW/m^2) and at a given distance (1 m) from a test piece. The burning conditions fit as close as possible the ramp of temperature of a hydrocarbon fire heating curve (about 200°C/min). Five thermocouples are used on the back side of each plate, and only the average temperature is reported on the plot. Plates were mounted vertically in the furnace.

however, the char falls off the plate (rapid change of slope at 400°C). When APP and boric acid are incorporated together in the thermoset resin (curve E), the time to failure increases very significantly, up to 29.5 min, and the resulting intumescent char adheres strongly to the plate, exhibiting a regular hemispherical shape.

The examples above show that a large improvement in flammability properties can be achieved using an intumescent system in bulk polymers and in coatings. Thermal protection is the main purpose of using intumescent materials; heat transfer is limited by the formation of an intumescent shield. Swelling is central to the fire-protective capabilities, and a fundamental understanding of the mechanisms that cause expansion is important. Temperature gradients and heat transfer play a central role in intumescent behavior. In particular, the effect on the temperature gradients of the growing bubbles cannot be neglected. The sizes of the bubbles may be quite different, due to the large temperature gradients within the intumescent melt. Considering this, a three-dimensional model was developed at the National Institute of Standards and Technology (NIST)[29] that incorporates bubble and melt hydrodynamics, heat transfer, and chemical reactions. In this model, the intumescent system is represented as a highly viscous incompressible fluid containing a large number of expanding bubbles. The bubbles obey equations of mass, momentum, and energy on an individual basis according to the values of the local parameters, and their collective behavior is responsible for the swelling and fire retardant properties of the material. This model provides a good basis for understanding and describing the physical aspect of intumescence.

In the previous discussion the chemical aspects were not addressed, but they are crucial. To make an intumescent system, three ingredients are necessary: an acid source (precursor for catalytic acidic species), a char-forming agent, and a blowing agent. Table 6.2 provides some example of components of intumescent formulations. In the case of PP–APP/PER, the reaction of the acidic species (APP and its degradation products into orthophosphates and phosphoric acid) with the char-former agent (PER) takes place in the first stage ($T < 280°C$) with formation of ester mixtures. The carbonization process then takes place at about 280°C (mainly via a free-radical process[33]). In the second step, the blowing agent decomposes to yield gaseous products (i.e. evolved ammonia from the decomposition of APP) which cause the char to swell ($280 \leq T \leq 350°C$). The intumescent material then decomposes at higher temperatures and loses its foamed character at about 430°C. Concurrently, the heat conductivity of the char decreases between 280 and 430°C, and the insulation of the substrate is enhanced.[34]

TABLE 6.2 Examples of Components of Intumescent Coatings

(a) Inorganic acid sources	(b) Polyhydric compounds
Phosphoric	Starch
Sulfuric	Dextrins
Boric	Sorbitol, mannitol
Ammonium salts	Pentaerythritol, monomer, dimer, trimer
Phosphates and polyphosphates	Phenol–formaldehyde resins
Borates	Methylol melamine
Sulfates	
Halides	(c) Amines and amides
Phosphates of amine or amide	Urea
Products of reaction of urea or Guanidyl urea with phosphoric acids	Urea–formaldehyde resins
	Dicyandiamide
	Melamine
Melamine phosphate	Polyamides
Product of reaction of ammonia with P_2O_5	(d) Others
Organophosphorus compounds	Charring polymers (PA6, PA6–clay nanocomposite PU, PC, ...)
Tricresyl phosphate	
Alkyl phosphates	
Haloalkyl phosphates	

Source: Date from Refs. 30 to 32.

The material resulting from the degradation of an intumescent formulation is a heterogeneous material. It is composed of “trapped” gaseous products in a phosphocarbonaceous cellular material (i.e., the condensed phase). This condensed phase is a mixture of solid phase and liquid phase (acidic tars) possessing the dynamic properties of interest which allows the trapping of gaseous and liquid products resulting from degradation of the polymer. The carbonaceous fraction of the condensed phase consists of polyaromatic species that are organized in stacks characteristic of a pregraphitization stage (Figure 6.3).[35]

The phosphocarbonaceous material constitutes crystalline macromolecular polyaromatic stacks bridged by polymer links and phosphate (poly-, di-, or orthophosphate) groups, crystalline additive particles, and an amorphous phase that encapsulates the crystalline domains. The amorphous phase is composed of small polyaromatic molecules, easily hydrolyzed phosphate species, alkyl chains formed via the degradation of additives, and the fragments of the polymer chain. It governs the protective behavior of the coating: This phase has to be voluminous enough to coat the crystalline domains perfectly and should exhibit appropriate viscoelasticity[36] (this aspect is discussed further below), which yields the dynamic properties of interest (avoiding dripping and accommodating the stress induced by solid particles and gas pressure).

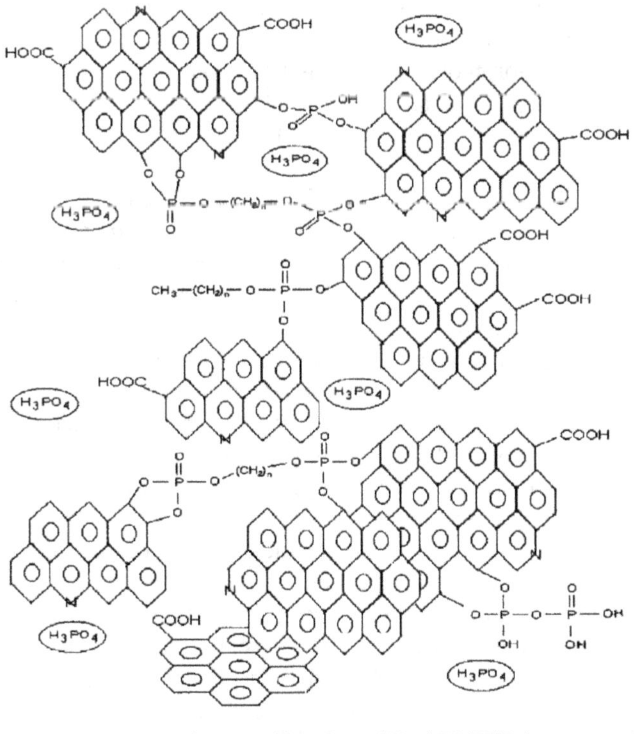

FIGURE 6.3 Intumescent coating resulting from PP–APP/PER heat treated at 350°C. (From Ref. 35, copyright © 1995, Elsevier, with permission.)

In this section we have demonstrated that intumescent systems provide efficient fire retardant properties to polymeric materials, both in bulk and as a coating. The mechanisms of action have been discussed and we have seen that the chemistry of such systems offers some flexibility in the synthesis of the char (e.g., potential reactivity of phosphates and oxidized functions). We can then expect to enhance the performance of the intumescent char by the addition of other reactive compounds in the formulation, which is the purpose of the next section.

6.3 ZEOLITES AS SYNERGISTIC AGENTS IN INTUMESCENT SYSTEMS

Several interesting developments have occurred recently that involved unexpected "catalytic" effects in various intumescent systems. Performance in terms of LOI, UL-94, or cone calorimetry was enhanced dramatically by adding small amounts of an additional compound, leading to a synergistic effect. In the following we use the following definition of *synergy*: A synergistic effect occurs when the combined effects of two chemicals are much greater than the sum of the effects of each agent given alone.

Work done in our laboratory has shown that by adding small amounts of minerals such as zeolites,[37,38] natural clays,[39] and zinc borates[40] to intumescent systems, the flame retardant performance can be enhanced enormously, Levchik et al. proposed the use of small amounts of talc and manganese dioxide combined with APP in PA6 to promote charring and to enhance the insulative properties of the intumescent coating, leading to a significant improvement in flammability performance.[41,42] Another approach using borosiloxane elastomer also shows a very large synergistic effect in intumescent systems.[43,44]

Here, only zeolites are considered. Zeolites are tectosilicates characterized by a three-dimensional framework of AlO_4 and SiO_4 tetrahedra (Table 6.3).[45] The framework contains channels and interconnected voids that are occupied by the cation and water molecules. Negative charges due to AlO_4 are balanced by cations. The size of the voids or the channels (apperture size in Table 6.3) is approximately the size of the usual organic molecules. Nevertheless, zeolites are not nanoparticles, but their internal nanostructure makes it reasonable to discuss their effect in this chapter. The ideal chemical formula is $M_{x/n}[(AlO_2)_x,(SiO_2)_y]\cdot zH_2O$. The part in brackets is the framework of the zeolite with a y/x ratio ≥ 1, and M is the charge-balancing cation.

In previous work[37] we have combined the intumescent APP–PER system with zeolites in an ethylene–butyl acrylate–maleic anhydride terpolymer (hereafter called LRAM3.5) and we have observed (in particular with 4A zeolite) a high degree of improvement in fire-proofing properties (Figure 6.4).

Table 6.3 shows that zeolite exhibits different structures, and we can expect that flammability properties might depend on this. In Figure 6.5 we investigate the influence of the Si/Al ratio using the sodium zeolites Y (Si/Al = 2.43), mordenite (Si/Al = 5), and ZSM-5 (Si/Al = 140). The trends observed are similar, with maxima for the LOI reached at 1.5 (Y) or 2 wt% (mordenite and ZSM-5) of

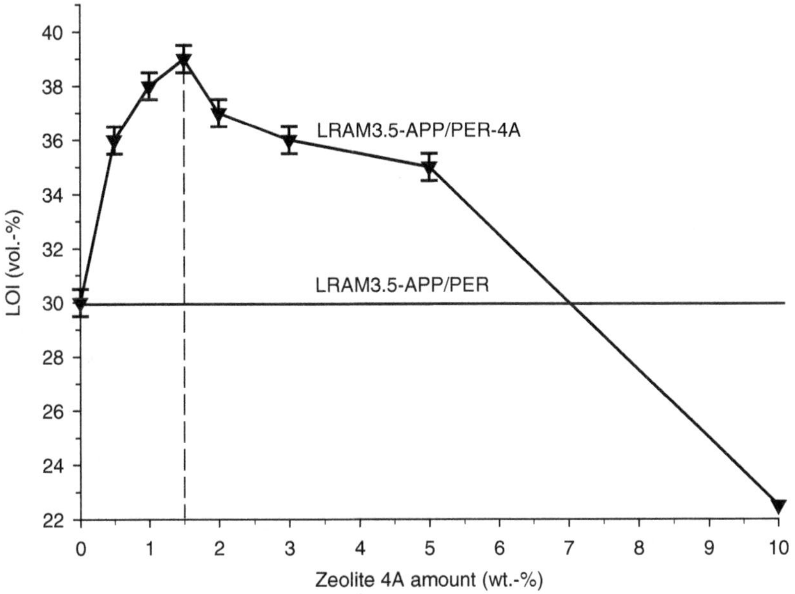

FIGURE 6.4 LOI versus zeolite level in a technical ethylenic terpolymer (ethylene–butylacrylate–maleic anhydride, Lotader P3200 from Elf-Atochem). The additive level is kept constant at 30 wt%, and the synergistic effect exhibits a maximum at 1.5 wt% 4A zeolite level.

TABLE 6.3 Chemical Characteristics of Zeolites

Building Units	Formula	Structure Type	Si/Al (at/at)	Aperture Size (nm)
D4R	$K_9Na_3[(AlO_2)_{12}(SiO_2)_{12}],27H_2O$	KA (3A)	1	0.32
D4R	$Na_{12}[(AlO_2)_{12}(SiO_2)_{12}],27H_2O$	NaA (4A)	1	0.35
D4R	$Ca_{4,5}Na_3[(AlO_2)_{12}(SiO_2)_{12}],27H_2O$	CaA (5A)	1	0.42
D6R	$Ca_{21,5}Na_{43}[(AlO_2)_{86}(SiO_2)_{106}]_2,76H_2O$	CaX (10X)	1.23	0.8
D6R	$Na_{56}[(AlO_2)_{86}(SiO_2)_{106}],276H_2O$	NaX (13X)	1.23	0.9–1
D6R	$Na_{86}[(AlO_2)_{56}(SiO_2)_{136}],250H_2O$	Y	2.43	1
T_6O_{16}	$Na_8[(AlO_2)_8(SiO_2)_{40}],24H_2O$	Mordenite	5	0.67–0.7
—	$Na_{0.7}[(AlO_2)_{0,7}(SiO_2)_{95.3}],16H_2O$	ZSM-5	140	0.52–0.58

Source: After Ref. [45].

zeolite. These values demonstrate significant differences (LOI = 40% with Y zeolite and LOI = 36% with ZSM-5 zeolite). It suggests to us that low Si/Al ratios might help achieve the best performance.

Zeolites are aluminosilicates, and considering this, an aluminosilicate composed of silica sheets (Si_2O_5) bonded to aluminum oxide/hydroxide layers [$Al_2(OH)_4$] (China clay with kaolinite composition) is compared with zeolites in LRAM3.56–PP/PER formulations (Figure 6.6). The curves exhibit the same

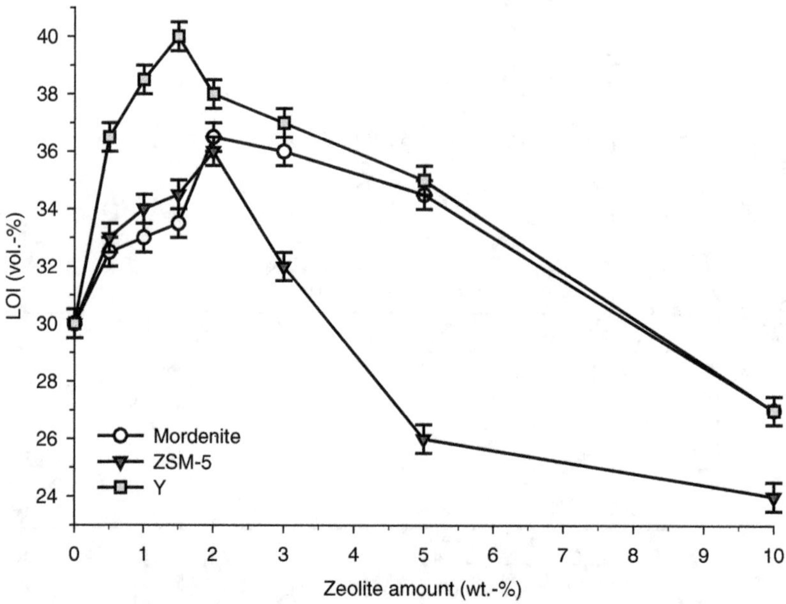

FIGURE 6.5 LOI values versus Y, mordenite, and ZSM-5 zeolite level of formulations LRAM3.5–APP/PER–zeolites (additive level remaining constant at 30 wt%).

behavior; an LOI maximum is always reached at 1.5 wt% of aluminosilicate in the formulations, but LOI measured with kaolinite is lower than that measured with Y and 4A zeolites. This result suggests therefore that an aluminosilicate with a zeolite structure should be used to obtain the best flame retardancy performance. In other work,[39] using different clays as synergistic agents, we proposed that the resulting LOI classification depended on the content of exchangeable cations of the clay (LOI values increase when the number of exchangeable cations per unit cell increases). Kaolinite-type minerals do not have such cations, whereas zeolites do. It was suggested that the flame retardant performance depends on the exchange properties of the aluminosilicates.

Cone calorimetry experiments confirm the enhancement of performance using zeolite.[37] Figure 6.7 shows the curves of RHR versus time of the formulation LRAM3.5–APP/PER with and without zeolite 4A. A significant decrease in the RHR maxima in the flame-retarded polymers is observed comparison with that of the virgin polymer. In the case of intumescent formulations, the RHR decreases strongly after ignition. It is noteworthy that the RHR behavior of the intumescent polymers is similar. The three maxima curves of intumescent materials are explained as follows:

- First, the formulations degrade and the intumescent shield forms.
- These coatings then degrade, and consequently, the residual materials degrade and form a new intumescent coating.
- Finally, the entire of mass residual material degrades.

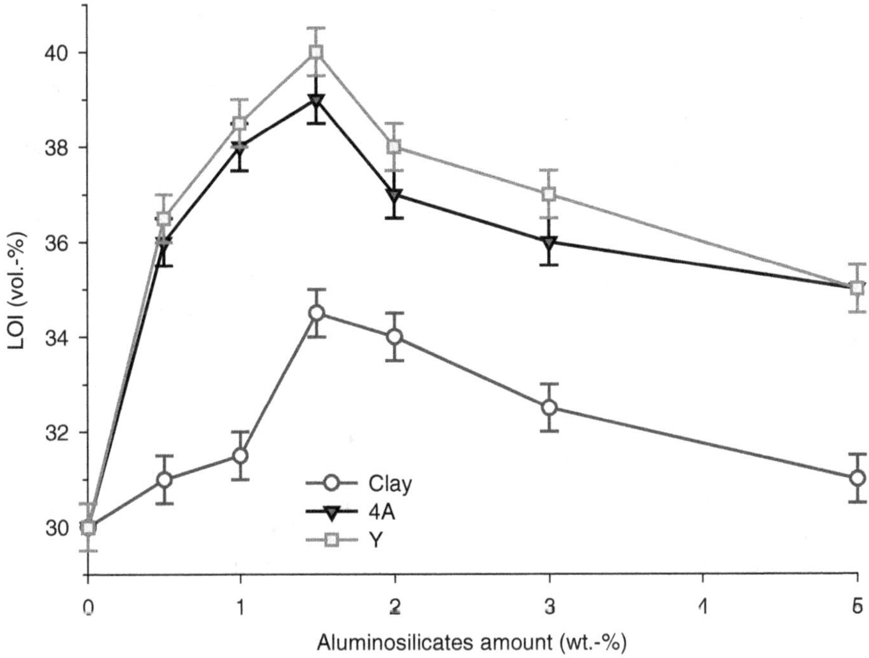

FIGURE 6.6 LOI values versus 4A, Y, and China clay's level of formulations LRAM3.5 –APP/PER–aluminosilicates (additive level remaining constant at 30 wt%).

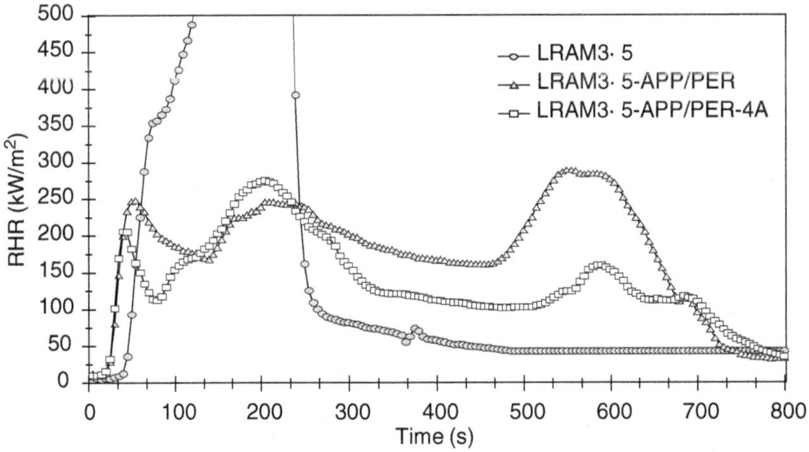

FIGURE 6.7 RHR versus time of the systems LRAM3.5, LRAM3.5–APP/PER, and LRAM3.5–APP/PER–4A (external heat flux = 50 kW/m^2). (From Ref. 37, copyright © 1996, Elsevier, with permission.)

The maxima of RHR are different. These differences become very significant at longer times. As an example, at $t = 600$ s the RHR value is only 150 kW/m^2 for the system with zeolite, whereas it is 300 kW/m^2 for the system without zeolite. This means that the intumescent shield developed with the system containing zeolite can resist longer in severe thermooxidative conditions than that without zeolite (visual observations of small flames at the surface of the samples as well as glowing spots indicate that the materials undergo thermooxidative degradation rather than pyrolysis).

The combination of zeolites with APP/PER or diammonium pyrophosphate (PY)/PER systems also leads to a high degree of improvement in the fire retardant properties in other polymers (Table 6.4).[37] The efficiency of zeolite depends on the polymer, but the synergistic effect is always observed, suggesting that zeolites should be used as an additional ingredient in intumescent formulations. The role played by the zeolite in the particular system LRAM3.5–APP/PER was explained earlier using spectroscopic studies of additive systems and of polymer-additive formulations.[46–50] When the temperature increases, an intumescent structure develops. At 280°C, stacks of polyaromatic species linked principally by phosphohydrocarbonated bridges are formed. These bridges provide dynamic properties to the structure to accommodate the stresses (assuming that the number of bridges and the dimension of polyaromatic structures are similar for the two systems). At this time, the structures developed from formulations with and without zeolite are distinguished by the organization of the carbon, the zeolite slowing its process of organization.[48]

The condensation of aromatic species and the decrease in phosphocarbonated species by scission of P–O–C bonds are observed when the temperature increases ($T > 280°C$). Consequently, enlargement in the size of the polyaromatic stacks drastically increases the viscosity of the material and thus leads to a loss of the properties of the coating. The addition of zeolite in the formulation maintains a large number of polyaliphatic links in the structure stabilized by

TABLE 6.4 FR Performance of Intumescent Formulations with and Without Zeolite[a]

System[b]	LOI (vol%)	UL-94 Rating (3.2 mm)	System[b]	LOI (vol%)	UL-94 Rating (3.2 mm)
PP–APP/PER	30	V-0	PP–APP/PER/13X	45	V-0
LDPE–APP/PER	24	V-0	LDPE–APP/PER/4A	26	V-0
PP–PY/PER	32	V-0	PP–PY/PER/13X	52	V-0
PS–APP/PER	29	V-0	PS–APP/PER/4A	43	V-0
LRAM3.5–APP/PER	29	V-0	LRAM3.5–APP/PER/4A	39	V-0

Source: After Ref. [37].

[a]The additives' loading is kept at a constant equaling 30 wt%, and the synergistic effect exhibits a maximum at 1 or 1.5 wt% zeolite loading.[13]

[b]PP, isotactic polypropylene; LDPE; low-density polyethylene; PS; polystyrene; LRAM3.5; ethylene–butylacrylate–maleic anhydride terpolymer (Lotader P3200, Elf-Atochem).

organic aluminosilicophosphate complexes and reduces the scission of P–O–C bonds and therefore the increase in the size of the polyaromatic stacks.[49]

Moreover, pyridinic nitrogen (it was reported that ammonia evolving from APP reacts with carbon–carbon double bonds to yield nitrogenated heterocycles) is observed at all temperatures in the case of zeolite and only up to 350°C in the case of a system without zeolite.[50] It was proposed that the delay in condensation of the polyaromatic network allows the retention of pyridinic nitrogen to participate in improvement in the mechanical properties and therefore to improvement in the fireproofing properties of the material.

Intumescent systems develop a phosphocarbonaceous structure which is thermally stabilized by zeolite. Blocks of polyethylenic units link by means of the formation of organic phosphates and/or organic aluminophosphates, and this limits depolymerization (i.e., the evolution of small flammable molecules able to feed the fire as "fuel"). Moreover, it was shown that zeolite allows the formation of a more "coherent" structure. The formation of a coherent macromolecular network and the participation of polyethylenic links seem to be favorable for obtaining fire retardancy. Indeed, development of an intumescent shield in a structure consisting of polyaromatic species creates a rigid material (e.g., the LRAM3.5–APP/PER formulation). On the other hand, the stabilization of polyethylenic links in an intumescent structure able to bridge polyaromatic species by means of alumino- and/or silicophosphate groups and/or because the free radicals in the intumescent structure may provide the mechanical properties of interest in the intumescent coating[48]—they provide flexibility to the carbonaceous shield. Under fire conditions, this shield delays the creation and propagation of cracks in which oxygen diffuses to the polymeric matrix and through which small molecules can be released as fuel.

Zeolite does not need to be a true catalyst, because it reacts (zeolite is a crucial reactant of the intumescent reaction) with the other ingredients of the intumescent formulation (reaction of phosphate with the aluminosilicate). The latter reaction permits the formation (in situ and in the conditions of fire) of species that stabilize the intumescent structure. So it can be expected that the use of species such as aluminosilicate, able to react with phosphate, should also provide enhanced flammability properties.

6.4 INTUMESCENTS IN POLYMER NANOCOMPOSITES

The char formers commonly used in intumescent formulations for thermoplastics are polyols such as pentaerythritol, mannitol, and sorbitol.[33,51] However, exudation and water solubility are problems associated with these additives.[51] Moreover, these additives are often not compatible with the polymeric matrix, and the mechanical properties of the formulations are then very poor. We have developed intumescent polyolefin-based formulations using charring polymers [thermoplastic polyurethane (TPU) and polyamide-6 (PA6)] as carbonization agents.[52–56] These formulated blends have improved mechanical properties compared with polymers loaded with classical flame retardants, and they avoid the problems of exudation and water solubility. As TPU–and PA6–clay hybrid nanocomposites exhibit superior performance in terms of mechanical properties, the idea has been to combine those polymer nanocomposites as an ingredient of intumescent formulation to improve both flame retardancy and mechanical properties of the polymer.

As mentioned above, montmorillonite (MMT) clay is one of the most commonly used nanofillers. It is a member of the general mineral group of the clays, and its general chemical formula is $(Ca,Na,H)(Al,Mg,Fe,Zn)_2(Si,Al)_4O_{10}(OH)_2-xH_2O$. It is an aluminosilicate, so we can expect an enhancement of flammability properties when it is incorporated in intumescent material. The concept is evaluated with a combination of APP–PA6 and APP–PA6–clay hybrid (hereafter called PA6nano: PA6 containing organomodified MMT clay nanodispersed in PA6 and exhibiting an exfoliated structure) as intumescent systems in poly(ethylene-co-vinyl acetate) (EVA) (EVA24: EVA containing 24 wt% vinyl acetate). Figure 6.8 shows the mechanical properties of EVA24-based materials in comparison with virgin EVA24 and EVA24 loaded with a classical flame retardant [EVA24 loaded with 60 wt% aluminum trihydroxide (ATH) coated with silanes]. Among the flame-retarded polymers, stress and elongation at break are highest for the formulation containing PA6nano.

When burning EVA24–APP/PA6 and EVA24–APP/PA6nano formulations, the formation of an intumescent char that smothers the flame is observed. Figure 6.9 shows that a synergistic effect on LOI is observed in both EVA24–APP/PA6 and EVA24–APP/PA6nano formulations with APP/PA6 mass ratios equaling 3. One can observe that the use of PA6nano improves the values from 32 vol% without exfoliated clay to 37 vol% with clay at APP/PA6 = 3 (wt/wt). A V-0 rating is achieved for $13.5 \leq \text{APP} \leq 34$ wt% without clay and for $10 \leq \text{APP} \leq 34$ wt% with clay (the total loadings in APP–PA6 and APP–PA6nano remain equaling 40 wt%). This result shows that the use of

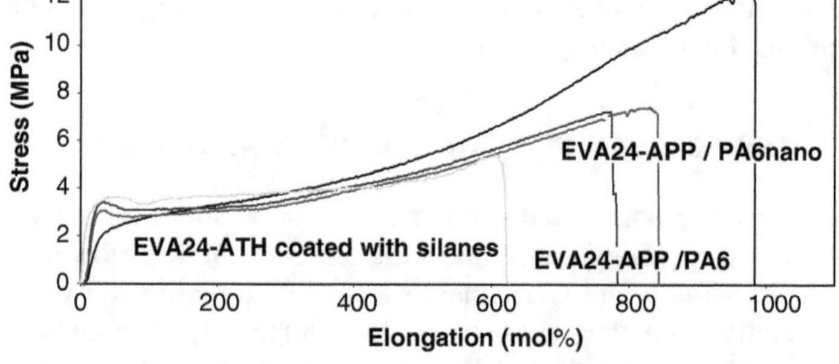

FIGURE 6.8 Mechanical properties of the formulations EVA24–ATH, EVA24–APP/PA6, and EVA24–APP/PA6nano (PA6nano = PA6 clay hybrid from UBE Industries; ratio APP/PA6 and APP/PA6nano = 3 wt/wt) compared with virgin EVA24. From PA6 clay nanocomposite hybrid as a char-forming agent in intumescent formulations. (From Ref. 32, copyright © 2000, John Wiley & Sons, Ltd., with permission.)

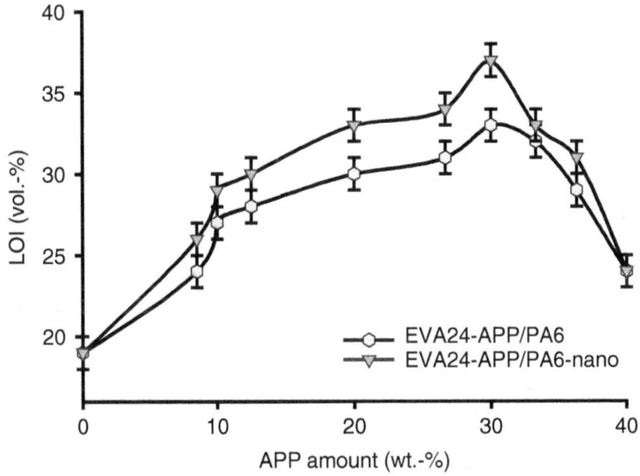

FIGURE 6.9 LOI values versus APP content in the intumescent formulations EVA24–APP/PA6 and EVA24–APP/PA6nano. From PA6 clay nanocomposite hybrid as a char-forming agent in intumescent formulations. (From Ref. 32, copyright © 2000, John Wiley & Sons, Ltd., with permission.)

PA6nano in the formulation allows a V-0 rating to be achieved at relatively low loading in APP (10 wt% in comparison with 13.5 wt%). This is a real advantage because it permits a decrease in the amount of APP in the formulation; additional APP can sometimes lead to a blooming effect and to migration throughout the polymer. The lowered amount of APP also permits the preservation of mechanical properties.

The RHR values of the intumescent EVA-based polymers are strongly reduced compared with the virgin EVA24 (Figure 6.10). The use of the PA6nano improves flame retardancy: the RHR peak = 320 kW/m^2 with PA6 and the RHR peak = 240 kW/m^2 with PA6nano. Visually, a char layer is formed and intumescences after ignition of the material. The height of the intumescent shield is about 1.5 cm, compared to 0.3 cm for the unburned material. Nevertheless, after combustion, the intumescent residue of the formulation containing PA6nano looks less fragile than that without PA6nano.

Using different spectroscopic techniques, the mechanism of action has been elucidated.[32,57] Upon heating in air, the two systems form a phosphocarbonaceous material; this structure is needed to get good fire performance.[57] In clay–nanocomposite formulations, the clay reacts with APP and forms aluminophosphate species (and probably silicophosphate species as well). These species thermally stabilize the phosphocarbonaceous structure up to 310°C. At higher temperatures, the phosphocarbonaceous structure is degraded because of the collapse of the aluminophosphate species. An amorphous "ceramiclike alumina" containing (or trapping) orthophosphoric and polyphosphoric acid species is then created. This can act as a protective barrier in addition to the intumescent shield, which can limit oxygen diffusion to the substrate and/or inhibit the migration of liquid

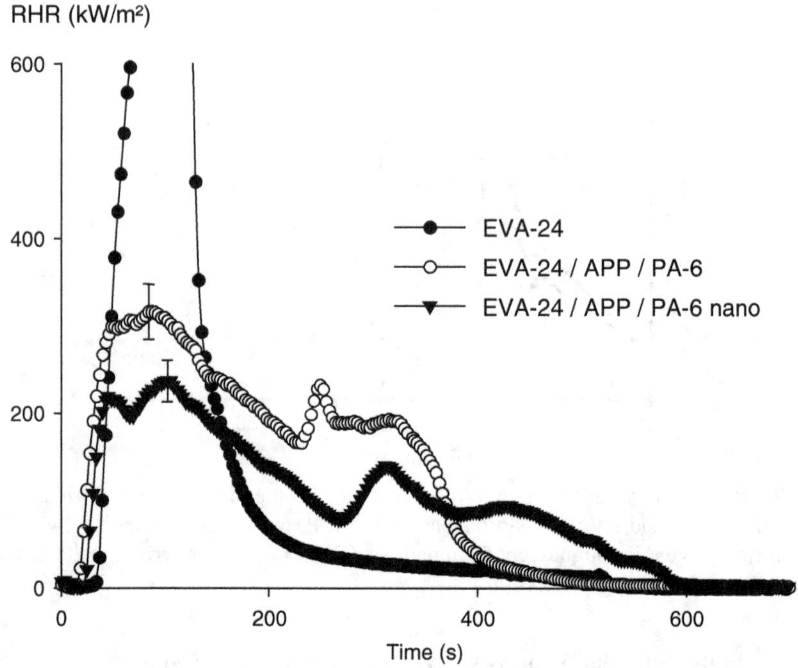

FIGURE 6.10 RHR values versus time of the formulations EVA24–APP/PA6 and EVA24–APP/PA6nano compared with virgin EVA24 (external heat flux = 50 kW/m^2). From PA6–clay nanocomposite hybrid as a char-forming agent in intumescent formulations. (From Ref. 32, copyright © 2000, John Wiley & Sons, Ltd., with permission.)

or gaseous decomposition products into the "hot zone". This barrier can also hinder the formation of cracks. For a formulation without nanocomposite, no phosphocarbonaceous structure is observed whatever the temperature, and only a layer of orthophosphoric acid is observed in the intumescent char.

As in the case of zeolite, the mechanism of action looks similar. No direct comparison can be made because MMT is a layered silicate compared to the cage structure of zeolite, and also because the carbonization agent is no longer a polyol but a char-forming polymer (PA6). Nevertheless, the main conclusion we can draw is that the action of the synergist (nanoclay or zeolite) is to stabilize in a first step the carbonaceous structure forming aluminophosphates and silicophosphates. With the nanoclay, this effect is only effective up to 310°C, whereas it is still efficient at 560°C with zeolite. To keep its protection efficient at high temperatures, the nanoclay permits the formation of protective ceramiclike material after collapse of the phosphocarbonaceous structure. Note that we did not detect any specific influence of the surfactant of the nanoclays, probably because of its low amount in the formulation.

In the study above, MMT was only incorporated in the char-forming polymer. The effect of MMT might be different if it is incorporated in the polymeric matrix. We have incorporated MMT both in EVA (EVA containing 19 wt% vinyl

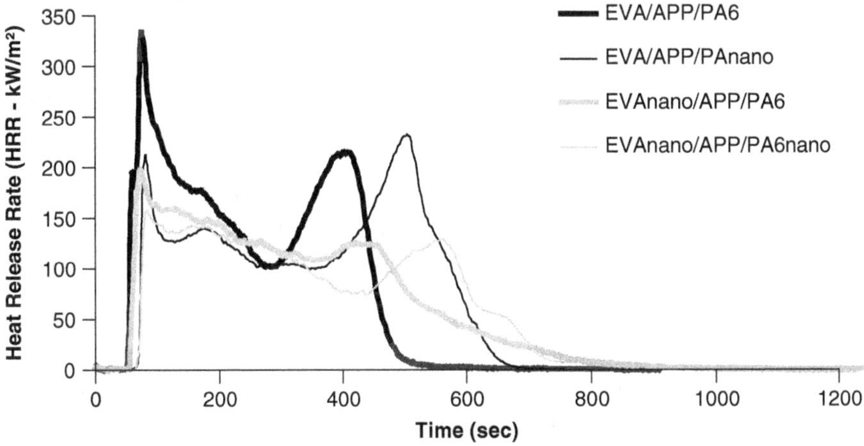

FIGURE 6.11 Heat release rate versus time for intumescent EVA–APP/PA6 formulations with and without clay.

acetate) and PA6 to make EVA nanocomposite (EVAnano) and PA6 nanocomposite (PA6nano), and then an EVA based intumescent system.[58] EVA and PA6 nanocomposites exhibit a mixed intercalated–exfoliated morphology and an exfoliated structure, respectively, as revealed by transmission electronic microscopy (TEM) and x-ray diffraction (XRD).

Curves of heat release rate (HRR) versus time for intumescent EVA-based formulations (Figure 6.11) exhibit two peaks assigned to the development of intumescence. The first corresponds to formation of a protective layer, and the second corresponds to its destruction or failure. It clearly appears that when a nanocomposite is included in the formulation (in the matrix, in the carbonization agent, or in both), the first peak heat release rate (PHRR) is reduced (from about 340 kW/m^2 to 200 kW/m^2). However, the second peak decreases only when EVAnano is used, suggesting the formation of a stronger char. Work is in progress to explain these phenomena.

6.5 NANOFILLERS AS SYNERGISTS IN INTUMESCENT SYSTEMS

According to our previous discussion and as a general rule, we suggest that all nanofillers able to react with phosphate could be used as a synergist in intumescent formulations. A rapid survey of the literature indicates that other nanofillers might be candidates as synergists in intumescent systems, including layered double hydroxides (LDHs), nanoparticles of TiO_2 and SiO_2, polyhedral oligomeric silsesquixonanes (POSSs), or carbon nanotubes. As a typical example, we will only examine the role of LDHs and nanoparticles of silica compared to organomodified MMT in intumescent EVA.

The intumescent system is a combination of APP–PA6 containing organomodified LDH (hereafter called OLDH; LDH was synthesized in our laboratory and the nitrate ions were exchanged by dodecyl sulfate according to the usual procedure[59]), nanoparticles of silica (hereafter called NPSi, Aerosil 200 supplied by Degussa, average primary particle size = 12 nm, SiO_2 content > 99.8%), and organo-modified MMT (hereafter called OMMT, Cloisite 30B supplied by Southern Clay Products San Antonio, Texas) in EVA (EVA containing 19 wt% vinyl acetate).

The thermooxidative degradation of the intumescent formulations takes place in four apparent steps (Figure 6.12). TGA curves of the intumescent formulations containing the nanofillers are similar to that of the formulation without nanofiller from ambient temperature up to 480°C. In this temperature range, three steps of degradation are observed. The first, between 250 and 400°C, is attributed to deacetylation of the EVA matrix, leading to the formation of unsaturated carbon–carbon bonds along the polymer chain. At the same time, thermal degradation of the ammonium polyphosphate begins. In the temperature range 420 to 480°C, two degradation steps overlap. Those two degradation steps lead to the formation of 35 wt% carbonaceous residues for EVA–APP/PA6 and EVA–APP/PA6–OLDH and 40 wt% for EVA–APP/PA6–OMMT and EVA–APP/PA6–NPSi. They result from phosphorylation reactions between the polymer or its degradation products (EVA and PA6) and the additives leading to the formation of a phosphocarbonaceous structure, as shown earlier (for OLDH and NPSi we assume a similar mechanism). The amount of residue for EVA–APP/PA6–OMMT and EVA–APP/PA6–NPSi is higher than the mineral content added in the formulation (2.2 wt% of OMMT and 3.2 wt% of NPSi). As a consequence, it may be assumed that the particles play a role in degradation of the formulation that leads to thermal stabilization of the systems.

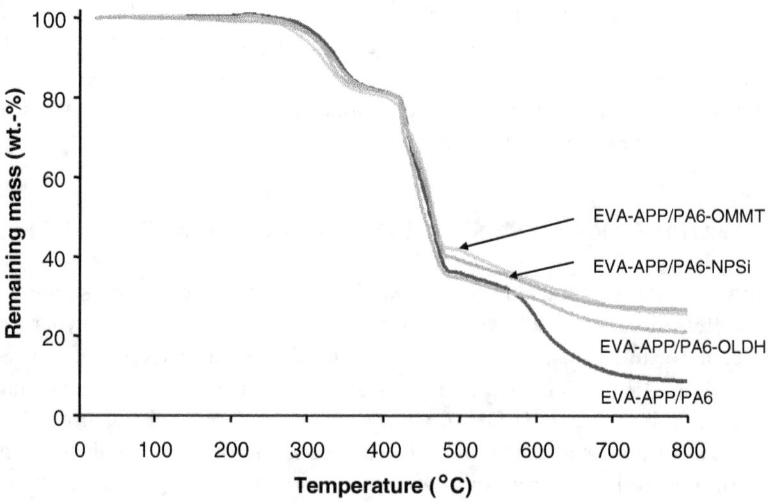

FIGURE 6.12 Thermogravimetric curves of intumescent formulations containing various nanofillers (airflow, 10°C/min).

TABLE 6.5 Cone Calorimeter Data of the Intumescent Formulations

	EVA–APP/PA6	EVA–APP/ PA6–OMMT	EVA–APP/ PA6–OLDH	EVA–APP/ PA6–NPSi
Peak HRR, t_1 (kW/m^2)	267	270	233	336
Peak HRR, t_2 (kW/m^2)	299	202	284	261
Total heat release (MJ/m^2)	68	69	74	68
Time to ignition (s)	36	76	44	61
Total CO emission (kg/kg)	0.04	0.03	0.03	0.03
Total CO_2 emission (kg/kg)	2.0	2.2	2.1	1.8
Total smoke release (m^2/m^2)	1422	1392	1367	1407

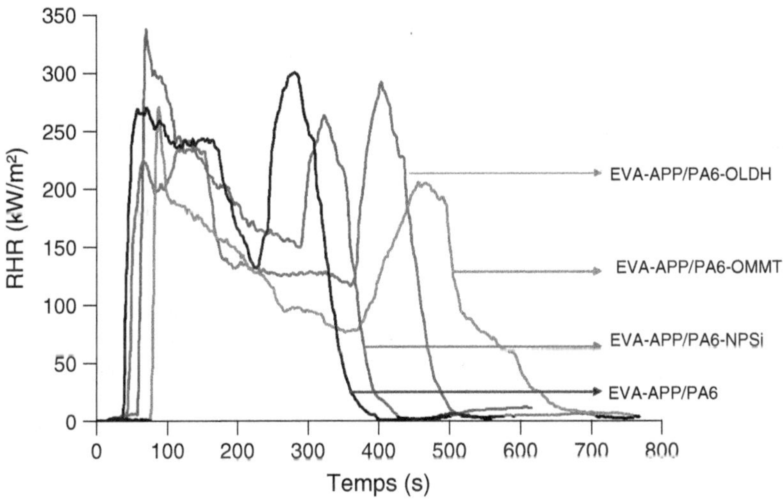

FIGURE 6.13 Heat release rate of intumescent formulations containing various nanofillers (external heat flux = 50 kW/m^2).

In the high-temperature range 480 to 800°C for EVA–APP/PA6–OMMT and EVA–APP/PA6–NPSi and between 600 and 800°C for EVA–APP/PA6–OLDH, an important stabilization of the system is observed (between 15 and 20 wt% compared to EVA–APP/PA6). This cannot be attributed to the mineral content. A reaction takes place between the mineral particles and the intumescent system. In this range of temperature, it may be assumed that a ceramiclike structure is formed from a reaction between the mineral (e.g., SiO_2, Al_2O_3, MgO) and the phosphate, as mentioned in Section 6.4.

The flame retardancy performance of intumescent formulations is evaluated by cone calorimetry (Table 6.5 and Figure 6.13). For each sample, an intumescence phenomenon is observed. HRR curves exhibit two peaks: the first (t_1) before 200 s and a second (t_2) between 300 s (EVA–APP/PA6) and 500 s (EVA–APP/PA6–OMMT); this behavior is typical of intumescent systems.[57]

The first peak is attributed to formation of a intumescent protective shield that leads to a decrease in heat and mass transfer between the flame and the material. When the intumescent shield is formed, HRR decreases and a pseudoplateau is observed. The second peak corresponds to destruction of the intumescent layer, leading to a sharp evolution of flammable gases: The higher the time for the second peak, the higher the thermal and mechanical stability of the intumescent shield. Then a thermally stable residue is formed (about 30 wt% of the initial mass of the sample).

When OMMT is incorporated in the intumescent system, the first peak HRR is narrower. However, its value is similar to that of the formulation without nanofiller. An increase in the time to ignition is observed. The second peak HRR is sharply reduced (decreased by 30% compared to the formulation without nanofiller) and appears around 200 s later. It suggests that formation of the intumescent shield is modified when OMMT is added in EVA–APP/PA6 and that the intumescent shield might be thermally and/or mechanically more stable. The formulation containing OLDH exhibits a decrease in the first peak HRR (decrease by 12% compared to the formulation without nanofiller), and the second HRR peak is delayed, appearing 150 s after that of EVA6–APP/PA6. The other parameters are not affected by the presence of the mineral filler. Finally, the addition of silica to the intumescent system leads to a sharp increase in the first peak HRR (26% increase compared to the formulation without nanofiller), showing a decrease in performance when using this mineral. The delay in the second peak HRR when inorganic particles are used in the intumescent system suggests that the thermal (as demonstrated in the high-temperature range during TGA experiments) and mechanical stability of the intumescent shield should be increased. Note that the dispersion of all fillers is reasonable.

The performance of intumescent formulations according to LOI and UL-94 standards is reported in Table 6.6. An intumescent phenomenon is observed whatever the materials. LOI values of the intumescent systems containing OMMT or OLDH are higher than that of the reference (EVA–APP/PA6). A UL-94 V-0 rating can be achieved using OMMT, and LOI increases from 28 to 32 vol% compared to the formulation without nanofiller. This confirms the results of Section

TABLE 6.6 Fire Retardant Performance of EVA–APP/PA6 Containing OMMT, OLDH, and NPSi According to LOI and UL-94 Tests

	EVA–APP/PA6	EVA–APP/ PA6–OMMT	EVA–APP/ PA6–OLDH	EVA–APP/ PA6–NPSi
Maximum after flame time (s)	60	4	28	>60
Total after flame time for five specimens (s)	126	15	60	—
Flaming drops	Yes	No	Yes	Yes
Time before first drop (s)	1	—	5	20
UL-94 rating	NC	V-0	V-2	NC
LOI (vol%)	28	32	29	26

6.4 even if the nanocomposite was prepared in a different way. In this case, the nanocomposite was prepared by melt blending in a Brabender mixer (dispersion of OMMT in EVA and PA6), whereas in Section 6.4, PA6nano was synthesized in situand further incorporated as an ingredient. OMMT therefore plays its role of synergist whatever the method of preparation.

Fire retardant performance is lower when LDH rather than OMMT is used. An explanation might be the different thermal stability between the two fillers, but further investigations are in progress to explain the difference. The addition of NPSi to an intumescent system leads to a dramatic decrease in performance. This result is the opposite of that in a study by Wei et al.,[60] in which a synergistic effect is observed for low silica loading (1 to 4 wt%) while an antagonist effect is observed for high loading (>6 wt%). However, in this study the carbon source was pentaerythritol, and the total additive amount was lower. Under the conditions of UL-94 testing, the sample burns totally after the first flame application, while for the formulation without nanofiller, the sample never burns totally, but because of a long combustion time, nonclassification (NC) is achieved. The change in the flammability properties of the intumescent system by addition of silica (NPSi) or of layered nanoparticles (OMMT and OLDH) can be explained partially by the viscoelastic properties of the char. The addition of mineral particles in EVA increases the viscosity of the material, as suggested by the increase in time before the first droplet observed during UL-94 tests (Table 6.6) and visual observation of the sample after UL-94 tests (Figure 6.14).

Viscosity is a key parameter of intumescence because the intumescence coating must adhere to the substrate and not drip. At the same time, the expansion (swelling) of the char has to be high enough to limit heat transfer (formation of a low-density foamed char), and the char strength has to accommodate stresses coming from the flame and internal pressure. The addition of nanofiller increases the viscosity of the char, but the expansion and the char strength should be

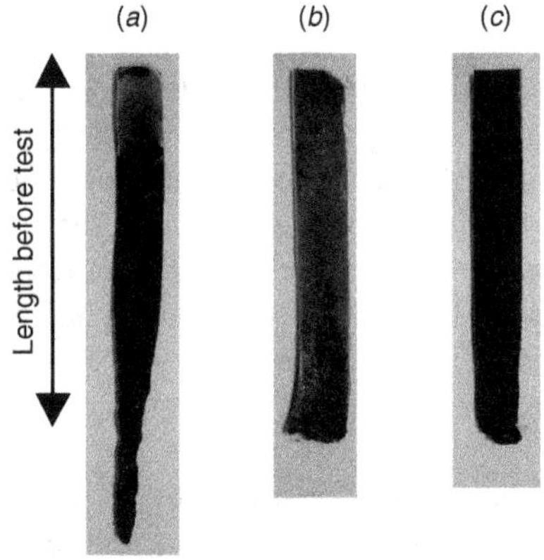

FIGURE 6.14 EVA–APP/PA6 (*a*), EVA–APP/PA6–OMMT (*b*), and EVA–APP/PA6 –OLDH (*c*) after UL-94 tests. (Note that swelling is not very high but real intumescent behavior was observed.)

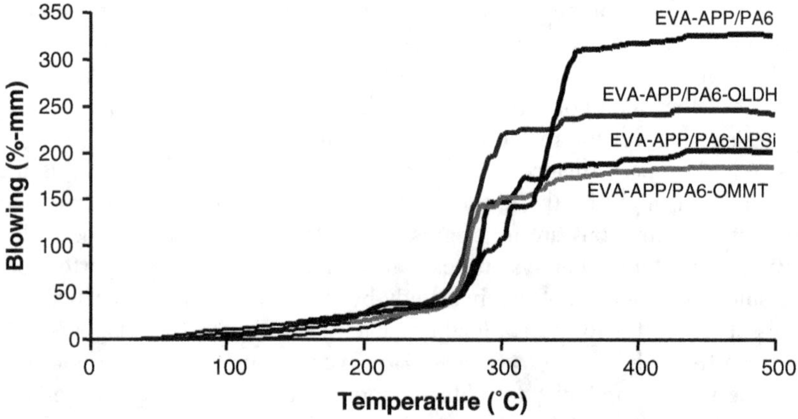

FIGURE 6.15 Expansion as a function of temperature of intumescent formulations. (Measurements were made with a rheometer in a parallel-plate configuration measuring the distance between the two plates when intumescent swelling pushed up the upper plate.)

measured to have a complete understanding. In our laboratory we have developed novel experiments to measure char strength and expansion of intumescent materials as a function of temperature using a high-temperature rheometer.[61] The expansion of the intumescent materials starts at 250°C for all intumescents and reaches 325% for the formulation without a nanofiller (Figure 6.15). It is lower for the formulations containing nanofillers and lies between 175% for EVA–APP/PA6–OMMT and 250% for EVA–APP/PA6–OLDH, with a value of 200% for EVA–APP/PA6–NPSi. The intumescent barrier begins to swell at 250°C and reaches its maximum expansion at 350°C for the formulation without a nanofiller, and at a lower temperature, 300°C, for the others. All expanded chars are stable up to 500°C. This result suggests that a certain level of expansion is required to get low flammability, but that the highest expansion does not provide the highest efficiency.

Char strength is a crucial parameter because it has to accommodate internal pressure and external stresses to avoid the formation of cracks. Figure 6.16 describes the evolution of the force applied on the top of the intumescent char as a function of the gap between the rheometer plates (Figure 6.16*a*). EVA–APP/PA6 –OMMT exhibits better char resistance than the other formulations, those of EVA–APP/PA6–NPSi and EVA–APP/PA6 are significantly lower, and EVA –APP/PA6–OLDH exhibits intermediate behavior. This suggests that high char strength should be required to get the best performance associated with reasonable expansion.

The expansion is related to the formation of an expanded foamed material with the role of limiting heat transfer, and the char strength permits cohesion of an insulative coating on the substrate and avoids the formation of cracks. The better flammability properties of formulations containing a nanofiller might be explained on the basis of the particular chemistry of the system. Reaction between

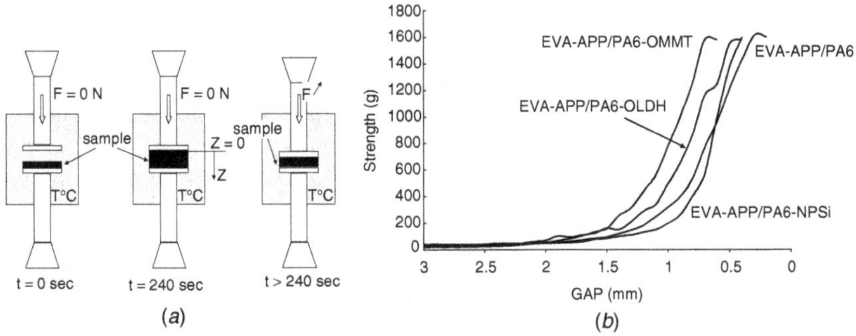

FIGURE 6.16 (*a*) Typical experiment for measuring the resistance of an intumescent char (sample ($h = 1$ mm), 10°C/min from 20 to 500°C, strength = 0, no strain). The upper plate is put in contact with the material and goes down linearly (0.02 mm/s) after reaching 500°C. The force is followed versus the distance between the two plates (gap). (*b*) Gap as a function of temperature of the intumescent formulations.

the phosphate and the nanofiller (aluminosilicate, silicate, and potentially other reactive particles) permits the stabilization of phosphate species at high temperatures as well as the intumescent char. It is not clear if the nanodispersion is absolutely necessary to obtain the best performance because our studies on intumescent systems combining zeolite (microdispersion) and organo-modified clay (nano-dispersion) show similar results in terms of fire performance. According to our results, nano-dispersion is essential only to get enhanced mechanical properties and/or to design further multifunctional materials. Further investigations on this topic are necessary to confirm the results.

6.6 CRITICAL OVERVIEW OF RECENT ADVANCES

Large synergistic effects are observed when nanofillers are incorporated in intumescent formulations. The presence of nanofillers modifies the chemical (reactivity of the nanofiller versus the ingredients of the intumescent system) and physical (expansion, char strength, and thermophysical properties) behavior of an intumescent char when undergoing flame or heat flux leading to enhanced performance. In a recent paper, Lewin describes this phenomenon as a catalytic effect.[62,63] It is noteworthy that the catalyst (the nanofiller) is a crucial ingredient (reactant) in the development of intumescence, forming additional species stabilizing the structure and enhancing the rheological behavior. The nanofiller is incorporated at amounts as low as 1 wt% (sometimes less, as in the case of the incorporation of nanoparticles of copper at an amount as low as 0.1 wt% in epoxy resin containing APP[64]), and it permits the formation of active species selecting chemical reactions in the condensed phase and yielding to char with the dynamic properties of interest. These are the typical roles of a catalyst. So we may speak of the catalytic mechanism or catalyst, but we must remember that the catalyst is also an essential reactant and reacts with the other ingredients of the formulation to form the final intumescent char. Of course, this statement cannot be extended to all nanocomposite systems, especially those composed of only polymer and clay, in which the mechanism of action is the accumulation of clay platelets, which form a protective barrier at the surface of the material.[6]

Many papers have been published on intumescent polypropylene because intumescent systems are well adapted to processing temperatures and are efficient (see Section 6.2).[21–23,26,27,31,43,60] Marosi et al.[65] combined an APP-based intumescent system in PP with OMMT and borosiloxane elastomer. The addition of a small amount of OMMT (1 wt%) in the formulation permits one to achieve a V-0 rating in the UL-94 protocol, but the increase in LOI is only 2 points. The authors also developed a novel approach using borosiloxane as a carrier of flame retardant and ceramic precursor. The combination of OMMT with borosiloxane increases the LOI by 8 points compared to the reference, and a V-0 rating is still achieved. These beneficial effects are confirmed by cone calorimetry. They explain the improvement in flame retardancy by the increase in viscosity under fire conditions, permitting to achieve a V-0 rating (addition of OMMT), which is consistent with our results in EVA–APP/PA6. No comment was offered on the potential reaction between APP and OMMT, as we have suggested above. A possible role for nanofillers in controlling the activity of flame retardants was also suggested: The flame retardant might promote exfoliation of the clay at the earliest stage of degradation and provide the first protective barrier (the concept of expandable nanocomposite).[66]. A similar explanation is reported with borosiloxane; here its main benefit is to make a flexible char rather than a fragile charred structure (Figure 6.17). It is claimed that borosiloxane-coated OMMT acts as a carrier of OMMT and delivers OMMT at the surface of the char, creating additional protection. No evidence of a reaction between APP and/or borosiloxane and/or OMMT is detected by heat treatment at 250 and 300°C. According to our previous work on intumescent systems,[48] phosphosilicate is formed only above 350°C, which may be a why it was not detected by Marosi et al. Our understanding is that borosiloxane should act as a carrier in the same

(a) (b)

FIGURE 6.17 Intumescent chars formed under the conditions of a cone calorimeter (*a*) without and (*b*) with borosiloxane. (From Ref. 65, copyright © 2003, Elsevier, with permission.)

way that zeolite forms phosphosilicate and probably borophosphate,[28] reinforcing the efficiency of the intumescent structure.

Tang et al.[67,68] also examined the incorporation of MMT in intumescent PP with a compatibilizer which is commonly used as a surfactant for making OMMT; evidence for nanocomposite formation is shown with and without the intumescent system. Cone calorimetry shows a large improvement in the flammability properties when using OMMT, similar to the results shown in Section 6.5. The mechanism of action postulated suggests the formation of an aluminophosphate structure, but no evidence was given. The same group[69] investigated the use EVA–PA6nano as a char former in a PP intumescent system,[52], as we have done in EVA.[32]. The benefit of using OMMT is once again proven by cone calorimetry.

Wilkie et al. prepared vinyl ester (PVE) nanocomposites using different OMMTs and POSSs.[70] As expected, significant reduction in peak HRR was observed. The goal was to strongly reduce the flammability of PVE. With the nanocomposite approach, the reduction was not enough for military applications on ships. They added phosphorus-containing flame retardants such as tricresylphosphate (TCP) and resorcinol diphosphate (RDP), selected using high-throughput techniques.[71] It is not mentioned in the paper whether the samples exhibit intumescent behavior, but according to the chemical nature of PVE and TGA results suggesting a condensed-phase mechanism, we may assume that there is formation of charred protective materials under the conditions of fire. Synergy between phosphorus-containing fire retardants and PVE nanocomposites (OMMT and POSS) is shown through cone calorimetry by reductions in the peak heat release rate, total heat release, and mass loss rate; there is no improvement in time to ignition. With this resin, the type of clay used showed different effects on the flammability of the nanocomposites formed. As far as we know, this is the first paper demonstrating the advantage of using POSS as a synergist. In this study, no mechanism of action is postulated, but we may assume that interactions should take place between the synergist (OMMT or POSS) and the phosphate that enhance the properties of the char.

One of the main applications of intumescent systems is as a coating to protect steel against collapsing in the case of fire.[61] Wang et al.[72] used OLDH as a nanofiller combined with an intumescent system in acrylate resin. The intumescent paints were evaluated on steel plate, measuring the temperature on the back side of the steel plate as a function of time as the temperature increased (standard ISO 834) (Figure 6.18). It can be observed that with the incorporation of 1.5% OLDH, the time to reach 300°C increases to 100 min, compared to 60 min without OLDH (virgin coating). It is also noteworthy that the thickness of the char layer of a formulation with 1.5% OLDH is similar to that of a formulation without. This confirms our results suggesting that the highest expansion is not necessary to get the best performance. The improvement in performance is partially explained by the char strength and the specific heat of the char layer.

The morphology of the char layer exhibits interesting features (Figure 6.19). Holes can be distinguished on the two pictures, but the diameters of holes in the char containing OLDH are much smaller (10 to 30 μm) than those of the

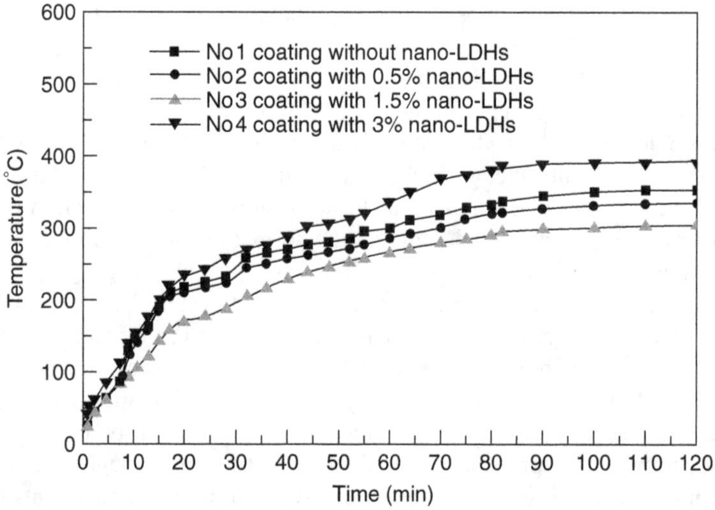

FIGURE 6.18 Temperature curves as a function of time for an intumescent coating containing different amounts of OLDH. (From Ref. 72, copyright © 2005, Elsevier, with permission.)

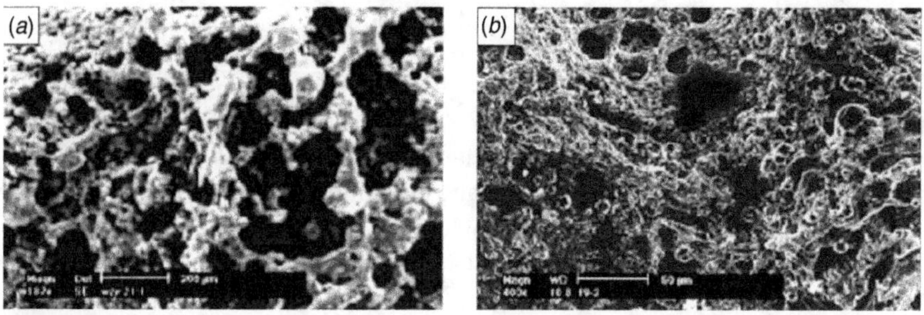

FIGURE 6.19 SEM image of the inner surface of a char layer of a formulation (*a*) without OLDH and (*b*) with 1.5 wt% OLDH. (From Ref. 72, copyright © 2005, Elsevier, with permission.)

char without OLDH. Wang suggests that small holes reinforce the char strength and avoid the formation of cracks at the surface of the char. Indeed, the formation of cells in the char structure, as in an evenly dispersed foam, reduces heat transfer and increases the efficiency of the char. When the cells are too large, the char strength is reduced and cracks can appear. The mechanism postulated is that OLDH catalyzes the esterification between the phosphate and the char former (polyol), but no evidence is given for this. From our previous work,[73] we know that metal hydroxides can also react with phosphate, and the formation of aluminophosphate and magnesium phosphate might be responsible for the development of an enhanced intumescent structure.

Intumescence also has a role in the flame retardancy of textiles.[74,75] Recently, Horrocks et al. investigated the inclusion of nanoparticles in combination with intumescent flame retardants in polyamide-6 and polyamide-6,6.[75,76] They suggest that the incorporation of nanoclay does not necessarily increase the LOI value when evaluated as film samples, but they showed the benefit of using nanoclay in polyamides when combined with APP-based formulations. The efficiency is increased by an average factor of 2.

From the literature, we demonstrated earlier that nanofillers are synergists for intumescent formulations. In this section we focused on the flammability of the materials—there was no mention of other properties. Nevertheless, we can expect enhancement in different ways to design multifunctional materials.

6.7 SUMMARY AND CONCLUSION

In this chapter we considered recent developments that have tried to increase the efficiency of intumescent system for polymeric materials using nanofillers, including organo-modified clays, layered double hydroxides, polyhedral silsesquioxane, and nanoparticles of silica as synergists. Intumescent nanocomposites exhibit superior flammability properties as well as enhancing properties, such as mechanical properties.

The mechanism of action is not completely elucidated, but we have identified a reaction that takes place between the nanofiller and the phosphate in order to stabilize the charred structure thermally. This reduces melt dripping and slows degradation of the material and thus the evolution of flammable molecules. This reaction does not significantly modify expansion of the intumescent coating, but it permits reinforcement of the char strength and avoids the formation of cracks. The char looks like a foam with evenly dispersed close-packed cells, ensuring the limitation of heat transfer between the flame and the substrate. Further investigations should study the influence of the chemical reaction and identify species such as alumino- or silicophosphate and their role on the physical properties of the intumescent char. Synergists such as nanofillers for making intumescent nanocomposites offer exceptional promise for making fire-safe polymers that meet legislative requirements. As a bonus, it should enable researchers to succeed in designing efficient multifunctional materials.

REFERENCES

1. Bourbigot, S.; Le Bras, M.; Troitzsch, J. Fundamentals: introduction, in: J. Troitzsch, Ed., *Flammability Handbook*. Care Hanser Verlag, Munich, Germany, 2003, pp. 3–7.
2. Lewin, M. Physical and chemical mechanisms of flame retarding of polymers, in: M. Le Bras, G. Camino, S. Bourbigot, and R. Delobel, Eds. *Fire Retardancy of Polymers: The Use of Intumescence*. Royal Chemical Society, Cambridge; England, 1998, pp. 3–32.
3. Gilman, J.W.; Kashiwagi, T.; Lichtenhan, J.D. Nanocomposites: a revolutionary new flame retardant approach. *SAMPE J.* **1997**, 33, 40–46.

4. Gilman, J.W. Flammability and thermal stability studies of polymer layered-silicate (clay) nanocomposites. *Appl. Clay Sci.* **1999**, 15(1–2), 31–49.
5. Gilman, J.W.; Kashiwagi, T.; Giannelis, E.P.; Manias, E.; Lomakin, S.; Lichtenhan, J.D.; Jones, P. Flammability studies of polymer layered silicate nanocomposites, in: M. Le Bras, G. Camino, S. Bourbigot, and R. Delobel, Eds., *Fire Retardancy of Polymers: The Use of Intumescence.* Royal Society, of Chemistry, London, 1998, pp. 203–221.
6. Zhu, J; Morgan, A.B.; Lamelas, F.J.; Wilkie, C.A. Fire properties of polystyrene–clay nanocomposites. *Chem. Mater.* **2001**, 13, 3774–3780.
7. Qin, H.; Su, Q.; Zhang, S.; Zhaoa, B.; Yan, M. Thermal stability and flammability of polyamide 66/montmorillonite nanocomposites. *Polymer* **2003**, 44, 7533–7538.
8. Bourbigot, S.; VanderHart, D.L.; Gilman, J.W.; Bellayer, S.; Stretz, H.; Paul, D.L. Solid state NMR characterization and flammability of styrene–acrylonitrile copolymer montmorillonite nanocomposite. *Polymer* **2004**, 45(22), 7627–7638.
9. Song, L.; Hu, Y.; Tang, Y.; Zhang, R.; Chena, Z.; Fa, W. Study on the properties of flame retardant polyurethane/organoclay nanocomposite. *Polym. Degrad. Stab.* **2005**, 87, 111–116.
10. Laachachi, A.; Leroy, E.; Cochez, M.; Ferriol, M.; Lopez Cuesta, J.M. Use of oxide nanoparticles and organoclays to improve thermal stability and fire retardancy of poly(methyl methacrylate). *Polym. Degrad. Stab.* **2005**, 89, 344–352.
11. Kashiwagi, T.; Morgan, A.B.; Antonucci, J.M.; VanLandingham, M.R.; Harris, R.H., Jr.; Awad, W.H.; Shields, J.R., Thermal and flammability properties of a silica–poly-(methylmethacrylate) nanocomposite. *J. Appl. Polym. Sci.* **2003**, 89(8), 2072–2078.
12. Lefebvre, J.; Le Bras, M.; Bourbigot, S. Lamellar double hydroxides/polymer nanocomposites: a new class of flame retardant material, in: M. Le Bras, C.A. Wilkie, S. Bourbigot, S. Duquesne, and C. Jama, Eds., *Fire Retardancy of Polymers: New Applications of Mineral Fillers.* Royal Society of Chemistry, London, 2005, pp. 42–53.
13. Zammarano, M.; Gilman, J.W.; Franceschi, M.; Meriani, S. In: M. Lewin, Ed., *Proceedings of the 16th BCC Conference on Flame Retardancy*, Business Communications Co., Norwalk, CT, 2005.
14. Kashiwagi, T.; Grulke, E.; Hilding, J.; Groth, K.; Harris, R.; Butler, K.; Shields, J.; Kharchenko, S.; Douglas, J. Thermal and flammability properties of polypropylene/carbon nanotube nanocomposites. *Polymer* **2004**, 45, 4227–4239.
15. Kashiwagi, T.; Du, F.; Winey, K.I.; Groth, K.M.; Shields, J.R.; Bellayer, S.P.; Kim, H.; Douglas, J.F. Flammability properties of polymer nanocomposites with single-walled carbon nanotubes: effects of nanotube dispersion and concentration. *Polymer* **2005**, 46, 471–481.
16. Devaux, E.; Bourbigot, S.; El Achari, A. Crystallisation behaviour of PA-6 clay hybrid nanocomposite. *J. Appl. Polym. Sci.* **2002**, 86, 2416–2423.
17. Bourbigot, S.; Le Bras, M.; Flambard, X.; Rochery, M.; Devaux, E.; Lichtenhan, J. Polyhedral oligomeric silsesquioxanes: application to flame retardant textile, in: M. Le Bras, C.A. Wilkie, S. Bourbigot, S. Duquesne, and C. Jama, Eds., *Fire Retardancy of Polymers: New Applications of Mineral Fillers*. Royal Society of Chemistry, London, 2005, pp. 189–201.

18. Jash, P.; Wilkie, C.A. Effects of surfactants on the thermal and fire properties of poly(methyl methacrylate)/clay nanocomposites. *Polym. Degrad. Stab.* **2005**, 88, 401–406.

19. Jama, C.; Quédé, A.; Goudmand, P.; Dessaux, O.; Le Bras, M.; Delobel, R.; Bourbigot, S.; Gilman, J.W.; Kashiwagi, T. Fire retardancy performance and thermal stability of materials coated by organosilicon thin films using a cold remote plasma process, in: G.L. Nelson and C.A. Wilkie, Eds. *Fire and Polymers: Materials and Solutions for Hazard Prevention.* American Chemical Society, Washington, DC, 2001, pp. 200–213.

20. Vandersall, H.L. Intumescent coating systems, their development and chemistry. *J. Fire Flammability* **1971**, 2, 97–140.

21. Camino, G.; Costa, L.; Trossarelli, L. Study of the mechanism of intumescence in fire retardant polymers; V: Mechanism of formation of gaseous products in the thermal degradation of ammonium polyphosphate. *Polym. Degrad. Stab.* **1985**, 12, 203–211.

22. Delobel, R.; Le Bras, M.; Ouassou, N.; Alistiqsa, F. Thermal behaviors of ammonium polyphosphate–pentaerythritol and ammonium pyrophosphate–pentaerythritol intumescent additives in polypropylene formulations. *J. Fire Sci.* **1990**, 8(2), 85–92.

23. Bourbigot, S.; Le Bras, M.; Delobel, R. Fire degradation of an intumescent flame retardant polypropylene. *J. Fire Sci.* **1995**, 13(1–2), 3–9.

24. Kandola, B.K.; Horrocks, A.R.; Myler, P.; Blair, D. The effect of intumescents on the burning behaviour of polyester-resin-containing composites. *Composites A Appl. Sci. Manuf.* **2002**, 33A(6), 805–817.

25. Zhang, S.; Horrocks, A.R. Substantive intumescence from phosphorylated 1,3-propanediol derivatives substituted on to cellulose. *J. Appl. Polym. Sci.* **2003**, 90(12), 3165–3172.

26. Jenewein, E.; Pirig, W.D. (Hoechst A.G.), *Eur. Patent* EP 0 735 119 A1, 1996.

27. Morice, L.; Bourbigot, S.; Leroy, J.M. Heat transfer study of polypropylene-based intumescent systems during combustion. *J. Fire Sci.* **1997**, 15, 358–374.

28. Jimenez, M.; Duquesne, S.; Bourbigot, S. Multiscale experimental approach for developing high-performance intumescent coatings.*Ind. Eng. Chem. Res.* **2006**, 45(13), 4500–4508.

29. Butler, K.M. Physical modeling of intumescent fire retardant polymers, in: K.C. Khemani, Ed., *Polymeric Foams: Science and Technology.* ACS Symposium Series, Vol. 669, American Chemical Society, Washington, DC, 1997, pp. 214–230.

30. Le Bras, M.; Bourbigot, S.; Félix, E.; Pouille, F.; Siat, C.; Traisnel, M. Characterization of polyamide-6–based intumescent additive for thermoplastic formulations. *Polymer* **2000**, 41(14), 5283–5298.

31. Bugajny, M.; Le Bras, M.; Bourbigot, S.; Poutch, F.; Lefebvre, J.-M. The use of thermoplastic polyurethanes as carbonization agents in intumescent blends, 1: Fire retardancy of polypropylene/thermoplastic polyurethane/ammonium polyphosphate blends. *J. Fire Sci.* **1999**, 17(11–12), 494–513.

32. Bourbigot, S.; Le Bras, M.; Dabrowski, F.; Gilman, J.W.; Kashiwagi, T. PA-6 clay nanocomposite hybrid as char forming agent in intumescent formulations. *Fire Mater.* **2000**, 24, 201–208.

33. Le Bras, M.; Bourbigot, S.; Delporte, C.; Siat, C.; Le Tallec, Y. New intumescent formulations of fire retardant polypropylene: discussion about the free radicals mechanism of the formation of the carbonaceous protective material during the thermo-oxidative treatment of the additives. *Fire Mater.* **1996**, 20, 191–202.
34. Bourbigot, S.; Duquesne, S.; Leroy, J.-M. Modeling of heat transfer study of polypropylene-based intumescent systems during combustion. *J. Fire Sci.* **1999**, 17(1), 42–49.
35. Bourbigot, S.; Le Bras, M.; Delobel, R.; Bréant, P.; Trémillon, J.-M. Carbonisation mechanisms resulting from Intumescence, II: Association with an ethylene terpolymer and the ammonium polyphosphate–pentaerythritol fire retardant system. *Carbon* **1995**, 33(3), 283–289.
36. Duquesne, S.; Le Bras, M.; Delobel, R. Visco-elastic behaviour of intumescent systems, in: M. Lewin, Ed., *Proceedings of the Conference on Recent Advances in Flame Retardancy of Polymeric Materials*, Norwalk, CT, 2002, pp. 50–64.
37. Bourbigot, S.; Le Bras, M.; Delobel, R.; Bréant, P.; Trémillon, J.-M. 4A zeolite synergistic agent in new flame retardant intumescent formulations of polyethylenic polymers: study of the constituent monomers. *Polym. Degrad. Stab.* **1996**, 54, 275–283.
38. Bourbigot, S.; Le Bras, M.; Trémillon, J.-M.; Bréant, P.; Delobel, R. Zeolites: new synergistic agents for intumescent thermoplastic formulations—criteria for the choice of the zeolite. *Fire Mater.* **1996**, 20, 145–158.
39. Le Bras, M.; Bourbigot, S. Mineral fillers in intumescent fire retardant formulations: criteria for the choice of a natural clay, filler for the ammonium polyphosphate/pentaerythritol. *Fire Mater.* **1996**, 20, 39–49.
40. Bourbigot, S.; Fontaine, G.; Duquesne, S.; Delobel, R. Synthesis and evaluation of new flame retardant phosphorus agents, in: M. Lewin, Ed., *Proceedings of the 18th BBC Conference on Recent Advances in Flame Retardancy of Polymeric Materials.* Business Communications Co., Stamford, CT, 2007.
41. Levchik, S.V.; Levchik, G.F.; Camino, G.; Costa, L. Mechanism of action of phosphorus-based flame retardants: II-Ammonium polyphosphate/talc. *J. Fire Sci.* **1995**, 13, 43–58.
42. Levchik, S.V.; Levchik, G.F.; Camino, G.; Costa, L.; Lesnikovich, A.I. Mechanism of action of phosphorus-based flame retardants in nylon 6, III: Ammonium polyphosphate/manganese dioxide. *Fire Mater.* **1996**, 20, 183–190.
43. Marosi, Gy.; Bertalan, Gy.; Anna, P.; Ravadits, I.; Bourbigot, S.; Le Bras, M.; Delobel, R. In: M. Lewin, Ed., *Recent Advances in Flame Retardancy of Polymeric Materials XI.* Business Communications Co., Norwalk, CT, 2000, pp. 115–128.
44. Anna, P.; Marosi, Gy.; Csantos, I.; Bourbigot, S.; Le Bras, M.; Delobel, R. Intumescent flame retardant systems of modified rheology. *Polym. Degrad. Stab.* **2002**, 77, 243.
45. Smith, J.V. Origin and structure of zeolites, In: J.A. Rabo, Ed., *Zeolite Chemistry and Catalysis.* Monograph 171. American Chemical Society, Washington, DC, 1976.
46. Bourbigot, S.; Le Bras, M.; Delobel, R. Carbonisation mechanisms resulting from intumescence: association with the ammonium polyphosphate–pentaerythritol fire retardant system. *Carbon* **1993**, 31(8), 1219–1226.
47. Bourbigot, S.; Le Bras, M.; Gengembre, L.; Delobel, R. XPS study of an intumescent coating: application to the ammonium polyphosphate/pentaerythritol fire-retardant system. *Appl. Surf. Sci.* **1994**, 81, 283–290.

48. Bourbigot, S.; Le Bras, M.; Delobel, R.; Décressain, R.; Amoureux, J.P. Synergistic effect of zeolite in an intumescent process: study of the carbonaceous structures using solid state NMR. *J. Chem. Soc. Faraday Trans.* **1996**, 92(1), 149–161.

49. Bourbigot, S.; Le Bras, M.; Delobel, R.; Trémillon, J.-M. Synergistic effect of zeolite in an intumescent process: study of the interactions between the polymer and the additives. *J. Chem. Soc. Faraday Trans.* **1996**, 92(18), 3435–3449.

50. Bourbigot, S.; Le Bras, M.; Gengembre, L.; Delobel, R. XPS study of an intumescent coating, II: Application to the ammonium polyphosphate/pentaerythritol/ethylenic terpolymer fire-retardant system with and without synergistic agent. *Appl. Surf. Sci.* **1997**, 120, 15–28.

51. Le Bras, M.; Bourbigot, S.; Le Tallec, Y.; Laureyns, J. Synergy in intumescence: application to β-cyclodextrin carbonisation agent in intumescent additives for fire retardant polyethylene formulations. *Polym. Degrad Stab.* **1997**, 56, 11–24.

52. Bourbigot, S.; Le Bras, M.; Siat, C. Use of polymer blends in flame retardancy of thermoplastic polymers, in: M., Lewin, Ed., *Recent Advances in Flame Retardancy of Polymeric Materials*, Vol. 7. Business Communications Co., Norwalk, CT, 1997, pp. 146–160.

53. Siat, C.; Bourbigot, S.; Le Bras, M. Combustion behaviour of ethylene–vinyl acetate copolymer-based intumescent formulations using oxygen consumption calorimetry. *Fire Mater.* **1998**, 22, 119–129.

54. Bugajny, M.; Le Bras, M.; Bourbigot, S. The use of thermoplastic polyurethanes as carbonization agents in intumescent blends, 2: Thermal behavior of a polypropylene/thermoplastic polyurethane/ammonium polyphosphate blend. *J. Fire Sci.* **2000**, 18(1–2), 7–27.

55. Le Bras, M.; Bugajny, M.; Lefebvre, J.-M.; Bourbigot, S. Use of polyurethanes as char forming agents in polypropylene intumescent formulations. *Polym. Int.* **2000**, 49, 1–10.

56. Almeras, X.; Dabrowski, F.; Le Bras, M.; Delobel, R.; Bourbigot, S.; Marosi, Gy.; Anna, P. Using polyamide-6 as charring agent in intumescent polypropylene formulations, I: Effect of the compatibilising agent on the fire retardancy performance. *Polym. Degrad Stab.* **2002**, 77, 205–214.

57. Le Bras, M.; Bourbigot, S.; Revel, B. Comprehensive study of the degradation of an intumescent EVA-based material during combustion. *J. Mater. Sci.* **1999**, 34, 5777–5789.

58. Duquesne, S.; Jama, C.; Delobel, R.; Le Bras, M. The use of nanocomposites in intumescent EVA, in: S. Finnegan, Ed., *Proceedings of Additives*. ECM, San Francisco, CA, 2003.

59. Miyata, S. Anion-exchange properties of hydrotalcite-like compounds. *Clays Clay Miner.* **1983**, 31, 305–313.

60. Wei, P.; Hao, J.; Du, J.; Han, Z.; Wang, J. An investigation on synergism of an intumescent flame retardant based on silica and alumina. *J. Fire Sci.* **2003**, 21, 17–28.

61. Jimenez, M.; Duquesne, S.; Bourbigot, S. Increasing intumescence efficiency by a systematic approach, in: M. Lewin, Ed., *Proceedings of the 16th BCC Conference on Flame Retardancy*, Business Communications Co., Norwalk, CT, 2005.

62. Lewin, M. Unsolved problems and unanswered questions in flame retardance of polymers. *Polym. Degrad Stab.* **2005**, 88, 13–19.

63. Miyata, S. U.S. Patent 5401442, 1995.

64. Antonov, A.; Yablokova, M.; Costa, L.; Balabanovich, A.; Levchik, G.; Levchik, S. The effect of nanometals on the flammability and thermooxidative degradation of polymer materials. *Mol. Cryst. Liquid Cryst. Sci. Technol. A Mol. Cryst. Liquid Cryst.* **2000**, 353, 203–210.

65. Marosi, G.; Marton, A.; Szep, A.; Csontos, I.; Keszei, S.; Zimonyi, E.; Toth, A.; Almeras, X.; Le Bras, M. Fire retardancy effect of migration in polypropylene nanocomposites induced by modified interlayer. *Polym. Degrad. Stab.* **2003**, 82, 379–385.

66. Marosi, G.; Keszei, S.; Marton, A.; Szep, A.; Le Bras, M.; Delobel, R.; Hornsby, P. Flame retardant mechanisms facilitating safety in transportation, in: M.; Le Bras, C.A.; Wilkie, S.; Bourbigot, S.; Duquesne, and C. Jama, Eds. *Fire Retardancy of Polymers: New Applications of Mineral Fillers*. Royal Society of Chemistry, London, pp. 347–360.

67. Tang, Y.; Hu, Y.; Wang, Y.S.; Gui, Z.; Chen, Z.; Fan, W. Intumescent flame retardant–montmorillonite synergism in polypropylene-layered silicate nanocomposites. *Polym. Int.* **2003**, 52, 1396–1400.

68. Tang, Y.; Hu, Y.; Li, B.; Liu, L.; Wang, Z.; Chen, Z.; Fan, W. Polypropylene/montmorillonite nanocomposites and intumescent, flame-retardant montmorillonite synergism in polypropylene nanocomposites. *J. Polym. Sci. A Polym. Chem.* **2004**, 42, 6163–6173.

69. Tang, Y.; Hu, Y.; Xiao, J.; Wang, J.; Song, L.; Fan, W. PA-6 and EVA alloy/clay nanocomposites as char forming agents in poly(propylene) intumescent formulations. *Polym. Adv. Technol.* **2005**, 16, 338–343.

70. Chigwada, G.; Jash, P.; Jiang, D.D.; Wilkie, C.A. Fire retardancy of vinyl ester nanocomposites: synergy with phosphorus-based fire retardants. *Polym. Degrad. Stab.* **2005**, 89, 85–100.

71. Gilman, J.W.; Bourbigot, S.; Schields, J.R.; Nyden, M.; Kashiwagi, T.; Davis, R.D.; VanderHart, D.L.; Demory, W.; Wilkie, C.A.; Morgan, A.B.; Harris, J.; Lyon, R.E. High throughput methods for polymer nanocomposites research: extrusion, NMR characterization and flammability property screening. *J. Mater. Sci.* **2003**, 38(22), 4451–4460.

72. Wang, Z.; Han E.; Ke, W. Influence of nano-LDHs on char formation and fire-resistant properties of flame-retardant coating. *Prog. Org. Coat.* **2005**, 53, 29–37.

73. Bourbigot, S.; Duquesne, S.; Sébih, Z.; Ségura, S.; Delobel, R. Synergistic aspects of the combination magnesium hydroxide and ammonium polyphosphate of ethylene–vinyl acetate copolymer, in: C.A.; Wilkie, and G.L., Nelson, Eds., *Fire and Polymers IV: Materials and Concepts for Hazard Prevention*. American Chemical Society, Washington DC, 2005, pp. 200–212.

74. Horrocks, A.R. Developments in flame retardants for heat and fire resistant textiles: the role of char formation and intumescence. *Polym. Degrad. Stab.* **1996**, 54, 143–154.

75. Horrocks, A.R.; Kandola, B.; Davies, P.J.; Zhang, S.; Padbury, S.A. Developments in flame retardant textiles: a review. *Polym. Degrad. Stab.* **2005**, 88, 3–12.

76. Padbury, S.A.; Horrocks, A.R.; Kandola, B. The effect of phosphorus-containing flame retardants and nanoclay on the burning behavior of polyamides 6 and 6.6, in: M., Lewin, Ed., *Proceedings of the 16th BCC Conference on Flame Retardancy*, Business Communications Co., Norwalk, CT, 2003.

7

FLAME RETARDANT PROPERTIES OF ORGANOCLAYS AND CARBON NANOTUBES AND THEIR COMBINATIONS WITH ALUMINA TRIHYDRATE

GÜNTER BEYER
Kabelwerk Eupen AG, Eupen, Belgium

7.1 INTRODUCTION

Fire risks and fire hazards are mainly the result of the combination of factors, including ignitability, ease of extinction, flammability of volatiles generated, amount of heat released on burning, rate of heat release, flame spread, smoke obscuration, and smoke toxicity. The most important fire risks and fire hazards are the rates of heat release, smoke production, and toxic gas release.[1] An early high rate of heat release causes fast ignition and flame spread, controls a fire's intensity, and is much more important than ignitability, smoke toxicity, or flame spread. The time for people to escape a fire is also controlled by the heat release rate.[2]

Once a fire starts in a room containing flammable materials, it will generate heat, which can ignite additional combustible materials. As a consequence, the rate at which the fire progresses increases, because more and more heat is released and a progressive increase in room temperature is observed. The radiant heat and the temperature can rise to such an extent that all materials within the room will be ignited easily, resulting in an extremely high rate of fire spread. This point in

Flame Retardant Polymer Nanocomposites, edited by Alexander B. Morgan and Charles A. Wilkie

time, termed *flashover*, leads to a fully developed fire. Escape from the room will then be nearly impossible, and spread of the fire to other rooms is very likely. When a fire goes to flashover, every polymer will release roughly 20% of its weight as carbon monoxide, resulting in too much toxic smoke. Therefore, most people die in big fires, and 90% of fire deaths are the result of fires becoming too big, resulting in too much toxic smoke.[3]

Each year about 5000 people are killed by fires in Europe and more than 4000 people in the United States. Direct property losses by fires in the United States are roughly 0.2% of the gross domestic product, and the total costs of fires are around 1% of the gross domestic product.[4] Therefore, it is important to develop well-designed flame retardant materials to decrease both fire risks and fire hazards.

Polymers are used in more and more fields of applications, and specific mechanical, thermal, and electrical properties are required. One further important property is the flame retardant behavior of polymers, which can be fulfilled traditionally by using intrinsically flame retardant polymers such as poly(vinyl chloride) (PVC) or fluoropolymers and by flame retardants like alumina trihydrate, magnesium hydroxide, organic brominated compounds, or intumescent systems based on nitrogen- or phosphorus-based compounds to prevent the burning of such polymers as polyethylene, polypropylene, or polyamide. These flame retardant systems sometimes exhibit serious disadvantages. Use of alumina trihydrate (ATH) and magnesium hydroxide (MDH) in flame retardant cables requires a very high portion of these fillers for the applied polymers poly(ethylene-co-vinyl acetate) (EVA), polyethylene (PE), or polypropylene (PP); filling levels of more than 60 wt% are necessary to achieve suitable flame retardancy. Clear disadvantages of these filling levels are the high density and the lack of flexibility of end products, the poor mechanical properties, and the problematic compounding and extrusion steps. In Europe, there are, at a minimum, reservations about the general use of brominated compounds as flame retardants. Intumescent systems are relatively expensive, and electrical requirements can restrict the use of these products.

A new class of materials, *nanocomposites*, avoids the disadvantages of traditional flame retardant systems. Generally, the term *nanocomposite* describes a two-phase material with a suitable nanofiller [usually modified layered silicates such as montmorillonite (organoclays) or carbon nanotubes] dispersed in the polymer matrix at the nanometer scale. Compared with pure polymers, the corresponding nanocomposites show tremendous improvements; the content of nanofillers within the polymer matrix is usually between 2 and 10 wt%. The most important improved properties are mechanical properties such as tension, compression, bending, and fracture, barrier properties such as permeability, and solvent resistance, translucence, and ionic conductivity. Ray and Okamoto's review[5] covers these improvements in detail. Other highly interesting properties exhibited by polymer nanocomposites concern their increased thermal stability and flame retardancy at very low filler levels.[6–8] The low filler content in nanocomposites for improved thermal stability is highly attractive for the industry because the end products can be made cheaper and are easier to process.

Several research groups have reported on the preparation and properties of EVA-based nanocomposites. EVA nanocomposites were prepared by Camino et al.[9] in a Brabender mixer and the thermal degradation was investigated. Hu et al.[10] prepared intercalated EVA nanocomposites; only 5% of filler content improved the flame retardancy of the nanocomposites. Camino et al.[11] described the synthesis and thermal behavior of layered EVA nanocomposites; the nanofiller was a synthetic modified fluorohectorite, which is a layered silicate, and protection against thermal oxidation and mass loss was observed in air. The modified silicates accelerated the deacetylation of the polymer but slowed thermal degradation of the deacetylated polymer due to the formation of a barrier at the surface of the polymer. Zanetti et al.[12] mixed modified fluorohectorite with EVA in an internal mixer and indicated that the accumulation of filler on the surface of a burning specimen created a protective barrier to heat and mass loss during combustion. There was suppressed dripping of burning particles in the vertical combustion in the case of nanocomposites, reducing the hazard of flame spread to surrounding materials. Melt-intercalated and additionally gamma-irradiated high-density (HDPE)/EVA nanocomposites were prepared by Hu et al.[13–14] based on a modified montmorillonite; increasing the clay content from 2% to 10% was beneficial, improving the flammability properties. Thermogravimetric analysis (TGA) data showed that nanodispersion of the modified montmorillonite within the polymer inhibited the irradiation degradation of the HDPE–EVA blend, which led to nanocomposites with better irradiation-resistant properties than those of the nonfilled blend. Other authors described the preparation of EVA-based nanocomposites in more detail. Sundararaj and Zhang[15] used a twin-screw extruder and found intercalation of modified montmorillonite with EVAs differing in melt flow index and vinyl acetate content. The use of maleated EVA obviously improved the exfoliation, probably due to chemical interaction between the maleated EVA and the filler. Camino et al.[16] studied the effect of the compounding apparatus on the properties of EVA nanocomposites. A discontinuous batch mixer, a single-screw extruder, and a counter-rotating and co-rotating intermeshing twin-screw extruder were used. Hu et al.[17] prepared EVA nanocomposites on a twin-screw extruder and a twin-roll mill. Morgan et al.[18] compared natural and synthetic clays to improve polymer flammability. The natural clay was a montmorillonite mined and refined in the United States; the synthetic clay was a fluorinated synthetic mica. Both clays were converted into organoclays by ion exchange with an alkyl ammonium salt and were then used to synthesize polystyrene(PS)-based nanocomposites by melt blending. Both nanocomposites showed very similar reductions in the peak of heat release rate. The major differences between the natural and synthetic clay were improved color and better batch-to-batch consistency, with higher costs for the synthetic clay.

Concerning the reaction mechanism of degradation and fire retardant behavior of EVA nanocomposites, Wilkie et al.[19] found that in the early EVA degradation, the loss of acetic acid seemed to be catalyzed by hydroxyl groups which were present on the edges of the montmorillonite. The thermal degradation of EVA in the presence and in the absence of the modified clay showed that the formation of reaction products differed in quantity and identity. He found that the products were formed as a result of radical recombination reactions that could occur because the degrading polymer was contained within the clay layers long enough to permit the reactions. The formation of these new products explained the variation of heat release rates. In cases with multiple degradation pathways, the presence of the modified montmorillonite could promote one of these at the expense of another and thus led to different products and hence a different rate of volatilization. From his investigations of polyamide-6 (PA6), PS, poly(methyl methacrylate)(PMMA), styrene–acrylonitrile (SAN), acrylonitrile–butadiene–styrene (ABS), high-impact PS (HIPS), PE and PP, Wilkie et al.[20] proposed a more general explanation of the fire retardant properties of nanocomposites. Since the clay layers acted as a barrier to mass transport and led to superheated conditions in the condensed phase, extensive random scission of the products formed by radical recombination was an additional degradation pathway of polymers in the presence of clay. The polymers that showed good fire retardancy upon nanocomposite formation exhibited significant intermolecular reactions, such as interchain aminolysis and acidolysis, radical recombination, and hydrogen abstraction. In the case of polymers that degraded through a radical pathway, the relative stability of the radicals was the most important factor for prediction of the effect that nanocomposite formation had on the reduction in the peak of the heat release rate. The more stable was the radical produced by the polymer, the better was the fire retardancy, as measured by the reduction of the heat release rates of the polymer–clay nanocomposites.

Other nanostructured fillers have also been described as flame retardants. Frache et al.[21] investigated the thermal and combustion behavior of PE–hydrotalcite nanocomposites. Hydrotalcites were synthesized and then intercalated with stearate anions, because of the compatibility of the long alkyl chain with the PE chains. The presence of the inorganic filler shielded PE from thermal oxidation, and a reduction of 55% for the peak of the heat release rate was observed. Nelson et al.[22] generated various nanocomposites using modified silica; PMMA–silica and PS–silica nanocomposites were obtained by single-screw extrusion. Although these nanocomposites exhibited higher thermal stabilities and oxygen indices, they burned faster than the virgin polymers according to horizontal burning tests, suggesting that nanocomposites themselves cannot be considered as flame retardant materials. In combination with traditional flame retardant additives, flame retardancy and better mechanical properties could be achieved using less flame-retardant additives in the presence of nanofillers. Zammarano et al.[23] studied the flame retardant properties of modified layered double hydroxide (LDH) nanocomposites, which can be more effective than modified montmorillonites in the reduction of heat release rates. This may be related to the layered structure of LDHs and their hydroxyl groups and water molecules. Zammarano et al.[24] reported on synergistic effects for LDHs in particular with ammonium polyphosphate.

Several papers from Kashiwagi et al.[25–27] described single- and multiwall carbon nanotubes enhancing the thermal stability of polymers without using any organic treatment or additional additives. The carbon nanotubes were at least as effective flame retardants as organoclays. PP and PMMA were investigated and the dispersion of the nanotubes within the polymer matrix was important for good flame retardancy. Kashiwagi reported that the ideal structure of a protective surface layer (consisting of clay particles and some char) was netlike and had sufficient physical strength not to be broken or disturbed by bubbling. The protective layer should remain intact over the entire burning period. The requested formation of a continuous, network-structured protective char was easiest with high-aspect-ratio nanoscale particles. Kashiwagi et al.[28] pointed out that in general, a variety of highly extended carbon-based nanoparticles, such as single- and multiwall carbon nanotubes as well as carbon nanofibers, will form this type of network if the nanofillers formed a jammed network structure in the polymer matrix such that the material as a whole behaves rheologically like a gel. Also, Schartel et al.[29] reported on the flame retardancy of multiwall carbon nanotubes in PA6; again, the increased melt viscosity of the nanocomposites and the fiber-network character were the dominant mechanisms influencing the fire performance.

Leroy et al.[30] investigated the influence of the aspect ratio of fillers on flame retardancy for the system EVA, magnesium hydroxide (MDH), and talc. Talc particles of different lamellarity and specific surface area were tested, leading to the conclusion that for highly lamellar talc particles the fire-retarding behavior became similar to that of EVA–MDH–modified montmorillonite–based nanocomposites but with a significant intumescence. This intumescence, which occurred during the preignition period in cone calorimeter tests, may to be related to three phenomena caused by the lamellar particles (modified montmorillonite or talc): heterogeneous bubble nucleation, increased viscosity, and charring promotion. Ferry et al.[31] described similar results on the intumescence effect of talc in EVA–MDH–zinc borate–talc; zinc borate acted as a binder in EVA–MDH–zinc borate formulations, as shown by Le Bras et al.[32]

Jho et al.[33] pointed out that modified montmorillonites alone are not sufficient as flame retardants used in cable applications. Also, Wilkie[34] gave a clear statement: "It is apparent that nanocomposite formation alone is not the solution to the fire problem, but it may be a component of the solution. We, and others, have been investigating combinations of nanocomposites with conventional fire retardants." Charring polymers such as PA6 and PA6 nanocomposites were used by Bourbigot et al.[35] to improve the flame retardancy of EVA. The organoclay increased the efficiency of the char as a protective barrier by thermal stabilization of a phosphorocarbonaceous structure in the intumescent char and additionally, the formation of a "ceramic" layer. Hu et al.[36] used a blend of PA6 and EVA–nanocomposite to improve the flame retardancy of PP.

Often, combinations of nanofillers with traditional micro-sized traditional flame retardants demonstrated synergistic effects. A halogen-free flame retardant nanocomposite was reported by Hu et al.[37] using PA6, modified montmorillonite,

MDH, and red phosphorus. This system showed higher mechanical and flame retardant properties than those of a classical flame-retarded PA6 and therefore a synergistic effect for all three fillers. Ferry et al.[38] partially substituted MDH in flame retardant EVA by organoclays; improvements in self-extinguishability were reported, and the main mechanism was connected to a phenomenon of intumescence leading to the formation of a foamlike structure during the preignition period. Horrocks et al.[39] demonstrated that the combination of organoclays with ammonium polyphosphate or polyphosphine oxide showed synergistic effects in flame retardancy for PA6. Whaley et al.[40] investigated blends of EVA and ethylene-co-octene copolymers with MDH and modified montmorillonites for cable compounds. The time dependence of the char layer formation in such systems suggested that optimal loadings for the montmorillonite could be different for applications with different cable jacketing thickness, and therefore the true performance of nanocomposite-based jacketing compounds needs to be assessed in actual cable constructions.

Shen et al.[41,42] reported on flame retardant improvements by using a filler combination of modified montmorillonites, MDH, and zinc borate in EVA. A modified montmorillonite with a smaller particle size gave a better UL-94 performance than one of a larger particle size, and a stronger char was formed during the presence of zinc borate with a very fine particle size.

Ristolainen et al.[43] used modified montmorillonite as a partial substitute for ATH in PP–ATH composites and observed enhanced flame retardancy with composites containing both fillers. Wilkie and Zhang[44] studied the fire behavior of PE combined with ATH and a modified montmorillonite. The combination of PE with 2.5% modified montmorillonite and 20% ATH gave a 73% reduction in the peak heat release rate, which was the same as that obtained when 40% ATH alone was used. A further increase in the montmorillonite loading did not improve the fire properties. Mechanical properties such as elongation at break could be improved in comparing compounds with or without montmorillonite at the same reduction in peak heat release rate.

Cusak et al.[45] found that zinc hydroxystannate greatly enhanced the performance of an ATH–organoclay synergistic fire retardant system in an EVA formulation that allowed reductions in the overall filler level with no or little compromise in terms of flame retardant or smoke suppressant properties.

In this chapter we review results of nanocomposites based on organoclays and carbon nanotubes and the synergistic effects of these fillers with micro-sized alumina trihydrate as a traditional flame retardant for cables.

7.2 EXPERIMENTAL PROCESS

7.2.1 Materials

Poly(ethylene-co-vinyl acetate)(EVA) from Exxon with different weight percentages of vinyl acetate were used. Such types of copolymers demonstrated their ability to promote nanocomposite formation by melt blending with organoclays.[46,47] Low-density polyethylene (LDPE) BPD 8063 from Innovene (formerly, BP Petrochemicals) was used as a nonpolar polymer matrix for carbon nanotubes. Alumina trihydrate (ATH) Martinal OL 104 LE from Albemarle was used as a conventional flame retardant. A commercially available organoclay (Nanofil 15) based on a layered silicate (montmorillonite modified by dimethyldistearylammonium cations) from Süd-Chemie AG was used. The content of the quaternary ammonium compound was 38 ± 2 wt%. Pure and crude multiwall carbon nanotubes (MWCNTs) and single-wall carbon nanotubes (SWCNTs) from the research group of Janos B. Nagy at the University of Namur and 1.5 kg of MWCNTs from Nanocyl S.A. (Belgium) were used.

7.2.2 Compounding

Mixing was performed on several different compounding machines. For EVA organoclay-based nanocomposites, a laboratory twin-roll mill and an internal mixer at 145°C were used and additionally a co-rotating twin-screw extruder from Leistritz in Germany with a 27-mm screw diameter and an aspect ratio of 40 L/D. The mass temperature was 190°C at the extruder die. A Brabender mixer as a discontinuous compounding machine was used at 45 rpm and 180°C processing temperature to generate LDPE filled with MWCNTs and SWCNTs as well as EVA filled with organoclay or MWCNTs at 160°C. A laboratory twin-roll mill at 160°C was used to compound different MWCNT/organoclay-based formulations to optimize their fire properties. A Buss Ko-kneader (with a rotating and simultaneously oscillating screw, 11 L/D, 46-mm screw diameter) was used as a continuous compounding machine to generate 60 kg of an optimized MWCNT–organoclay compound and different EVA–organoclay nanocomposites; in both cases the processing temperature was 160°C. A 20-L/D single-screw cable extruder with a 80-mm screw diameter was used to produce an insulated flexible 2.5-mm^2 copper wire with a wall thickness of 0.86 mm for flame retardant insulations.

7.2.3 Analyses

Cone calorimeter tests were carried out at 35 or 50 kW/m^2 heat flux with horizontal orientation of the samples [plates ($100 \times 100 \times 3$ mm) or cut cables] according to ASTM E1354. The data reported were the average of three measurements for each sample, with a standard uncertainty of the measured heat release rates of $\pm 5\%$. Thermogravimetric analyses were done under helium or airflow at a 20°C/min heating rate.

7.3 ORGANOCLAY NANOCOMPOSITES

7.3.1 Processing and Structure of EVA/Organoclay-Based Nanocomposites

Depending on the nature of the filler distribution within the matrix, the morphology of nanocomposites can evolve from the intercalated structure with a regular alternation of layered silicates and polymer monolayers to the exfoliated (delaminated) structure with layered silicates randomly and distributed homogeneously within the polymer matrix. The easiest and technically most attractive way to produce these types of materials is kneading a polymer in the molten state with a modified layered silicate such as montmorillonite, with the native Na^+ interlayer cation exchanged by a quaternary alkylammonium cation; the modified filler, called an *organoclay*, is much more compatible with the polymer matrix.

Information on the nanocomposite morphology was obtained by transmission electron microscopy (TEM) and x-ray diffraction (XRD) observation. Compounding was done on a twin-roll mill, and exfoliated silicate sheets were observed together with small stacks of intercalated sheets.[47] This structure may be described as a semi-intercalated semi-exfoliated structure that did not change principally with the vinyl acetate content of the EVA matrix, even a larger number of stacks were observed for EVA with lower vinyl acetate contents.[47] There were no great differences within the morphology of these nanocomposites.

7.3.2 Thermal Stability of EVA/Organoclay-Based Nanocomposites

Thermogravimetric analysis (TGA) is widely used to characterize the thermal stability of polymers. The mass loss of a polymer due to volatilization of products generated by thermal decomposition is monitored as a function of a temperature ramp. Nonoxidative decomposition occurs when the heating of the material is done under an inert gas flow such as helium or nitrogen, while the use of air or oxygen allows investigation of oxidative decomposition reactions. The experimental conditions highly influenced the reaction mechanism of the degradation. EVA is known to decompose in two consecutive steps. The first step was identical under both oxidative and nonoxidative conditions, occurring between 350 and 400°C and involving the loss of acetic acid. The second step involved the thermal decomposition of the unsaturated backbone either by further radical scissions (nonoxidative decomposition) or by thermal combustion (oxidative decomposition) (Figure 7.1).

In helium, the EVA nanocomposite showed a negligible reduction in thermal stability compared to pure EVA or EVA filled with Na-montmorillonite (microcomposite). In contrast, when decomposed in air, the same nanocomposite exhibited a rather large increase in thermal stability because the maximum of the second degradation peak was shifted 40°C to higher temperatures while the maximum of the first decomposition peak remained unchanged[47] (Table 7.1). In this case the explanation for the improved thermal stability was char formation occurring under oxidative conditions. The char acted as a physical barrier between the polymer and the superficial zone where the combustion of the polymer occurred. The optimum thermal stabilization was obtained at an organoclay level of 2.5 to 5.0 wt%, as indicated by the results in Table 7.1 on the maximal temperatures for the main degradation peak for EVA nanocomposites.

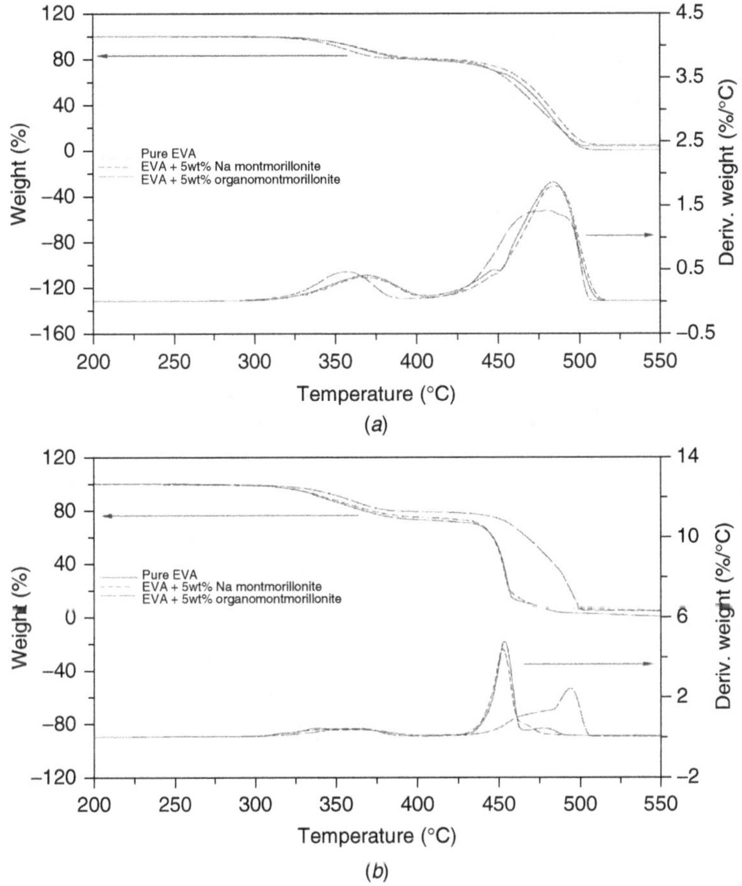

FIGURE 7.1 TGA of EVA, EVA microcomposite with 5 wt% Na^+-montmorillonite, and EVA nanocomposite with 5 wt% organoclays under (*a*) helium and (*b*) airflow. Heating rate, 20°C/min; EVA, Escorene UL-00328 with 28 wt% vinyl acetate content. Organoclay, Nanofil 15. (From Ref. 48, copyright © 2001, John Wiley & Sons, Ltd., with permission.)

7.3.3 Flammability Properties of EVA/Organoclay-Based Nanocomposites

From an engineering point of view, it is important to identify the principal fire hazard associated with products so that relevant properties can be measured. Extensive research at the National Institute for Standards and Technology in the United States led to an important conclusion that allows significant simplification of the problem. The heat release rate, in particular the peak of heat release rate, is the single most important parameter in a fire and can be viewed as the "driving force" of the fire.[49] Therefore, today the universal choice of an engineering instrument to measure flame retardant properties of polymers is the cone calorimeter. The measuring principle is the oxygen depletion with a relationship

TABLE 7.1 Maximum Temperature at the Main Degradation Peak Measured by TGA Under Airflow at 20°C/min for EVA and EVA-Based Nanocomposites with Different Organoclay Contents[a]

Organoclay Content (wt%)	Maximum Temperature at the Main Degradation Peak (°C)
0	452.0
1	453.4
2.5	489.2
5	493.5
10	472.0
15	454.0

[a]EVA, Escorene UL-00328 with 28 wt% vinyl acetate content; organoclay, Nanofil 15

between the mass of oxygen consumed from the air and the amount of heat released. In a typical cone calorimeter experiment, polymer samples placed in aluminum dishes are exposed to a defined heat flux (typically, 35 or 50 kW/m^2). Simultaneously, the following properties are measured: heat release rate, peak heat release rate (PHRR), time to ignition, total heat released, mass loss rate, mean CO yield, mean specific extinction area, and others.

The flame retardant properties of the EVA nanocomposites were determined by a cone calorimeter with a heat flux of 35 kW/m^2 (Figure 7.2). Under such

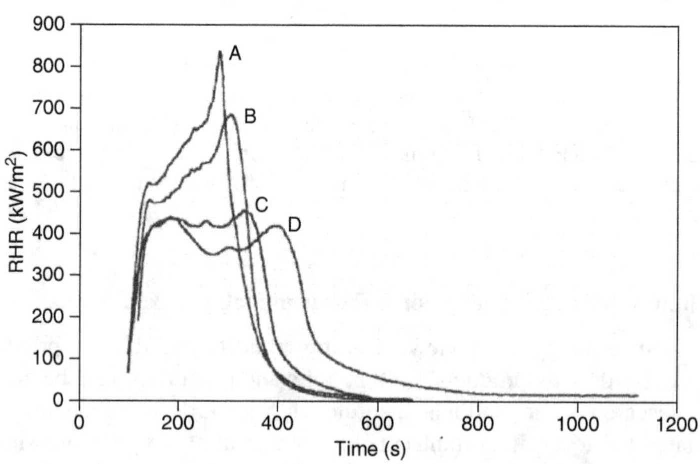

FIGURE 7.2 Heat release rate at heat flux = 35 kW/m^2 for various EVA (Escorene UL-00328 with 28 wt% vinyl acetate content)–based materials: A, pure EVA matrix and EVA matrix with 5 wt% Na$^+$-montmorillonite; B, EVA + 3 wt% of organoclay Nanofil 15; C, EVA + 5 wt% of organoclay Nanofil 15; D, EVA + 10 wt% of organoclay Nanofil 15. (From Ref. 48, copyright © 2001, John Wiley & Sons, Ltd., with permission.)

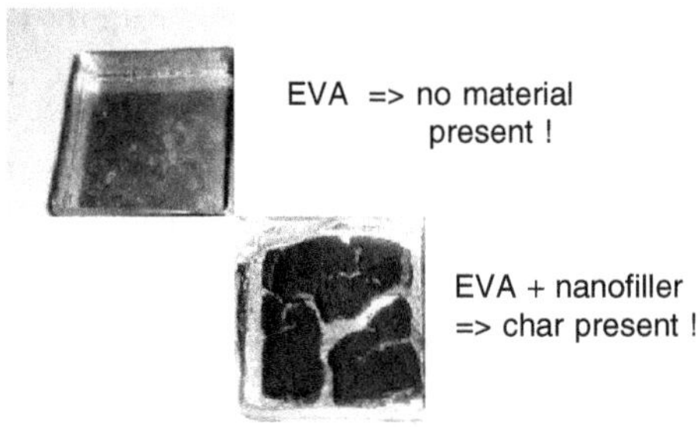

FIGURE 7.3 Samples of pure EVA and EVA nanocomposite with 5 wt% organoclays by cone calorimeter combustion after 200 sec; heat flux = 35 kW/m^2; polymer plates of 100 × 100 × 3 mm within aluminium dishes. EVA, Escorene UL-00328 with 28 wt% vinyl acetate content; organoclay, Nanofil 15. (From Ref. 48, copyright © 2001, John Wiley & Sons, Ltd., with permission.)

conditions, simulating a developing fire scenario, the effect of the organoclay was already observed at 3 wt% clay. A decrease by 47% of PHRR compared to virgin EVA, as well as a shift toward longer times, was detected for a nanocomposite containing 5 wt% of the organoclay. Increasing the filler content to 10 wt% did not significantly improve the PHRR reduction. As a decrease in PHRR indicates a reduction in burnable volatiles generated by degradation of the polymer matrix, such a decrease clearly showed the flame retardant effect due to the presence of the organoclays and their "molecular" distribution throughout the matrix. Furthermore, the flame retardant properties were improved by the fact that the heat release was spread over a much longer period. The flame retardant properties were due to the formation of a char layer during nanocomposite combustion. This char acted as an insulating and nonburning material and reduced the emission of volatile products (fuel) into the flame area. The silicate layers of the organoclay played an active role in the formation of the char but also strengthened it and made it more resistant to ablation.

Cone calorimeter experiments at a heat flux of 35 kW/m^2 showed that virgin EVA was burned completely, without any residue. In contrast, very early strong char formation was found for an EVA nanocomposite with an analogous cone calorimeter experiment; this char was stable and did not disappear by combustion (Figure 7.3). Finally, compared to a pure EVA matrix, the nanocomposite burned without producing burning droplets (UL-94 vertical procedure),[50] a characteristic feature that also limits the propagation of a fire. Burning droplets will be an important additional characteristic for cables to be classified within the Euroclasses B1ca, B2ca, Cca, or Dca, defined by a draft for the European standard EN 50399.

7.3.4 NMR Investigation and Fire Retardant Mechanism of EVA Nanocomposites

The degradation of EVA and EVA nanocomposites was investigated by solid-state cross-polarization magic angle opening ^{13}C-NMR spectroscopy; Bourbigot et al.[51] described the measurement method in detail. EVA (Escorene UL-00112 with 12 wt% vinyl acetate content) and a nanocomposite of EVA with 5 wt% of the organoclay were degraded in a cone calorimeter with a heat flux of 50 kW/m^2. Samples were removed from the heat flux after 50, 100, 150, 200, and 300 s, and the presence of EVA and the char formation was measured by the shift (in ppm) of the signals related to the signal from the standard tetramethysilane. The following results were obtained[52]:

- *Before irradiation of EVA and EVA nanocomposite:* 33 ppm => $-CH_2-$, polymer backbone of EVA; 75 ppm => $-CH_3$, acetate group of EVA; 172 ppm => $-C-O$, acetate group (small signal) of EVA
- *After irradiation of EVA:* 50 s, new signals at 130 ppm (char: aromatics and polyaromatics) and 180 ppm ($-C{=}O$ with start of oxidation), EVA signals still present; 150 s, no signals => no organic material present
- *After irradiation of EVA nanocomposite:* 50 s, new signals at 130 ppm (char: aromatics and polyaromatics) and 180 ppm ($-C{=}0$ with start of oxidation), EVA signals still present; 100 s, char formation and EVA signals still present; 200 s, char formation and EVA signals still present; >300 s, no signals => no organic material present

Obviously, the formation of nanocomposites clearly promoted char formation and delayed the degradation of EVA.

7.3.5 Intercalation Versus Exfoliation of EVA Nanocomposites

Often, it is reported in the literature that exfoliation is the most effective structure for maximal enhancements of properties improved by nanocomposites. Therefore, it was of interest to shift the ratio of the mixed intercalated–exfoliated structure that was observed within EVA nanocomposites by twin-roll mill compounding[47] to the only exfoliated structure. This was done by melt compounding EVA (Escorene UL-00328) with 5 phr (parts per hundred parts resin) of organoclay by a co-rotating twin-screw extruder (27-mm screw diameter, 40 L/D). Two screw designs were used: a first screw for maximal mixing using mixing elements, and the second screw for maximal dispersion using kneading blocks. The screws were used from 300 up to 1200 rpm. TEM and XRD demonstrated that for the highest shear rate (1200 rpm) and highest friction (second screw), the mixed structures were shifted to the exfoliated structure. However, cone calorimeter data showed that there were no changes in the PHRRs for all the melt-compounded nanocomposites. Obviously, the mixed intercalated–exfoliated structures within the EVA nanocomposites already had maximal reductions in PHRR. This result is important for companies using EVA-based nanocomposites as a flame retardant system, because it simplifies the task of organoclay dispersion as a main processing step.

7.3.6 Combination of the Classical Flame Retardant Filler Alumina Trihydrate with Organoclays

Cable compounds must be flame retardant to achieve a low flame spread, defined by the widely used international cable fire test (IEC 60332-3-24).[53] A combination of 65 wt% ATH and 35 wt% of a high filler level accepting polymer matrix such as EVA must often be used for cable outer sheaths.[54]

The performances of two compounds were compared. Both compounds were prepared on a Buss Ko-kneader (46-mm screw diameter, 11 L/D). One compound was made from 65 wt% ATH and 35 wt% EVA (28% vinyl acetate content), and a second compound was made from 60 wt% ATH, 5 wt% organoclay, and 35 wt% EVA. Both compounds were investigated with TGA in air and by a cone calorimeter at 50 kW/m^2 heat flux. TGA in air clearly showed a delay in degradation by the small amount of organoclays (Figure 7.4).

The char of the EVA–ATH–organoclay compound generated by the cone calorimeter was very rigid und showed only very few small cracks; but the char of the EVA–ATH compound was much less rigid (lower mechanical strength) and with many big cracks. This could be why for the nanocomposite the PHRR was reduced to 100 kW/m^2, compared to 200 kW/m^2 for the EVA–ATH compound. To obtain the same decrease for the PHRR by the flame retardant filler ATH only, the content of ATH must be increased to 78 wt% within the EVA–ATH system.

The great improvements in flame retardancy by the organoclays also opened the possibility of decreasing the level of ATH within the EVA–polymer matrix. To maintain 200 kW/m^2 as a sufficient peak heat release level, the content of ATH could be decreased from 65 wt% to 45 wt% by the presence of only 5 wt% organoclay within the EVA–polymer matrix. The reduction in the total amount of these fillers resulted in improved mechanical and rheological properties of the EVA-based nanocomposite.

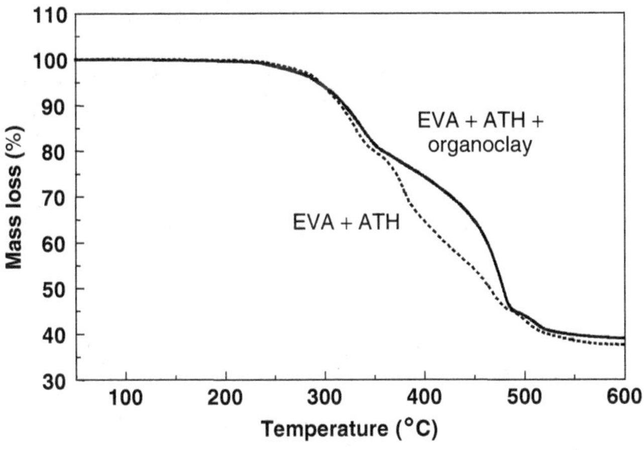

FIGURE 7.4 TGA in air of a compound with 35 wt% EVA and 65 wt% ATH in relation to a nanocomposite compound with 35 wt% EVA, 60 wt% ATH, and 5 wt% of organoclays. EVA, Escorene UL-00328 with 28 wt% vinyl acetate content; organoclay, Nanofil 15; ATH, Martinal OL 104 LE. (From Ref. 48, copyright © 2001, John Wiley & Sons, Ltd., with permission.)

7.3.7 Coaxial Cable Passing the UL-1666 Fire Test with an Organoclay/ATH-Based Outer Sheath

There are many applications for indoor cables passing the large-scale fire test UL-1666 (riser test for cables) with a 145-kW burner in a two-story facility. This very severe fire test defines the following important points of measurements: (1) maximal temperature of fire gases at 12 ft: 850°F, and (2) maximal height for flames: 12 ft. Compounds with halogenated flame retardants are often used to pass this test, but more and more flame retardant nonhalogen (FRNH) cables are requested by the market for the riser test. Cables based on nanocomposite compounds demonstrate their promising performance for this fire test.

An example of FRNH cables passing UL-1666 is shown in Figure 7.5. The outer sheath was based on a nanocomposite with an industrial EVA–ATH–organoclay composition. The analogous coaxial cable was tested with an outer sheath based on EVA–ATH. In both compounds the relation of polymer to filler was the same and Table 7.2 presents the results. The improved flame retardant properties were due to the formation of a char layer during

FIGURE 7.5 FRNH coaxial cable (1/2-in. diameter) with a nanocomposite-based outer sheath passing the UL-1666 cable fire test. (From Ref. 52, copyright © 2005, Sage Publications, with permission.)

TABLE 7.2 Fire Performance of FRNH Coaxial Cables with EVA–ATH and EVA–ATH–Organoclay Outer Sheathes

UL-1666 Requirement	EVA–ATH Compound	EVA–ATH–Organoclay Compound
Maximal temperature at 12 ft: <850°F	1930°F	620°F
Maximal flame height: <12 ft	>12 ft	6 ft

nanocomposite combustion. This insulating and nonburning char reduced the emission of volatile products from polymer degradation into the flame area, and thus minimized the maximal temperature and height of the flames.

7.4 CARBON NANOTUBE NANOCOMPOSITES

7.4.1 General Properties of Carbon Nanotubes

Carbon nanotubes (CNTs) are tubular derivatives of fullerenes. They were first observed in arc discharge experiments and exhibit properties which are quite different from those of closed-cage fullerenes such as C_{60}, C_{70}, and C_{76}. Special topologies are responsible for the unique and interesting properties of CNTs. As novel carbon materials the CNTs are of great interest within the field of material science research. Due to their high mechanical strength, capillary properties, and remarkable electronic structures, a wide range of potential uses is reported. Typical applications for CNTs are supports for metals in the field of heterogeneous catalysis, material for hydrogen storage, as composite materials in polymer science,[55,56] and for immobilization of proteins and enzymes. Several techniques, such as arc discharge, laser ablation, catalytic methods, and others, have been developed for the production of CNTs.[57] Many material science researchers are working on the development of methods for large-scale production of CNTs to realize their speculated applications.

CNTs can consist of one (SWCNTs) or more (MWCNTs) cylindrical shells of graphitic sheets. Each carbon is completely bonded to three neighboring carbon atoms through sp^2 hybridization to form a seamless shell. CNTs can have a very high aspect ratio, above 1000. It is reported[58] that polymer degradation can be retarded by CNTs as indicated by thermogravimetric analysis. In PVOH a loading level of 20 wt% MWCNTs shifts the beginning of the polymer degradation and the single degradation peak to higher temperatures.

7.4.2 Synthesis and Purification of Carbon Nanotubes

Crude MWCNTs and SWCNTs were produced by catalytic decomposition of acetylene on $Co{-}Fe/Al(OH)_3$ catalysts. The CNTs contained used catalysts and other by-products. The catalysts and support contents of the CNTs samples are shown in Table 7.3. Crude MWCNTs contain Co, Fe, and alumina. Purified MWCNTs were synthesized from crude MWCNTs by dissolution of the catalyst support in concentrated NaOH, dissolution of the metal catalyst in concentrated HCl, drying at 120°C in an air oven and additional drying at 500°C under vacuum. Crude SWCNTs contain Co and MgO. Purified SWCNTs were synthesized from crude SWCNTs by dissolution of the catalyst support in concentrated HCl, purification by air oxidation at 300°C and then drying at 120°C in an air oven.

7.4.3 Flammability of EVA–MWCNT and EVA–MWCNT–Organoclay Compounds

It was possible to investigate for the first time worldwide the flame retardant properties of carbon nanotube compounds by cone calorimeter at 35 kW/m^2.[59,60]

TABLE 7.3 Properties of MWCNTs

	Nanotubes		Catalyst		Support	
Sample	Length (μm)	Diameter (nm)	Co (wt%)	Fe (wt%)	Al_2O_3 (wt%)	MgO (wt%)
Crude MWCNTs	ca. 50	5–15	0.3	0.3	19	—
Purified MWCNTs	ca. 50	5–15	0.2	0.3	0.2	—

TABLE 7.4 Peak of Heat Release Rates at Heat Flux = 35 kW/m² for Various Compounds with Organoclays and Multiwall Carbon Nanotubes

		MWCNT (phr)			
Sample	EVA[a] (parts resin)	Purified	Crude	Organoclay[b] (phr)	PHRR (kW/m^2)
EVA[a]	100.0	—	—	—	580
1[c]	100.0	2.5	—	—	520
2	100.0	5.0	—	—	405
3	100.0	—	—	2.5	530
4	100.0	—	—	5.0	470
5[c,d]	100.0	2.5	—	2.5	370
6a[c]	100.0	—	5.0	—	403
6b[e]	100.0	—	5.0	—	405

[a] Escorene UL-00328 with 28 w% vinyl acetate content.
[b] Nanofil 15.
[c] The screw velocity was 45 rpm and the mass temperature was 136°C.
[d] The nanotubes and the organoclay were premixed before addition.
[e] The screw velocity was 120 rpm and the mass temperature was 142°C.

All compounds were melt-blended in a Brabender mixing chamber. It is evident from the results in Table 7.4 that all the filled polymers had improved flame retardant properties. For EVA and EVA-based nanocomposites containing 2.5 phr of filler, the PHRR decreased as follows: EVA > organoclays ~ purified MWCNTs. For EVA and EVA-based composites containing 5.0 phr of filler, the PHRR decreased as follows: EVA > organoclays > purified MWCNTs = crude MWCNTs. Crude MWCNTs were as effective in the reduction of PHRR as purified MWCNTs! Increasing the filler content from 2.5 phr to 5.0 phr caused an additional flame retardant effect that was most significant when purified or crude MWCNTs were used.

A synergistic effect for flame retardancy between MWCNTs and organoclays was observed for a nanocomposite containing 2.5 phr of purified MWCNTs and 2.5 phr of organoclays (Figure 7.6). The latter sample was found to be the best flame retardant compound. The variation of the screw velocity from 45 rpm (sample A) to 120 rpm (sample B) did not change the flame retardant properties for composites containing 5.0 phr of crude MWCNTs. There was also no reduction in time to ignition for the MWCNT-based EVA composite, in

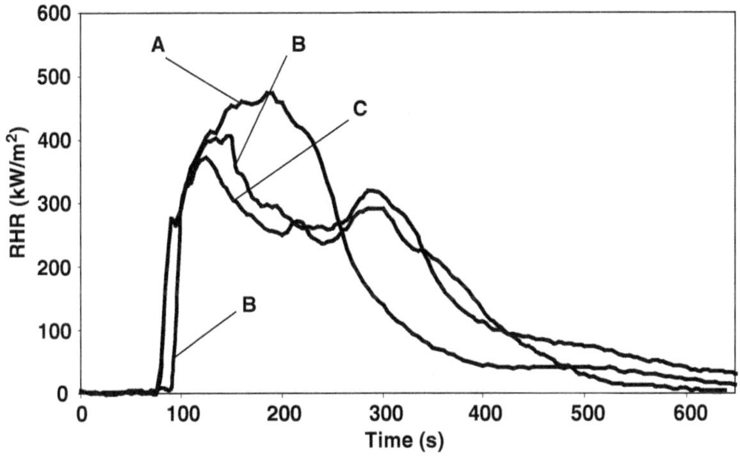

FIGURE 7.6 Heat release rates at heat flux = 35 kW/m^2 for various EVA-based materials: A, EVA + 5.0 phr organoclays; B, EVA + 5.0 phr pure MWCNTs; C, EVA + 2.5 phr organoclays + 2.5 phr pure MWCNTs. EVA, Escorene UL-00328 with 28 wt% vinyl acetate content; organoclay, Nanofil 15. (From Ref. 59, copyright © 2002, John Wiley & Sons, Ltd., with permission.)

contrast to the organoclay-based EVA composite, which undergoes an early thermal degradation of the quaternary ammonium compound within the galleries of the organoclay.[61]

7.4.4 Crack Density and Surface Results of Charred MWCNT Compounds

For flame retardant EVA-based composites containing 5 phr of fillers (Table 7.4), the crack density increased in the order purified MWCNTs < organoclays. A very important synergistic effect reducing the crack density to a very low level was observed for nanocomposite containing both 2.5 phr of purified MWCNTs and 2.5 phr of organoclays (Figure 7.7). The synergistic effect for improved flame retardancy by the filler combination MWCNTs and organoclays can be explained by the improved closed surface. The char acted as an insulating and nonburning material with reduction of emission of volatile products (fuel) into the flame area. The fewer cracks present, the better was the reduction in fuel emission and therefore the reduction of PHRR. The fillers played an active role in the formation of this char, but obviously the MWCNTs, with their long aspect ratio, could strengthen it and make it more resistant to mechanical cracking.

7.4.5 Flammability of LDPE Carbon Nanotube Compounds

Compounds of SWCNTs and MWCNTs in LDPE BPD 8063 were melt blended in a Brabender mixing chamber according to the formulations indicated in Tables 7.5 and 7.6. The corresponding cone calorimeter measurements are shown

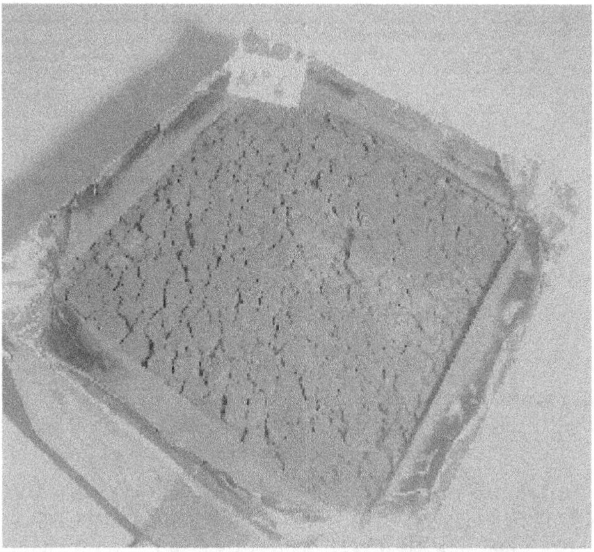

(a)

(b)

FIGURE 7.7 (*a*) EVA with 5 phr organoclays after combustion; (*b*) EVA with 5 phr pure MWCNTs after combustion; (*c*) EVA with 2.5 phr pure MWCNTs and 2.5 phr organoclays after combustion. EVA, Escorene UL-00328 with 28 wt% vinyl acetate content; organoclay, Nanofil 15. (From Ref. 59, copyright © 2002, John Wiley & Sons, Ltd., with permission.)

(c)

FIGURE 7.7 (*continued*)

TABLE 7.5 SWCNTs in LDPE

Sample	LDPE (wt%)	SWCNT (wt%)	
		Purified	Crude
BPD 8063	100.0	—	—
5 SWCNT	95.0	5	—
10 SWCNT	90.0	10	—
5 crude SWCNT	95.0	—	5
10 crude SWCNT	90.0	—	10

TABLE 7.6 MWCNTs in LDPE

Sample	LDPE (wt%)	MWCNT (wt%)	
		Purified	Crude
BPD 8063	100.0	—	—
5 MWCNT	95.0	5	—
10 MWCNT	90.0	10	—
5 crude MWCNT	95.0	—	5
10 crude MWCNT	90.0	—	10

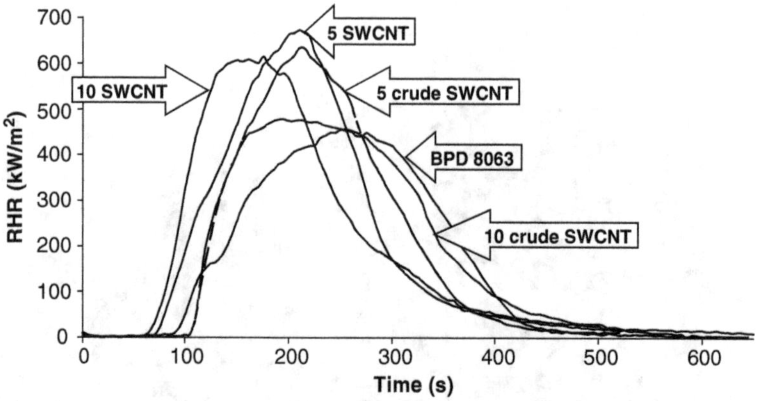

FIGURE 7.8 Heat release rates for SWCNTs in LDPE, Heat flux = 35 kW/m^2. (From Ref. 62, copyright © 2005, John Wiley & Sons, Ltd., with permission.)

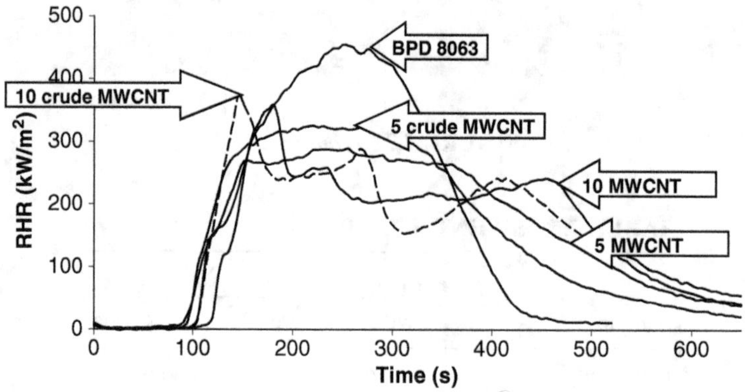

FIGURE 7.9 Heat release rates for MWCNTs in LDPE. Heat flux = 35 kW/m^2. (From Ref. 62, John Wiley & Sons, Ltd., with permission.)

in Figures 7.8 and 7.9. The results from the cone measurements of different carbon nanotubes in LDPE demonstrated that SWCNTs did not act as flame retardants in LDPE. MWCNTs acted as flame retardants in LDPE with no reductions in time to ignition (in contrast to organoclays), and crude MWCNTs showed similar reductions for the PHRR as purified MCNTs.

7.4.6 Cable with the New Fire Retardant System MWCNTs–Organoclays–ATH

It was of interest to transform the results for the synergistic filler-based fire retardant system MWCNTs–organoclays to real products.[62] Therefore, the carbon nanotube supplier Nanocyl S.A. synthesized 1.5 kg of MWCNTs. To produce a flame retardant insulated wire by a real cable production extruder, a minimum of approximately 60 kg of compound was needed to fill the extruder and

TABLE 7.7 Blends of MWCNTs and Organoclays in Cable Compound Formulations

Compound	Composition
1	Technical cable compound (EVA–PE–ATH–organoclay –processing additives)
2A	Same formulation as compound 1, but substitution of 50% organoclay by the same amount of MWCNTs
2B	Same formulation as compound 1, but substitution of 100% organoclay by the same amount of MWCNTs

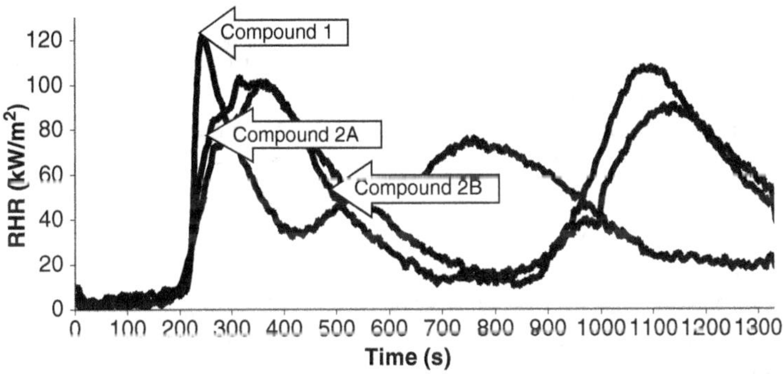

FIGURE 7.10 Heat release rates for compounds made by twin-roll mill with various filler blends of MWCNTs and organoclays. Heat flux = 35 kW/m^2. (From Ref. 62, copyright © 2005, John Wiley & Sons, Ltd. with permission.)

to run a small insulated wire production. It was checked whether the MWCNTs–organoclay filler blend system can be transformed to a real cable compound without problems from the polymer EVA investigated.[59] A well-running organoclay-based cable compound named compound 1 was used, and the weight ratio between these two fillers was changed stepwise. The sum of both fillers always remained constant (Table 7.7) within the cable compounds. Compounding was done on a twin-roll mill, and the reductions in PHRRs for the three compounds were measured (Figure 7.10). The results clearly indicated that for the filler blend and the MWCNT-only-based compounds, the first PHRR was reduced maximally. The second PHRR was observed at the longest times for the two compounds 2A and 2B, indicating that the chars were less cracked (more stable in time). Therefore, a 1:1 blend of MWCNTs and organoclays (compound 2A; see Table 7.7) for the cable compound was used. This allowed production of the required quantity of the cable compound. Compounding of the formulation 2A was done on an 11-L/D Buss Ko-kneader with a 46-mm screw diameter, and 60 kg was produced without any processing problems. Processing on the Buss

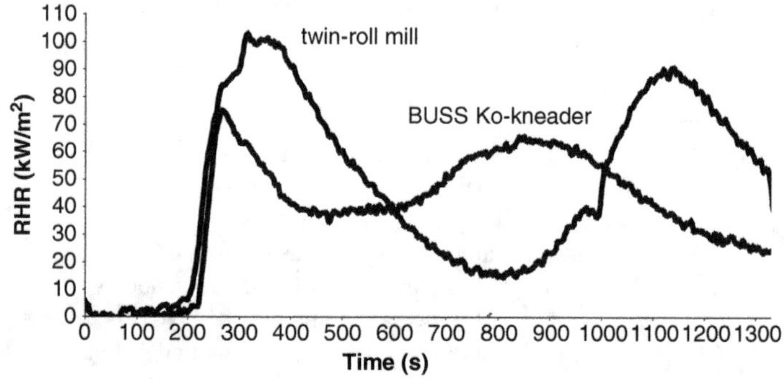

FIGURE 7.11 Heat release rates for cable compound 2A with the filler blends of MWCNTs and organoclays by twin-roll mill and Buss Ko-kneader. Heat flux = 35 kW/m^2. (From Ref. 62, copyright © 2005, John Wiley & Sons, Ltd., with permission.)

Ko-kneader improved the PHRRs compared to those with the corresponding twin-roll mill compounding (Figure 7.11).

Two insulated wires with identical geometric parameters were produced on a 20-L/D single-screw cable extruder with a 80-mm screw diameter. The diameter of the copper wire was 1.78 mm, and the wall thickness for the insulation was 0.8 mm. For one wire the insulation was compound 1 (filler combination by ATH–organoclay) and for the other wire the optimized compound 2A (filler combination by ATH–organoclay–MWCNTs) was used as insulation. The MWCNT-based compound 2A showed a remarkably increased viscosity over that of the standard nanocomposite compound 1, as indicated by reduced revolutions per minute of the screw and increased power takeup by the extruder motor; a high-pressure capillary viscosimeter showed a higher viscosity for compound 2A for all shear rates up to 3000 s^{-1}.

A small-scale fire test according to IEC 60332-1 (the insulation was exposed to a Bunsen burner ignition source) was very similar for both insulated wires. No dripping of burning polymer was noted, and the charred lengths were identical. But the char of the insulation made with compound 2A was much less cracked than the char generated from compound 1. This may be the result of the strengthening effect by the MWCNT, with its very large L/D ratio; a proposal of such a mechanism for the function of carbon nanotubes was published recently.[63]

Heat release rates and times to ignition of the two insulated wires were measured using a cone calorimeter. The wires were cut in samples of 10 cm, and the standard cone sample holder was filled with the cables pieces. Twenty-six wires were mounted by building up a single layer of wires with no gap between them,[64] and the ends of the wires were not sealed; this mounting was called a *single-layer design*. Also, four cut wires with no sealing of the ends, simulating an unjacketed cable,[64] were put together. An aramid fiber binder was used to maintain the integrity of the bundles; this mounting with 24 wires per layer and a total of two layers was called a *bundle design* (Figure 7.12). Both mounting designs demonstrated quite different cone calorimeter results (Figures 7.13 and 7.14). For

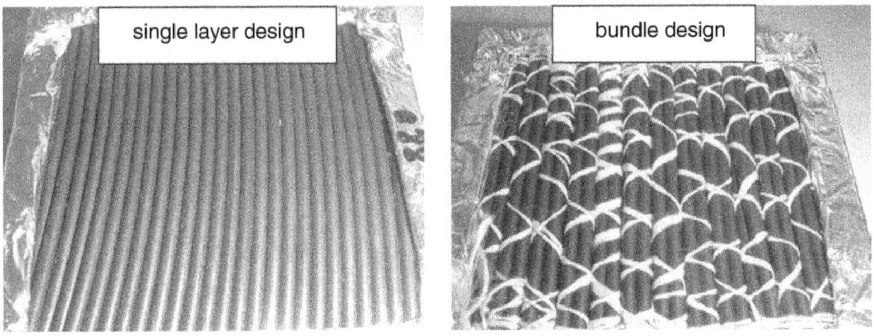

FIGURE 7.12 Various insulated wire mounting designs for cone calorimeter tests: single-layer design and bundle design (From Ref. 62, copyright © 2005, John Wiley & Sons, Ltd., with permission.)

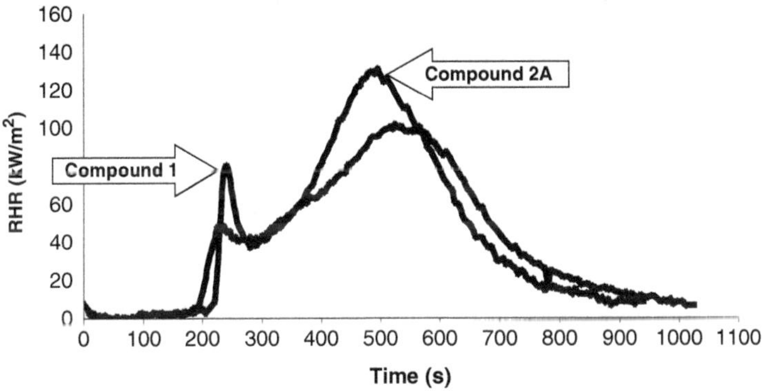

FIGURE 7.13 Heat release rates for a single-layer design. Heat flux = 35 kW/m^2. (From Ref. 62, copyright © 2005 John Wiley & Sons, Ltd., with permission.)

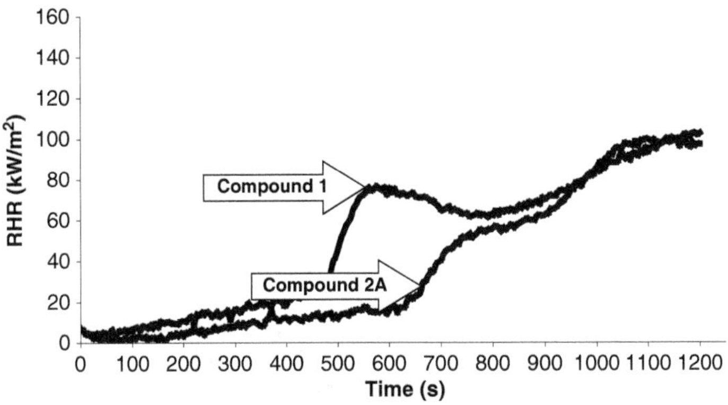

FIGURE 7.14 Heat release rates for a bundle design. Heat flux = 35 kW/m^2. (From Ref. 62, copyright © 2005 John Wiley & Sons, Ltd., with permission.)

the single-layer design the cone tests were finished within 20 minutes. This is the flame application time for cables as defined in the new European proposal for flame tests of cables (EN 50399). Wires insulated with compound 1 had a higher PHRR value within the first 5 min. For the bundle design, the duration of the cone calorimeter tests was very long, due to the insulation effect of the first charred layer on those located underneath in the second layer; therefore, the tests were stopped after 20 minutes. This mounting design is representative for many end-product applications. Wires insulated with the filler-blended compound 2A did not show any increase for the PHRR within the first 10 minutes compared to the wires insulated with the only organoclay-based compound, 1.

7.5 SUMMARY AND CONCLUSIONS

Nanocomposites made by melt-blending poly(ethylene-co-vinyl acetate) (EVA) with modified layered silicates are used for halogen-free flame retardant cables. Combinations of organoclays and traditional flame retardants on basis metal hydroxides must be used for proper flame retardancy, as outlined for a coaxial cable fulfilling the UL-1666 riser test. MWCNTs act as very efficient flame retardants at low filler contents in EVA. An optimized formulation for flame retardant insulated wires was developed based on filler-blend multiwall carbon nanotubes/organoclays. Small-scale fire tests according to IEC 60 332-1, with flames of a Bunsen burner attacking the insulation of the wire, showed that the char was strengthened by the long L/D ratio of the MWCNTs. The improved char results in better flame retardant performance by the wires.

REFERENCES

1. Hirschler, M.M. Fire performance of organic polymers, thermal, decomposition, and chemical decomposition. *Polym. Mater. Sci. Eng.*, **2000**, ACS Meeting, Washington, DC, Aug., 83, 79–80.
2. Babrauskas, V. The generation of CO in benchscale fire tests and the prediction for realscale fires. *Fire Mater.* **1995**, 19, 205–213.
3. Hirschler, M.M. Fire safety, smoke toxicity and halogenated materials. Commentary in: *Flame Retard. News* 2005, Apr., Business Communications Co., Norwalk, CT.
4. Stevens, G.C. Countervailing risks and benefits in the use of flame retardants, in: *Proceedings of the Flame Retardants 2000 Conference*, London, Feb. 8–9, 2000, pp. 131–145.
5. Ray, S.S.; Okamoto, M. Polymer/layered silicate nanocomposites: a review from preparation to processing. *Prog. Polym. Sci.* **2003**, 28, 1539–1641.
6. Beyer, G. Nanocomposites offer new way forward for flame retardants. *Plast. Add. Compound.* **2005**, Sept.–Oct., 7, 32–35.
7. Gilman, J.W.; Kashiwagi, T.; Giannelis, E.P.; Manias, E.; Lomakin, S.; Lichtenhan, J.D.; Jones, P. Nanocomposites: radiative gasification and vinyl polymer flammability, in: M. Le Bras, G. Camino, S. Bourbigot, and R. Delobel, Eds., *Fire Retardancy of*

Polymers: The Use of Intumescence. Royal Society of Chemistry, London, 1998, pp. 203–221.

8. Gilman, J.W.; Kashiwagi, T.; Lichtenhan, J.D. A revolutionary new flame retardant approach. *SAMPE J.* **1997**, 33–40.
9. Camino, G.; Zanetti, M.; Riva, A.; Braglia, M.; Falqui, L. Thermal degradation and rheological behaviour of EVA/montmorillonite nanocomposites. *Polym. Degrad. Stab.* **2002**, 77, 299–304.
10. Hu, Y.; Tang, Y.; Wang, S.; Gui, Z.; Fan, W. Preparation and flammability of ethylene–vinyl acetate copolymer/montmorillonite nanocomposites. *Polym. Degrad. Stab.* **2002**, 78, 555–559.
11. Camino, G.; Mülhaupt, R.; Zanetti, M.; Thomann, R. Synthesis and thermal behaviour of layered silicate–EVA nanocomposites. *Polymer* **2001**, 42, 4501–4507.
12. Zanetti, M.; Camino, G.; Mülhaupt, R. Combustion behaviour of EVA/fluorohectorite nanocomposites. *Polym. Degrad. Stab.* **2001**, 74, 413–417.
13. Hu, Y.; Lu, H.; Kong, Q.; Cai, Y.; Chen, Z.; Fan, W. influence of gamma irradiation on high density polyethylene/ethylene–vinyl acetate/clay nanocomposites. *Polym. Adv. Technol.*, **2004**, 15, 601–605.
14. Hu, Y.; Lu, H.; Kong, Q.; Chen, Z.; Fan, W. Gamma irradiation of high density poly(ethylene)/ethylene–vinyl acetate/clay nanocomposites: possible mechanism of the influence of clay on irradiated nanocomposites. *Polym. Adv. Technol.*, **2005**, 16, 688–692.
15. Sundararaj, U.; Zhang, F. Nanocomposites of ethylene–vinyl acetate copolymer (EVA) and organoclay prepared by twin-screw melt extrusion. *Polym. Compos.* **2004**, 25, 535–542.
16. Camino, G.; Gianelli, W.; Dintcheva, N.T.; Verso, S.L.; La Mantia, F.P. EVA–montmorillonite nanocomposites: effect of processing conditions. *Macromol. Mater. Eng.* **2004**, 289, 238–244.
17. Hu, Y.; Tang, Y.; Wang, J.; Gui, Z.; Chen, Z.; Zhuang, Y.; Fan, W.; Zong, R. influence of organophilic clay and preparation methods on EVA/montmorillonite nanocomposites. *J. Appl. Polym. Sci.*, **2004**, 91, 2416–2421.
18. Morgan, A.B.; Chu, L.; Harris, J.D. A flammability performance comparison between synthetic and natural clays in polystyrene nanocomposites. *Fire Mater.* **2005**, 29, 213–229.
19. Wilkie, C.; Costache, M.C.; Jiang, D.D. Thermal degradation of ethylene–vinyl acetate copolymer nanocomposites. *Polymer* **2005**, 46, 6947–6958.
20. Wilkie, C.; Costache, M.C.; Jang, B.N. The relationship between thermal degradation behavior of polymer and the fire retardancy of polymer/clay nanocomposites. *Polymer* **2005**, 46, 10678–10687.
21. Frache, A.; Constantino, U.; Gallipoli, A.; Nchetti, M.; Camino, G.; Bellucci, F. New nanocomposites constituted of polyethylene and organically modified ZnAl-hydrotalcites. *Polym. Degrad. Stab.*, **2005**, 90, 586–590.
22. Nelson, G.L.; Yngard, R.; Yang, F. Flammability of polymer–clay and polymer–silica nanocomposites. *J. Fire Sci.* **2005**, 23, 209–226.
23. Zammarano, M.; Franceschi, M.; Bellayer, S.; Gilman, J.W.; Meriani, S. Preparation and flame resistance properties of revolutionary self-extinguishing epoxy nanocomposites based on layered double hydroxides. *Polymer* **2005**, 46, 9314–9328.

24. Zammarano, M.; Franceschi, M.; Mantovani, F.; Minigher, A.; Celotto, M.; Meriani, S. Flame resistance properties of layered-double-hydroxides/epoxy nanocomposites; in: M. Le Bras et al., Eds., *Proceedings of the 9th European Meeting on Fire Retardancy and Protection of Materials*, Villeneuve d'Ascq., France, Sept. 17–19 2003.
25. Kashiwagi, T.; Grulke, E.; Hilding, J.; Harris, R.; Awad, W.; Douglos, J. Thermal degradation and flammability properties of poly(propylene)/carbon nanotube composites. *Macromol. Rapid Commun.* **2002**, 23, 761–765.
26. Kashiwagi, T.; Grulke, E.; Hilding, J.; Groth, K.; Harris, R.; Butler, K.; Shields, J.; Kharchenko, S.; Douglas, J. Thermal and flammability properties of polypropylene/carbon nanotube nanocomposites. *Polymer* **2004**, 45, 4227–4239.
27. Kashiwagi, T.; Du, F.; Winey, K.; Groth, K.; Shields, J.; Bellayer, S.; Kim, H.; Douglas, J. Flammability properties of polymer nanocomposites with single-walled carbon nanotubes: effects of nanotube dispersion and concentration. *Polymer* **2005**, 46, 471–481.
28. Kashiwagi, T.; Du, F.; Douglas, J.; Winey, K.I.; Harris, R.; Shields, J. Nanoparticle networks reduce the flammability of polymer nanocomposites. *Nat. Mater.*, **2005**, 4, 928–933.
29. Schartel, B.; Pötschke, P.; Knoll, U.; Abdel-Goad, M. Fire behaviour of polyamide 6/multiwall carbon nanotube nanocomposites. *Eur. Polym. J.* **2005**, 41, 1061–1070.
30. Leroy, E.; Lopez-Cuesta, J.-M.; Clerc, L. Influence of talc physical properties on the fire retarding behaviour of (ethylene–vinyl acetate copolymer/magnesium hydroxide/talc) composites. *Polym. Degrad. Stab.* **2005**, 88, 504–511.
31. Ferry, L.; Durin-France, A.; Lopez-Cuesta, J.-M.; Crespy, A. Magnesium hydroxide/zinc borate/talc compositions as flame-retardants in EVA copolymer. *Polym. Int.*, **2000**, 49, 1101–1105.
32. Le Bras, M.; Bourbigot, S.; Carpentier, F.; Delobel, R. Rheological investigations in fire retardancy: application to ethylene–vinyl-acetate copolymer–magnesium hydroxide/zinc borate formulations. *Polym. Int.* **2000**, 49, 1216–1221.
33. Jho, J.Y.; Hong, C.H.; Lee, Y.B.; Bae, J.W.; Nam, B.U.; Nam, G.J.; Lee, K.J. Tensile and flammability properties of polypropylene-based RTPO/clay nanocomposites for cable insulating material. *J. Appl. Polym. Sci.*, **2005**, 97, 2375–2381.
34. Wilkie, C. Nanocomposite formation as a component of a fire retardant system, in: *Proceedings of the 10th European Meeting on Fire Retardancy and Protection of Materials*, Berlin, Sept. 7–9, 2005.
35. Bourbigot, S.; Le Bras, M.; Dabrowski, F.; Gilman, J.W.; Kashiwagi, T. PA-6 clay nanocomposite hybrid as char forming agent in intumescent formulations. *Fire Mater.*, **2000**, 24, 201–208.
36. Hu, Y.; Tang, Y.; Xiao, J.; Wang, J.; Song, L.; Fan, W. PA-6 and EVA alloy/clay nanocomposites as char forming agents in poly(propylene) intumescent formulations. *Polym. Adv. Technol.*, **2005**, 16, 338–343.
37. Hu, Y.; Song, L.; Lin, Z.; Xuan, S.; Wang, S.; Chen, Z.; Fan, W. Preparation and properties of halogen-free flame-retarded polyamide 6/organoclay nanocomposite. *Polym. Degrad. Stab.* **2004**, 86, 535–540.
38. Ferry, L.; Gaudon, P.; Leroy, E.; Lopez Cuesta, J.-M. Intumescence in ethylene–vinyl acetate copolymer filled with magnesium hydroxide and organoclays, in: *Fire Retardancy of Polymers*. Royal Society of Chemistry, London 2005, pp. 302–312.

39. Horrocks, A.R.; Kondola, B.K.; Padbury, S.A. Interactions between nanoclays and flame retardant additives in polyamide 6, and polyamide 6,6 films, in: *Fire Retardancy of Polymers*. Royal Society of Chemistry, London 2005, pp. 223–238.

40. Whaley, P.D.; Cogen, J.M.; Lin, T.S.; Bolz, K.A. Nanocomposite flame retardant performance: laboratory testing methodology in: *Proceedings of the 53rd International Wire and Cable Symposium*, Philadelphia, PA, Nov. 2004, pp. 605–611.

41. Shen, K.K.; Olsen, E. Borates as FR in halogen-free polymers, in: *Proceedings of the 15th Annual BCC Conference on Flame Retardancy*, Stamford, CT, June 6–9, 2004.

42. Shen, K.K.; Olsen, E. Recent advances on the use of borates as fire retardants in halogen-free systems, in: *Proceedings of the 16th Annual BCC Conference on Flame Retardancy*, Stamford, CT, May 22–25, 2005.

43. Ristolainen, N.; Hippi, U.; Seppälä, J.; Nykänen, A.; Ruokolainen, J. Properties of polypropylene/aluminum trihydroxide composites containing nanosized organoclay. *Polym. Eng. Sci.* **2005**, 45, 1568–1575.

44. Wilkie, C.; Zhang, J. Fire retardancy of polyethylene–alumina trihydrate containing clay as a synergist. *Polym. Adv. Technol.* **2005**, 16, 549–553.

45. Cusack, P.A.; Cross, M.S.; Hornsby, P.R. Effects of tin additives on the flammability and smoke emission characteristics of halogen-free ethylene–vinyl acetate copolymer. *Polym. Degrad. Stab.*, **2003**, 79, 309–318.

46. Beyer, G.; Alexandre, M.; Henrist, C.; Cloots, R.; Rulmont, A.; Jérôme, R.; Dubois, P. Poster presentation: Preparation, morphology, mechanical and flame retardant properties of EVA/layered silicate nanocomposites, presented at the World Polymer Congress, IUPAC Macro 2000, 38th Macromolecular IUPAC Symposium, Warsaw, Poland, 2000.

47. Beyer, G.; Alexandre, M.; Henrist, C.; Cloots, R.; Rulmont, A.; Jérôme, R.; Dubois, P. Preparation and properties of layered silicate nanocomposites based on ethylene vinyl acetate copolymers. *Macromol. Rapid Commun.* **2001**, 22, 643–646.

48. Beyer, G. Flame retardant properties of EVA-nanocomposites and improvements by combination of nanofillers with aluminum trihydrate, *Fire Materi.* **2001**, 25, 193–197.

49. Babrauskas, V.; Peacock, R.D. Heat release rate: the single most important variable in fire hazard. *J. Fire Safety* **1992**, 18, 255–272.

50. UL-94, *Test for Flammability of Plastic Materials for Parts in Devices and Appliances*. 1966-10-00. Underwriters Laboratories, Inc., Chicago, **1966**.

51. Bourbigot, S.; Le Bras, M.; Leeuwendal, R.; Shen, K.K.; Schubert, D. Recent advances in the use of zinc borates in flame retardancy of EVA. *Polym. Degrad. Stab.*, **1999**, 64, 419–425.

52. Beyer, G. Flame retardancy of nanocomposites: from research to technical products. *J. Fire Sci.* **2005**, 23, 75–87.

53. IEC 60332-3-24, *Tests on Electrical Cables Under Fire Conditions*, Part 3–24; Test for vertical flame spread of vertically-mounted bunched wires or cables; Category C, 2000-10-00. International Electrotechnical Commission, 2000.

54. Herbert, M.J.; Brown, S.C. New developments in ATH technology and applications, in: *Proceedings of the Flame Retardants '92 Conference*, London, Jan. 12–13, 1992, pp. 100–119.

55. Breuer, O.; Sundararaj, U. Big returns from small fibers: a review of polymer/carbon nanotube composites. *Poly. Compos.* **2004**, 25, 630–645.

56. Tang, B.Z.; Xu, H. Preparation, alignment, and optical properties of soluble poly(phenylacetylene)-wrapped carbon nanotubes. *Macromolecules* **1999**, 32, 2569–2576.
57. Bernier, P.; Journet, C. Production of carbon nanotubes. *Appl. Phys. A Mater. Sci. Process.*, **1998**, 67, 1–9.
58. Coleman, J.N.; Blau, W.J.; Dalton, A.B.; Munoz, E.; Collins, S.; Kim, B.G.; Razal, J.; Selvidge, M.; Vieiro, G.; Baughman, R.H. Improving the mechanical properties of single-walled carbon nanotube sheets by intercalation of polymeric adhesives. *Appl. Phys. Lett.*, **2003**, 82, 1602.
59. Beyer, G. Carbon nanotubes as flame retardants for polymers. *Fire Mater.* **2002**, 26, 291–293.
60. Beyer, G. Improvements of the fire performance of nanocomposites, in: *Proceedings of the 13th Annual BCC Conference on Flame Retardancy*, Stamford, CT, June 3–6, 2002.
61. Gilman, J.W.; Jackson, C.L.; Morgan, A.B.; Harris, R.; Manias, E.; Giannelis, E.P.; Wuthenow, M.; Hilton, D.; Philipps, H. Flammability properties of polymer–layered-silicate nanocomposites: polypropylene and polystyrene nanocomposites. *Chem. Mater.*, **2000**, 12, 1866–1873.
62. Beyer, G. Filler blend of carbon nanotubes and organoclays with improved char as a new flame retardant system for polymers and cable application. *Fire Mater.*, **2005**, 29, 61–69.
63. Beyer, G.; Gao, F.; Yuan, Q. A mechanistic study of fire retardancy of carbon nanotube/ethylene vinyl acetate copolymers and their clay composites. *Polym. Degrad. Stab.*, **2005**, 89, 559–564.
64. Elliot, P.J.; Whiteley, R.H. A cone calorimeter test for the measurement of flammability properties of insulated wire. *Polym. Degrad. Stab.* **1999**, 64, 577–584.

8

NANOCOMPOSITES WITH HALOGEN AND NONINTUMESCENT PHOSPHORUS FLAME RETARDANT ADDITIVES

Yuan Hu and Lei Song

State Key Lab of Fire Science, University of Science and Technology of China, Anhui, China

8.1 INTRODUCTION

8.1.1 Polymer–Organoclay Nanocomposites

Since the twentieth century, natural and synthetic polymeric materials have been used to replace many natural materials in several applications, such as construction, electrical and electronics components, household, and transportation. Although these polymeric materials provide many benefits, most of them are unfortunately more flammable than the materials that have been replaced. Therefore, the use of flame retardants to reduce the flammability of these replacement polymers and production of smoke or toxic products during their combustion has become an important aspect of the research, development, and application of new materials.[1,2]

In the past two decades, many academic or industrial researchers have paid attention to polymer/layered-silicate nanocomposites (PLSs), especially to polymer/organically modified clay (organoclay) nanocomposites. These PLS materials often exhibit remarkable improvement in properties over those of both virgin polymer and conventional filled systems, and many represent a better choice

Flame Retardant Polymer Nanocomposites, edited by Alexander B. Morgan and Charles A. Wilkie

than the latter system. Therefore, nanotechnology has been described as the next great frontier of materials research because nanocomposite formation brings about improved material performance, including enhanced mechanical, thermal, optical, dimensional, and barrier performance properties.[3–7] Moreover, these PLS nanocomposites have considerably improved flame retardant properties.[8–16] It has become clear that one of the most promising aspects of PLS materials is the concomitant improvement in both flammability and physical properties,[3–16] most notably attributed to specific interactions occurring on the nanometer scale. Therefore, some researchers suggest that PLS nanocomposites will provide a totally new and promising approach to flame retardancy of polymers without compromising their desirable service properties.

8.1.2 Conventional Halogen and Nonintumescent Phosphorus-Containing Flame Retardants

Conventional flame retardants act either in the vapor or condensed phase through a chemical and/or physical mechanism to interfere with the combustion process at various stages of the process (e.g., during heating, decomposition, ignition, or flame spread).[2] Halogenated flame retardants, arguably one of the most widely used classes of flame retardants, act in the vapor phase through a radical chain reaction to interrupt the combustion reaction.[17] Brominated or chlorinated flame retardants can be employed, although the former is more useful for a variety of reasons, such as cost versus performance. Also, the addition of metal oxides, such as antimony trioxide (AO), as coadditives to halogenated flame retardants further increases halogenated additive efficiency through the formation of antimony trihalide.[18] The antimony trihalide possesses even greater volatility under conditions of combustion whereby it can interfere with and retard the propagating radical chain reactions in the flame, even though the oxide itself has no effect.[15] The polybrominated aromatics, such as decabromdiphenyloxide (DB) and AO, are the most commonly used additives that provide a good flame retardant effect (i.e., high-temperature resistance, good processability, and excellent compatibility with the polymer formulation, such as fiber-reinforced materials).[19–21] Although halogenated flame retardants are used widely, in particular for polymers composites or in materials for electronics, they can impart dangerous effects. Specifically, they may give rise to toxic, acidic, and dense smoke,[22] which can threaten people and damage costly equipment. Both the European Union (EU) and the United States (U.S.) government have expressed concern about the toxicity and environmental impact of the primary halogenated additives in used today. The EU has proposed to restrict the use of brominated diphenyloxide flame retardants because highly toxic and potentially carcinogenic brominated furans and dioxins may form during combustion.[23] The World Health Organization (WHO) and the U.S. Environmental Protection Agency (EPA) also recommend exposure limit and risk assessment of dioxins and similar compounds.[24,25] As a result of the growing concern regarding the adverse effects inherent during thermal degradation in the halogenated additives used currently, a significant amount of research has been devoted to the development of nonhalogenated additives for flame retardant polymer formulations.

Phosphorus-containing flame retardants (PFRs) have been used as effective flame retardants; the range of PFRs is extremely wide and diverse, since the element exists in several oxidation states. Phosphines, phosphine oxides, phosphonium compounds, phosphonates, elemental red phosphorus, and phosphates have all been studied as flame retardants,[2,26–30] with varying degree of success. The PFRs are usually divided into two categories: (1) inorganic derivatives: for example, ammonium polyphosphate (APP) and red phosphorus; and (2) organic derivatives such as aromatic phosphates: for example, triphenyl phosphate; alkyl-substituted triaryl phosphates such as cresyl diphenyl phosphate, isopropylphenyl diphenyl phosphate, *tert*-butylphenyl diphenyl phosphate, tricresyl phosphate, trixylyl phosphate, and so on, and oligomeric phosphates such as resorcinol bis(diphenyl phosphate) (RDP).

By analogy, nitrogen-containing flame retardants may also be considered as environmental friendly since they are less toxic, do not produce dioxin- and halogen acid–containing by-products, and give a low release of smoke during combustion. The most important nitrogen containing flame retardants (NFRs) are melamine and its derivatives.[31,32] Melamine phosphates and pyrophosphates are another distinct type of flame retardant combining a nitrogen element with a phosphorus element, which may give rise to synergistic effects in flame retardant polymeric materials. The action of both PFRs and NFRs in the vapor phase is best described as a radical mechanism to interrupt and thus suppress the exothermic processes of combustion. Both may also act in the condensed phase by promoting carbonaceous residue or char formation. To date, two char-forming mechanisms have been reported: (1) redirection of the chemical reactions involved in decomposition in favor of reactions yielding carbon rather than CO or CO_2, and (2) formation of a surface layer of protective char. The char acts as a barrier to inhibit gaseous products from diffusing to and feeding the flame and to shield the polymer surface from heat and air.[2,26–32]

Several patents and literature since the 1990s have reported the combination of thermoplastic polymer–organoclay nanocomposites and commercial flame retardant additives to obtain simultaneously reduced flammability and improved mechanical properties. The commercial flame retardant additives used include halogenated,[33–37,41,46,49,56–59] phosphorus,[36–40,55,60] nitrogen,[38,50,60] metal hydroxide,[61–65] intumescent,[66–70] and other flame retardant additives.[71,72]. The goal in this chapter is to review the recent development of thermoplastic polymer–organoclay nanocomposites with halogenated, nonintumescent phosphorus and nitrogen flame retardant additives.

8.2 PREPARATION METHODS AND MORPHOLOGICAL STUDY

Generally, a polymer–organoclay with flame retardant additives may be prepared either by a blending process through melt or solution blending, or by an in situ polymerization process, depending on starting materials and processing techniques. Prior to preparation, the clay is organically modified with an appropriate surfactant (e.g., alkyl ammonium salt) by ion exchange in water to obtain organically modified clay: namely, organoclay. These preparation methods are described briefly here in the context of different polymer resin formulations.

8.2.1 Melt Compounding and Solution Blending

In the solution blending method, the organoclay is first swollen in a solvent. Then a flame retardant additive and polymer are added to the organoclay solution. The polymer dissolves in the solution, and the polymer molecule chain can intercalate into the gallery of the organoclay. Upon solvent removal, a nanocomposite is obtained. The solvent must be compatible with polymer, organoclay, and flame retardant additives. It can be difficult to select an appropriate solvent for this method. Therefore, solution blending is not often used to prepare flame retardant nanocomposites. In the melt-blending method, the flame retardant polymer–organoclay nanocomposites are prepared by mixing the flame retardant additive, the organoclay, and the polymer at above the softening or melting point of the polymer through mechanical mixing and shearing introduced by an internal mixer or twin-screw extruder. Generally, organic modification of the clay can be achieved prior to the nanocomposite preparation. However, it can also be achieved by adding both pristine clay and surfactant to the flame retardant polymer formulation during melt blending, thus eliminating some preparation steps.[73] This method is environmentally benign, due to the absence of organic solvents, is compatible with current industrial processing, and provides a potentially economical and convenient way of imparting and promoting superior flame retardancy for commercial polymers.

Recently, our group reported several types of flame retardant polymer–organoclay nanocomposites with either halogenated-antimony or halogen-free flame retardant additives[34–36] prepared using melt-blending methods. For example, flame retardant acrylonitrile–butadiene–styrene (ABS)/organoclay nanocomposites were prepared by blending ABS, organoclay (OMT), and flame retardant additives [decabromodiphenyloxide (DB)–antimony oxide (AO)] using a twin-roll mill to obtain ABS/OMT/DB–AO nanocomposite.[35] Flame retardant polyamide-6 (PA6)/organoclay nanocomposite was prepared by blending PA6, OMT, and either DB–AO or magnesium hydroxide–red phosphorus (MH–RP) using a twin-screw extruder to obtain PA6/OMT/DB–AO and PA6/OMT/MH–RP nanocomposites.[34,36]

X-ray diffraction (XRD) and transmission electron microscopy (TEM) analyses revealed the morphology of each melt-blended nanocomposites formulation.[34–36] Figure 8.1 shows that the *d*-spacing of the OMT increases after organic modification by hexadecyltrimethylammonium bromide (C16) from 1.3 nm to 2.2 nm. The *d*-spacing increase suggests that the C16 intercalates into the gallery between the individual silicate sheets of MMT. Upon formation of the ABS–OMT and PA6–OMT nanocomposites, the *d*-spacing increases further, from 2.2 nm to 3.1 and 3.3 nm, respectively, and the diffuse diffraction peak for the nanocomposite indicates the formation of an intercalated–exfoliated morphology. Clearly, the

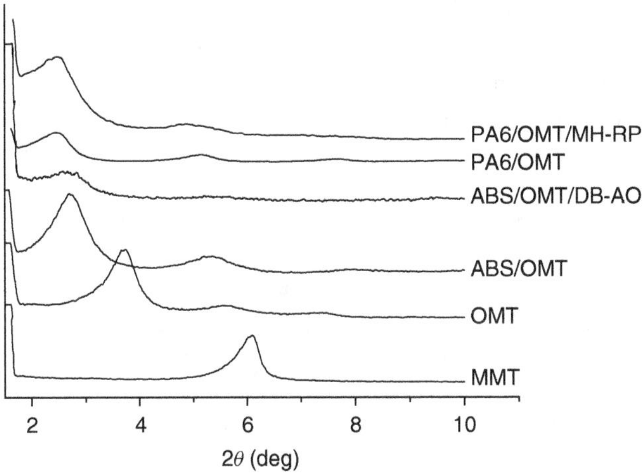

FIGURE 8.1 XRD patterns for pristine clay (MMT), organoclay (OMT), ABS–OMT (ABS/5% OMT), ABS/OMT/DB–AO (ABS/5% OMT/15% DB–3% AO), PA6–OMT (PA6/2% OMT), and PA6/OMT/MH–RP (PA6/2% OMT/6% MH–5% RP). (Adapted from Refs. 34 to 36.)

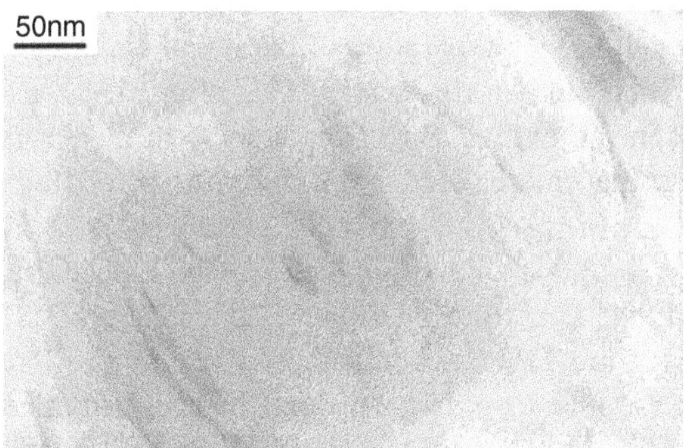

FIGURE 8.2 HREM of ABS–OMT nanocomposite. (From Ref. 35, copyright © 2004, Elsevier, with permission.)

addition of flame retardant additives, either DB–AO or MH–RP, in the nanocomposite formulation during melt blending has a minimal effect on the position and shape of the (001) plane diffraction peaks, and therefore no further exfoliation occurs. This result is probably attributable to both DB–AO and MH–RP being stable solid particles. High-resolution electron microscopy (HREM) micrographs (Figure 8.2) for ABS–OMT further confirm that a mixed morphology forms in the polymer matrix during melt processing,[35] yielding some tactoids containing only two to four clay sheets along with single clay sheets. The generally accepted opinion is that both exfoliated and intercalated morphologies impart measurable improvement in both mechanical and flame retardant properties, but the exfoliated morphology imparts greater improvement in mechanical properties than does the intercalated morphology.[13] There does not appear to be a significant difference between the intercalated and exfoliated morphologies in flame retardant performance for polymer nanocomposites.[41] When both intercalated and exfoliated morphologies exist in a polymer matrix, increasing the exfoliation level probably benefits the improvement of the mechanical properties for polymer nanocomposites. The exfoliated morphology formation depends on the nature and interactions between the surfactant, the clay, and the polymer matrix. Until now, the relationship between the morphology and properties of the polymer–organoclay nanocomposites has not been completely revealed, but with further research and development on the nanocomposites, this may be resolved.

Zanetti et al.[33] reported the preparation of brominated-antimony flame retardant polypropylene–graft–maleic anhydride/organoclay (PP-g-MA/OMT/DB–AO) nanocomposites by mixing the polymer, the commercial organoclay, and DB–AO in a Brabender internal mixer. It was found from analysis of this system that intercalated and exfoliated morphologies coexist in the polymer matrix.[33] The literature[56,57] discloses that flame retardant polypropylene (PP)–organoclay nanocomposites containing either DB–AO or chlorinated paraffin wax (CPW) were prepared by melt blending with a twin-screw extruder. Organoclay is difficult to exfoliate in the PP matrix because of the poor compatibility between PP matrix and organoclay. In the preparation of these flame retardant PP–organoclay nanocomposites, maleic anhydride–grafted PP (PP-g-MA) was used as a compatibilizer to assist in the exfoliation of organoclay in the PP matrix. XRD and TEM analyses show that intercalated and exfoliated morphologies coexist in the polymer matrix.

Kim[37] reported a complicated melting blend method much different from those mentioned above. First, a mixture of triphenyl phosphate (TPP) and organoclay was mixed to obtain nano TPP.[37] Then an epoxy resin and silane-coupling agent were added to the mixture to obtain the final flame retardant mixture. Finally, this mixture and ABS was blended in a Haake Plasti-Corder mixer to obtain the flame retardant ABS–OMT–TPP nanocomposites. The epoxy resin and silane-coupling agent act as flame retardant synergists. XRD analysis shows that the *d*-spacing of organoclay increases by intercalation of TPP.[37] The intercalated TPP can escape from nano TPP upon heating to 300°C[37]; there is no chemical bond between TPP and clay. However, the intercalation of TPP into the gallery space of the organoclay could reduce the volatility and suppress the evaporation rate of TPP. Alkyl-substituted triaryl phosphates (ArPs) such as cresyl diphenyl phosphate, isopropylphenyl diphenyl phosphate, *tert*-butylphenyl diphenyl phosphate, tricresyl phosphate, and trixylyl phosphate have very limited application in engineering plastics because of their high volatility and relatively low flame retardancy. The organoclay can probably be used to increase the thermal stability and flame retardant properties of these aromatic phosphates by forming an intercalated ArP–organoclay hybrid.

The literature[40] described a type of flame retardant organoclay combining flame retardant elements with organoclay. The flame retardant organoclay (OLP–clay) was modified by oligomeric phosphate ammonium salt. In a two-step synthesis, the oligomeric phosphate was prepared from the polymerization of diphenyl 4-vinylphenyl phosphate (DPVPP), styrene, and 4-vinylbenzylchloride followed by ammonium salt introduction through reaction of pendant benzyl chloride groups with dimethylhexadecylamine.[40] Two OLP–clays were reported, with the oligomeric phosphate ammonium cation containing 55 and 75% DPVPP, respectively, and the corresponding polystyrene–organoclay nanocomposites (PS/OLP–clay) were prepared by melt blending. The literature also provided several OLP–clays with the oligomeric phosphate ammonium cation containing different substituted vinyl phosphates, such as diphenyl-4-vinylbenzylphsophate (DPVBP), 1-vinylphosphonic acid (VPA), and 1-phenylvinylphosphonic acid (PVPA). In another report,[41] the flame retardant organoclay (OLB–clay) with an oligomeric bromine ammonium cation (OLB) and the corresponding polystyrene–organoclay nanocomposites (PS/OLB–clay) were prepared. XRD and TEM analysis[40,41] reveals that an intercalated morphology is obtained in these PS nanocomposites. Typically, a significant amount of either bromine or phosphorus is incorporated in the clay, and these flame retardant elements will be well mixed with the clay. If the clay is nano-dispersed in the polymer matrix, these flame retardant elements also will be well dispersed throughout the polymer matrix along with the clay, and this should facilitate the improvement in the flame retardancy of the polymer.

Two patents reported flame retardant polymeric materials derived from the incorporation of flame retardant clays.[50,60] A Japanese patent[50] reported that PA6, poly(butylene terephthalate) (PBT), poly(oxymethylene) (POM), and poly (phenylene sulfide) (PPS), respectively, were prepared as polymer/clay–triazine nanocomposites using the synthetic silicate, fluorohectorite (FSM), by melt blending. Various melamine salts were used either to treat the FSM directly or added to the polymer formulation in the range 8 to 15 wt%. The inventors state that the nanometer-scale dispersion of FSM additive is very important for the flame retardant properties of the resulting nanocomposite. Another patent[60] described flame retardant polypropylene–clay nanocomposites in which the clay was preinterclated with an amine-based flame retardant melamine phosphate (MP) either alone or together with an alkyl ammonium compound. The flame retardant clay was then mixed with a polypropylene–graft–maleic anhydride material using a solution blending technique to give a flame retardant master batch. In a final step, the flame retardant master batch was melt-blended with polypropylene to prepare the corresponding flame retardant polypropylene-clay nanocomposites. It is obvious that clay can be modified not only by organic surfactants but also by inorganic amine-based flame retardant additives through ion exchange in water to obtain flame retardant clays. These clays can either be used separately or combined with other flame retardant additives to obtain the flame retardant polymer formulation.

8.2.2 In Situ Polymerization Method

In the in situ polymerization method, flame retardant additives and organoclay are dispersed within a liquid monomer or monomer solution. Then the mixture is polymerized and initiated either by heat, radiation, or the addition of a suitable initiator. An initiator that may diffuse into the clay can be incorporated into the clay galleries as a result of a cation exchange. In this method, miscibility between the monomer and the organoclay is better than that between the polymer macromolecule and organoclay. Therefore, this method facilitates organoclay dispersion and exfoliation in the polymer matrix.

Our group recently reported the formation of halogen-free flame retardant polyurethane nanocomposites (PU) [polyurethane–OMT–MPP] synthesized via in situ polymerization.[38] The polyurethane is polymerized with monomers [polyether and toluene diisocyanate (TDI)], a chain extender (diglycol), and a cross-linking agent (glycerin). Upon PU nanocomposite formation, the organoclay (OMT) and melamine polyphosphate (MPP) mixed with polyether form an intercalated polyether–OMT–MPP mixture, and the reaction with TDI gives a prepolymer. The prepolymer is then mixed with glycerin and diglycol to synthesize the flame retardant PU–OMT–MPP nanocomposites. The XRD patterns of MMT, OMT, PU–OMT, and PU–OMT–MPP are shown in Figure 8.3. The basal spacing (d_{001}) of MMT and OMT are 1.3 and 2.1 nm, respectively. Figure 8.3 shows that the (001) diffraction peaks of PU–OMT and PU–OMT–MPP both

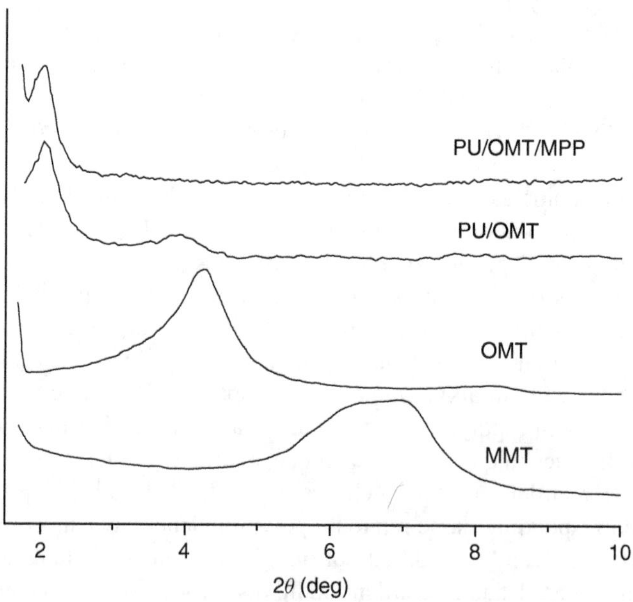

FIGURE 8.3 XRD patterns for pristine clay (MMT), organoclay (OMT), PU–OMT (PU/5% OMT), PU–OMT–MPP (PU/5% OMT/6% MPP). (From Ref. 38, copyright © 2005, Elsevier, with permission.)

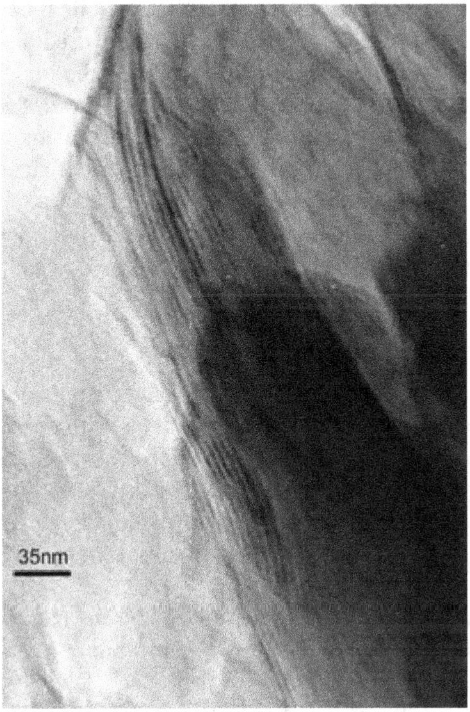

FIGURE 8.4 HREM for PU–OMT–MPP (PU/5% OMT/6% MMP) nanocomposite. (From Ref. 38, copyright © 2005, Elsevier, with permission.)

shift from 2.1 nm to 4.8 nm, thus increasing the gallery height by 2.7 nm. This result indicates an intercalated morphology. Apparently, addition of MPP neither alters this intercalated morphology nor facilitates any further exfoliation, as supported by HREM (Figure 8.4).

Chigwada[39] reported the preparation of polystyrene–organoclay nanocomposites containing various aromatic phosphates [PS–OMT–aromatic phosphate (ArP)]. The nanocomposites were prepared via in situ polymerization by mixing styrene, ArP, organoclay, and initiator. An intercalated morphology is formed in the polystyrene nanocomposite.[39] Moreover, addition of ArP again neither affected the *d*-spacing nor facilitated further clay sheet exfoliation in these polystyrene nanocomposites, a result consistent with other reports employing similar small-molecule flame retardants.

A halogenated flame retardant ABS–organoclay nanocomposite was prepared using a two-step method.[59] Initially, the organoclay was swelled with both monomers, acrylonitrile and styrene followed by polymerization to give SAN–organoclay nanocomposites. Then the SAN–organoclay nanocomposite was blended with polybutadiene rubbers and halogenated flame retardant additives in a twin-screw extruder to give flame retardant ABS–organoclay nanocomposites; XRD analysis and TEM observation confirm an intercalated morphology.

In the polymerization method, the organoclay is swollen with the liquid monomer or a monomer solution. The clay is a type of layered aluminosilicate, so it is hydrophilic and possesses high polarity. Although the organoclay is modified organically, it still shows high polarity. The attractive forces between the organoclay sheets and the monomer can facilitate the intercalation of monomer into the gallery space of the organoclay. The intercalation efficiency depends on the nature of the organic groups of the surfactant on the surface of the clay sheet and the chemical structure of the monomer used. This method facilitates organoclay dispersion and exfoliation in the polymer matrix.

8.2.3 Summary of Synthetic Methods

In summary, flame retardant polymer–organoclay nanocomposites can be prepared using melt blending, solution blending, or the in situ polymerization method. Intercalated, exfoliated, or intercalated–exfoliated morphology is formed in the polymer matrix. Addition of flame retardant additives hardly affects nanocomposite formation. Specifically, flame retardant additives are introduced into the polymer nanocomposite formulation by three approaches:

1. A flame retardant additive can be added during nanocomposite preparation. Generally, both inorganic and organic flame retardant additives can be added to the formulation by this approach; and their presence is not expected to affect the nanocomposite morphology.
2. Flame retardant additives are incorporated into the organoclay through ion exchange or physical adsorption before nanocomposite preparation.
3. An organic group or oligomer containing flame retardant elements such as bromine or phosphorus can be grafted to the surfactant used to modify the clay.

It is well known that the miscibility between flame retardant additives and the polymer matrix in a flame retardant formulation restricts the properties of the flame retardant polymer. Poor miscibility between flame retardant additives and the polymer matrix often causes a decrease in the mechanical and other properties of the base polymer used in a flame retardant formulation. However, the addition of organoclay may bring about simultaneous improvement in the flame retardant and mechanical properties, thereby offsetting any adverse effects imparted by the flame retardant additives. Moreover, the latter two approaches both impart a flame retardant performance to organoclay by incorporating the flame retardant additive into the gallery space of the organoclay. When the organoclay is nano-dispersed, the flame retardant additives will also be well dispersed in the polymer matrix during nanocomposite preparation, which facilitates dispersion of the flame retardant additives in the polymer nanocomposite. Moreover, addition of the organoclay may lead to a significant reduction in the amount of the flame retardant needed, while maintaining or even improving the flammability performance and physical properties of the flame retardant polymer.

8.3 THERMAL STABILITY

Thermogravimetric analysis (TGA) is used routinely to study the thermal stability of polymeric materials that have been carried out in inert atmosphere or in air. Generally, the addition of organoclay to the polymer enhances the thermal stability by acting as a superior insulator and mass transport barrier to volatile by-products generated during decomposition, thereby suppressing combustion.[2] TGA analysis of MPP, PU, OMT, PU–OMT, PU–MPP, and PU–OMT–MPP nanocomposites are shown in Figure 8.5. The PU degradation process usually includes three stages[48]:

1. Depolymerization of polyurethane leads to the formation of alcohols and isocyanates.
2. Isocyanates dimerize to carbodiimides, which react with the alcohol groups to give relatively stable cross-linked substituted ureas. Trimerization of isocyanates may also occur under certain conditions to yield thermally stable isocyanurate rings.
3. High-temperature degradation of these stable cross-linking structures yields volatile products and charred carbonaceous structure. From TGA, the onset of MPP thermal degradation occurs at about 360°C, volatilizing melamine and generating polyphosphoric acid.[68]

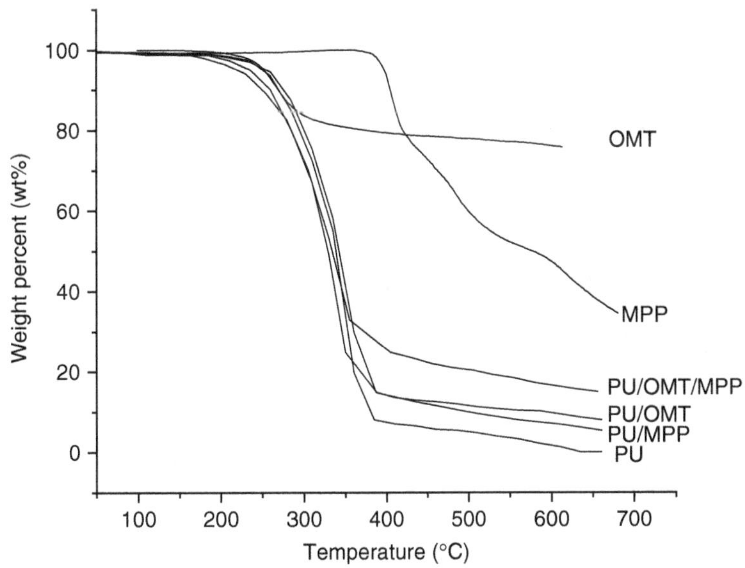

FIGURE 8.5 Comparison of the TGA curves of PU, PU–MPP (PU/6% MPP), PU–OMT (PU/5% OMT), PU–OMT–MPP (PU/5% OMT/6% MPP), OMT, and MPP. N_2 atmosphere, a heating rate of 15°C/min. (From Ref. 38, copyright © 2005, Elsevier, with permission.)

The thermal degradation of OMT begins at about 200°C and proceeds according to the generally agreed upon Hofmann degradation mechanism[42] typical of ammonium-based systems. The ammonium cation (C16) loses an olefin through β-hydrogen elimination, thus producing a free amine and an acid proton. To balance charge, the acid proton remains on the surface of the MMT while the amine and olefin volatilize. The acid proton probably has a catalytic effect on the initial stages of decomposition of polyurethane within the OMT. Therefore, the onset degradation temperature of PU–OMT is slightly lower than that of pure PU. In the presence of MPP, the degradation temperature and rate of PU increase, due to the catalytic effect of the polyphosphoric acid.[32] In the presence of both OMT and MPP, the onset degradation temperature is decreased further. These results show that both OMT and MPP impart catalytic effects during the initial stages of PU decomposition. In contrast, the formation of carbonaceous char residue is enhanced (Table 8.1), which is probably attributable to the barrier effect generated by clay sheets, which delays the escape of volatile decomposition products from the nanocomposite. Also, a possible synergistic effect occurs between the OMT and MPP to enhance the formation of carbonaceous char residue.

Table 8.2 shows TGA data for PS–OMT–ArP nanocomposites prepared by the in situ polymerization method, including the onset degradation temperature at which 10% of the sample was lost ($T_{10\%}$), the midpoint of the degradation ($T_{50\%}$), and the fraction of material that was nonvolatile at 600°C (char). Addition of organoclay to PS led to an increase in $T_{10\%}$ from 351°C to 401°C. However, the addition of both tricresylphosphate (TCP) and trixylylphosphate (TXP) led to a decrease in $T_{10\%}$ and a smaller decrease in the corresponding $T_{50\%}$ value.[39] The reduced thermal stability of the flame retardant nanocomposite may be attributed to the low thermal stability and high volatility of the TCP and TXP flame retardants as described in the literature.[39] The authors think that the $T_{50\%}$ value is less depressed because TCP and TXP are degraded and volatilized completely by the time the midpoint of the degradation is reached. Because the thermal stability

TABLE 8.1 TGA Results for PU, PU–OMT, PU–MPP, and PU–OMT–MPP[a]

	Solid Residue at 630°C (wt%)	
Sample	Total	Char Residue
PU	0	0
OMT	74	74
MPP	42	42
PU/5% OMT	9	5
PU/6% MPP	6	4
PU/5% OMT/6% MPP	16	9

Source: After Ref. 38.

[a]PU, polyurethane; OMT, organoclay modified with hexadecyltrimethylammonium bromide; MPP, melamine polyphosphate.

TABLE 8.2 TGA Data for Polystyrene–Organoclay Nanocomposites Containing Various Aromatic Phosphates[a]

Sample	$T_{10\%}$ (°C)	$T_{50\%}$ (°C)	Char (%)	P Content (%)
PS	351	404	0	0
PS + 3% organoclay	401	454	4	0
15% TCP	353	419	2	1.3
15% TCP, 3% organoclay	374	439	6	1.3
15% TXP	370	437	3	1.1
15% TXP, 3% organoclay	376	443	6	1.1
15% RDP	417	447	2	1.6
15% RDP, 3% organoclay	387	438	8	1.6
15% RDP, 5% organoclay	404	446	8	1.6

Source: After Ref. 39.

[a]TCP, Tricresylphosphate; TXP, trixylylphosphate; RDP, resorcinol diphosphate; organoclay, organically modified montmorillonite, dimethylbenzyl hydrogenated tallow ammonium (hydrogenated tallow is a mixture of 65% C18, 30% C16, 5% C14) substituted clay.

of resorcinol diphosphate (RDP) is greater than that of TCP and TXP, the $T_{10\%}$ value is correspondingly increased. The presence of both ArP and organoclay apparently does not have an observable effect on char formation.

The thermal stability was also determined by TGA for a polymer–organoclay nanocomposite in which the organoclay contains oligomeric phosphate or bromine materials. Specifically, the TGA results for OLP–clays with 55 and 75% DPVPP and the corresponding PS nanocomposites are shown in Table 8.3. The OLP–clays show good thermal stability, with a $T_{10\%}$ value ranging from 330 to 340°C, higher thermal stability than that of traditional organoclay modified with C16 ammonium cations that degrade at about 200°C. Furthermore, the onset temperature, $T_{10\%}$, increased from 331°C to 345°C, with the content of DPVPP increasing from 55% to 75%. Moreover, the melt blending of OLP–clays with PS

TABLE 8.3 TGA Data for PS/55% DPVPP–Modified Clay and PS/75% DPVPP–Modified Clay Nanocomposites Prepared by Melt Blending[a]

Sample	$T_{10\%}$ (°C)	$T_{50\%}$ (°C)	Char (%) at 600°C	P Content (%)
Commercial PS	389	434	0	0
55% DPVPP modified clay	331	—	60	5.6
PS/55% DPVPP modified clay (5%)	425	465	11	0.28
75% DPVPP modified clay	345	455	40	7.7
PS/75% DPVPP modified clay (3%)	430	470	9	0.23
PS/75% DPVPP modified clay (5%)	421	472	12	0.39

Source: After Ref. 40.

[a]PS, polystyrene; DPVPP, diphenyl-4-vinylphenylphosphate.

led to increases in both the $T_{10\%}$ and $T_{50\%}$ values and the char upon nanocomposite formation; and increasing clay content slightly decreases the $T_{10\%}$ values. The thermal stability of oligomeric phosphate is higher than that of aromatic phosphate. The corresponding flame retardant nanocomposite shows the same regularity. TGA studies[41] of the oligomeric brominated ammonium–modified clay (OLB–clay) and corresponding polymerized nanocomposite show that the bromine content does not affect the thermal stability of OLB–clay regardless of the presence of various ammonium salts. Clearly, the thermal stability of PS/OLB–clay nanocomposites increases with increasing OLB–clay content. It appears that the degree of thermal stability of the nanocomposite is dependent on the identity of the flame retardant additive used.

It is concluded, therefore, that the combination of organoclay with oligomeric phosphate or bromine leads to an increase in thermal stability of both the modified clay and the corresponding polymer nanocomposite. The role of the flame retardant element is entirely different. The combination of a phosphate flame retardant with OMT enhances char formation. However, the addition of brominated flame retardant additives appears not to affect char formation. Thus, a general trend may not exist (i.e., in some cases bulk polymerization apparently gives enhanced thermal stability, whereas in others, melt blending may give a better result).[40,41]

8.4 MECHANICAL PROPERTIES

Table 8.4 lists the data of the mechanical properties of PP-g-MA/OMT/DB–AO nanocomposites that were prepared by melt blending. Included in the table are the stress yield, elongation at break, and storage modulus. Consistent with polymer–clay nanocomposites, the addition of only 5% OMT leads to a 100% increase in the modulus, a 1% decrease in the elongation at break, and a 19% increase in the yield stress.[33] The subsequent and sequential addition of both DB and AO in the nanocomposite formulation does not have any obvious effect, beneficial or adverse, on the mechanical properties of the material after processing. The addition of DB leads to a storage modulus increase, but storage

TABLE 8.4 Mechanical Properties of PP-g-MA/OMT/DB–AO Nanocomposites[a]

Sample	Yield Stress (MPa)	Elongation at Break (%)	Storage Modulus (MPa)
PP-g-MA	16.9	5.4	462
PP-g-MA/22% DB	15.1	4.2	628
PP-g-MA/5% OMT	20.1	4.2	955
PP-g-MA/5% OMT/22% DB/6% AO	23.3	3.8	950

Source: After Ref. 33.

[a]PP-g-MA, polypropylene–graft–maleic anhydride; DB, decabromodiphenyl oxide; OMT, organoclay modified with the ammonium cation on the clay contained a methyl group, tallow (containing 70, 25, 4, and 1 mol% of C18, C16, C14, and C12 carbon chains, respectively), and two hydroxyethyl groups; AO, antimony trioxide.

modulus increases are governed more by OMT than by DB. In a recent patent report, the addition of both a halogenated flame retardant and organoclay to an ABS matrix led to excellent flame retardancy while maintaining the desirable mechanical properties in the product nanocomposites.[59]

While evaluating the mechanical properties of the PS–OMT–TCP nanocomposites prepared by polymerization,[39] a content of 15 wt% TCP appeared to be a transition point for the mechanical properties derived. Both the remarkable plasticization effect of TCP and the enhancement of OMT probably affected the mechanical properties. Below 15 wt% TCP, the strength was enhanced by the addition of organoclay, whereas above 15 wt% TCP, the strength was not affected by the presence of organoclay. In contrast, below 15% TCP the elongation was not significantly dependent on the presence of organoclay, whereas above 15%, the elongation was slightly improved by the presence of organoclay. Up to about 15% TCP, the mechanical properties of the nanocomposite were not affected significantly by the presence of a flame retardant. Above 15 wt% TCP, the mechanical properties were sufficiently compromised that the flame retardant nanocomposite could not be utilized successfully in typical applications,[39] due primarily to the excessive plasticization effect of such a large amount of TCP. In the case of PS–organoclay nanocomposites in which oligomeric phosphate was incorporated in the clay by ion exchange, different mechanical property behavior was observed, including Young's modulus, stress at break, and strain at break of PS/OLP–clay nanocomposites. As the OLP–clay content increased there was some decrease in mechanical properties, probably due to the excessive plasticization effect of such a large amount of oligomeric phosphate.[40]

The mechanical properties of pure PA6, PA6/OMT, PA6/MH–RP, and PA6/OMT/MH–RP are given in Table 8.5.[36] The tensile strength of PA6/MH–RP decreased with the addition of MH–RP. This reduction was attributed primarily to agglomeration and phase separation of flame retardants from the bulk PA6. However, in the presence of OMT, the tensile strength of both PA6/OMT (PA6-1)

TABLE 8.5 Mechanical Properties of PA6 and PU Materials[a]

Sample	Composition	Tensile Strength (MPa)
PA6	Pure nylon 6	80.5
PA6-1	PA6 + OMT 2 wt%	103.4
PA6-2	PA6 + MH 8 wt% + RP 5 wt%	71.1
PA6-3	PA6 + OMT 2 wt% + MH 6 wt% + RP 5 wt%	98.2
PU	Pure PU	1.53
PU–OMT	PU + OMT 5 wt%	2.91
PU–MPP	PU + MPP 6 wt%	2.11
PU–OMT–MPP	PU + OMT 5 wt% + MPP 6 wt%	3.58

Source: Data from Refs. 36 and 38.

[a]PA6, nylon-6; PU, polyurethane; OMT, organoclay modified with hexadecyltrimethylammonium bromide; MPP, melamine polyphosphate; MH, magnesium hydroxide; RP, red phosphorus.

and PA6/OMT/MH–RP (PA6-3) increased compared to that of pure PA6. Specifically, more than a 25% increase in strength was achieved at only a 2 wt% loading of the OMT to the formulation. The reinforcement effect of the OMT appears to offset any adverse effects brought about by the addition of flame retardant and to enhance tensile properties even at very low loading levels, as shown in Table 8.5.

The mechanical properties of PU, PU–OMT, PU–MPP, and PU–OMT–MPP are also given in Table 8.5.[38] Interestingly, the tensile strength of PU–OMT, PU–MPP, and PU–OMT–MPP is higher than that of pure PU. In the PU case, more than a 90% increase in the tensile strength was achieved by the addition of only 5 wt% of the OMT. Apparently, the OMT and the MPP together interacted synergistically to enhance the tensile strength of PU. As a result, the tensile strength of PU–OMT–MPP increased from 1.53 MPa to 3.58 MPa, about 134% increase compared with that of pure PU. The observable positive effect of MPP on tensile strength was probably due to hydrogen bond formation between MPP and the polyurethane molecule and also to chemical bonds formed in the polymerization process between the amino group of MPP and the isocyanate group of TDI.

Clearly, the effects of flame retardant additives on the mechanical properties of polymer–clay nanocomposites are highly dependent on the identity of the flame retardant additive and polymer matrix. In general, the addition of a flame retardant usually decreases the mechanical properties, due to the poor miscibility between the flame retardant additives and the polymer matrix.[36,39–41] For a given clay content, the mechanical properties decrease as the content of flame retardants increases. Therefore, a good flame retardant polymer formulation should employ the minimal content of the flame retardant additive. Moreover, some organic flame retardant additives cause plasticization of some polymers.[40] An appropriate plasticization effect enhances the mechanical properties, but excessive plasticization compromises the mechanical properties of the flame retardant polymer. In the case of these flame retardant additives, a good flame retardant polymer formulation should employ the appropriate content of the flame retardant. However, incorporation of reactive flame retardants into the polymer matrix through typical organic transformation may facilitate completely different behavior.[38]

8.5 FLAMMABILITY PROPERTIES

Without question, the cone calorimeter instrument and technique is one of the most effective bench-scale methods used for systematic study of the flammability properties of materials under firelike conditions.[8–16] Sample plates 100 × 100 mm in size are investigated under forced-flaming conditions.[74] The sample size is of the smallest order of magnitude discussed in fire engineering and of the largest used in polymer analysis.[74] It is also a universal approach to ranking and comparing the fire behavior of materials. Therefore, the cone calorimeter is used as a characterization tool in the research and development of flame retardant polymeric materials. Typical parameters of analysis include the time to ignition (t_{ign}), heat release rate (HRR), peak heat release rate (PHRR) and time to the peak heat release rate, total heat release (THR), mass loss rate (MLR), and specific extinction area (SEA). The heat release rate, in particular the peak HRR (PHRR), has been found to be the most important parameter to evaluate the fire safety of natural and synthetic materials.[21] A cone calorimeter provides comprehensive insight into not only fire risks, such as the heat release rate, total heat release, and time to ignition, but also fire hazards such as smoke release and CO production. Polymer–organoclay nanocomposites[8–16] are reported to possess improved flame retardant properties through characterization of the flammability performance as measured by a cone calorimeter. A reduction in PHRR values from that of base polymer is a typical characteristic of polymer–organoclay nanocomposites, a property that is widely believed to be universal across many polymer classes.[8–16] These results demonstrate primarily a significant decrease in the peak heat release rate, a change in the char structure, and a decrease in the rate of mass loss during combustion in a cone calorimeter.[8–16] However, for polymer–organoclay nanocomposites, the time to ignition, t_{ign}, is usually lower than that of base polymer, while the total heat released is unchanged.[41] The decrease in t_{ign} suggests that formation of a nanocomposite upon the incorporation of organically modified clay into a base polymer facilitates ignition and degradation of the polymer; a constant THR value for the base polymer and the nanocomposite indicates complete combustion of the polymer, even in the presence of completely exfoliated clay sheets.[41] Under conditions of complete combustion, the solid residue left is merely the inorganic clay added after complete degradation of any organic modifier of clay sheet surfaces. These studies demonstrate that polymer organoclay nanocomposites eventually burn and cannot be used alone as a flame retardant system.

The goal in research of polymer–organoclay nanocomposites is to advance nanocomposites into commercial end products that can pass various fire safety regulations by means of certain fire standard tests without reducing other desirable material properties critical to practical application. In addition to the cone calorimeter, other conventional bench-scale fire tests are the limiting oxygen index (LOI) and the UL-94 test, widely used as industrial standards to determine the ignitability and flammability of polymeric materials. Therefore, to impart relevance to our discussion of conditions of industrial evaluation and application, we also evaluated these samples according to the LOI and UL-94 tests as well as the prototypical, more fundamental evaluation methods, which include TGA, XRD, and cone calorimetry. However, experiments show that nanocomposites did not perform better than virgin polymer in these tests and sometimes even showed poorer performance.[15] Therefore, a need exists for a routine experimental method that can assess accurately whether the combination of conventional flame retardants and polymer nanocomposites can give rise to beneficial flame retardant performance, such as the cone calorimeter test supported by other conventional fire tests.

The LOI measures the minimum oxygen concentration required to support candlelike combustion. The LOI method simply tests the ease at which combustion is extinguished for downward-burning materials. Therefore, the method describes the effectiveness of flame retardant additives during early stage of research and development. Many researchers consider the LOI to be very useful for research work and in connection with small fire situations. On the other hand, the UL-94 vertical test gives insight into the ease with which a polymer may burn upward or self-extinguish. There is a difference in the measuring method. The LOI is measured by downward burning, whereas the UL-94 vertical test is by upward burning. Generally, the LOI values of UL-94 V-1 and V-0 rated materials were higher than those of materials rated HB. However, it does not mean that higher LOI values will give better UL-94 V ratings.[74,75] These three methods (cone calorimeter, LOI, UL-94) give very different information, however, this information is important to describe fully and properly the flame retardant behavior of a given material. Therefore, given the added complexity of flame retardant nanocomposite systems, the question arises of how to ascribe flame retardancy properly to nanocomposite-based systems based on cone calorimeter, LOI, and UL-94 tests. To solve this problem depends not only on the identity of base polymer and flame retardant system but also on application of the end product. A study[74] showed that LOI and UL-94 were not closely related, since upward and downward burning are quite different, especially with regard to heat transfer. Weil et al.[75] reported that the LOI value might be leveled to some degree in certain conditions with UL-94 or cone calorimeter data, but it was difficult to show close relations between LOI and UL-94 or cone calorimeter data because of the downward-burning test condition of the LOI.[75] Further effort is needed to determine the correlation between these tests.

8.5.1 Cone Calorimetry

Figure 8.6 shows the plots of heat release rate (HRR) data for pure PA6, PA6-n (PA6–OMT), PA6/DB–AO, and PA6-n/DB–AO (PA6/OMT/DB–AO). Also, Figures 8.7 and 8.8 show the plot of the HRR data for pure ABS, ABS–DB, ABS–DB–AO, and ABS–DB/OMT, and ABS–DB–AO/OMT nanocomposites. The data in Figures 8.6 and 8.8 show the effect of the addition of DB and AO on pure polymer and polymer–organoclay nanocomposites on the HRR values observed. The addition of 18% DB to pure ABS reduces the PHRR to 534 kW/m^2 from 1078 kW/m^2 for pure ABS, while combining 18% DB and 5% OMT with ABS reduces the PHRR further, to 350 kW/m^2. In the case of DB–AO flame retardant systems, addition of 15% DB–3% AO to ABS reduces the PHRR value from 1078 kW/m^2 to 349 kW/m^2, while combining 15% DB–3% AO and 5% OMT with ABS further reduces the PHRR to 235 kW/m^2. Owen and Harper previously reported that a synergistic effect exists between Sb_2O_3 and halogenated flame retardant additives such as DB.[43] Analogously, the PHRR values obtained for PA6/DB–AO/OMT and ABS/DB–AO/OMT are lower than those for the respective flame retardant polymer in the absence of OMT. Using cone calorimetry as the evaluation tool, Zanetti et al.[33] showed that the flame retardant halogen–antimony–clay synergism also exists in PP-g-MA/OMT/DB–AO nanocomposites. However, in their system the presence of DB–AO did not significantly affect the HRR behavior observed for pure PP-g-MA, a result that

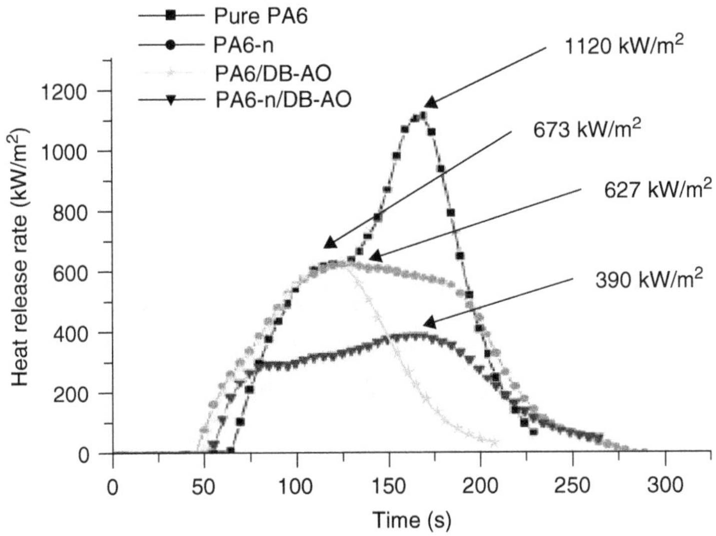

FIGURE 8.6 Comparison of the heat release rate plots for PA6, PA6-n (PA6/5% OMT), PA6/DB–AO (PA6/15% DB–5% AO), and PA6-n/DB–AO (PA6/5% OMT/15% DB–5% AO). Heat flux: 50 kW/m^2. (From Ref. 34, copyright © 2003, Wiley-VCH Verlag GmbH & Co., with permission.)

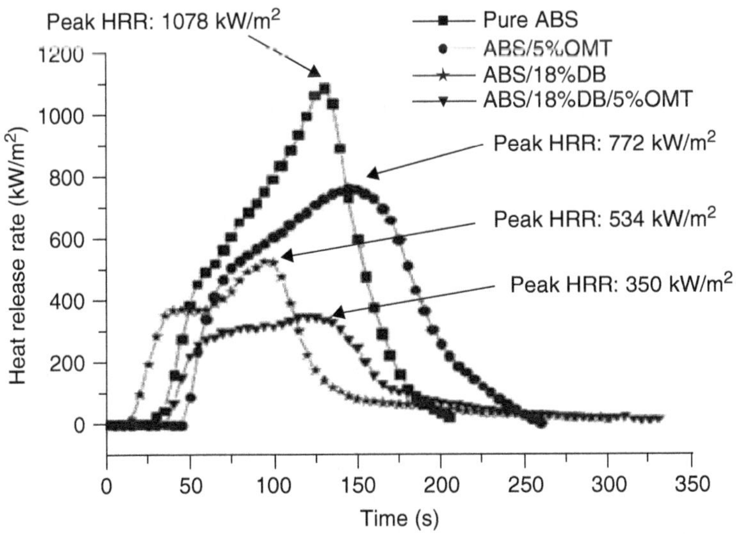

FIGURE 8.7 Comparison of HRR plots for ABS, ABS–OMT (ABS/5% OMT), ABS–DB (ABS/18% DB), and ABS/OMT/DB (ABS/15% OMT/18% DB). Heat flux: 50 kW/m^2. (From Ref. 35, copyright © 2004, Elsevier, with permission.)

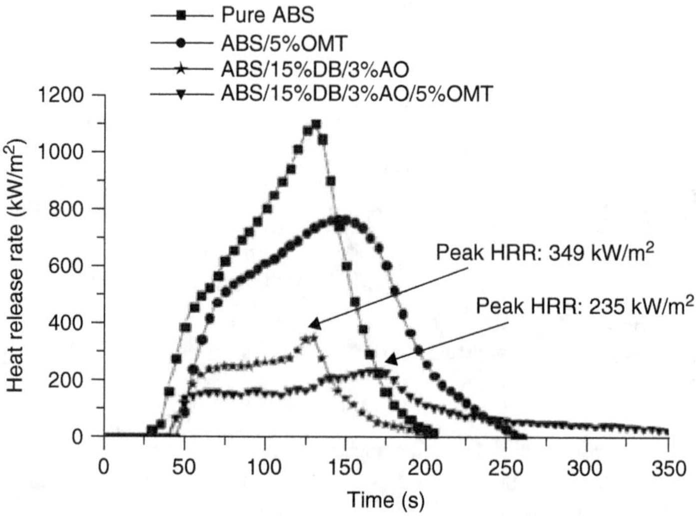

FIGURE 8.8 Comparison of HRR plots for ABS, ABS–OMT (ABS/5% OMT), ABS/DB–AO (ABS/15% DB–3% AO), and ABS/OMT/DB–AO (ABS/5% OMT/15% DB–3% AO). Heat flux: 50 kW/m^2. (From Ref. 35, copyright © 2004, Elsevier, with permission.)

contradicts those obtained with the above-mentioned PA6 and ABS[34,35] flame retardant nanocomposites. In general, those polymer–organoclay nanocomposites show a lower PHRR value than that for pure polymer, whose value is reduced further when DB is added to the formulation. However, both DB and AO are present in the nanocomposites, the PHRR decreases even more.

Analysis of PA6,[34] ABS,[35] PP-g-MA,[33] and PP[56,57] nanocomposites shows that addition of both organoclay and halogenated flame retardants to flame retardant polymer formulation leads to superior flame retardant performance in the cone calorimeter that can meet minimal criteria for flame retardancy. It is an enabling technology that gives materials with far superior properties and performance. Clearly, a synergistic effect exists between nanocomposite formation and halogenated flame retardants additives. It is likely that the organoclay could be used to replace a part of the halogenated flame retardant needed to reach a satisfactory degree of retardancy. From a practical point of view, the negative effects introduced by the addition of halogenated flame retardant additives (i.e., toxicity, smoke, and corrosiveness) would also be reduced significantly.

Melamine cyanurate (MCA) is an efficient flame retardant for polyamide, as it facilitates thermal decomposition of polyamide, probably because it not only interferes with the hydrogen-bonding network of the polymer but also catalyzes base hydrolysis of the polymer macromolecule chains.[44,45] Figure 8.9 shows plots of HRR data for PA6, PA6–MCA, PA6–OMT, and PA6–MCA–OMT materials.[34] The addition of MCA to PA6 caused a small decrease in the PHRR and hardly changed the MLR. However, when MCA was added to PA6–OMT,

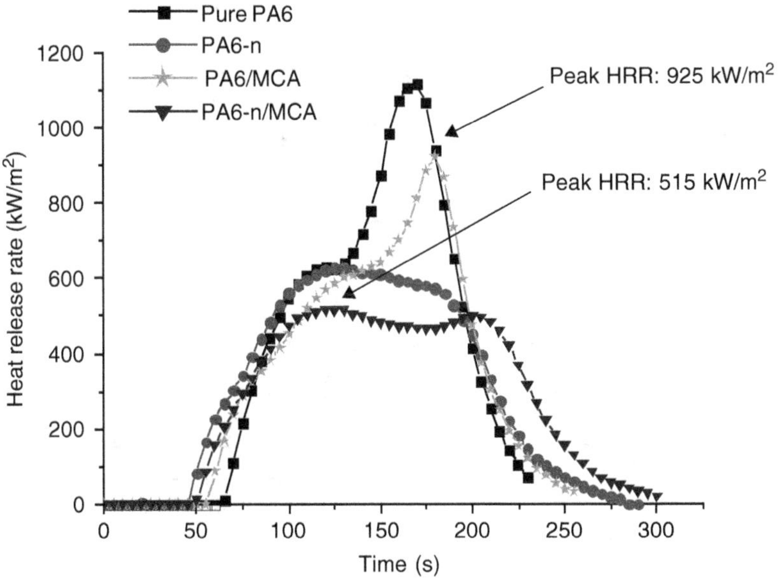

FIGURE 8.9 Comparison of HRR plots for for PA6, PA6-n (PA6/5% OMT), PA6/15% DB–5% AO, and PA6-n/DB–AO (PA6/5% OMT/15% DB–5% AO). Heat flux: 50 kW/m^2. (From Ref. 34, copyright © 2003, Wiley-VCH Verlag GmbH & Co., with permission.)

the PHRR decreased significantly. The most likely explanation is the formation of carbonaceous–silicate char in the condensed phase that retards degradation of the PA6 matrix.

Cone calorimetric data for PA6 materials is shown in Figure 8.10 and Table 8.6.[36] The addition of either 2% OMT or 8% MH (magnesium hydroxide)–5% RP (red phosphorus) to pure PA6 decreased the PHRR values by 39% or 59% from that of pure PA6 (Figure 8.10 and Table 8.6). Moreover, when 6% MH–5% RP was added to PA6/2% OMT, the PHRR decreased further, again suggesting a synergistic effect between the additives. The difference between PA6/2% OMT/6% MH–5% RP and PA6/8% MH–5% RP was replacement of 2 wt% MH with an equivalent amount of OMT, with the PHRR value decreasing by 73% from that of pure PA6. It is likely that more of the MH can be replaced by OMT to improve properties further. That is dependent on the synergistic effect between the organoclay and flame retardant additives and the polymer class. Similar trends are obtained for the MLR and HRR data, thus suggesting that the MH–RP–OMT additives impart their flame retardant effects primarily in the condensed phase. The observed flame retardant behavior in these PA6 nanocomposites is attributed to a barrier effect arising from the formation of char in the presence of organically modified exfoliated clay sheets. Apparently, synergistic effects exist between the clay, the MH–RP, and the PA6 components, as suggested by the data in Table 8.6.

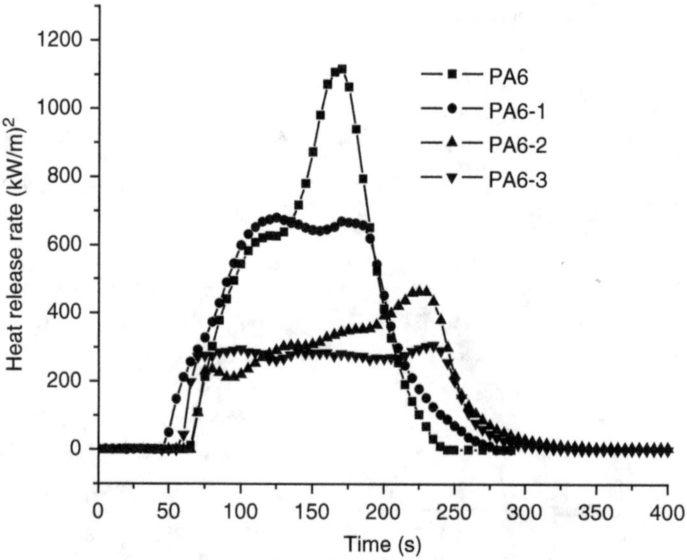

FIGURE 8.10 Comparison of HRR plots of PA6, PA6-1 (PA6/2% OMT), PA6-2 (PA6/8% MH–5% RP), and PA6-3 (PA6/2% OMT/6% MH–5% RP). Heat flux: 50 kW/m^2. (From Ref. 36, copyright © 2004, Elsevier, with permission.)

TABLE 8.6 Peak Heat Release Rate and Peak Mass Loss Rate of PA6 Materials[a]

Sample	Peak HRR (kW/m^2)	Char Residue (wt%)	Peak MLR (g/m^2·s)
PA6	1120	0.65	0.38
PA6-1	681	4.80	0.20
PA6-2	463	6.28	0.18
PA6-3	308	9.88	0.14

Source: After Ref. 36.

[a]PA6, polyamide 6; PA6-1, PA6/2% OMT; PA6-2, PA6/8% MH–5% RP; PA6-3, PA6/2% OMT/6% MH–5% RP; OMT, organoclay modified with hexadecyltrimethylammonium bromide; MH, magnesium hydroxide; RP, red phosphorus.

Figure 8.11 shows that the addition of either OMT or MPP to PU results in a sharp decline in the HRR. For PU/5% OMT material with 5 wt% OMT, the PHRR value was reduced by nearly 57% from that of pure PU. On the other hand, when 6% MPP was added to PU, the HRR value was reduced by only 39%, and a plot of the data shows a curve that has two peaks. In the literature, MPP has been reported to alter the thermooxidative degradation mechanism through which degradation of PU occurs.[38] For PU/5% OMT/6% MPP material with 5 wt% OMT and 6 wt% MPP, the PHRR value was reduced by nearly 74% from that of pure PU, and a plot of the HRR data for the PU/5% OMT/6% MPP

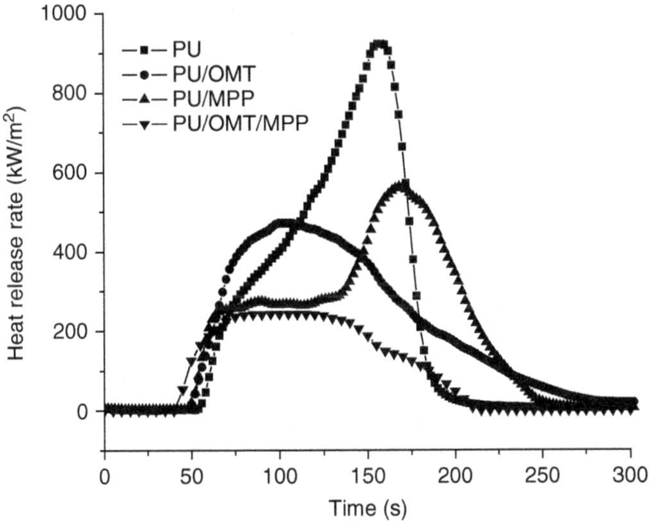

FIGURE 8.11 Comparison of HRR plots of PU, PU–OMT (PU/5% OMT), PU–MPP (PU/6% MPP), and PU–OMT–MPP (PU/5% OMT/6% MPP). Heat flux: 35 kW/m^2. (From Ref. 38, copyright © 2005, Elsevier, with permission.)

material shows only one broad peak, in contrast to the two peaks observed for the PU–MPP material, suggesting that OMT alters the PU/6% MPP combustion process. Consistency is obtained between the PHRR and the MLR data, a result that also indicates that the flame retardancy of MPP–OMT occurs predominantly via a condensed-phase mechanism.

Cone calorimetric data for PU and corresponding flame retardant nanocomposites are listed in Table 8.7. The data include the peak heat release rate (PHRR), mass loss rate (MLR), specific extinction area (SEA), amount of CO released,

TABLE 8.7 Cone Data for Flammability of PU, PU–OMT, PU–MMP, and PU–OMT–MMP[a]

	Sample			
	PU	PU–OMT	PU–MMP	PU–OMT–MPP
Peak heat release rate (kW/m^2)	923	472	563	243
Mass loss rate (g/·sm^2)	0.4	0.24	0.27	0.14
Specific extinction area (m^2/kg)	1399	473	488	415
CO release amount (kg/kg)	2.33	0.37	2.33	0.33
CO_2 release amount (kg/kg)	3.67	1.91	3.67	1.71

Source: After Ref. 38.

[a]PU, polyurethane; OMT, organoclay modified with hexadecyltrimethylammonium bromide; MMP, melamine polyphosphate.

TABLE 8.8 Cone Calorimetric Data for PS–OMT–TCP Nanocomposites[a]

Sample	t_{ign} (s)	t_{PHRR} (s)	PHRR [kW/m^2 (% reduction)]	THR (MJ/m^2)	MLR (g/m^2·s)	Average SEA (m^2/kg)
PS	62	124	1419	109.7	17	1097
PS + 3% clay	57	85	610 (56)	85.5	14	1695
15% TCP + PS	59	108	1122 (20)	63.4	14	1560
15% TCP + 3% clay	59	109	495 (65)	59.1	14	1803
30% TCP + 3% clay	43	60	378 (74)	49.5	14	2401
30% TCP + 5% clay	53	87	342 (76)	45.8	14	2310
30% TCP + 10% clay	55	119	324 (79)	47.3	14	2285
10% TCP + 3% clay	49	101	485 (65)	62.4	15	2159

Source: After Ref. 39.

[a]PS, polystyrene; TCP, tricresylphosphate; clay, organically modified montmorillonite, dimethylbenzyl hydrogenated tallow ammonium (hydrogenated tallow is a mixture of 65% C18, 30% C16, 5% C14) substituted clay.

and amount of CO_2 released. The maximum values of HRR, MLR, and SEA decrease in the presence of OMT or MPP. With the addition of both OMT and MPP, the HRR, MLR, and SEA decrease even more. The maximum amount of CO and CO_2 released decreases with the addition of OMT regardless of whether or not MPP is added, whereas they remain invariant with MPP addition alone. The addition of OMT reduces the amount of flammable small decomposition products released during combustion. Clearly, synergistic effects between clay, MPP, and PU exist that tend to retard the heat release rate, suppress the release of smoke, and decrease the amount of potentially toxic gas released during the combustion of PU-derived materials.

Cone data for PS–OMT–ArP nanocomposites are shown in Table 8.8, again showing synergistic effects. The author used three types of ArP: TCP, TXP, and RDP.[39] Table 8.8 lists cone data only for PS–OMT–TCP composition. When TCP and organoclay are both present, the time to ignition is highly variable, with no clear trend emerging. The THR is reduced significantly with the addition of either TCP or clay, but more so with the addition of both TCP and organoclay. The THR decreases correspondingly, but the SEA increases significantly. The presence of both clay and TCP appears to give rise to a substantial reduction in the PHRR and a significant reduction in the THR compared to the values obtained for pure polymer and a nanocomposite containing only clay. Similar behavior is observed for PS blends of either TXP or RDP as that of TCP discussed here.[39] The combination of conventional flame retardants, aromatic phosphate, and clay again appears to interact synergistically to improve flame retardancy and thermal degradation above that obtained for polystyrene.

Table 8.9 gives the cone data for flame retardant polystyrene nanocomposites (PS/OLP–clay) prepared using clay modified with an oligomeric phosphate ammonium salt (OLP) containing both ammonium salt and phosphate-pendant groups placed randomly along the polymer backbone. The HRR data for the

TABLE 8.9 Cone Calorimetric Data for PS Nanocomposites Prepared Using 55% DPVPP-Modified Clay and 75% DPVPP-Modified Clay, Respectively, via Melt Blending[a]

	Composition				
	PS	5% clay	3% clay	5% clay	10% clay
t_{ign} (s)	36 ± 5	40 ± 5	54 ± 2	43 ± 3	44 ± 3
PHRR (kW/m^2)	1411 ± 18	837 ± 32	638 ± 10	416 ± 12	268 ± 1
t_{PHRR} (s)	87 ± 4	93 ± 7	71 ± 3	69 ± 6	100 ± 4
Average HRR (kW/m^2)	755 ± 11	571 ± 20	380 ± 4	234 ± 2	158 ± 2
THR (MJ/m^2)	102 ± 1	58 ± 11	76 ± 3	58 ± 5	54 ± 0
SEA (m^2/kg)	1134 ± 24	1323 ± 28	1481 ± 11	1492 ± 46	1475 ± 27
Average MLR (g/m^2)	29 ± 0	25 ± 1	20 ± 1	13 ± 1	10 ± 1

Source: After Ref. 40.

[a]DPVPP, diphenyl-4-vinylphenylphosphate.

various compositions employ two loading levels of the phosphate component, 55% and 75% (Table 8.9). Upon incorporation of the OLP–clay into PS, the THR is reduced by nearly 50% from that obtained for pure PS. In general, the time to ignition for polymer–organoclay nanocomposites is shorter than that for pure polymer. Interestingly, addition of OLP–clay to a PS nanocomposite increases the time to ignition, possibly attributable to the high thermal stability of the OLP. It was suggested that the OLP may behave differently from classic clay surfactants and phosphate flame retardants and may be intimately involved in retarding combustion in this system. The absolute increase is greater than that which one could obtain with addition of the components at the same loading level as small aromatic phosphates (ArPs). The reduction in PHRR suggests that this system has a potential for the formation of flame retardant polymeric systems, with practical applications. The MLR decreases and the SEA increases, which is similar to that of PS–OMT–ArP. The literature describes the effect that increasing phosphorus content typically has on the results of a cone calorimeter test.[40] A reduction in PHRR correlates rather well with increasing phosphate content in PS systems.[40] For constant clay content, the reduction in PHRR is greater for increasing phosphorus content.

PS/OLB–clay nanocomposites prepared using novel oligomeric bromine ammonium-modified clay (OLB–clay) were discussed earlier in the chapter.[41] The authors studied the relationship between nanocomposite formation and low levels of bromine on the flame retardancy of polystyrenes (PSs) using cone calorimetry. Specifically, the effects of bromine content and the identity of the oligomeric ammonium on the cone calorimetry behavior of PS nanocomposites were studied.[41] Consistent with most other polymer nanocomposite materials, addition of the OLB–clay to PS decreased the t_{ign}. The OLB–clay contained only 30% inorganic clay, whereas commercial organically modified clay contained about 70% inorganic clay. No reduction in the PHRR and the average MLR occurred until 10% organically modified clay was added. The THR was also not affected significantly, probably due to the low amounts of added inorganic clay and to the fact that complete combustion still occurred. Above 10% OLB–clay, there is a significant reduction in PHRR and THR and a slight reduction in average MLR for the nanocomposites. The THR clearly decreases as the bromine content increases. The poor results observed for PS/OLB–clay nanocomposites at less than 10 wt% loading are probably due to the low amounts of inorganic clay present. The authors suggest that nanocomposite formation affects the PHRR and that the presence of bromine is important in the reduction of the THR.[41] The addition of each component contributes to a different effect on the cone calorimetric parameters. A comparison between the oligomeric trimethyl, oligomeric triethyl, and oligomeric dimethylhexadecylammonium salts shows that the first salt gives a larger reduction in the PHRR observed.[41] Interestingly, the bromine content at less than 4 wt% is far less than that of the conventional halogenated flame retardant additives. Therefore, it appears that bromine-containing nano-dispersed clays may be useful as flame retardant additives.[41]

A Japanese patent issued to Sekisui Chemical Co. Ltd.[47] describes cone calorimeter data for polyethylene (PE)-derived nanocomposites in which organically modified layered silicates (OMTs) were combined with a variety of conventional flame retardant additives. The ammonium salt–modified silicate (SBAN-400) was incorporated into the PE compositions at 10 phr (Table 8.10). The PHRR results for the PE nanocomposite materials are 50% lower than the PHRRs for pure PE and PE with 10 phr pristine clay (MMT). These results suggest that without suitable organic treatment of the clay, the addition of MMT has little effect on flame retardant properties. Due to the favorable interactions arising from the organic surface treatment of the clay, the OMT forms a nano-dispersed structure in the PE matrix, while the PE–pristine MMT material is an immiscible or conventional composite. At 10-phr loading levels, where the OMT reduces the PHRR substantially for PE nanocomposite, DB–AO has little effect and APP alone has only a weak effect on PHRR of PE, respectively. Addition of either 15 phr of DB–AO or 15 phr of APP to pure PE decreases the PHRR values by only 10 to 20% from that of pure PE. However, addition of 5 phr of APP to a PE–SBAN N-400 nanocomposite with 10 phr of SBAN N-400 causes a 63% reduction in the PHRR. The addition of 5 phr of phenylphosphate and 10 phr of SBAN N-400 also reduces the PHRR by 60% (Table 8.10).

8.5.2 LOI and UL-94 Tests

Table 8.11 data show that addition of 5 wt% OMT in the various flame retardant ABS formulations with DB–AO increased the LOI about 0.5. The LOI increased by 0.5 with the addition of 2 wt% OMT to PA6. However, the LOI increased by 2.0 with the combination of 2 wt% OMT with MH–RP. The LOI of the PU–OMT–MPP increased by 3.5 from that of PU–MPP, while the LOI of PU–OMT increased by 1.5 compared to pure PU. Also, the ABS–DB/OMT nanocomposite achieves an UL-94 V-0 rating, whereas the ABS–DB material

TABLE 8.10 Cone Calorimeter Data for PE Materials[a,b]

	Example							
	1	2	3	4	5	6	7	8
Composition (phr) Polyethylene	100	100	100	—	100	100	100	100
SBAN N-400	—	10	—	—	—	—	10	10
Bengal (pristine MMT) A	—	—	10	—	—	—	—	—
APP	—	—	—	10	—	15	5	—
DBDPO	—	—	—	—	7.5	—	—	—
Sb_2O_3	—	—	—	—	2.5	—	—	—
Phenyl phosphate	—	—	—	—	—	—	—	5
PHRR (kW/m^2)	1327	687	1067	1272	1309	989	493	543

Source: After Ref. 47.

[a] Sample size, 100 × 100 × 3 mm; heat flux; 50 kW/m^2; phr, parts per hundred parts resin.
[b] SBAN N-400, type of organically modified clay provided by Hojun Kogyo Co., Ltd.; APP, ammonium polyphosphate; DBDPO, decabromodiphenyl oxide; Sb_2O_3, antimony oxide.

does not under UL-94 test conditions. LOI and UL-94 test results suggest that beneficial synergistic effects may exist between the polymer, the flame retardant (DB–AO, MH–RP, and MPP), and the OMT. As discussed previously, the improved performance of the flame retardant nanocomposites may probably be attributed to the barrier effects imparted by the OMT.

Interesting, unexpected results are obtained for both organoclay and MCA incorporated in PA6 matrix prepared using melt blending in the UL-94 test.[34] Table 8.11 shows data in which the PA6–MCA composite obtains a V-0 rating in the UL-94 test, whereas the PA6–OMT–MCA material burns and therefore fails under the same conditions. Even when 25 wt% MCA is added to the PA6–OMT nanocomposite, a V-0 rating cannot be obtained. Increased MCA content, which aids heat removal by accelerating melt dripping in the vertical burning test, becomes antagonistic to the nano-dispersed clay layers, which enhance the formation of char, which in turn suppresses melt dripping. The mechanism by which each of these two components contributes to flame retardancy explains the result observed in the UL-94 test.

Wilkie used a UL-94 test to study whether any synergy exists between aromatic phosphate flame retardants and PS nanocomposite formation,[39] the results of which are shown in Table 8.12. At 30% ArP content, various V ratings are obtained that improve as the clay content increases.

A patent[46] discloses glass fiber–reinforced poly(1,4-butylene terephthalate) (PBT) with organically modified clay (organoclay) and flame retardant additives. The PBT, glass fibers, brominated flame retardant, and Sb_2O_3 were processed using a laboratory Ko-kneader. In a second step, the organoclay and an alkali metal salt were then blended with the PBT mixture prepared in the first step using a gyratory mixer to prepare the final samples. The flammability and mechanical properties of the samples were then evaluated, the results of which are shown in Table 8.13. The results of these two flammability tests show the samples to be

TABLE 8.11 Flame Retardant Properties of ABS, PA6, and PU Materials[a]

Sample	Composition	LOI	UL-94 Test
ABS	Pure ABS	18.7	Burning
ABS/OMT	ABS + OMT 5 wt%	21.5	Burning
ABS/DB	ABS + DB 18 wt%	22	Burning
ABS/DB/OMT	ABS + DB18 wt% + 5 wt%	22.6	V-0
ABS/DB–AO	ABS + DB 15 wt% + AO 3 wt%	27	V-0
ABS/DB–AO/OMT	ABS + OMT 5 wt% + DB 15 wt% + AO 3 wt%	27.5	V-0
PA6	Pure nylon-6		Burning
PA6-n	PA6 + OMT 5 wt%		Burning
PA6/DB–AO	PA6 + DB 15 wt% + AO 5 wt%		V-0
PA6-n/DB–AO	PA6 + OMT 5 wt% + DB 15 wt% + AO 5 wt%		V-0
PA6/MCA	PA6 + 15 wt% MCA		V-0
PA6-n/MCA	PA6 + OMT 5 wt% + 15 wt% MCA		Burning
PA6	Pure nylon-6	21	Burning
PA6-1	PA6 + OMT 2 wt%	21.5	Burning
PA6-2	PA6 + MH 8 wt% + RP 5 wt%	29	V-0
PA6-3	PA6 + OMT 2 wt% + MH 6 wt% + RP 5 wt%	31	V-0
PU	Pure PU	19.0	
PU/OMT	PU + OMT 5 wt%	20.5	
PU/MPP	PU + MPP 6 wt%	24.0	
PU/OMT/MPP	PU + OMT 5 wt% + MPP 6 wt%	27.5	

Source: Data from Refs. 34 to 36 and 38.

[a]ABS, acrylonitrile–butadiene–styrene; PA6, polyamide-6; PU, polyurethane; OMT, organoclay modified with hexadecyltrimethylammonium bromide; MPP, melamine polyphosphate; MH, magnesium hydroxide; RP, red phosphorus; MCA, melamine; DBDPO, decabromodiphenyl oxide; AO, antimony oxide.

TABLE 8.12 UL-94 Results for PS–OMT–ArP Materials[a]

Sample	UL-94 Test
30% TCP + 5% organoclay + PS	V-1
30% TCP + 10% organoclay + PS	V-1/V-0?
30% TCP + 3% organoclay + PS	V-2
30% RDP + 5% organoclay + PS	V-2?
30% RDP + 10% organoclay + PS	V-0/V-1?
30% TXP + 5% organoclay + PS	V-2

Source: After Ref. 39.

[a]TCP, tricresylphosphate; TXP, trixylylphosphate; RDP, resorcinol diphosphate; organoclay, organically modified montmorillonite, dimethylbenzyl hydrogenated tallow ammonium hydrogenated tallow is a mixture of 65% C18, 30% C16, 5% C14) substituted clay.

TABLE 8.13 Properties of Brominated Flame Retardant PBT–Organoclay Nanocomposites[a]

	Example				
	1	2	3	4	5
Composition (wt%) PBT	55.0	55.1	55.0	55.2	55.2
Glass fibers	30.0	30.0	30.0	30.0	30.0
Decabromodiphenyl ether	9.0	9.0	9.0	9.0	9.0
Sb_2O_3	4.5	4.5	4.5	4.5	4.5
Claytone 34	1.0	—	—	—	—
Claytone 40	—	1.0	—	—	—
Bentone 27	—	—	1.0	—	—
Bentone SD-1	—	—	—	1.0	—
Bentone 500	—	—	—	—	1.0
Potassium oleate	0.5	0.4	0.5	0.3	0.3
UL-94 (1.6 mm thick)	V-0 non-dripping	V-0 non-dripping	V-0 non-dripping	V-0 non-dripping	V-0 non-dripping
Impact strength (kJ/m^2)	25.9	25.8	22.5	32.2	29.9

Source: After Ref. 46.

[a]PBTP, poly(1,4-butylene terephthalate); Claytone 34 and Claytone 40, dimethyldioctadecylammonium betonite; Bentone SD-1, smectite modified with organic compounds; Bentone 27 and Bentone 500, montorillonite modified with organic compounds. Claytone and Bentone are registered brand names of China-Clay Handelsgesellschaft and NL Chemicals, respectively.

self-extinguishing with a V-0 rating and nondripping. It is likely that the identity of the organoclay affects the synergistic effect on the flame retardant properties between the organoclay and the brominated flame retardants, probably due to the degree of dispersion and exfoliation of the organoclay in the polymer matrix.

In another patent, a flame retardant poly(butylene terephthalate) (PET) nanocomposite[49] was prepared using PBT, an organically modified clay (organoclay), a fluorocarbon polymer, polytetrafluoroethylene (PTFE), a brominated flame retardant, and a stabilizer by the melt blending method. The patent states that a combined organoclay and PTFE, dispersed in a styrene–acrylonitrile (SAN) copolymer (50% PTFE), could replace up to 40% of the brominated polycarbonate–Sb_2O_3 flame retardant in the PBT (Table 8.14). The flame retardant blends containing either the organoclay or PTFE do not obtain a V-0 rating, but when both are added together to the nanocomposite, a V-0 rating can be achieved (example 4). The addition of PTFE and organoclay together enables a significant reduction in the brominated flame retardant content required to achieve a V-0 rating. The fact that the behavior is observed only when they are added together indicates a synergistic effect between PTFE and organoclay on flame retardant properties. Without both PTFE and organoclay present, the PBT nanocomposite

TABLE 8.14 PBT Fire Retardant Blend Data[a]

	Example					
	1	2	3	4	5	6
Composition (%) Valox 315	76.74	81.32	81.32	84.37	81.4	84.37
BC-58	15	12	12	10	12	10
Sb_2O_3	7.88	6.3	6.3	5.25	6.3	5.25
T-SAN	0.08	0.08	0.08	0.08	—	0.08
Zn phos.	0.3	0.3	0.3	0.3	0.3	0.3
Clayton HY (% of PBT)	—	—	2	1	2	0.25
UL-94 test	V-0	F	V-0	V-0	F	V-0
Total flame-out time (s)	10.6	—	10.2	15.9	—	—

Source: Data from Refs. 49 and 74.

[a] Valox 315, poly(butylene terephthalate) (weight average molecular weight = 105, 000); BC-58, brominated bisphenol A polycarbonate oligomers (58% bromine); T-SAN, polytetrafluoroethylene dispersion in styrene–acrylonitrile copolymer (>50% PTFE); Zn phos., zinc phosphate, ester-interchange inhibitor; Clayton HY, MMT [dimethyl di(hydrogenated tallow) ammonium ion]; PBT, poly(1,4-butylene terephthalate).

material cannot be classified as flame retardant polymer according to the UL-94 test, even if either the PTFE or organoclay content increase two- to threefold. Moreover, the data in Table 8.14 indicate the benefits of adding both PTFE and organoclay, even at very low levels, for the latter to give rise to a synergistic effect that further improves flame retardancy over that obtainable for any individual component alone.

A Canadian patent[61] discloses the application of organoclay on the flame retardant formulation used in telecommunication cable based on polyolefin. In general, either chlorine or bromine flame retardants are used in the telecommunication cable even though they can release significant amounts of toxic smoke during combustion. The addition of only 3 to 8% added organoclay and 0.5 to 40% PTFE microparticles leads to a significant reduction in the amount of halogenated flame retardants added while providing equivalent flame retardant properties and improved performance in terms of smoke emissions in a UL-910 test.

Inoue and Hosokawa of Showa Denko reported in a patent[50] the use of silicate–triazine intercalation compounds in the flame retardant polymeric composites listed in Table 8.15. Various melamine salts were used to treat a synthetic silicate [fluorinated synthetic mica (FSM)] prior to flame retardant nanocomposite formation. Specifically, 8 to 15 wt% of the melamine as well as other melamine salts were also added to the polymer using melt blending. FSM is chemically similar to MMT, but the aspect ratio of the individual FSM silicate layers is 5 to 10 times larger than that of MMT. By combining the melamine and the polymer–organoclay nanocomposites, a V-0 rating is obtained in the UL-94 test, while simultaneously increasing both the modulus and the heat distortion temperature. The inventors state that the flame retardant properties of the nanocomposites derived are highly dependent on the ability to achieve nanoscale

TABLE 8.15 UL-94 Data for FSM Nanocomposites

Example	Composite[a]	Silicate (%)	Additive[b]	% Added	Exfoliation	UL-94
1	PA6/O-FSM	5	MA	3.3	80	HB
2	PA6/M-FSM	5	—	—	80	V-2
3	PA6/M-FSM	5	MA	3.3	80	V-2
4	PA6/M-FSM	5	MA	10	80	V-0
5	PA6/M-FSM	5	MCA	3.3	>50	V-2
6	PA6,6/O-FSM	5	MCA	3.3	>50	HB
7	PA6,6/M-FSM	5	MA	3.3	>50	V-0
8	PA6,6/M-FSM	5	MCA	3.3	—	V-0
9	PA6,6/M-FSM	3	MA	5	>50	V-0

Source: After Ref. 50.

[a]PA6, nylon-6 (relative viscosity is 2.37); PA6,6: nylon-6,6 (relative viscosity is 2.61); O-FSM, dioctadecyl dimethyl ammonium fluorinated synthetic mica; M-FSM, fluorinated synthetic mica modified by melamine.

[b]MA, melamine; MCA, melamine cyanurate.

TABLE 8.16 Properties of Polypropylene–Clay Nanocomposites

	Example					
	1	2	3	4	5	6
Composition[a] (%) Polypropylene	86.6	86.6	79.9	100	93.3	94.6
MMT1	6.7	—	6.7	—	—	—
MMT2	—	6.7	—	—	6.7	—
MMT3	—	—	—	—	—	5.4
PP-g-MA	6.7	6.7	13.4	—	—	—
d-spacing (nm)	5.5	4.8	5.8	—	1.5	1.2
Dispersibility	good	good	good	—	poor	poor
Flexural modulus (MPa)	2270	2150	2440	1360	1760	1720
Limited oxygen index (LOI)	22.3	23.0	21.9	17.5	19.9	18.4

Source: After Ref. 50.

[a]MMT1, octadecyl ammonium and melamine phosphate–modified montmorillonite (*d*-spacing: 1.8 nm); MMT2, melamine phosphate–modified montmorillonite (*d*-spacing: 1.5 nm); MMT3, sodium montmorillonite.

dispersion of the FSM clay. They found that without a uniform dispersion of the clay layers in the PA6,6 (>50% exfoliation; Table 8.16, example 6), only an HB rating (self-extinguishing in a horizontal burn) could be obtained in the UL-94 test. They also established that the melamine compounds had to be added to both the FSM and the resin to obtain the V-0 ratings (Table 8.15). Such dual-addition behavior contrasts the results obtained for melt-blending PA6–OMT–MCA,[34] probably due to the correlation between the OMT and melamine. In the PA6 case, the organoclay modified with C16 and MCA and PA6 are blended simultaneously to prepare the PA6–OMT–MCA. The interaction between tethered MCA and the clay probably gives rise to the flame retardant behavior observed.

Another patent[60] describes a preparation of flame retardant polypropylene–clay nanocomposites. Prior to nanocomposite preparation, the clay was modified with either melamine phosphate (MP) or a mixture of MP and alkylammonium salt to obtain flame retardant clay. The flame retardant clay was then incorporated into a polypropylene–graft–maleic anhydride copolymer prior to blending with pure polypropylene. The results of mechanical and flame retardant properties of the virgin polymer and various polypropylene–clay nanocomposites are given in Table 8.16. Apparently, PP–clay nanocomposites in which the clay modified with MP alone or together with alkylammonium compound combined with PP-g-MA possess better dispersion, mechanical, and flame retardant properties than those of pure polymer (example 4), PP–clay nanocomposites without PP-g-MA (example 5), or PP–sodium MMT composite (example 6). XRD analysis suggests that MP can intercalate in the clay and addition of PP-g-MA can cause nanocomposite formation. The maleic anhydride group and the amino group of the MP on the clay surface probably form a chemical bond during nanocomposite formation. Moreover, the presence of both PP-g-MA and organic ammonium facilitate the greater clay sheet exfoliation and improve mechanical properties. Clearly, exfoliation of the organoclay must give rise to a nano-dispersed structure in the polymer matrix in order to achieve significant improvement in both the mechanical and flame retardant properties of the resulting materials. The MP, organic ammonium and the PP-g-MA may all impart positive effects through favorable interactions with either inorganic or organic cations on the clay sheet surfaces. Similar results were observed by Okada et al. in flame retardant polyolefin nanocomposites.[47]

In summary, the presence of both appropriate organoclay and flame retardants can facilitate a significant reduction in the flammability of the parent polymers. However, the organoclay must form either highly intercalated or exfoliated nanoscale structures in the polymer matrix to achieve significant improvement in the flame retardant properties discussed in this review. Across several polymer families, several groups observe a synergistic effect between the various organic cations, flame retardants, both standard and novel, and the clay itself used in the preparation of these enhanced flame retardant materials. Moreover, recent data by several of these same authors suggest that the addition of an organoclay to a flame retardant composition may also help in reducing the amount of the conventional flame retardants required to achieve a satisfactory level of flame retardant property, thus rendering such flame retardant nanocomposite materials feasible for a myriad of potentially significant industrial applications.

8.6 FLAME RETARDANT MECHANISM

The generally accepted mechanism that has been suggested[40] to explain the reduction in flammability for polymer–organoclay nanocomposites containing intercalated or exfoliated clays is based on barrier effects. During combustion, the polymer matrix is heated to thermal degradation temperature, and volatile thermal degradation products are generated in the polymer matrix. Because the boiling temperatures of most of the volatile products are much lower than the thermal degradation temperatures of the polymer, the products are superheated as they are generated.[29] Then bubbles nucleate and grow below the heated polymer surface and are released into the gas phase as fuel vapor. These bubbles agitate the melt polymer surface and can interfere with the formation of a carbonaceous char layer, heat-transfer barrier at the polymer surface.[29] The literature[8–16] suggests that the reduced flammability of polymer–organoclay nanocomposites can best be explained by enhanced char formation in the condensed phase. Specifically, the presence of nanoscale clay sheets in the polymer matrix retards the vigorous bubbling process and facilitates the formation of a continuous protective charred layer made of clay and carbonaceous char on the burning polymer surface that insulates the underlying polymeric substrate. Specifically, the carbonaceous–silicate char slows heat and mass transfer between the gaseous and condensed phases, thereby retarding further the thermooxidative degradation of the polymer. Therefore, successful nano-scaled dispersion or exfoliation of the individual clay sheets appears to influence reaction kinetics, thermal decomposition by-product migration, volatilization, and char formation in the polymer matrix under conditions of combustion. The presence of the protective layer is clearly important in the flammability reduction by these additives, but these clay sheets tend to develop large lateral surface cracks in which vigorous bubbling still occurs.[66] The individual organoclay is not enough to be used as a flame retardant additive. The combination of organoclay and flame retardant additives improves the effectiveness of flame retardant systems.

If the ammonium cation that was used to modify the clay sheet surface has β-hydrogens, the onset thermal decomposition of organoclay occurs around 200°C via the generally accepted Hofmann elimination process.[42] The ammonium cations will lose an olefin and an amine in a base-catalyzed process. To maintain charge balance, a proton is left on the surface of the MMT, as illustrated in reaction (8.1). The proton probably serves as a catalyst that facilitates further decomposition of polymer within the gallery space of the OMT. The amine and olefin are combustibles that fuel the combustion in the flame. As an example, thermal degradation of polyamide can be efficiently catalyzed by acids, even in trace amounts. This catalytic effect may contribute to greater heat release rates in the early stage of nanocomposite combustion than observed for pure polyamide, such as PA6,[16] and decrease the thermal stability of the polyamide.

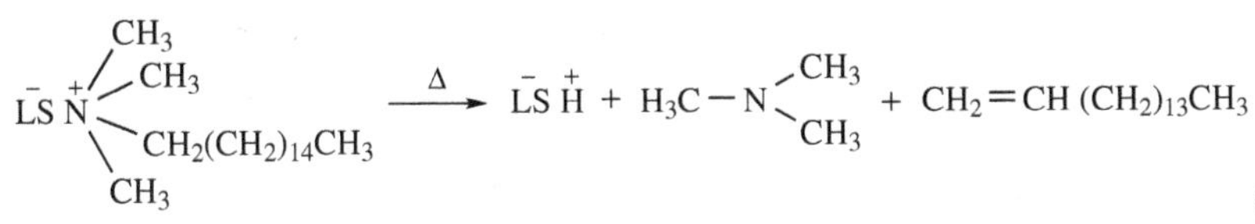

(8.1)

8.6.1 Combination of Nanocomposites and Halogen Flame Retardant Additives

Decabromodiphenyl oxide is one of the halogenated flame retardant additives that act in the vapor phase through a radical chain reaction to interrupt the combustion reaction. Also, the addition of metal oxides such as antimony trioxide (AO) as coadditives to halogenated flame retardants further increases the efficiency of the halogenated additive through the formation of antimony trihalide, a volatile product that can interfere with and slow the propagating radical chain reactions in the flame, even though the oxide itself has no effect.[15] During combustion, DB decomposes and releases hydrogen bromide. Then the HBr reacts with AO to form antimony tribromide, a vapor-phase flame retardant that inhibits the radical chain reaction in the flame [reaction (8.2)]. When DB–AO is combined with a nanocomposite, the reaction between AO and sodium bromide[34] present as an impurity in the OMT as a residue of clay modified with ammonium cation yields antimony tribromide [reaction (8.3)]. This reaction may be catalyzed by the proton formed during reaction (8.1) on the clay sheets dispersed in the polymer matrix:

$$6HBr + Sb_2O_3 \longrightarrow 2SbBr_3 + H_2O \tag{8.2}$$

$$3\overset{-}{LS}\;\overset{+}{H} + 6NaBr + Sb_2O_3 \xrightarrow{\Delta} 3\overset{-}{LS}\;\overset{+}{Na} + 2SbBr_3 + 3H_2O \tag{8.3}$$

When DB–AO was incorporated into PA6, three distinct processes have been suggested[19,20]: (1) generation of free-radical chain-terminating agents (i.e., antimony tribromide, (2) promotion of char formation through dehydrogenation reactions, and (3) formation of a blanket of hydrogen bromide, which acts as a barrier between the fuel gas and condensed phases. However, even though DB–AO is a very effective flame retardant additive, it can facilitate polyamide chain degradation and give rise to combustible monomers or similar species [reaction (8.4)][20,21]. The thermal degradation of PA6 can be catalyzed efficiently by the protons formed during reaction (8.1) on the clay sheets dispersed in the polymer matrix as illustrated in reaction (8.5). While combining DB–AO with a PA6–OMT nanocomposite, AO may react with any amine hydrobromide present to form an antimonate complex [reaction (8.6)] that is believed to be responsible for the condensed-phase activity of the bromine–Sb_2O_3 flame retardant additives[34].

$$\sim NH-\underset{\underset{O}{\|}}{C}\sim \;+\; HBr \longrightarrow \sim NH_3^+Br^- \;+\; \sim\underset{\underset{O}{\|}}{C}-OH \tag{8.4}$$

$$\overset{-}{LS}\,\overset{+}{H} + \sim NH-\underset{\underset{O}{\|}}{C}\sim + NaBr \xrightarrow{\Delta} \overset{-}{LS}\,\overset{+}{Na} + \sim NH_3^+Br^- + \sim\underset{\underset{O}{\|}}{C}-OH \tag{8.5}$$

$$x\,SbBr_3 \;+\; y \sim NH_3^+Br^- \longrightarrow x\,SbBr_3,\, yNH_4Br \tag{8.6}$$

The slower thermal decomposition observed for the PA6/OMT/DB–AO nanocomposite has been ascribed to barrier effects arising from the nano-dispersed clay sheets. Synergetic effects between the OMT and DB–AO probably occur to augment PA6 flame retardancy [reactions (8.1)–(8.6)].[19,20,33,34,51] Moreover, evaluation of flame retardant nanocomposites by conventional testing methods such as the UL-94 vertical burning test indicates that compatibility and synergy exists between DB–AO, OMT, and the polymer matrix. In the case of PP and ABS, similar synergetic effects between DB–AO, OMT, and the polymer matrix occur and increase flame retardancy.

8.6.2 Combination of Nanocomposites and Nonintumescent Phosphorus Flame Retardant Additives

Considering the flame retardant behavior observed for an MH,[52,53] OMT, and phosphorus species flame retardant composition,[54] synergistic effects probably take place between the OMT and the MH–RP.[36] If the ammonium cation used to modify the clay sheet surface contains β-hydrogen atoms, the alkylammonium cation will lose an olefin and an amine and generate a proton adsorbed on the clay surface to maintain a charge balance around 200°C. These deposited, catalytically active protons promote the endothermic conversion of magnesium hydroxide (MH) to magnesium oxide and water.

Simultaneously, the dispersed silicate sheets of the clay tend to retard volatilization of the water vapor liberated from the MH conversion. Both the water vapor and the acid can then accelerate the thermooxidative degradation of RP, which leads to the formation of a highly cross-linked polyphosphoric acid (PPA). This PPA may then facilitate either further thermooxidation degradation of the PA6, indiscriminate cross-linking reactions, or charring. During combustion, the polyphosphoric acid reacts with thermal degradation products of PA6, MgO, and OMT to form a stable carbonaceous charred layer with a glassy coating. This stable, physically protective barrier on the surface of the polymer material can then insulate the underlying polymer substrate from further combustion events and subsequently, slow the heat and mass transfer between the gaseous and condensed phases.

SEM studies[36] show that in the case of PA6/MH–RP containing only flame retardants, the charred residue forms a loose porous network structure after burning (Figure 8.12*a*). The charred residue of PA6/OMT/MH–RP containing both organoclay and MH–RP (Figure 8.12*b*) is more compact than that of PA6/MH–RP (Figure 8.12*a*). Moreover, the microstructure of the PA6/OMT/MH –RP charred residue shows greater homogeneity and compactness (Figure 8.12*d*) than that of PA6/MH–RP (Figure 8.12*c*). It is well known that the presence of the protective charred layer is clearly important in flammability reduction during combustion. As discussed above for the clay sheet barrier effect, the presence of nanoscale clay sheets in the polymer matrix inhibits the vigorous bubbling process and facilitates the formation of a continuous compact protective charred layer made of clay and carbonaceous char on the burning polymer surface. This

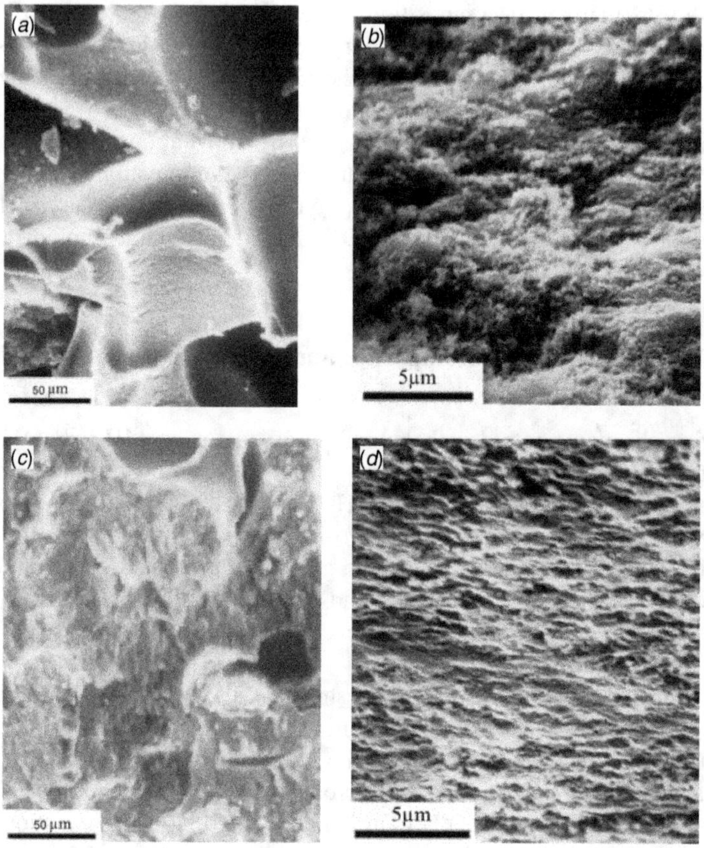

FIGURE 8.12 Scanning electron micrograph of charred residue of (*a*, *b*) PA6/MH–RP (PA6/8% MH–5% RP) and (*c*, *d*) PA6/OMT/MH–RP (PA6/2% OMT/6% MH–5% RP). (From Ref. 36, copyright © 2004, Elsevier, with permission.)

protective layer more effectively protects the underlying polymeric substrate from combustion and reduces heat and mass transfer between the gaseous and condensed phases. The scanning electron microscopy (SEM) observation demonstrates this result, and subsequently, observable and significant changes occur in the experimental HRR and MLR values.

Consistent with the behavior observed, discussed previously, a mechanism through which flame retardancy is imparted by OMT–MPP addition may be[38] that during the combustion of the PU–OMT–MPP composition, the MPP can degrade to form highly cross-linked polyphosphoric acid with concomitant release of melamine, CO_2, and H_2O. The polyphosphoric acid derivatives and the protons deposited on the clay sheets can then promote thermooxidative degradation, cross-linking, and charring of the PU matrix. The exfoliated H^+-bearing clay sheets both retard the release of melamine, CO_2, and H_2O and improve the cross-linking reaction of the PU matrix. Similarly, during combustion, the polyphosphoric acid again reacts with the decomposition products of PU and OMT to form a stable carbonaceous char with a glassy coating. This stable protective barrier on the surface of partially combusted polymer materials that may insulate the underlying polymeric substrate from further compromising combustion and mass transfer events as described in prior cases such as PA6. The presence of the OMT serves to augment the barrier effect imparted by the char residues formed on the surface of the PU matrix. Again, as mentioned previously for other systems, synergistic effects probably exist between the organoclay, the flame retardants, and the polymer matrix that tend to retard the heat release rate, suppress the release of smoke, and decrease the release of toxic gas during the combustion of the polymeric material.

The organophosphate flame retardants, including aromatic phosphate and ogli-meric phosphate, have both vapor- and condensed-phase flame retardant effects. When organophosphate was used in noncharring polymer such as a polyolefin, it can be considered that vapor-phase flame retardant action is the main fire retardant mechanism, which is identical to the free-radical termination mechanism of halogen flame retardants, capturing the hydrogen and hydroxy radicals that contribute to combustion reactions.[77] If phosphates are used in char-forming polymers such as polyester and polyurethane, the condensed-phase mechanism plays a significant role. The condensed-phase mechanisms are explained primar ily by the migration of organophosphates to the surface, then decomposition to phosphoric acid and these acids, forming a molten viscous surface layer that protects the underlying polymer substrate from flame and oxygen; organophosphates can react with the degraded products to promote and enhance char formation.[78] By incorporation of organophosphate and OMT, the nanoscale clay sheets can increase the efficiency of vapor-phase flame retardant action, enhance the molten viscous surface layer formed by phosphoric acid, and enhance char formation through retarding the volatilization of organophosphate, migrating to the molten polymer surface, and inhibiting the degraded products.

In summary, a possible mechanistic explanation has been set forth to explain the synergistic effect observed in several flame retardant polymer nanocomposite formulations as reviewed herein. Specifically, barrier and catalytic effects arising from the incorporation of both an organoclay and flame retardants change the chemical reactions to favor those that yield protective char residues rather than CO or CO_2. Moreover, those additives facilitate those favorable chemical reactions to occur at earlier stages of decomposition, thereby suppressing decomposition and imparting enhanced flame retardancy to such materials. The char layer that forms is facilitated by the incorporation of the organoclay and flame retardants, acting as a barrier to insulate the underlying polymeric substrate and inhibiting heat and mass transfer between the gaseous and condensed phases.

8.7 SUMMARY AND CONCLUSIONS

The choice of an appropriate flame retardant polymer formulation frequently depends on the end-product application. The variables that one must consider in choosing flame retardant compositions, specific additives, and means of incorporation include the material to be flame retarded, the fire safety standards with which the product must comply, toxicity, and cost considerations. The literature and patents reviewed herein reveal that the combination of polymer–organoclay nanocomposites and conventional flame retardant allows successful synergistic improvements of both the flammability and mechanical properties of many polymer systems. Many factors influence the improvement in mechanical and flame retardant properties, such as the formation of a nanostructure, the dispersibility of nanoscale clay, and the interaction between nanoscale clay and flame retardants. In fact, fundamental to enhancing flame retardancy successfully in nanocomposite materials, organoclay must form either an intercalated or an exfoliated nanoscale structure in the polymer matrix prior to exposure to the combustion conditions. Synergistic effects have been observed to occur between the organoclay and conventional flame retardants on the flame retardant properties of materials that drastically alter the decomposition kinetics, mass and energy transfer, and the volatilization of the degraded product in the polymer matrix during combustion. Moreover, due to the barrier and char formation–promoting effects observed upon incorporation of OMT, the amount of the necessary conventional flame retardants added, to achieve a satisfactory flame retardancy can be reduced significantly.

Therefore, it appears that a combination of organoclays and conventional flame retardants possesses significant potential to be useful flame retardant systems. A more detailed understanding of the flame retardant mechanism by which such an additive combination exerts its positive effects may further improve its performance and safety and reduce overall additive loading and cost. Moreover, the development of feasible and relevant manufacturing methods based on the intercalated flame retardant clays described here which facilitate the dispersion of flame retardant additives and increase flame retardant efficiency foretells of a promising future for flame retardant polymer nanocomposite materials in everyday applications.

REFERENCES

1. Irvine, D.J.; McCluskey, J.A.; Robinson, I.M. Fire hazards and some common polymers. *Polym. Degrad. Stab.* **2000**, 67, 383–396.
2. Lu, S.Y.; Hamerton I. Recent developments in the chemistry of halogen-free flame retardant polymers. *Prog. Polym. Sci.* **2002**, 27, 1661–1712.
3. Hu, Y.; Song, L.; Xu, J.; Yang, L.; Chen, Z.Y.; Fan, W.C. Synthesis of polyurethane/clay intercalated nanocomposites. *Colloid Polym. Sci.* **2001**, 279, 819–822.
4. Wang, S.F.; Hu, Y.; Song, L.; Wang, Z.Z.; Chen, Z.Y.; Fan, W.C. Preparation and thermal properties of ABS/montmorillonite nanocomposite. *Polym. Degrad. Stab.* **2002**, 77, 423–426.
5. Ray, S.S.; Yamada, K.; Okamoto, M.; Ueda, K. Polylactide-layered silicate nanocomposite: a novel biodegradable material. *Nano Lett.* **2002**, 2, 1093–1096.
6. Alexandre, M.; Dubois, P.; Sun, T.; Garces, J.M.; Jerome, R. Polyethylene–layered silicate nanocomposites prepared by the polymerization-filling technique: synthesis and mechanical properties. *Polymer* **2002**, 43, 2123–2132.

7. Ray, S.S.; Okamoto, M. Polymer/layered silicate nanocomposites: a review from preparation to processing. *Prog. Polym. Sci.* **2003**, 28, 1539–1641.
8. Song, L.; Hu, Y.; Li, B.G.; Wang, S.F.; Fan, W.C.; Chen, Z.Y. A study on the synthesis and properties of polyurethane/clay nanocomposites. *Int. J. Polym. Anal. Char.* **2003**, 8, 317–326.
9. Bourbigot, S.; Le Bras, M.; Dabrowski, F.; Gilman, J.W.; Kashiwagi, T. PA-6 clay nanocomposite hybrid as char forming agent in intumescent formulations. *Fire Mater.* **2000**, 24, 201–208.
10. Bourbigot, S.; Devaux, E.; Flambard, X. Flammability of polyamide-6/clay hybrid nanocomposite textiles. *Polym. Degrad. Stab.* **2002**, 75, 397–402.
11. Zhu, J.; Wilkie, C.A. Thermal and fire studies on polystyrene–clay nanocomposites. *Polym. Int.* **2000**, 49, 1158–1163.
12. Gilman, J.W.; Jackson, C.L.; Morgan, A.B.; Harris, R.; Manias, E.; Giannelis, E.P.; Wuthenow, M.; Hilton, D.; Phillips, S.H. Flammability properties of polymer–layered-silicate nanocomposites: polypropylene and polystyrene nanocomposites. *Chem. Mater.* **2000**, 12, 1866–1873.
13. Gilman, J.W. Flammability and thermal stability studies of polymer layered-silicate (clay) nanocomposites. *Appl. Clay Sci.* **1999**, 15, 31–49.
14. Hartwig, A.; Putz, D.; Schartel, B.; Bartholmai, M.; Wendschuh-Josties, M. Combustion behaviour of epoxide based nanocomposites with ammonium and phosphonium bentonites. *Macromol. Chem. Phys.* **2003**, 204, 2247–2257.
15. Zhu, J.; Uhl, F.M.; Wilkie, C.A. Recent studies on thermal stability and flame retardancy of polystyrene–nanocomposites. In: G.L. Nelson and C.A. Wilkie, Eds., *Fire and Polymer: Materials and Solutions for Hazard Prevention.* Oxford University Press, Oxford, England, 2001, pp. 24–33.
16. Tang, Y.; Hu, Y.; Wang, S.F.; Gui, Z.; Chen, Z.Y.; Fan, W.C. Intumescent flame retardant–montmorillonite synergism in polypropylene–layered silicate nanocomposites. *Polym. Int.* **2003**, 52, 1396–1400.
17. Camino, G. Fire retardant polymeric materials, in: G. Scott, Ed., *Atmospheric Oxidation and Antioxidants*. Elsevier, Amsterdam, The Netherlands, 1993.
18. Lewin, M. Unsolved problems and unanswered questions in flame retardance of polymers. *Polym. Degrad. Stab.* **2005**, 88(1), 13–19.
19. Levchik, S.V.; Weil, E.D.; Lewin, M. Thermal decomposition of aliphatic nylons. *Polym. Int.* **1999**, 48(7), 532–557.
20. Levchik, S.V.; Weil, E.D. Combustion and fire retardancy of aliphatic nylons. *Polym. Int.* **2000**, 49(10), 1033–1073.
21. Babrauskas, V.; Peacock, R.D. Heat release rate: the single most important variable in fire hazard. *Fire Saf. J.*, **1992**, 18(3), 255–261.
22. Camino, G.; Costa, L.; Luda, M.P. Overview of fire retardant mechanisms. *Polym. Degrad. Stab.* **1991**, 33(2), 131–154.
23. Stevens, G.C.; Mann, A.H. *Risks and Benefits in the Use of Flame Retardants in Consumer Products*. DTI Report, London, 1999.
24. Van Esch, G.J. Tris(2-butoxyethyl) phosphate, tris(2-ethylhexyl)-phosphate and tetrakis(hydroxymethyl)phosphonium salts. in: *Environmental Health Criteria 218—Flame Retardants*. WHO, Geneva, Switzerland, 2000.

25. U.S. Environmental Protection Agency. *EPA/600*, Vols. 1–3. U.S. GPO, Washington, DC, 1994.
26. Lomakin, S.M.; Zaikov, G.E. *Ecological Aspects of Polymer Flame Retardancy*. VSP, Utrecht, The Netherlands, 1999.
27. Green, J. A review of phosphorus-containing flame retardants. *J. Fire Sci.* **1996**, 14(5): 353–366.
28. Annakutty, K.S.; Kishore, K. Synthesis and properties of flame-retardant polyphosphate esters: a review. *J. Sci. Ind. Res.* **1989**, 48(10), 479–493.
29. Kashiwagi, T. Polymer combustion and flammability: role of the condensed phase. *Proc. Combust. Inst.* **1994**, 28, 1423–1437.
30. Troitzsch, J. Flame retardant polymers current status and future-trends. *Makromol. Chem. Macromol. Symp.* **1993**, 74: 125–135.
31. Horacek, H.; Grabner, W. Nitrogen based flame retardants for nitrogen-containing polymers. *Makromol. Chem. Macromol. Symp.* **1993**, 74, 271–276.
32. Chen, W.Y.; Wang, Y.Z.; Chang, F.C. Thermal and flame retardation properties of melamine phosphate-modified epoxy resins. *J. Polym. Res. Taiwan* **2004**, 11(2), 109–117.
33. Zanetti, M.; Camino, G.; Canavese, D.; Morgan, A.B.; Lamelas, F.J.; Wilkie, C.A. Fire retardant halogen–antimony–clay synergism in polypropylene layered silicate nanocomposites. *Chem. Mater.* **2002**, 14(1), 189–193.
34. Hu, Y.; Wang, S.F.; Ling, Z.H.; Zhuang, Y.L.; Chen, Z.Y.; Fan, W.C. Preparation and combustion properties of flame retardant nylon 6/montmorillonite nanocomposite. *Macromol. Mater. Eng.* **2003**, 288(3), 272–276.
35. Wang, S.F.; Hu, Y.; Zong, R.W.; Tang, Y.; Chen, Z.Y.; Fan, W.C. Preparation and characterization of flame retardant ABS/montmorillonite nanocomposite. *Appl. Clay Sci.* **2004**, 25(1–2), 49–55.
36. Lei, S.; Yuan, H.; Lin, Z.H.; Xuan, S.Y.; Wang, S.F.; Chen, Z.Y.; Fan, W.C. Preparation and properties of halogen-free flame-retarded polyamide 6/organoclay nanocomposite. *Polym. Degrad. Stab.* **2004**, 86(3), 535–540.
37. Kim, J.; Lee, K.; Lee, K.; Bae, J.; Yang, J.; Hong, S. Studies on the thermal stabilization enhancement of ABS: synergistic effect of triphenyl phosphate nanocomposite, epoxy resin, and silane coupling agent mixtures. *Polym. Degrad. Stab.* **2003**, 79(2), 201–207.
38. Song, L.; Hu, Y.; Tang, Y.; Zhang, R.; Chen, Z.Y.; Fan, W.C. Study on the properties of flame retardant polyurethane/organoclay nanocomposite. *Polym. Degrad. Stab.* **2005**, 87(1), 111–116.
39. Chigwada, G.; Wilkie, C.A. Synergy between conventional phosphorus fire retardants and organically-modified clays can lead to fire retardancy of styrenics. *Polym. Degrad. Stab.* **2003**, 80(3), 551–557.
40. Zheng, X.; Wilkie, C.A. Flame retardancy of polystyrene nanocomposites based on an oligomeric organically-modified clay containing phosphate. *Polym. Degrad. Stab.* **2003**, 81(3), 539–550.
41. Chigwada, G.; Jash, P.; Jiang, D.D.; Wilkie, C.A. Synergy between nanocomposite formation and low levels of bromine on fire retardancy in polystyrenes. *Polym. Degrad. Stab.* **2005**, 88(3), 382–393.

42. Xie, W.; Gao, Z.M.; Pan, W.P.; Hunter, D.; Singh, A.; Vaia, R. Thermal degradation chemistry of alkyl quaternary ammonium montmorillonite. *Chem. Mater.* **2001**, 13(9), 2979–2990.

43. Owen, S.R.; Harper, J.F. Mechanical, microscopical and fire retardant studies of ABS polymers. *Polym. Degrad. Stab.* **1999**, 64(3), 449–455.

44. Casu, A.; Camino, G.; De Giorgi, M.; Flath, D.; Morone, V.; Zenoni, R. Fire-retardant mechanistic aspects of melamine cyanurate in polyamide copolymer. *Polym. Degrad. Stab.* **1997**, 58(3), 297–302.

45. Shimasaki, C.; Watanabe, N.; Fukushima, K.; Rengakuji, S.; Nakamura, Y.; Ono, S.; Yoshimura, T.; Morita, H.; Takakura, M.; Shiroishi, A. Effect of the fire-retardant, melamine, on the combustion and the thermal decomposition of polyamide-6, polypropylene and low-density polyethylene. *Polym. Degrad. Stab.* **1997**, 58(1–2), 171–180.

46. Breitenfellner, F.; Kainmuelle, T. (Ciba Geigy Corp.). Flame-retarding, reinforced moulding material based on thermoplastic polyesters and the use thereof. U.S. Patent 4546126, Oct. 8, 1985.

47. Okada, K. (Sekisui Chemical Co. Ltd.). Flame-retardant polyolefin resin composition. Jpn. Patent JP11228748, Aug. 24, 1999.

48. Duquesne, S.; Le Bras, M.; Bourbigot, S.; Delobel, R.; Camino, G.; Eling, B.; Lindsay, C.; Roels, T. Thermal degradation of polyurethane and polyurethane/expandable graphite coatings. *Polym. Degrad. Stab.* **2001**, 74(3), 493–499.

49. Takekoshi, T.; Fouad, F.; Mercx, F.P.M.; De Moor, J.J.M. (General Electric). Fire retardant blends. U.S. Patent 5773502, June 30, 1998.

50. Inoue, H.; Hosokawa, T. (Showa Denko KK). Silicate–triazine complex and flame retardant resin composite containing the same complex. Jpn. Patent JP10081510, Mar. 31, 1998.

51. Sallet, D.; Mailhoslefievre, V.; Martel, B. Flame retardancy of polyamide 11 with a decabromodiphenyl–antimony trioxide mixture: a bromine–antimony–nitrogen synergism. *Polym Degrad. Stab.* **1990**, 30(1), 29–39.

52. Lewin, M. Synergism and catalysis in flame retardancy of polymers. *Polym. Adv. Technol.* **2001**, 12(3–4), 215–222.

53. Wu, Q.; Lu, J.P.; Qu, B.J. Preparation and characterization of microcapsulated red phosphorus and its flame-retardant mechanism in halogen-free flame retardant polyolefins. *Polym. Int.* **2003**, 52(8), 1326–1331.

54. Dabrowski, F.; Le Bras, M.; Cartier, L.; Bourbigot, S. The use of clay in an EVA-based intumescent formulation: comparison with the intumescent formulation using polyamide-6 clay nanocomposite as carbonisation agent. *J. Fire Sci.* **2001**, 19(3), 219–241.

55. Ito, K.; Hayashi, H.; Nakajima, H. (Idemitsu Petrochemical Co., Calp Corp.). Flame-retardant polyolefin resin composition and its molded article. Jpn. Patent JP2005029628, Feb. 3, 2005.

56. Lee, S.G.; Won, J.C.; Lee, J.H.; Choi, K.Y. Flame retardancy of polypropylene/montmorillonite nanocomposites. *Polymer-Korea* **2005**, 29(3), 248–252.

57. Lee, J.H.; Nam, J.H.; Lee, D.H.; Kim, M.D.; Kong, J.H.; Lee, Y.K.; Nam, J.D. Flame retardancy of polypropylene/montmorillonite nanocomposites with halogenated flame retardants. *Polymer-Korea* **2003**, 27(6), 569–575.

58. Vexler, G.; Cornibert, J. (Nordx Cdt., Inc.). Flame retardant dual insulation design and cable construction for plenum applications. Can. Patent CA2421440, Sept. 10, 2004.
59. Kaku, B.; Lee, M.S.; Ko, S.; Go, K.; Sho, S. (Industrial Technical Research Institute). ABS nano composite material and method for producing the same. Jpn. Patent JP2001200135, July 24, 2001.
60. Lee, Y.H.; Jung, W.B.; Yang J.H. (Samsung General Chemicals Co.). Polypropylene–clay composite having excellent flame-resistance and producing method thereof. Kr. Patent WO0206388, Jan. 24, 2002.
61. Vexler, G.; Cornibert, J. (Nordx Cdt., Inc.). Flame retardant foam under layer having refined micro-cellular structure. Can. Patent CA2435719, Jan. 21, 2005.
62. Goodman, H.; Legrix, A.H.R. (Imerys Minerals, Ltd.). Flame retardant polymer compositions comprising a particulate clay mineral. G.B. Patent WO03082965, Oct. 9, 2003.
63. Hashimoto, M.; Watanabe, T.; Tokuda, S. (Furukawa Electric Co., Ltd.). Flame-retardant resin composition and insulated electric wire coated therewith. Jpn. Patent JP2004075993, Mar. 11, 2004.
64. Kausch, C.; Verrocchi, A.; Pomeroy, J.E.; Peterson, K.M.; Payne, P.F. (Omnova Solutions, Inc.) Flame resistant polyolefin compositions containing organically modified clay. U.S. Patent 6414070, July 2, 2002.
65. Zhang, J.; Wilkie, C.A. Fire retardancy of polyethylene–alumina trihydrate containing clay as a synergist. *Polym. Adv. Technol.* **2005**, 16(7), 549–553.
66. Kashiwagi, T.; Harris, R.H.; Zhang, X.; Briber, R.M.; Cipriano, B.H.; Raghavan, S.R.; Awad, W.H.; Shields, J.R. Flame retardant mechanism of polyamide 6–clay nanocomposites. *Polymer* **2004**, 45(3), 881–891.
67. Gilman, J.W.; Kashiwagi, T.; Giannelis, E.P.; Manias, E.; Lomakin, S.; Lichtenham, J.D.; Jones, P. Flammability studies of polymer layered silicate nanocomposites: polyolefin, epoxy, and vinylester resins. In: M. Le Bras, G. Camino, S. Bourbigot, and R. Delobel, Eds., *Fire Retardancy of Polymers: The Use of Intumescence*. Royal Society of Chemistry, London, 1998, pp. 203–221.
68. Jahromi, S.; Gabrielse, W.; Braam, A. Effect of melamine polyphosphate on thermal degradation of polyamides: a combined x-ray diffraction and solid-state NMR study. *Polymer* **2003**, 44(1), 25–37.
69. Tang, Y.; Hu, Y.; Xiao, J.F.; Wang, J.; Song, L.; Fan, W.C. PA-6 and EVA alloy/clay nanocomposites as char forming agents in poly(propylene) intumescent formulations. *Polym. Adv. Technol.* **2005**, 16(4), 338–343.
70. Tang, Y.; Hu, Y.; Li, B.G.; Liu, L.; Wang, Z.Z.; Chen, Z.Y.; Fan, W.C. Polypropylene/montmorillonite nanocomposites and intumescent, flame-retardant montmorillonite synergism in polypropylene nanocomposites. *J. Polym. Sci. A Polym. Chem.* **2004**, 42(23), 6163–6173.
71. Ogoshi, M.; Kondo, Y.; Kanbara, H. (Kansai Research Institute). Flame-retardant resin composition and resin composite material using the same. Jpn. Patent JP2003138072, May 14, 2003.
72. Kai, T. (Canon KK). Flame-retardant thermoplastic resin composition and flame-retardant resin molding. Jpn Patent JP2002309106, Oct. 23, 2002.
73. Gilman, J.W.; Kashiwagi, T. Use of polymer layered-silicate nanocomposites with conventional flame retardants, in: T.J. Pinnavaia and G. Beall, Eds., *Polymer–Clay Nanocomposites*. Wiley, Chichester, West Sussex, England 2000, pp. 193–206.

74. Schartel, B. Some comments on the use of cone calorimeter data. *Polym. Degrad. Stab.* **2005**, 88(3), 540–547.
75. Weil, E.D.; Hirschler, M.M.; Patel, N.G.; Said, M.M.; Shakir, S. Oxygen index: correlations to other fire tests. *Fire Mater.* **1992**, 16(4), 159–167.
76. Hong, S.; Yang, J.; Ahn, S.; Mun. Y.; Lee, G. Flame retardancy performance of various UL94 classified materials exposed to external ignition sources. *Fire Mater.* **2004**, 28(1), 25–31.
77. Murashko, E.A.; Levchik, G.F.; Levchik, S.V.; Bright, D.A.; Dashevsky, S. Fire-retardant action of resorcinol bis(diphenyl phosphate) in PC–ABS blend, II: Reactions in the condensed phase. *J. Appl. Polym. Sci.* **1999**, 71(11), 1863–1872.
78. Jang, B.N.; Wilkie, C.A. The effects of triphenylphosphate and resorcinol bis(diphenyl phosphate) on the thermal degradation of polycarbonate in air. *Thermochim. Acta* **2005**, 433, 1–12.

FIGURE 3.5 Image of a PS/mass fraction 10% MMT nanocomposite pyrolized for 82 s. The image on the right shows a cross section and has had the surface char partially removed.

FIGURE 3.6 Image of a PS/mass fraction 10% MMT nanocomposite pyrolized for 95 s. The image on the right shows a cross section and partial removal of the surface char.

FIGURE 3.7 Image of a PS/mass fraction 10% MMT nanocomposite pyrolized for 200 s. The image on the right shows a cross section and partial removal of the surface char.

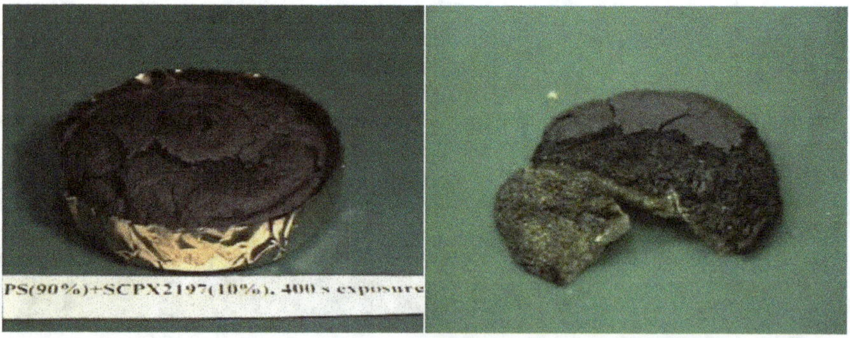

FIGURE 3.8 Image of a PS/mass fraction 10% MMT nanocomposite pyrolized for 400 s. The image on the right shows a cross section and partial removal of the surface char.

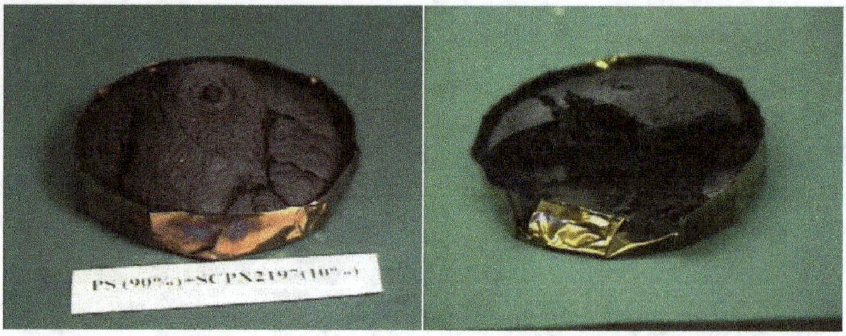

FIGURE 3.9 Image of a PS/mass fraction 10% MMT nanocomposite pyrolized for 1150 s. The image on the right shows a cross section and partial removal of the surface char.

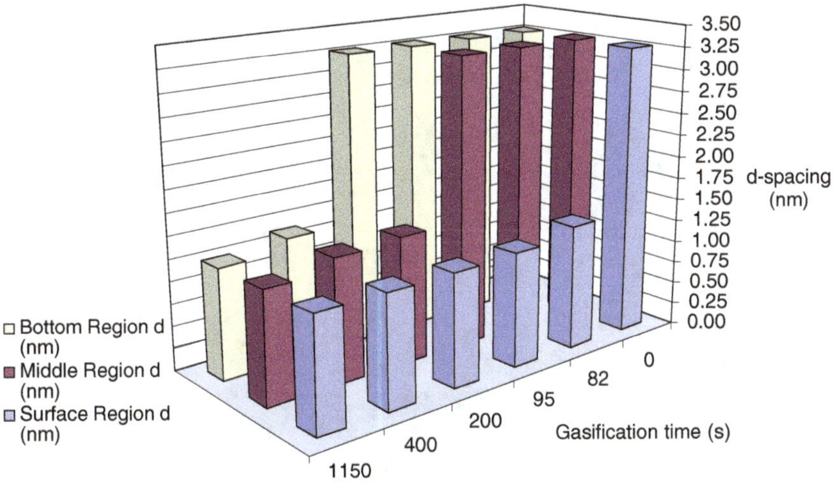

FIGURE 3.11 Clay–char *d*-spacings versus gasification time and sampling region, revealing a step decrease in *d*-spacing, from 3.27 nm (the *d*-spacing of the PS AMMT nanocomposite) to 1.3 nm with increased exposure time.

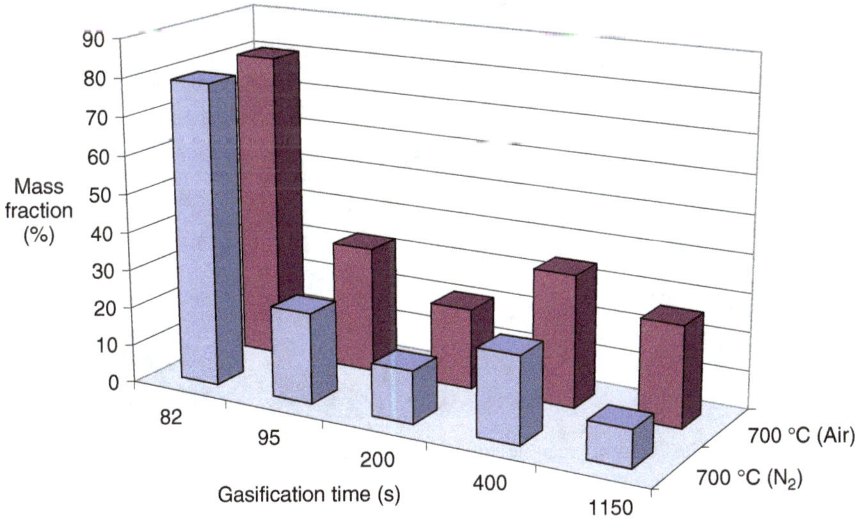

FIGURE 3.12 Mass fraction (%) loss in TGA (in N_2 and air) for surface region clay–chars at various gasification exposure times.

FIGURE 3.16 Gasification residue of PP/PP-g-MA/MMT with 7.7% PP-g-MA and MMT.

FIGURE 3.17 Gasification residue of PP/PP-g-MA/synthetic hectorite with 7.7% PP-g-MA and hectorite.

FIGURE 3.18 Gasification residue of PP/PP-g-MA/synthetic mica with 7.7% PP-g-MA and mica.

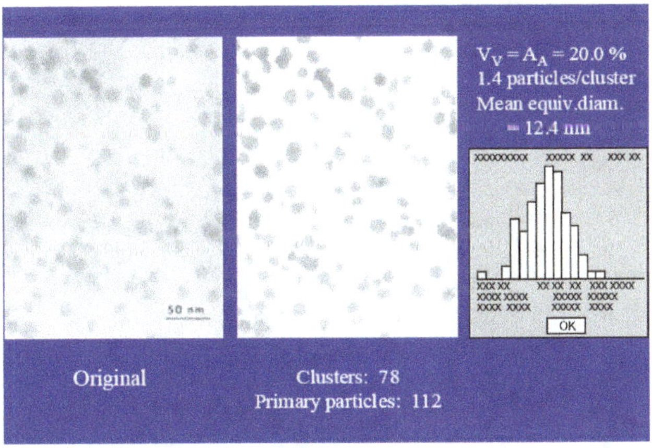

FIGURE 10.1 TEM image of a PMMA–nanosilica nanocomposite (*left*), analyzed image (*middle*), and a histogram distribution of diameter (*right*). (From Ref. 15.)

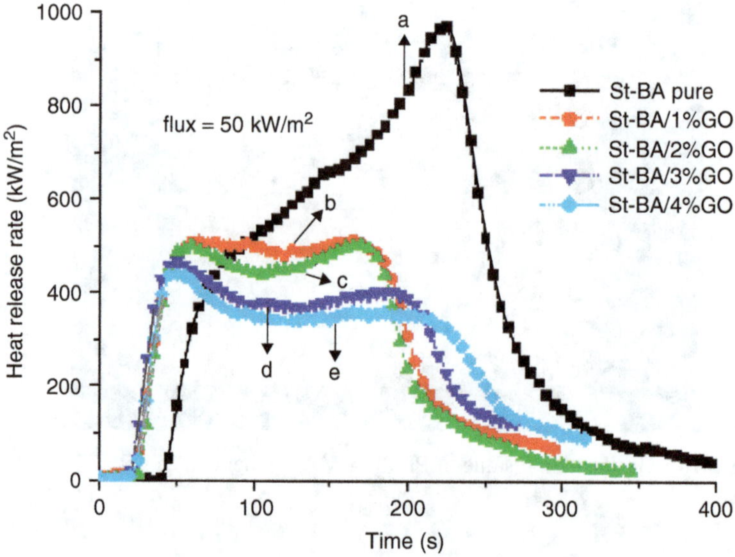

FIGURE 10.14 Heat release rate curves of St–BA and St–BA/GO nanocomposites. (From Ref. 41, copyright © 2004, Elsevier, with permission.)

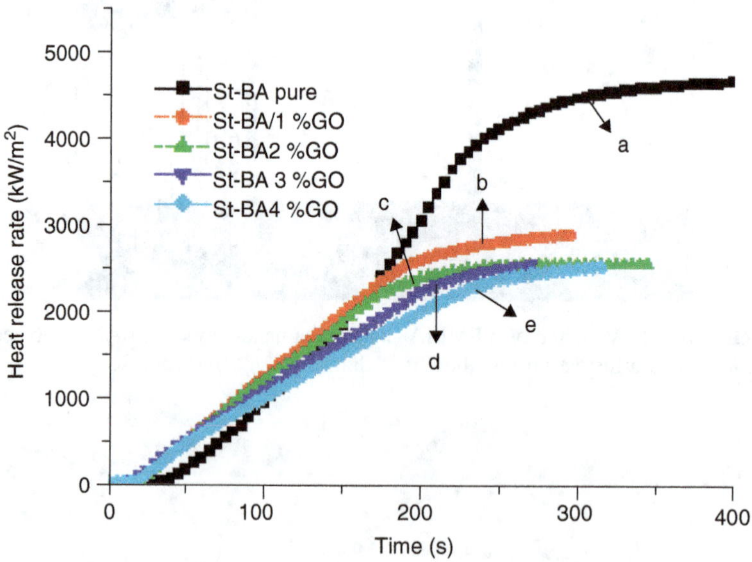

FIGURE 10.15 Total heat released for St–BA and St–BA nanocomposites at 50 kW/m^2. (From Ref. 41, copyright © 2004, Elsevier, with permission.)

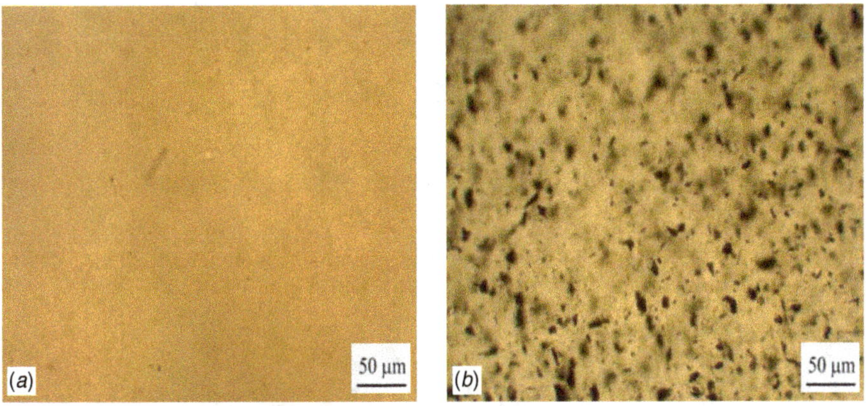

FIGURE 10.17 Optical microscopy images of PMMA–SWNT(0.5%) with two different dispersion of nanotubes with numerous agglomerates: (*a*) good dispersion; (*b*) poor dispersion. (From Ref. 58.)

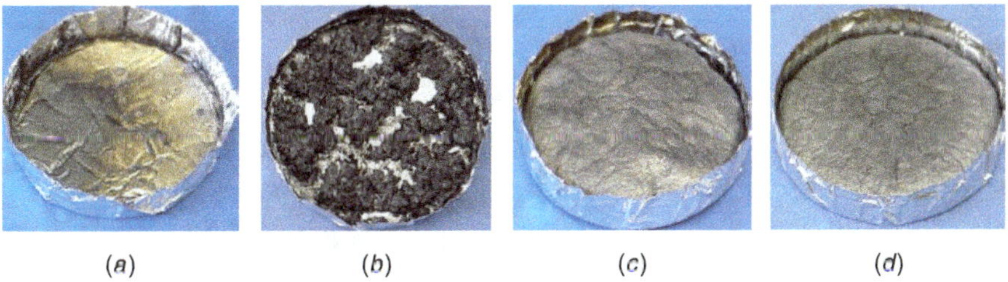

FIGURE 10.21 Residues of PMMA–SWNT after gasification tests in a nitrogen atmosphere at 50 kW/m^2: (*a*) PMMA; (*b*) PMMA–SWNT(0.2%); (*c*) PMMA–SWNT(0.5%); (*d*) PMMA–SWNT(1%).

FIGURE 10.27 Cross section of the residue of PP–MWNT (1%).

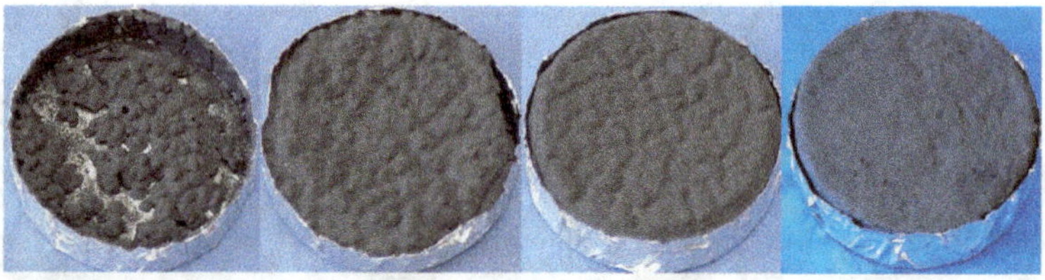

(a) (b) (c) (d)

FIGURE 10.33 Residues after the gasification tests at 50 kW/m^2 in nitrogen: (*a*) with PR-24(1%); (*b*) with PR-24(2%); (*c*) with PR-24(4%); (*d*) with PR-1(4%).

FIGURE 12.5 Polyamide-6 nanocomposite after 1 min of processing (purple, intercalated) and 7 min of processing (red, exfoliated). (From Ref. [98].)

9

THERMOSET FIRE RETARDANT NANOCOMPOSITES*

Mauro Zammarano

Building and Fire Research Laboratory, National Institute of Standards and Technology, Gaithersburg, Maryland

9.1 INTRODUCTION

Thermoset polymers are used in a variety of applications, including construction and transportation, where they are generally preferred to thermoplastics because of their greater strength and better resistance to creep, especially at high operative temperatures. However, like all organic materials, they are flammable, and their use in replacing traditionally nonflammable materials (e.g., metals or ceramics) has increased the fire hazard in the past years. Evolved smoke and toxic gases from these fires create hazards for both people and the environment. For these reasons it is necessary to develop thermoset resins with reduced flammability. This can be achieved by a modification of the polymer (e.g., increase in the cross-link density or insertion of flame retardant moieties in the network) or by the use of fire retardant additives. Halogen-based flame retardants are extremely effective, but there are many concerns about their use, especially

*This work was carried out by the National Institute of Standards and Technology (NIST), an agency of the U.S. government and by statute is not subject to copyright in the United States. The identification of any commercial product or trade name does not imply endorsement or recommendation by NIST. The policy of NIST is to use metric units of measurement in all its publications, and to provide statements of uncertainty for all original measurements. In this document, however, data from organizations outside NIST are shown, which may include measurements in nonmetric units or measurements without uncertainty statements.

Flame Retardant Polymer Nanocomposites, edited by Alexander B. Morgan and Charles A. Wilkie

in Europe, due to the potential release of corrosive and toxic chemicals.[1,2] Therefore, there is a trend to replace them with halogen-free flame retardants. Unfortunately, the efficiency of traditional halogen-free flame retardants such as aluminum or magnesium hydroxide is low, and filling levels of 50 wt%* or even more are usually required to comply with the current fire safety standards. The high filler load dramatically decreases the mechanical properties, and due to the elevated viscosity, also the processability of the formulations. Polymer nanocomposites, on the other hand, have attracted research attention because 5 to 10 wt% loading of nanofiller is sufficient to reduce the maximum rate of heat release by 50 to 70%.[3] Furthermore, nanocomposites exhibit better physical and performance properties than do pristine polymers and conventional composites.[4] Improved tensile and thermal properties,[5,6] reduced permeability,[7] reduced solvent uptake,[8] and increased heat distortion temperature have been reported for these materials in the literature. The improvements are mainly a consequence of the unique interfacial effects that result from the dispersion of nanoparticles with high specific surface area and high aspect ratio.[9–11] A few weight percent of reinforcing agent that is properly distributed in the polymer matrix creates significantly more surface area and polymer–filler interactions than do conventional composites.[12]

This characteristic also greatly affects the flammability of polymeric nanocomposites through three main mechanisms:

1. Extended particle–particle and polymer–particle interactions increase the viscosity of the melt through the formation of a jammed network structure in the polymer matrix, such that the material as whole behaves rheologically like a gel.[13] This phenomenon limits flame propagation through the inhibition of dripping[14] and the decrease in the rate of release of combustible gases emitted by bubbling.[15]
2. High-aspect-ratio nanoparticles, reassembling on the polymer surface during combustion, create an intercalated carbonaceous–silicate residue[3] that lowers the rate of diffusion of the degradation products by a "labyrinth" effect.[16]
3. The large surface area for nanofiller–polymer contact enhances catalytic effects such as the catalysis of charring reactions[17,18] or radical trapping mechanisms.[19]

At present, clays are by large the most investigated nanofillers in flame retardancy. In this chapter we focus on thermoset nanocomposites based on layered silicates and a lately emerging class of layered crystals known as *layered double hydroxides* (LDHs). The use of polyhedral oligomeric silsesquioxanes and nanotubes nanocomposites is discussed in Chapter 10. The preparation of thermoset nanocomposites based on spherical nanosilica is also reported in the literature. It is shown that while being heated in the nanocomposite, nanosilica, migrates to the surface of the material, due to the relatively low surface potential energy

*wt%, meaning mass fraction percent, is used as such throughout this chapter.

of silicon.[20] The silica accumulated on the surface creates a protective barrier that greatly improves char stability in oxidative conditions. Thermoset nanocomposites based on nanosilica are not further discussed here, due to the lack of flammability data for these systems.

Epoxy nanocomposites are used here as a case study. This is because epoxy resins are, by large, the most investigated resins for the preparation of nanocomposite systems. The principles illustrated for this specific system (e.g., parameters affecting the dispersion and the strategies for its improvement, and the influence of the nanofiller and its surface modifier on cross-linking, thermal stability, and flame retardancy) can be generalized for other thermosetting resins. The data available on flammability of other thermoset nanocomposites are barely enough to discuss. However, some data on polyurethane and vinyl ester nanocomposites are reviewed briefly in Sections 9.6 and 9.7, respectively.

9.2 CLAYS

Clays are among the most common minerals on Earth's surface. They may be divided into two broad groups: cationic clays and anionic clays. Both of them have a layered structure in which the layers exhibit a residual electrical charge that is compensated by intercalated exchangeable ions. In particular, cationic clays have negatively charged silicate layers with cations in the interlayer space to balance the charge, whereas anionic clays have positively charged metal hydroxide layers with charge-balancing anions in the gallery region. Cationic clays are widespread in nature and are generally isolated from the minerals, whereas anionic clays are rarer in nature but relatively simple and inexpensive to synthesize.

9.2.1 Cationic Clays

Cationic clay layers are obtained by the combination of two basic building blocks, the Si(O,OH) tetrahedra and the $M(O,OH)_6$ octahedra (where M = Al^{3+}, Mg^{2+}, Fe^{3+}, or Fe^{2+}). The combination of a tetrahedral sheet with an octahedral sheet such way that the oxygen ions of the octahedral sheet also belong to the tetrahedral sheets gives rise to 1 : 1 minerals (e.g., kaolinite or serpentinite M = Al^{3+} or Mg^{2+}, respectively) with layers about 0.7 nm thick.[21] Similarly, 2 : 1 phyllosilicate minerals are obtained with an octahedral sheet sandwiched between two tetrahedral sheets (Figure 9.1*a*). The most commonly used clays for the preparation of nanocomposites (i.e., montmorillonite, hectorite, and saponite) belong to this group. They exhibit a layered structure with 1 nm thickness and extremely high aspect ratios (i.e., 50 to 1000).

9.2.2 Anionic Clays

Anionic clays are a family of lamellar mixed metal hydroxides, also called hydrotalcite-like compounds (HTlcs)* or, more often, layered double hydroxides

*Hydrotalcite is actually the name of a specific mineral whose formula is $Mg_6Al_2(OH)_{16}CO_3{}^{2-} \cdot nH_2O$.

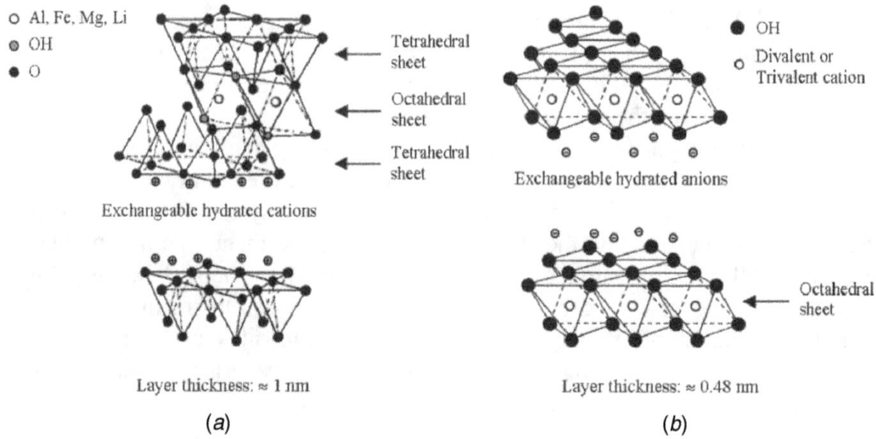

FIGURE 9.1 Comparison between the layered structures of (*a*) 2 : 1 phyllosilicates and (*b*) layered double hydroxides.

(LDHs). LDH lamellae are formed by edge-sharing octahedral units where each octahedron is formed by a cation coordinated with six hydroxyl groups (Figure 9.1*b*). The general formula to describe the chemical composition for an octahedral unit is

$$\underset{\text{intralayer composition}}{[M_{1-x}M_x^{III}(OH)_2]^q} \quad \underset{\text{interlayer composition}}{[A^{n-}_{q/n} \cdot mH_2O]}$$

where $M = M^{2+}$, a divalent cation (Mg^{2+}, Zn^{2+}, Ca^{2+}, Co^{2+}, Cu^{2+}, Mn^{2+}), or $M = M^+$ monovalent cation (Li^+)†; M^{3+} is a trivalent cation (Al^{3+}, Cr^{3+}, Fe^{3+}, Co^{3+}, Ni^{3+}, Ga^{3+}, Mn^{3+}); A^{n-} is an exchangeable interlayer anion; and q is the value of the layer charge for an octahedral unit: $q = x$ for M bivalent cation and $q = 2x - 1$ for M monovalent cation. Important features of LDHs are the highly tunable intra- and interlayer compositions that allow one to fit the properties of the clay to applications in a large number of fields (e.g., catalysis and their supports, adsorbents, ceramic precursors, electrochemical reactions, stabilizers, gene therapy).[22]

The layer charge q and therefore the anion exchange capacity (AEC) can be tuned by adjusting the ratio $\rho = (1 - x)/x$ between the monovalent–divalent cation M and the trivalent cation M^{3+}. Typically, LDH exchange capacities vary in the range 200 to 470 meq per 100 g and are higher than the corresponding cation exchange capacity (CEC) of silicate clays such as sodium montmorillonite (exchange capacity = 80 to 145 meq per 100 g). The aspect ratios of lamellae are similar to, or even higher than, those observed for cationic clays. LDH layers are 0.48 to 0.49 nm thick,[23,24] and their planar dimensions can be tuned between 0.06 and 20 μm by properly adjusting the synthesis conditions.[25,26]

†The Li–Al LDH is the only known example of a M^+–M^{3+} LDH. Li–Al LDH structure is obtained by insertion of Li^+ ions in the octahedral vacancies of gibbsite [γ-$Al(OH)_3$].

LDHs are the most investigated anion-exchangeable layered materials, but other examples of crystals belonging to this family are layered hydroxide (or basic) salts[27] and layered hydroxy double salts.[28] Layered phosphate[29] and phosphonate[30] can also be used for the preparation of layered organo-modified derivatives through topotactic exchange reactions with anionic moieties.

9.3 THERMOSET NANOCOMPOSITES

There are several factors that must be taken into account when investigating thermoset nanocomposites:

- Their properties are related to the complex interaction between the level of dispersion of the nanoparticles and the network formation of the polymer. In fact, the spatial arrangement of the nanoparticles disrupts the cross-link density of the matrix, and this effect is much more severe as the dispersion of the particle improves.[31] The reduced cross-link density may hasten the thermal degradation and decrease the char yield.[32]
- The nanoparticles can play a catalytic role[33] and/or modify the chemistry of the resin. For example, the epoxy monomer diglycidyl ether of bisphenol A (DGEBA) self-polymerizes due to the catalytic effect of protonated alkylammonium cation in the first stage of curing (Figure 9.2).[34] This phenomenon, together with the restricted accessibility to the confined polymer in the intergallery region, promotes the formation of a linear rather than cross-linked structure and the presence of unreacted primary amine that may act as a plasticizer.
- The different affinity between the clay surface and the components of the formulation (i.e., curing agent, monomer, catalyst, initiator, accelerator, additives, etc.) can cause segregation and heterogeneous distribution of the components by preferential intercalation or adsorption on the clay surface.[35] This phenomenon can affect the polymeric network formation by

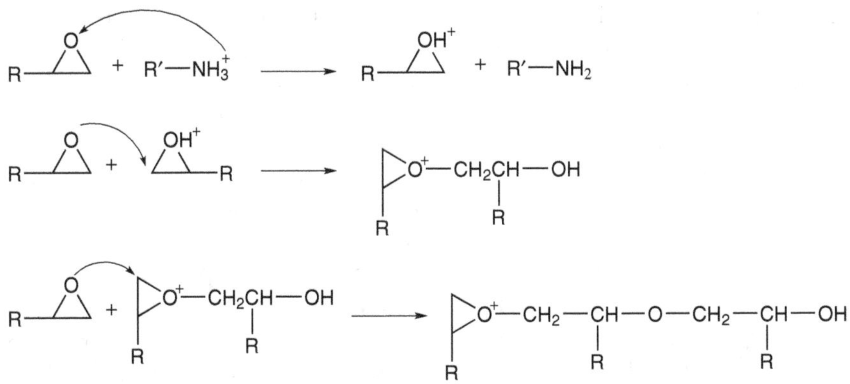

FIGURE 9.2 Catalytic effect of alkylammonium in the curing of DGEBA. (From Ref. 34, copyright © 1996, Elsevier, with permission.)

changing the curing kinetics[14] or by promoting a plasticizing effect due to the presence of unreacted monomers.[34]

- The thermal history during curing dramatically affects the glass transition temperature T_g; for fully cured DGEBA system differences up to 30°C in T_g where observed.[36] This effect is much more relevant than the one due to the presence of layered silicate and/or the variation in its level of dispersion in epoxy nanocomposites.[31]

Therefore, one should be very careful comparing the properties between the pure thermosetting resins and their relative nanocomposites, and all the previously effects must be taken in account.

9.4 EPOXY NANOCOMPOSITES BASED ON CATIONIC CLAYS

Epoxy resins are, by and large, the most investigated resins for the preparation of nanocomposite systems, and for this reason they are used as a case study in this chapter. The principles illustrated for this specific system can be generalized for other thermosetting resins.

9.4.1 Preparation Procedures

Epoxy–clay nanocomposites are prepared by in situ polymerization with organo-modified clays. In this technique, the prepolymer intercalates between the layered crystals during swelling, a process in which heat, stirring, and/or sonication is applied to favor prepolymer diffusion into the clay gallery. Then polymerization can be initiated between the intercalated sheets either by heat, radiation, or a suitable initiator. A variant of in situ polymerization is the exfoliation–adsorption technique, in which the organo-modified clay is first dispersed in a solvent before starting the standard in situ polymerization.

Alkylammonium ions are the most common organomodifiers, but their use is limited by their low thermal stability.[37,38] In addition to promoting intercalation in the intragallery region of the polymer precursor, the organic compatibilizer can react and promote the polymerization of the polymer precursor. Both reactive and unreactive compatibilizers are useful in preloading the gallery region with epoxide monomer, but only reactive compatibilizer can catalyze the intragallery polymerization. Based on work by Lan et al.,[39] the ratio between intra- and extragallery polymerization rates must be tuned in order to exfoliate the clay. Indeed, if this ratio is too low, the polymerization rate outside the layers (extragallery polymerization) will be faster than the polymerization rate between the layers (intragallery polymerization) and only intercalated nanocomposites will form. On the other hand, if the ratio is too high (i.e., the intergallery polymerization rate is much larger than the extragallery rate), the powders are not processable, due to extremely fast curing and phase segregation. The importance of intragallery catalysis is shown by the decrease in layer exfoliation with decreasing Brønsted acidity of the ammonium salt passing from monoalkyl to tetraalkyl onium ions. Catalyzed homopolymerization of DGEBA is also obtained by base-catalyzed oxirane ring-opening reaction of hydroxyl groups in the organic surfactant [e.g., bis(2-hydroxyethyl)methyl tallow alkylammonium], as shown in Figure 9.3. [7] However, as observed in previous work,[40] the catalytic cross-linking activity of alkyl primary ammonium salt is higher than the activity of hydroxyl-containing surfactants and leads to a higher degree of exfoliation. Similar results were found by Camino et al.[33]

The length of the chain of the surfactant and the exchange capacity of the clay control the quantity of epoxy monomer that can be intercalated before the polymerization starts. In fact, the aliphatic chain of alkylammonium salts in the range C8 to C18 assumes a vertical or nearly vertical orientation. So the longer the chain, the more volume is available for epoxide monomers preloaded in the gallery.[39] Furthermore, high-CEC clays require a large number of gallery onium ions to balance the layer charge. Therefore, because of the increasing population density of gallery onium ions, fewer epoxide monomers will be accommodated inside the clay galleries as the layer charge density is increased. The mechanism is illustrated in Figure 9.4. A high density of the surfactant layer can also limit the intragallery diffusion of epoxy and curing agent, and tend to form intercalated nanocomposites rather than exfoliated nanocomposites. This was speculated observing the different behavior of vermiculite, fluorohectorite, montmorillonite, and hectorite with a CEC of 1.6, 1.2, 0.86, and 0.66 meq/g, respectively. Hectorite and montmorillonite gave an ordered exfoliated structure, whereas vermiculite and fluorohectorite gave an intercalated structure.[39]

A comprehensive description of the forces acting during exfoliation was reported by Park et al.[41] They assumed that the driving force for exfoliation is the elastic force due to conformational entropy developed in the clay galleries during epoxy curing. Polymer chains store elastic energy to recoil, and more energy is stored with growing molecular weight (i.e., with the degree of curing). The recoil of the cross-linked epoxy chain in the gallery, and therefore exfoliation, is hindered by (1) the electrostatic attractive forces between the intercalated cations and the negatively charged layers, (2) the van der Waals forces between the organic

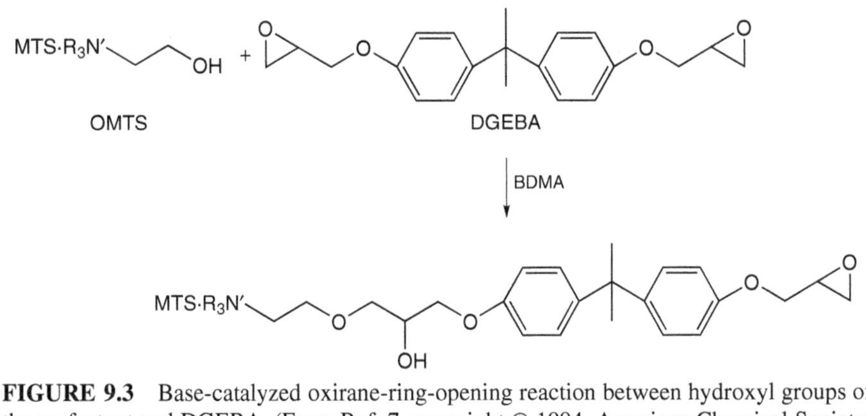

FIGURE 9.3 Base-catalyzed oxirane-ring-opening reaction between hydroxyl groups of the surfactant and DGEBA. (From Ref. 7, copyright © 1994, American Chemical Society, with permission.)

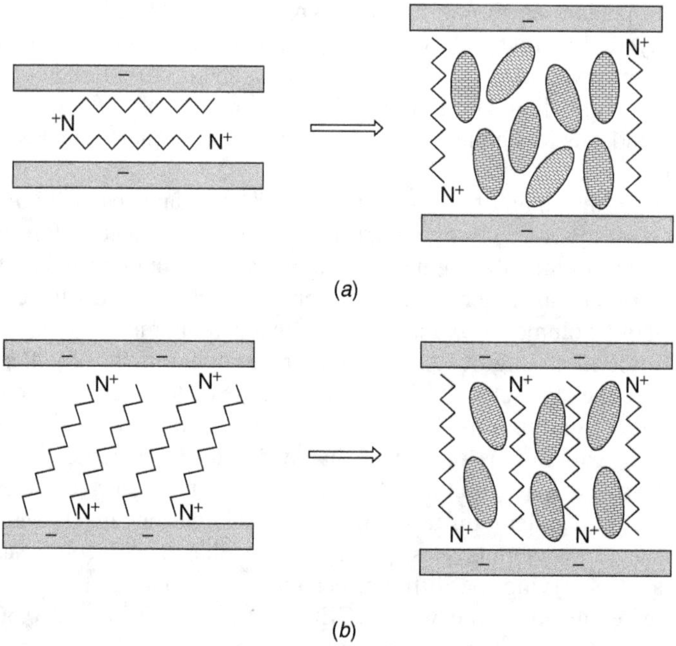

FIGURE 9.4 Influence of charge density on the epoxide monomer preload in the clay gallery: (*a*) low-charge-density clay; (*b*) high-charge-density clay. Note that the *d*-spacing after swelling remains constant despite the charge density. (From Ref. 39, copyright © 1995, American Chemical Society, with permission.)

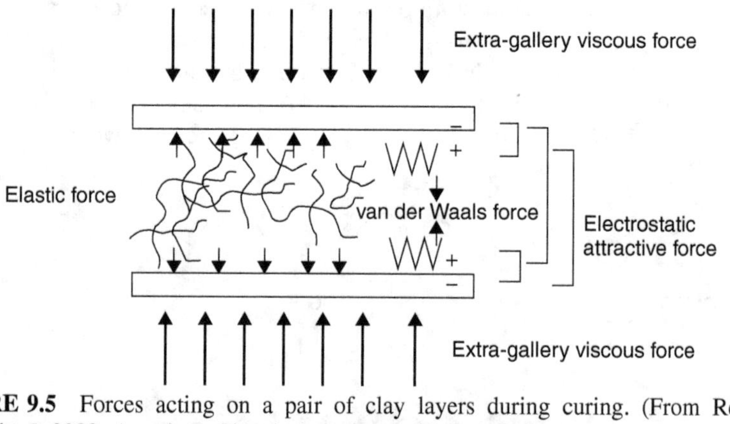

FIGURE 9.5 Forces acting on a pair of clay layers during curing. (From Ref. 41, copyright © 2003, American Chemical Society, with permission.)

aliphatic chains of the alkylammonium salts, and (3) the viscous force offered by the extragallery polymer network (see Figure 9.5). The relaxation process of curing epoxy chains and the possibility that the clay layers can be pushed out of the tactoids to yield exfoliation depend strongly on how the viscosity of the extragallery epoxy molecules evolves with time, as the extragallery epoxy offers viscous resistance to separation of the clay layers. Clay layer exfoliation is difficult if the viscosity of the epoxy in the extragallery region rises quickly. Fast curing conditions lead to a fast increase in viscosity and do not allow enough time for recoiling of the cross-linking epoxy molecules.

Epoxy–clay nanocomposites were studied extensively, but only ordered exfoliated nanocomposites were reported with in situ polymerization. Two recent works showed that disordered and highly exfoliated epoxy–clay nanocomposites can be prepared using an exfoliation–adsorption process.[42,43] In this method the organo-modified clay is first dispersed in a solvent. It is well known that due to the weak forces that stack the layers together, layered silicates can easily be swelled and eventually delaminated in an adequate solvent. The polymer then adsorbs onto the delaminated sheets, and when the solvent is evaporated, a highly disordered structure is obtained before starting the standard in situ polymerization.

Under this process are also gathered the nanocomposites obtained through emulsion polymerization where the layered silicate is dispersed in the aqueous phase. This approach was used by Ma et al.[43] To catalyze intragallery polymerization, they prepared a montmorillonite (MMT) modified with a reactive surfactant, a protonated *m*-xylylenediamine. A water emulsion of clay and DGEBA was prepared, and after evaporation of water at 105°C in vacuo, the curing agent 4-aminophenyl sulfone was added in a stoichiometric amount. Despite the high CEC (120 meq per 100 g) of the clay, they were able to prepare disordered exfoliated nanocomposites. The x-ray diffraction (XRD) pattern of the uncured clay–epoxy system after evaporation of water (epoxy/DM-clay) is shown in Figure 9.6. Only a broad weak peak is observed at about 5.8°, showing that the

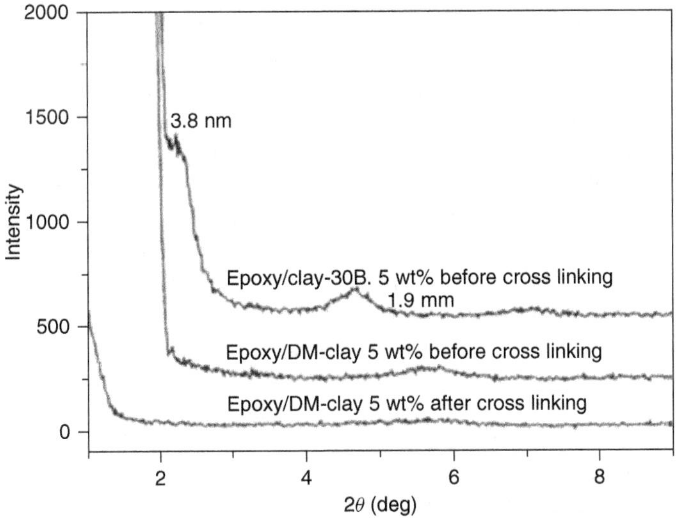

FIGURE 9.6 XRD pattern of a clay–epoxy system (epoxy/DM-clay) before and after curing prepared by the exfoliation–adsorption process, and for comparison, bis(2-hydroxyethyl)methyl tallow alkylammonium–modified MMT (Epoxy/clay-30B). (From Ref. 43, copyright © 2004, American Chemical Society, with permission.)

exfoliation-adsorption process allows obtaining a high degree of disorder before curing. For comparison, the XRD pattern of a MMT modified with a reactive modifier [bis(2-hydroxyethyl)methyl tallow alkylammonium] prepared according to Ref. 40, is also shown in Figure 9.6. In this case, first and second orders of diffraction are observed at $2\theta = 2.3°$ and $2\theta = 4.7°$, respectively, indicating that there is an increase in d-spacing up to 3.8 nm, but the clay maintains a regular parallel alignment.

Chen et al.[42] used a similar exfoliation–adsorption process, but acetone was used instead of water, for preparing the clay slurry. The epoxy monomer (DGEBA) was cured with methyltetrahydrophtalic anhydride and the clay (MMT with a CEC of 92.6 meq per 100 g) was modified with an accelerator for epoxy curing, 2,4,6-tris(dimethylaminomethyl)phenol. In Figure 9.7 transmission electron microscopy (TEM) micrographs show the final structure obtained after curing. For comparison, the typical structure of an epoxy–clay nanocomposite prepared by standard in situ polymerization[7] is illustrated in the TEM micrographs of Figure 9.8. It is evident that in this case the clay tactoids are mostly preserved, since the layers are swelled (d-spacing between 8 and 12 nm) but maintain their original parallel alignment, whereas the nanocomposite prepared by the exfoliation–adsorption technique is highly disordered and exfoliated.

9.4.2 Characterization of the Composite

Epoxy resin–based nanocomposites display a totally different behavior depending on their glass transition temperature. The best improvements in mechanical properties are obtained for systems with subambient glass transition temperature. In elastomeric epoxy matrices a stunning combination of increased tensile stress at break, elongation, and Young's modulus is achieved.[44] In Figure 9.9 is shown the strain at break values for an exfoliated epoxy–magadiite* nanocomposite prepared from magadiite modified with methyloctadecylammonium ion (C18A1M), an intercalated nanocomposite prepared from magadiite modified with trimethyloctadecylammonium ion (C18A3M), and a conventional composite prepared from magadiite modified with octadecylammonium ion (C18A). As expected, the tensile properties improve with increasing degree of nanolayer separation.

In high-T_g epoxy thermosets, neither intercalated nor exfoliated nanosilicates lead to an improvement in the tensile stress at break.[39,45] However, even brittle epoxy matrices exhibit an increase in Young's modulus, and the stress intensity factor causes K_{Ic}, a measure of the energy dissipation at a crack tip during fracture, to increase. This effect appears generally to be more pronounced for exfoliated structures than for intercalated ones. Becker et al.[31] investigated the properties of octadecylammonium-modified montmorillonite-based nanocomposites in di-, tri-, and tetrafunctional epoxy. They observed an increase in K_{Ic} and reduced T_g compared to neat epoxy for all the systems. The reduction in T_g is

*Magadiite is a cationic clay of formula $Na_2Si_{14}O_{29} \cdot nH_2O$ belonging to the family of layered silicic acids.

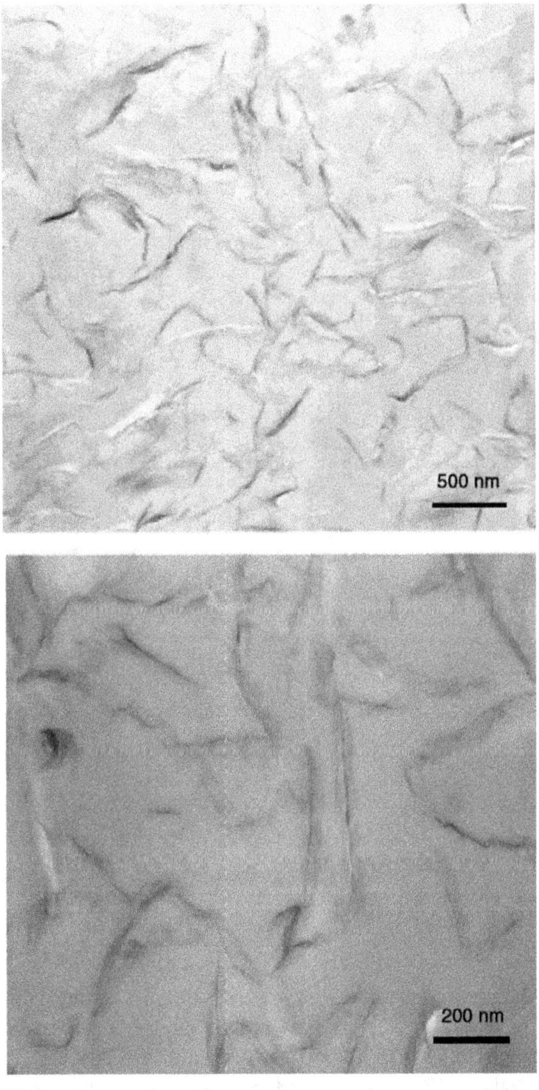

FIGURE 9.7 TEM micrographs of a nanocomposite prepared by the exfoliation–adsorption technique. The clay is highly disordered and exfoliated. (From Ref. 42, copyright © 2004, American Chemical Society, with permission.)

higher for epoxies with high cross-link density. It suggests that the spatial arrangement of the nanoparticles disrupts the cross-link density of the matrix and that this effect is more evident for high cross-link density of the matrix and improved dispersion of the clay layers. Another possible explanation for the decrease in T_g is due to the homopolymerization and plasticizing effect catalyzed by the protonated alkylammonium, as discussed previously (Figure 9.2). Messersmith et al.[7] obtained a broadening and slight increase in T_g (about 4°C) for a glassy epoxy

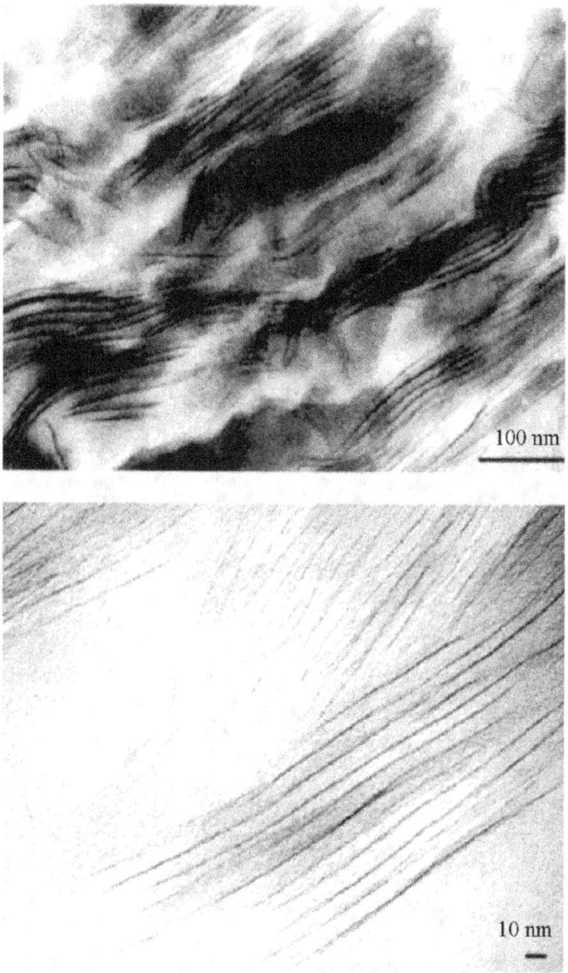

FIGURE 9.8 TEM micrographs of an epoxy–clay (4 vol%) nanocomposite obtained by standard in situ polymerization. The clay tactoids are mostly preserved since the layers are swelled (*d*-spacing between 8 and 12 nm) but maintain their original parallel alignment. (From Ref. 7, copyright © 1994, American Chemical Society, with permission.)

resin by using a mica-type silicate modified with bis(2-hydroxyethyl)methyl tallow alkylammonium. It suggests that by using a proper reactive compatibilizer, the chemical bonding at the interface of the silicate–epoxy matrix can lead to hindered relaxational mobility in the polymer segments near the interface that exceeds the effect of cross-link disruption, due to the spatial arrangement of the nanoparticles.

Dynamic mechanical analysis (DMA) of epoxy nanocomposites shows that the elastic modulus E' appears to be substantially enhanced, especially at temperatures above T_g. For example, DGEBA cured with benzyldimethylamine, below T_g

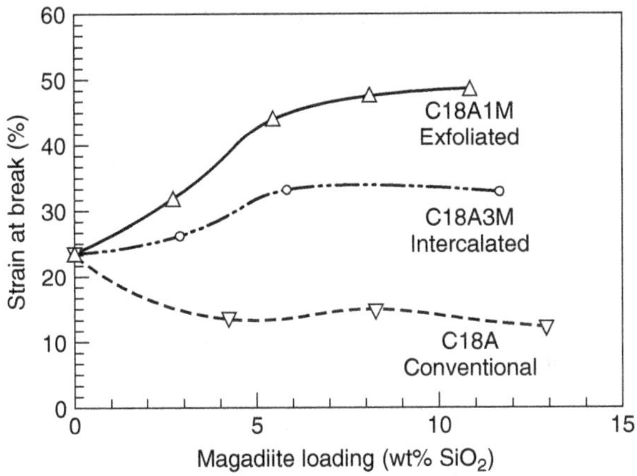

FIGURE 9.9 Comparison of the strain at break values for an exfoliated epoxy–magadiite nanocomposite prepared from magadiite modified with methyloctadecylammonium ion (C18A1M), an intercalated nanocomposite prepared from magadiite modified with trimethyloctadecylammonium ion (C18A3M), and a conventional composite prepared from magadiite modified with octadecylammonium ion (C18A). (From Ref. 44, copyright © 1998, American Chemical Society, with permission.)

shows a 58% increase in modulus by the dispersion of 4 vol% montmorillonite.[7] At 40°C, E' equals 2.44 and 1.55 GPa for the nanocomposite and unfilled cross-linked matrix, respectively. But above T_g (e.g., at 150°C) the elastic modulus improvement reaches a 4.5 factor, with E' values of 11 and 50 MPa for the unfilled and filled epoxy, respectively. Other interesting features of epoxy nanocomposite are improved chemical stability and solvent resistance,[46] dimensional stability,[47] and optical transparency.[48]

9.4.3 Thermal Stability and Combustion Behavior

9.4.3.1 Thermal Decomposition of Neat Epoxy Resins The thermal decomposition of neat epoxy resins was investigated by several authors by means of thermogravimetric analysis and spectroscopic analysis.[49–51] In nitrogen two main steps of decomposition are observed. The first process with a maximum rate at about 300°C causes the main loss in weight. It involves elimination of water through dehydration of secondary alcohol groups and the formation of unsaturated structures. The unsaturations generate weak aliphatic C–O and C–N bonds in the β-position which break down, giving formation of phenolic chain ends and secondary amine-terminal functions, respectively. The following fragmentation of the cross-linked structure implies the volatilization of chain fragments, which is more likely for bifunctional than polyfunctional epoxies. Volatilization is, however, limited by the development of an antagonistic process that leads to the condensation of polyaromatic structures. Elemental nitrogen may be implicated in the formation of heat-resistant cyclized structures, which are more stable than carbonaceous structures. In the second step of thermal decomposition, with a maximum rate at about 400°C, scission of the residual aliphatic C–N and C–O, together with C–phenyl bonds of bisphenol A, leads to almost complete volatilization. A small amount of stable charred residue is the product of aromatization reactions involving cyclic and heterocyclic polymerized structures created in chain scissions occurring at lower temperature and the aromatic ring of the pristine resin. The amount of residue varies typically between 4 and 30 wt% and increases with the cross-link density.

In air a thermooxidative mechanism of degradation leads to a higher amount of polyaromatic structures up to about 500°C. However, at higher temperatures, complete volatilization of material occurs, and no residue is observed.

9.4.3.2 *Thermal Decomposition of Organically Modified Layered Silicates*

The thermal stability of organically modified clays plays a key role in the synthesis and processing of polymer-layered silicate (PLS) nanocomposites.[38] The thermal decomposition of ammonium-treated clays proceeds, schematically, in four steps:

1. Evolution of absorbed water and gaseous species below 180°C
2. Evolution of organic substances from 200 to 500°C
3. Dehydroxylation of the aluminosilicate between 500 and 700°C
4. Evolution of products associated with residual organic carbonaceous residue between 700 and 1000°C

The onset temperature of decomposition of alkyl quaternary ammonium–modified montmorillonite, in nonoxidative thermal degradation, is about 180°C.[37] Initial degradation of the surfactant follows either a Hofmann elimination or an S_N2 nucleophilic substitution mechanism.[37,52,53] Both mechanisms can affect the performance of high-processing-temperature nanocomposites and, in general, the thermal stability and combustion behavior of nanocomposites. In particular, Hofmann elimination generates acidic sites on the layered silicate that can act as a protonic acid catalyst on polymer decomposition.[33,54,55] Imidazolium and phosphonium salts exhibit improved thermal stability compared to ammonium salts.[14,38] Alkylimidazolium salt–modified layered silicates were used successfully to prepare organoclays that exhibit an onset of decomposition temperature up to 392°C.

9.4.3.3 ***Thermal Decomposition of Polymeric Nanocomposites Based on Cationic Clays*** The low onset of decomposition temperature of onium salts in most cases limits the thermal stability of polymeric nanocomposites. Figure 9.10 shows the TGA curves in a nitrogen atmosphere for magadiite modified with methyloctadecylammonium ion (C18A1M), magadiite modified with trimethyloctadecylammonium ion (C18A3M), and the respective nanocomposites containing 20 wt% clay.[44] The lower-temperature weight loss for the C18A3M nanocomposite is indicative of the decomposition and volatilization of the quaternary

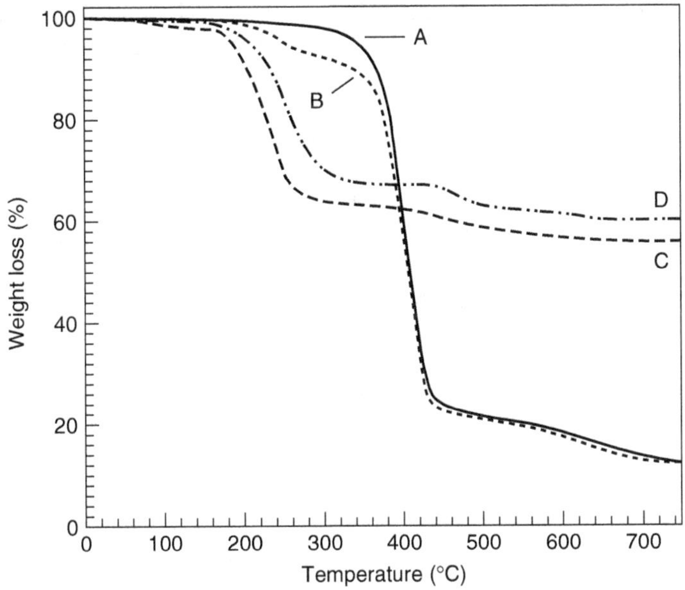

FIGURE 9.10 Thermogravimetric analysis curves for epoxy nanocomposites containing 20 wt% C18A1M–magadiite (curve A) C18A3M–magadiite (curve B). Curves C and D are for the pristine C18A1M–and C18A3M–magadiites, respectively (N_2 atmosphere). (From Ref. 44, copyright © 1998, American Chemical Society, with permission.)

alkylammonium cations in the clay, as is suggested by the analogous weight loss observed for neat C18A3M–magadiite. By comparison, the C18A1M nanocomposite exhibits higher thermal stability. No low-temperature weight loss for the decomposition of surfactant is observed because the secondary onium ions react with the oxirane rings and become part of the polymer network. Therefore, the thermal stability of the epoxy matrix is not affected by the presence of the organoclay in tethered epoxy nanocomposites. For the preparation of this nanocomposite, an elastomeric epoxy resin with a relatively low thermal stability of about 300°C was used.

Different results are obtained for high-thermal-stability epoxies. In this case the nanocomposite shows a decrease in the onset of decomposition temperature compared to the neat polymer. Hussain et al.[56] showed that the onset of decomposition of an amino-cured epoxy is 420°C, whereas a nanocomposite containing 5 wt% octadecylammonium montmorillonite exhibits, at the same temperature, 420°C, a weight loss of 10%. It is obvious that in this case the weight loss in the nanocomposite is not only a direct effect of volatilization of the surfactant but is attributed to the catalytic effect of onium decomposition. Camino et al.[33] compared the effect of different organo-modified montmorillonites on the thermal stability of a DGEBA cured with methyl tetrahydrophthalic anhydride. They observed that the octadecylammonium montmorillonite nanocomposite has the lowest onset at 5 wt% loss (288°C), compared to bis(2-hydroxyethyl)ammonium montmorillonite

(329°C) and dimethyl hydrogenated-tallow (2-ethylexyl)ammonium montmorillonite (342°C). They concluded that this is due to the larger catalytic activity related to the monoalkyl structure of its organic modifier compared to the di-, tri-, and tetraalkyl substitution of the other clays.

In nitrogen, a comparison of the residue yields from clay nanocomposites reveals little improvement in the carbonaceous char yield once the presence of the silicate in the residue is accounted for.[44,57] Instead, in air the char yield increase is more relevant.[33] Hussain and co-workers[56] studied the effect of clay (octadecylammonium montmorillonite) and an organophosphorus epoxy modifier [9,10-dihydro-9-oxa-10-phosphaphenanthrene-10-oxide (DOPO)] on DGEBA and tetrafunctional tetraglycidyldiaminodiphenylmethane (TGDDM) epoxies cured with amine curing agent (an isomer mixture of 3,5-diethyltoluene-2,4-diamine and 3,5-diethyltoluene-2,6-diamine). They prepared 3 wt% P containing epoxies by reacting DGEBA or TGDDM with DOPO. Standard and modified epoxies were used for preparing clay nanocomposites that show a mixed intercalated–exfoliated structure. The clay enhanced char formation in air significantly. For example, the char yield at 600°C increases from 14 wt% in the neat DGEBA to 38 and 42 wt% for 5 wt% clay–DGEBA nanocomposite and 3 wt% P-modified epoxy, respectively. Similar results were observed for TGDDM-based formulations. The results are summarized in the histogram of Figure 9.11. The presence of clay increased the oxidation resistance of char during degradation in air. An increase in the yield and stability of char in oxidative conditions was observed in epoxy nanocomposites by Camino and co-workers.[33] In this case, the maximum char yield and barrier effect to oxygen were maximum for exfoliated

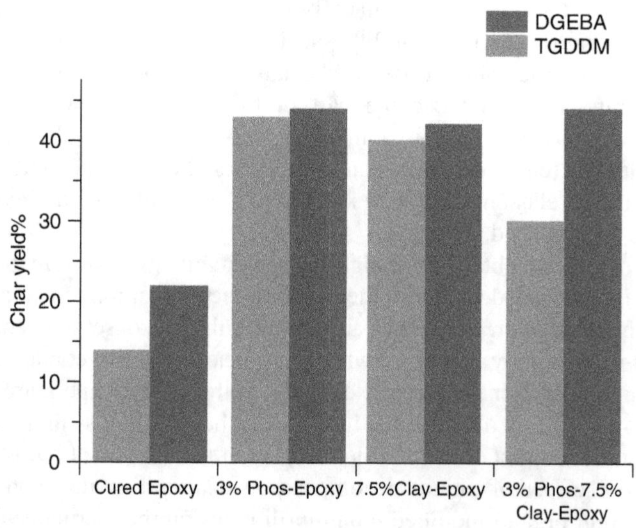

FIGURE 9.11 Histogram of the char yields at 600°C for DGEBA and TGDDM control samples, 3% P-modified epoxies, and 7.5% clay nanocomposites. (From Ref. 56, copyright © 2004, John Wiley & Sons, Inc., with permission.)

structures and negligible for intercalated or microstructures. Gilman et al.[3] characterized polymer nanocomposites by means of a radiative gasification apparatus. They observed that independent of the chemical structure (thermoplastic or thermoset) and nanostructure (delaminated or intercalated) of the nanocomposite, the clay layers reassemble on the surface of the polymer, forming a carbon–silicate multilayered structure with an interlayer spacing of about 1.3 nm. The intercalated carbon–silicate structure may act as a mass transport barrier, slowing the escape of volatiles generated by the polymer degradation and the diffusion of oxygen. It explains the increased oxidation resistance reported in polymeric nanocomposites.[17,58]

9.4.3.4 Combustion Behavior of Polymeric Nanocomposites Based on Cationic Clays Layered silicates dispersed at a nanoscale in a polymeric matrix reduce the maximum rate of heat release up to 50 to 70% with a clay loading of only 5 wt%. The degree of clay dispersion affects the performance observed. On the one hand, a nano-dispersion is necessary, as no significant improvement in terms of the heat of release rate is observed when layered silicates are dispersed in the polymeric matrix at a micro level.[3] On the other hand, exfoliation of clay is not necessary, and intercalated nanocomposites perform as well as delaminated nanocomposites.[59]

Polymer nanocomposites are attracting research attention in the field of flame retardancy primarily for their low combustion rate. Preliminary results showed that clays can reduce the amount of traditional flame retardants required to comply with the standards, but so far, the approach to nanocomposites itself is not sufficient to facilitate passing commercial tests. For example, in terms of the limiting oxygen index (LOI), the improvement achieved by using clays is generally not relevant. A typical increase between 1 and 2% in LOI values is seen with 5 wt% organoclay.[14] Better results were observed by Hussain et al.,[56] who found an increase in LOI values from 25.0% for neat DGEBA to 32.7 and 34.5% for 5.0 and 7.5 wt% nanocomposites, respectively. Similarly, for a tetrafunctional TGDDM epoxy, the LOI values increase from 26.3% in neat polymer to 35.1 and 36.7% in nanocomposites containing 5 and 7.5 wt% clay, respectively. These outcomes are quite surprising and could be related to the unusually high char yield observed in these epoxy systems (see Section 9.4.3.3).

More impressive results and a better comprehension of the mechanism of action of clays in terms of combustion behavior can be achieved by means of cone calorimetry. The first study reported on the flammability of epoxy clay nanocomposites investigated the effect of dimethyl-ditallow-ammonium montmorillonite in DGEBA cured with either methylenedianiline (MDA) or benzyldimethylamine (BDMA).[3] The XRD characterization of the MDA- and BDMA-cured nanocomposites indicates an intercalated structure with a d-spacing of 3.5 and 4.3 nm, respectively. Cone calorimeter data show that the peak heat release rate (PHRR) and the average heat release rate (HRR) are significantly improved with 6 wt% silicate. The HRR plots for DGEBA–MDA and the DGEBA–MDA clay nanocomposites are compared in Figure 9.12. The PHRR, HRR, and average mass loss rate (MLR) decrease by about 40%. The heat of combustion, smoke obscuration,

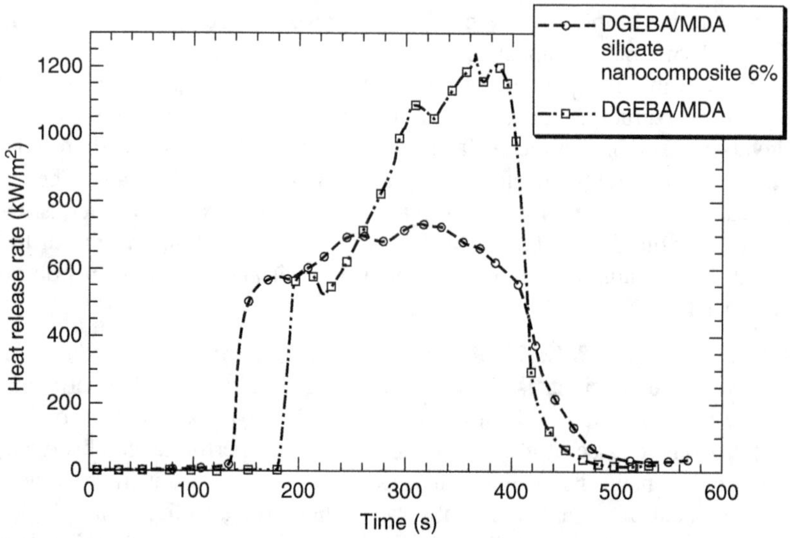

FIGURE 9.12 Comparison of HRR curves for neat DGEBA cured with methylenedianiline (MDA) and a relative nanocomposite containing 6 wt% dimethyl ditallow ammonium montmorillonite. (From Ref. 3.)

and carbon monoxide yields are unchanged. These results suggest that the mechanism of action of clay nanocomposites is operating primarily in the condensed phase. A shorter ignition time is observed in the nanocomposite. This may be due to the low stability of the unreactive organic modifier.

The difference between tethered and nontethered epoxy nanocomposites prepared with reactive hydroxyethyl-substituted quaternary alkylammonium modifier and unreactive quaternary alkylammonium modifier, respectively, was investigated.[59] A tether was formed by reaction of the anhydride curative (hexahydro-4-methylphthalic anhydride) with the alcohol group on the hydroxyethyl-substituted quaternary alkylammonium treatment used on the montmorillonite. The HRR plots (Figure 9.13) do not show any significant difference between the tethered and nontethered nanocomposites, also in terms of ignition time. The reduction in the PHRR was in the range 10 to 20%. Furthermore, the same authors found even a slight increase in the PHRR for a nanocomposite containing 5 wt% quaternary alkylammonium–modified montmorillonite over neat DGEBA when an aromatic amine (Curative W, Shell) was used as the curing agent. These results were lower than expected. In all cases the presence of clay caused a decrease in T_g of about 14°C. It suggests that nanoparticles can interfere with the network formation of a polymer and also hasten the combustion behavior.

The reduction in the PHRR observed by Camino et al.[31] in a similar DGEBA–anhydride system is considerably higher. They used methyl tetrahydrophthalic anhydride as the curing agent and 10 wt% organoclays. The resulting nanocomposites had an ordered intercalated–exfoliated structure. A reduction of 68 and 38% in the PHRR compared to neat epoxy was observed in a nanocomposite

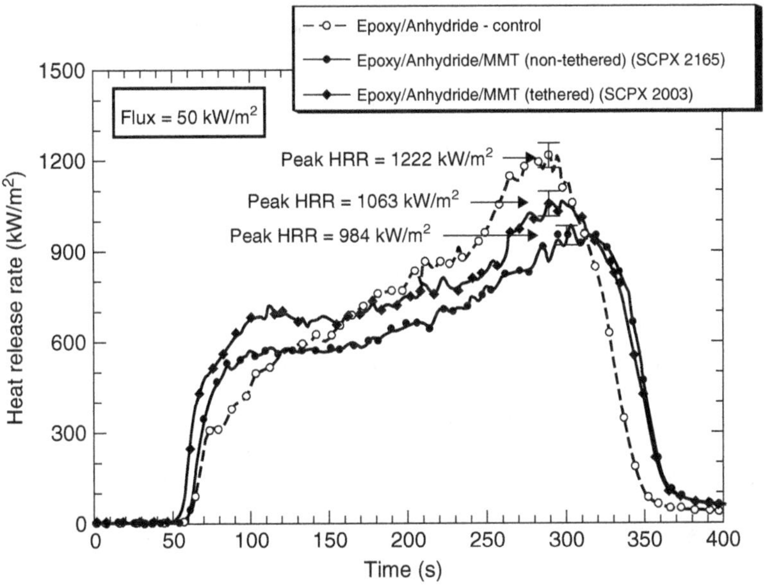

FIGURE 9.13 HRR plots for epoxy–anhydride control sample, epoxy–anhydride (non-tethered) nanocomposite (SCPX2165), and epoxy-anhydride (tethered) nanocomposite. (From Ref. 60.)

prepared with bis(2-hydroxyethyl)ammonium montmorillonite and octadecylammonium montmorillonite, respectively. It must be noted that curing in pure epoxy used as a control sample was different from curing in nanocomposites. In fact, 1 wt% of imidazole was added as catalyzer to the epoxy–anhydride mixture in the neat resin, whereas no imidazole was used for preparation of the nanocomposites since the clay itself exhibits a catalytic effect on the oxirane-ring-opening reaction. Therefore, it cannot be excluded that the unusually high reduction of the PHRR was partially related to the different curing conditions. In particular, as elsewhere reported[14] and as confirmed by the results in this work, use of an acidic initiator decreases the onset of decomposition temperature.

Hartwig et al.[14] investigated the influence of external heat fluxes on the fire behavior of a cycloaliphatic epoxy (3,4-epoxycyclohexylmethyl-3′,4′-epoxycyclohexane carboxylate) nanocomposite based on ammonium (hexadecyltrimethylammonium)- and phosphonium (hexadecyltriphenylphosphonium)-modified layered silicates. The resin was cured with poly(tetrahydrofuran) containing hydroxyl end groups in the presence of benzyltetrahydrothiophenium as initiator. The clay content in the nanocomposites was 4.7 wt%. In the presence of the clay, the kinetic of reaction was slower than that of pure polymer, and higher curing temperatures were used. The authors suggest that the superacid $HSbF_6$, formed through decomposition of the initiator or reaction of the primarily formed carbocation with a proton donor, may be adsorbed on the clay surface. This leads to a reduced amount of superacid available to initiate the reaction. The pure

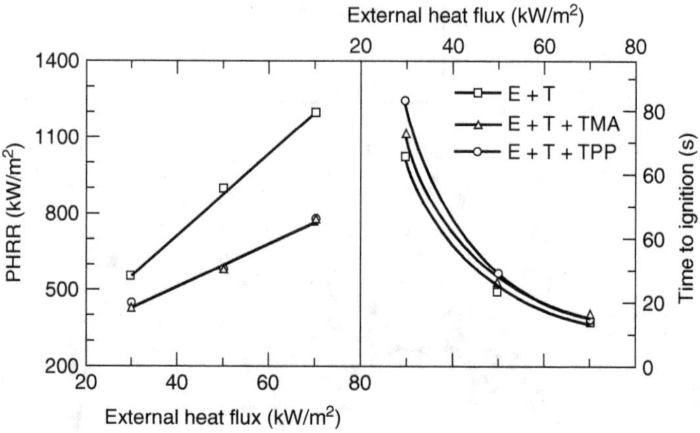

FIGURE 9.14 Influence of external heat flux on the peak heat release rate and time to ignition (t_{ign}). Pure epoxy (E + T), a nanocomposite containing hexadecyltrimethylammonium silicate (E + T + TMA), and a nanocomposite containing hexadecyltriphenylphosphonium silicate (E + T + TPP) were characterized by cone calorimetry at different heat fluxes (i.e., 30, 50, and 70 kW/m^2). (From Ref. 14, copyright © 2003, Wiley-VCH Verlag GmbH & Co., with permission.)

epoxy (E + T), the nanocomposite containing hexadecyltrimethylammonium silicate (E + T + TMA), and the nanocomposite containing hexadecyltriphenylphosphonium silicate (E + T + TPP) were characterized with cone calorimetry at different heat fluxes (i.e., 30, 50, and 70 kW/m^2). The results for the PHRR and t_{ign} (time to ignition) are summarized in Figure 9.14. The reduction in the PHRR was more evident at a high external heat flux: 33 and 20% for an external heat flux of 70 and 30 kW/m^2, respectively. The t_{ign} value was higher for nanocomposites than for neat polymer, especially at low heat flux. This is an unusual result that can be explained by the particular curing chemistry of the system. In fact, adsorption of the acidic initiator on the clay surface leads to an increase in thermal stability for the nanocomposites.

The reduced rate of combustion in nanocomposites can also be observed easily by a horizontal Bunsen burner test such as the UL-94 horizontal burning test or the FAR 25.853. The samples are ignited in a horizontal configuration with a methane flame, and the propagation speed of the flame is recorded. As illustrated in Figure 9.15, the time necessary for the front flame to reach the 150-mm mark on the samples increases by up to 78%[14] in epoxy nanocomposites containing 4.7 wt% of organoclay. However, all samples burned completely, whereas as discussed in Section 9.5.3.3, anionic clay–based nanocomposites may also exhibit self-extinguishing behavior.

The total heat evolved from clay nanocomposites reveals little improvement once the reduction in combustible organic material due to the presence of silicate is accounted for. A reduction of about 1 to 3% compared to pure epoxy is observed by cone calorimetry.[14] This effect is far below that usually resulting from adding traditional flame retardants. For this reason, researchers turned

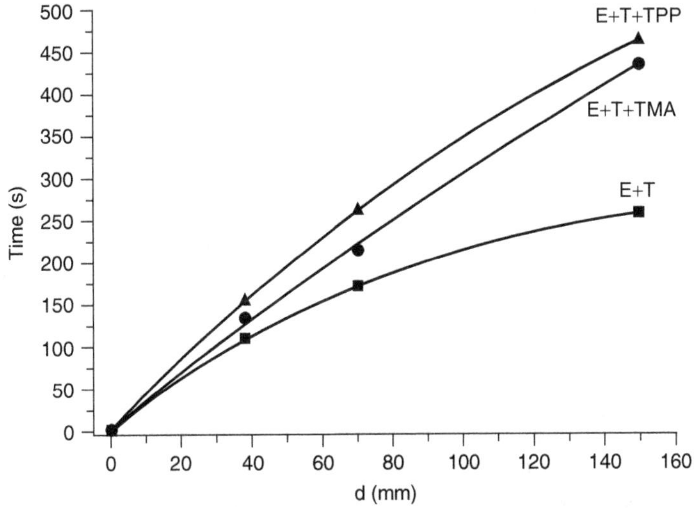

FIGURE 9.15 Time required for flame spread in a horizontal burner test for pure epoxy (E + T), a nanocomposite containing hexadecyltrimethylammonium bentonite (E + T + TMA), and a nanocomposite containing hexadecyltriphenylphosphonium bentonite (E + T + TPP). The clay content is 4.7 wt%. The resin is a cycloaliphatic epoxy (3,4-epoxycyclohexylmethyl-3′,4′-epoxycyclohexane carboxylate) cured with poly(tetrahydrofuran) containing hydroxyl end groups in the presence of benzyltetrahydrothiophenium as initiator. (From Reg. 14, copyright © 2003, Wiley-VCH Verlag GmbH & Co., with permission.)

their attention to the identification of possible synergistic effects between clays and conventional flame retardants that might combine the low combustion rate of nanocomposites with the reduction in the evolved heat of traditional flame retardants. Formulations containing phosphorus flame retardants (e.g., ammonium polyphosphate) and organoclays had already been investigated and showed interesting results in intumescent thermoplastic systems.[60] The combination of organophosphorus and clay was also studied in epoxy resins.[56] DGEBA and tetrafunctional TGDDM epoxies were modified with P-containing organic moieties. Standard and P-modified resins were used for preparing 7.5 wt% clay nanocomposites. For both TGDDM and DGEBA nanocomposites, P-containing epoxies showed an antisynergistic effect with clays. For example, the reduction in the PHRR compared to neat DGEBA was 40, 50, and 38% for the clay–DGEBA nanocomposite, 3 wt% P-modified epoxy, and 3 wt% P-modified epoxy nanocomposite, respectively. The same trend was also observed for TGDDM-based formulations, but in this case the effectiveness of clay was lower (17% reduction in the PHRR).

9.5 EPOXY NANOCOMPOSITES BASED ON ANIONIC CLAYS

As discussed in Section 9.2.2, the anion exchange capacities of LDH are about three to four times higher than the corresponding cation exchange capacities of silicate clays such as sodium montmorillonite. The electrostatic stacking forces between layers and intercalated anions increase with the exchange capacity; this is unfavorable for the exfoliation process[61,62] and may explain the relatively low number of LDH-based nanocomposites reported in the literature.

9.5.1 Preparation Procedures

Usually, only intercalated structures with low *d*-spacings are obtained with LDH, as found for poly(styrene sulfonate), poly(vinyl sulfonate), poly(acrylic acid), poly(ethylene oxide), and dioctyl sulfosuccinate LDH nanocomposites.[63–66] Thermoplastic nanocomposites based on LDH are usually prepared by solution intercalation or in situ polymerization. In our previous work we showed that it is also possible to get delamination of LDH in thermoplastic polymers by melt compounding when a suitable inter-or intralayer composition of LDH and proper processing conditions are selected.[67]

The preparation of thermosetting nanocomposites based on anionic clays is achieved by in situ polymerization with procedures and principles similar to those discussed in Section 9.4.1 for cationic clays. In situ polymerization by an exfoliation–adsorption technique in which organo-modified clay is dispersed in a solvent before beginning standard in situ polymerization does not seem to be effective for LDH. This is because anionic clays are not as easily delaminated in solvents such as cationic clays. A complete delamination in formamide of a high-aspect-ratio LDH in nitrate form has been reported.[68] However, the high boiling temperature (210°C) of formamide makes extraction of this solvent inconvenient. It is reported that alcohols or ethylene glycol can also intercalate between LDH layers and promote swelling and delamination.[69] The use of alcohols may even be deleterious and prevent further expansion of the clay gallery after swelling, as discussed later.

Hsueh and Chen[70] reported the preparation of a LDH–epoxy nanocomposite by standard in situ polymerization. They synthesized an aminolaurate-modified LDH by the coprecipitation method at a constant pH. The clay (filler content 3 to 7 wt%) was swelled in DGEBA at 55°C for 3 h; mixed 2 h at room temperature with the curing agent, a commercial polyoxypropylene diamine (Jeffamine D400, Huntsman Corp.); and cured at 75°C for 3 h and 135°C for an additional 3 h. XRD patterns showed that during the swelling the *d*-spacing of LDH increased from 2.1 nm to about 3.0 nm. A further increase in *d*-spacing was observed during curing at 75°C, and after postcuring at 135°C, no XRD could be detected. TEM micrographs show stacks of a few ordered exfoliated layers with a *d*-spacing of about 8 nm.

Work in our laboratory showed that complete intercalation of DGEBA in organo-modified LDH can be obtained without using long-chain aliphatic surfactants or swelling agents.[40] Specifically, three different LDHs modified with 3-aminobenzenesulfonate, 4-toluenesulfonate, and hydroxybenzenesulfonate, respectively, were synthesized. The organo-modified LDHs were obtained by ionic exchange in an acid medium of a magnesium–aluminum carbonate LDH (Pural MG61HT, Sasol Germany GmbH). Its chemical formula for octahedral unit is:

$$Mg_{1-x}Al_x(OH)_2(CO_3{}^{2-})_{x/2} \cdot nH_2O$$

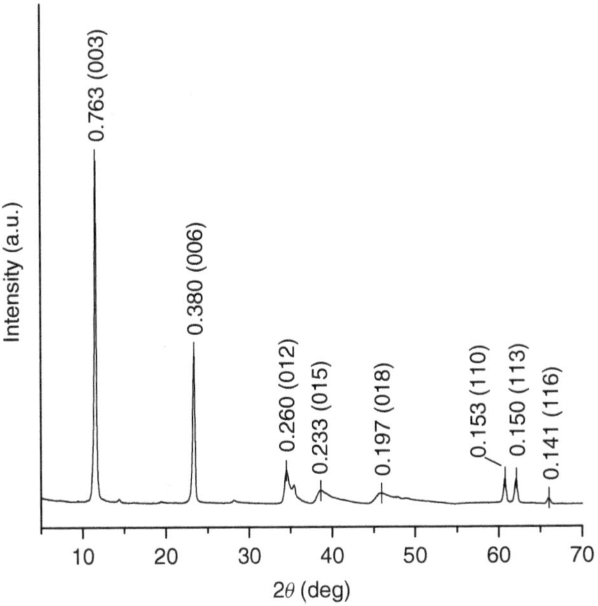

FIGURE 9.16 Diffraction pattern of the carbonate LDH precursor (Mg61) with its Miller indexing. The $d_{003} = 0.76$ nm is the characteristic basal d-spacing of a carbonate magnesium–aluminum LDH, in which carbonate is the intercalated exchangeable anion in the clay gallery. (From Ref. 40, copyright © 2005, Elsevier, with permission.)

where $x = 0.33$ and $n \approx 0.5$. Its diffraction pattern with characteristic d-spacings and Miller indexing is shown in Figure 9.16. Despite the high affinity between carbonate anions and LDH layers, the exchange with organic sulfonates was massive. No peak related to the carbonate LDH (d-spacing 0.76 nm) could be detected in the XRD patterns of the organo-modified LDH as reported in Figure 9.17. Actually, residue carbonate anions are co-intercalated with the organic sulfonates as indicated by FTIR and elemental analysis data.[40] The lack of a basal peak at 0.76 nm does not mean that carbonate anions are completely removed. In fact, a low amount of small carbonate anions can be solubilized into the interlayer region of the large organic anions without any detectable change in d-spacing. This behavior is expected for a layered solid with sufficiently rigid layers.

The organo-modified LDHs (6.4 wt%) were swelled in DGEBA 12 h at 80°C and 2 h at 120°C. The two epoxyphilic clays Mg61–ABS (modified with amino-substituted benzene sulfonate) and Mg61–HBS (modified with hydroxy-substituted benzenesulfonate) showed different results. In fact, no intercalation was observed for Mg61–HBS, whereas a complete intercalation of the epoxy monomer was observed for Mg61–ABS. The lack of intercalation with Mg61–HBS is probably due to thermally activated grafting by condensation reactions between the OH group of the organic anion (i.e., hydroxybenzenesulfonate) and the hydroxyl groups of the LDH layer. With Mg61–ABS, instead, the x-ray

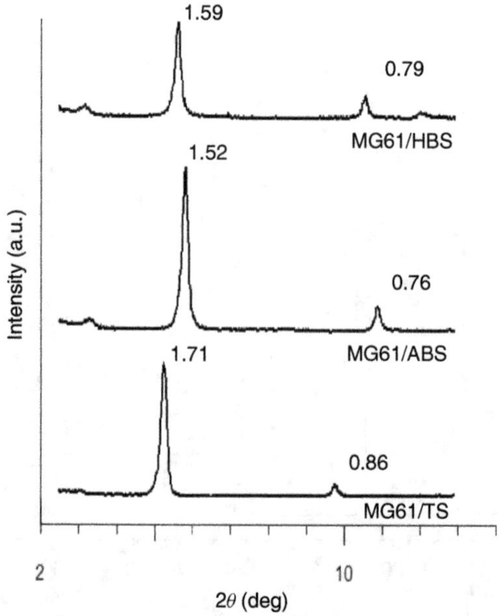

FIGURE 9.17 Powder diffraction patterns of the carbonate LDH precursor after anionic exchange with 4-toluenesulfonate (Mg61–TS), 3-aminobenzensulfonate (Mg61–ABS), and hydroxybenzenesulfonate (Mg61–HBS).

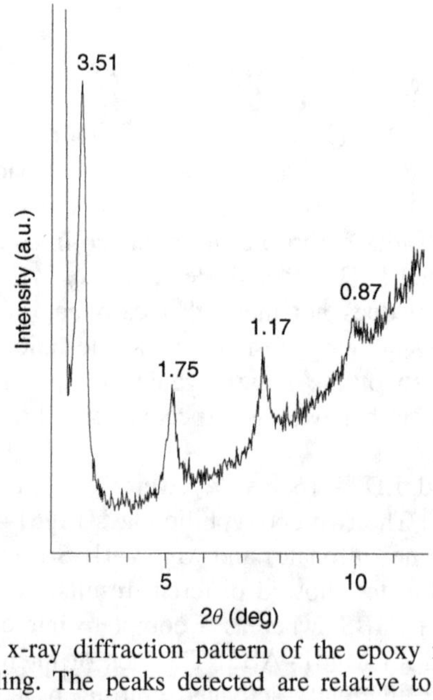

FIGURE 9.18 Liquid x-ray diffraction pattern of the epoxy monomer with 6.4 wt% Mg61–ABS after swelling. The peaks detected are relative to the first four orders of basal reflections. (From Ref. 40, copyright © 2005, Elsevier, with permission.)

diffraction of the liquid sample shows four orders of basal reflections (Figure 9.18), indicating a highly ordered intercalated structure with a d-spacing of 3.51 nm. A thin layer of the same liquid epoxy sample was spread on a grid and observed by TEM (Figure 9.19*a*). The layers are clearly visible and TEM outcomes in terms of d-spacing are in agreement with XRD.

After swelling, the formulations were cured with a stoichiometric amount of poly(propylene glycol)bis(2-aminopropyl) ether (Jeffamine D230, Huntsman Corp.). The clay content in the final composite was 5 wt%. The x-ray diffraction patterns of the samples after curing are shown in Figure 9.20. An intercalated structure was observed with Mg61–TS (modified with 4-toluenesulfonate). The d-spacing increased from 1.71 nm in the pristine organo-LDH to 2.33 nm. For the sample containing Mg61–ABS, two weak peaks at 2.84 and 1.86 nm were detected. It is proposed that these peaks are a second and a third order of diffraction, respectively. This hypothesis allows an estimated d-spacing of about 5.6 nm, which is in qualitative agreement with the nanostructure observed by TEM (Figure 9.19*b*).

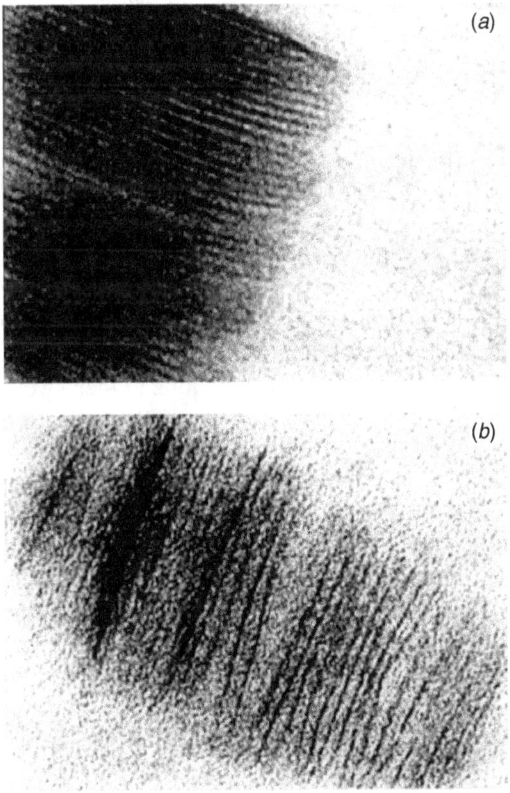

FIGURE 9.19 TEM micrographs of (*a*) a liquid epoxy monomer intercalated in MG61–ABS (6.4 wt%) after swelling and (*b*) the same sample after curing with D230. (From Ref. 40, copyright © 2005, Elsevier, with permission.)

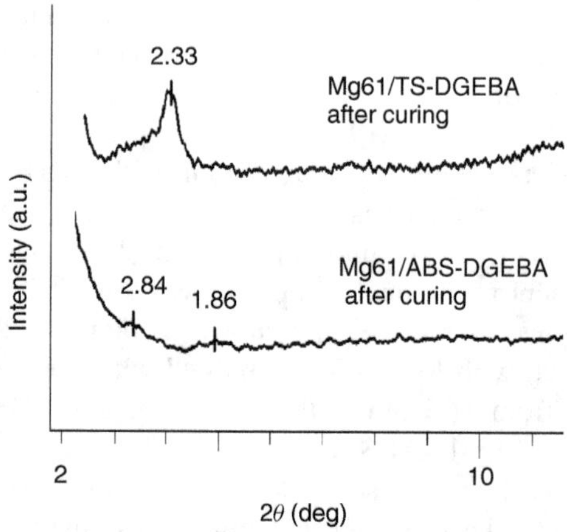

FIGURE 9.20 X-ray diffraction patterns of the epoxy nanocomposites cured with Jeffamine D230 containing 5 wt% Mg61–TS or Mg61–ABS.

The importance of the organic modifier and, in particular, of its reactivity with the epoxy monomer is pointed out by the different increase in d-spacing in the nanocomposites based on Mg61–TS ($\Delta d \approx 0.6$ nm) and Mg61–ABS ($\Delta d \approx 4.0$ nm) compared to that of the pristine organoclays. Mg61–TS is modified with a nonreactive surfactant, whereas Mg61–ABS is modified with an amine-substituted benzenesulfonate that catalyzes the intragallery polymerization and enhances the dispersion. Similar outcomes were discussed in Section 9.4.1 for cationic clay–based nanocomposites.

The influence of the swelling conditions and curing agents on the morphology of the nanocomposite in the system DGEBA–Mg61/ABS have also been investigated. n-Butanol was used as a swelling agent, and triethylene tetraamine (TTA) or 4,4′-diaminodiphenyl sulfone (DDS) was used in a stoichiometric amount as a curative. Mg61–ABS was dispersed in n-butanol and then swelled with DGEBA 2 h at 120°C and 2 h at 145°C in vacuo. The LDH content was 5 wt%. Samples cured with both TTA and DDS showed a d-spacing after curing (3.3 and 3.4 nm, respectively) that was lower than the d-spacing observed after swelling (3.5 nm). According to Chen et al.,[71] a d-spacing contraction of about 0.2 nm may be related to the stiffness of the extragallery polymer, which during curing increases faster than the stiffness of the intragallery polymer, such that the intragallery material becomes compressed. We proposed that bridging reactions (i.e., the reaction of one epoxy monomer with the amine groups of two facing lamellae) prevented further expansion of the clay gallery. Indeed, n-butanol during swelling promotes bridging reactions due to a chain transfer mechanism that increases the reactivity of oxirane rings.[72] The different behavior of the sample cured with TTA and DDS from the sample cured with Jeffamine D230 (where a large increase in the basal distance was observed) could also be explained by the different diffusion rate (i.e., polarity, mobility) and reactivity of the curing agents.[39,73]

9.5.2 Characterization of the Composite

Hsueh and Chen[70] showed that the use of aminolaurate-modified LDH in DGEBA cured with polyoxypropylene diamine (Jeffamine D400) enhances the thermomechanical properties of a matrix. Dynamical mechanical analysis (DMA) pointed out a gradual increase in the glass transition temperature, from 48°C for neat epoxy to 53°, 55°, 58°, and 61°C for a nanocomposite with a clay content of 1, 3, 5, and 7 wt%, respectively. The tensile properties were also enhanced, as reported in Figure 9.21. The tensile strength and Young's module were monotonically increasing with nano-additive content, whereas the strain at break exhibited a maximum at 3 wt%. The coefficient of thermal expansion was reduced significantly both below and above T_g. The optical transparency of the nanocomposite was comparable to that of neat polymer.

Gensler et al.[74] investigated the water vapor permeability of LDH–epoxy nanocomposites. They synthesized an LDH modified with stearylic acid anion. The clay was first dispersed at room temperature for 2 h in a bisphenol A diglycidyl ether monomer at high shear. After swelling, an anhydride hardener (hexahydrophthalic anhydride) and an accelerator (not specified) were added to the mixture. The samples were cured for 2 h at 120°C and a further 5 h at 140°C. During swelling and curing the *d* spacing of the LDH increased from 1.65 nm in the neat organo-modified LDH to about 6.5 nm in the nanocomposite. The water vapor permeability of epoxy nanocomposite films was measured at 23°C and 85% RH according to DIN 53122-1. As shown in Figure 9.22, the water vapor permeability was reduced by factors of 5 and 10 for an LDH contents of 3 and 5 wt%, respectively.

9.5.3 Thermal Stability and Combustion Behavior

In previous work,[40] a magnesium–aluminum carbonate LDH (Pural MG61HT, Sasol Germany GmbH) was used as a precursor for the synthesis of organo-modified LDH. This anionic clay was modified by exchange in an acid medium with 4-toluenesulfonate and 3-aminobenzenesulfonate, and nanocomposites based on 5 wt% of organo-modified LDH and DGEBA were prepared. In this section the thermal stability and combustion behavior of these systems are discussed.

9.5.3.1 Thermal Decomposition of Layered Double Hydroxides The decomposition mechanism of a carbonate LDH is a three-step process (Figure 9.23). The first step, at low temperature, corresponds to the loss of interlayer water; this step is reversible.[75] The second and third steps, at higher temperatures, correspond to the condensation of hydroxyls of the octahedral layer, together with decomposition of the anion CO_3^{2-}. The inorganic residue (≈56.0 wt%) is composed of magnesium oxide and a spinel-like structure, $MgAl_2O_4$. The decomposition is strongly endothermic: the enthalpies of reactions are 356 J/g for the first step and 594 J/g for the second and third steps, respectively. The total heat absorption capacity (950 J/g) is comparable with that of magnesium (1200 J/g) and aluminum hydroxides (1190 J/g).[76]

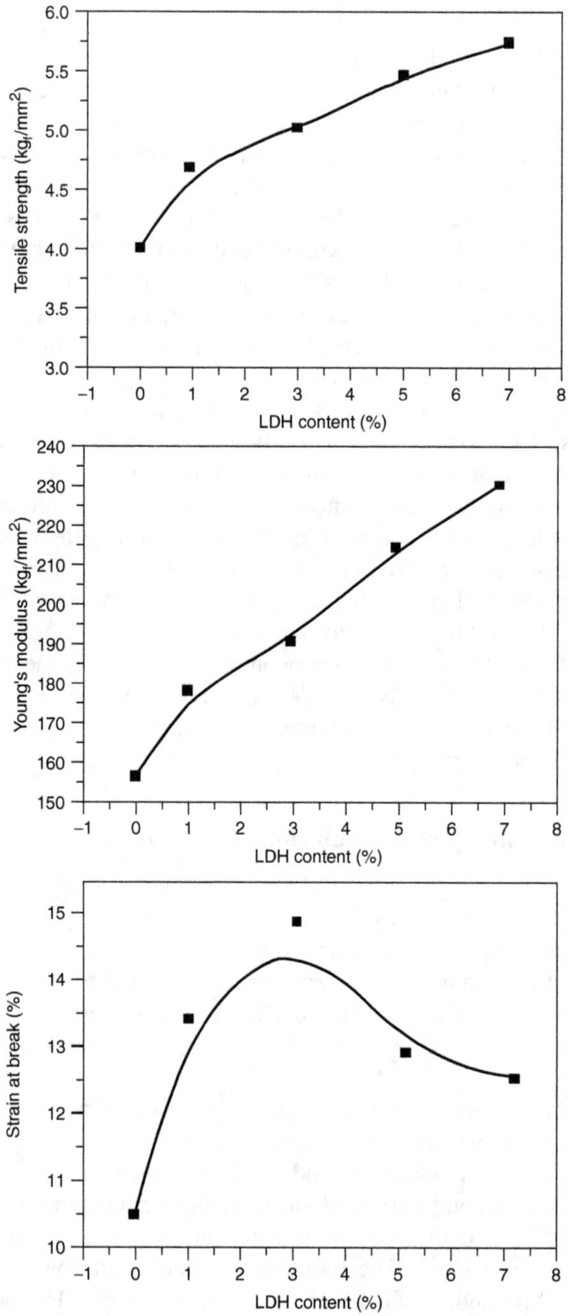

FIGURE 9.21 Tensile properties (tensile strength, Young's modulus and strain at break) versus filler content in a LDH epoxy nanocomposite. (From Ref. 70, copyright © 2003, Elsevier, with permission.)

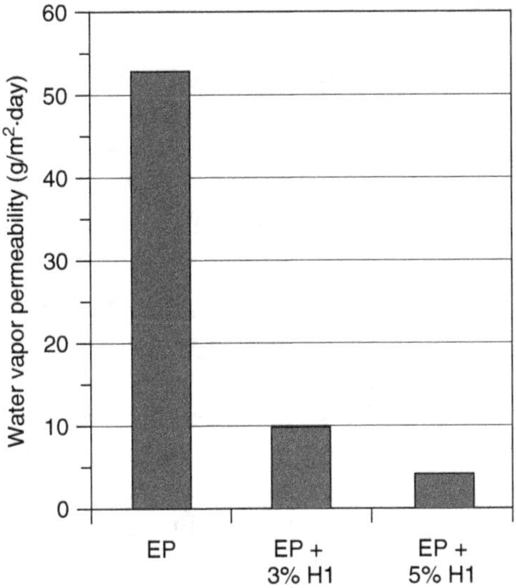

FIGURE 9.22 Water vapor permeability for pure epoxy (EP) and nanocomposites containing 3 wt% (3% H1) or 5 wt% (5% H1) of LDH modified with stearylic acid anion. (From Ref. 74, copyright © 2002, Wiley-VCH Verlag GmbH & Co., with permission.)

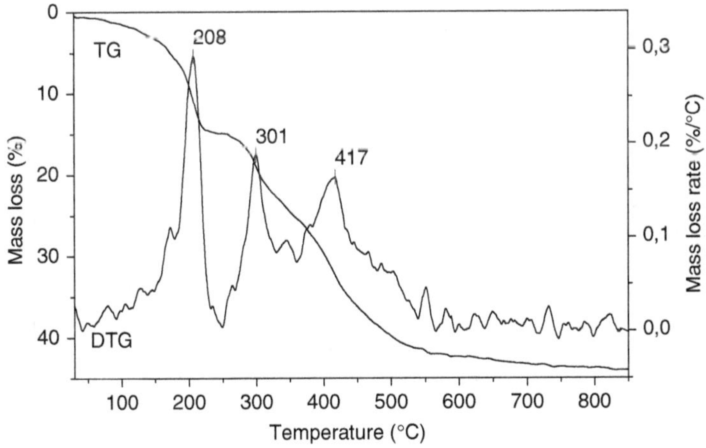

FIGURE 9.23 Thermogravimetrical analysis of aluminum–magnesium LDH in carbonate form (Pural MG 61HT, Sasol). (From Ref. 40, copyright © 2005, Elsevier, with permission.)

After anionic exchange, the organic-modified LDH presents two main steps in decomposition. The first step is related to the release of adsorbed and intercalated water; in the second step the endothermic decomposition of hydroxyl layers is overlapped with the exothermic decomposition of organic anions. The residue is composed mostly by magnesium–aluminum oxides, but the surfactant can also contribute to form a high-temperature stable char. As an example, the thermogravimetric analysis in nitrogen of 3-aminobenzensulfonate/LDH exhibits a residue of 40 wt% at 900°C. The theoretical inorganic residue (i.e., MgO and $MgAl_2O_4$) is only 37.6 wt% which means that other species, such as sulfate- and sulfur-containing organic moieties, are still retained at 900°C. This is further supported by an FTIR absorption spectrum of the residue.[40]

9.5.3.2 Thermal Decomposition of Polymeric Nanocomposites Based on Anionic Clays The thermal decomposition of DGEBA nanocomposites cured with polyoxypropylene diamine (Jeffamine D230) and containing 4-toluenesulfonate/LDH was investigated by simultaneous thermal analysis (STA) in air. The LDH nanocomposite (TS/LDH) is compared to the neat epoxy and to a bis(2-hydroxyethyl)ammonium montmorillonite nanocomposite (30B). The clay content was 5 wt% for both nanocomposites. In Figure 9.24, differential thermal analyses obtained by STA are shown. A main exothermic peak is observed at about 550°C for neat epoxy. In the LDH nanocomposite this peak is split in two parts, so the heat release rate is decreased and the heat evolution delayed, where as no relevant difference is observed between neat epoxy and the cationic clay nanocomposite.

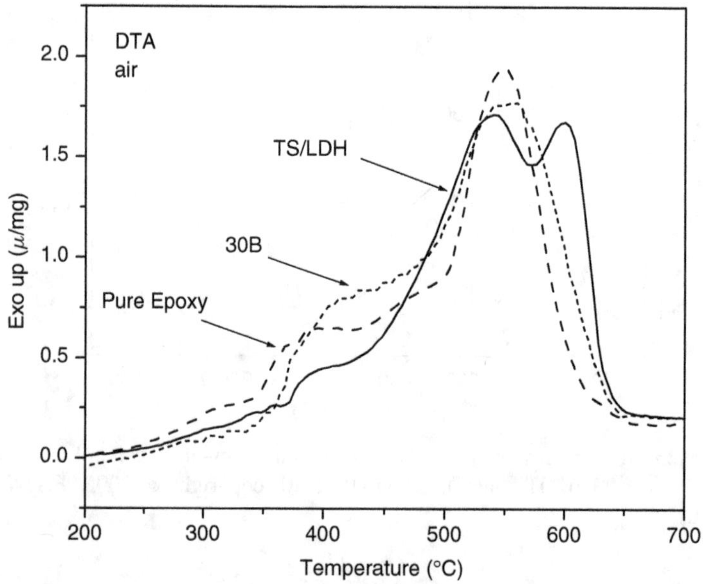

FIGURE 9.24 Differential thermal analysis in air of neat epoxy resin (pure epoxy), a 4-toluenesulfonate/LDH nanocomposite (TS/LDH), and a bis(2-hydroxyethyl)ammonium montmorillonite nanocomposite (30B). The nanoadditive content is 5 wt%. The main exothermic peak observed in the neat epoxy is split in two parts in the LDH nanocomposite.

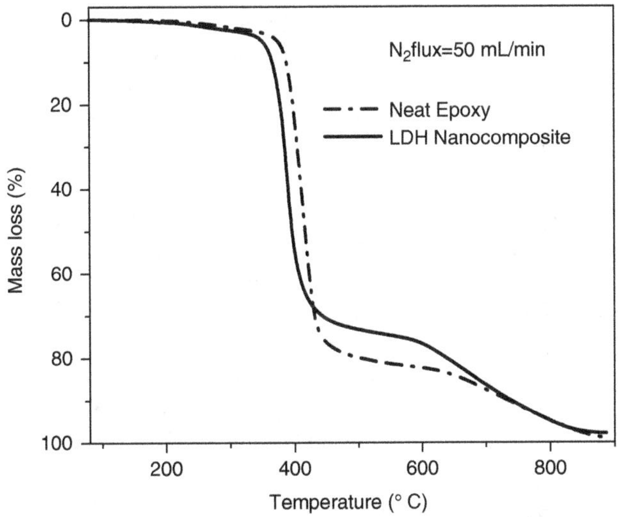

FIGURE 9.25 Thermogravimetrical analysis of a neat epoxy sample (DGEBA cured with diaminodiphenyl sulfone) and a nanocomposite containing 5 wt% LDH modified with 3-aminobenzenesulfonate. Reprinted (From Ref. 40, copyright © 2005, Elsevier, with permission.)

TGA of organosulfonate-modified LDH-based nanocomposites pointed out that the presence of clay promotes char formation. As an example, Figure 9.25 shows the TGA of a neat epoxy sample (DGEBA cured with diaminodiphenyl sulfone) and a nanocomposite containing 5 wt% of LDH modified with 3-aminobenzenesulfonate. Compared to pure epoxy, the nanocomposite exhibits a reduction in the onset of decomposition temperature of about 15°C and an increase in the residue at 550°C of about 36%. Char promotion and a decrease in thermal stability are also observed for epoxy resins containing phosphorus-containing flame retardants such as ammonium polyphosphate (APP). APP decomposes at about 200°C into polyphosphoric acid and ammonia. The polyphosphoric acid catalyzes dehydration of the polymer, which is the first step in degradation of the epoxy network and promotes the formation of unsaturated compounds with subsequent charring.[49,77] The charring agent precursor in intumescent systems is usually a phosphorus compound, in most cases APP, but sulfur compounds can also be used. When heated, sulfonates decompose to form strong mineral acids that catalyze dehydration reactions.[78] For this reason we propose that the organosulfonate used as a surfactant in LDH promotes charring reactions in epoxy nanocomposites. A further clue that charring is related to the nature of the surface modifier rather than the LDH itself comes from previous studies. Hsueh and Chen[70] prepared an epoxy nanocomposite based on LDH modified with an aminocarboxylate (12-aminolaurate), and observed, compared to the pristine polymer, an increase in thermal stability due to the typical barrier effect of clays but no increase in residue at higher temperatures.

On the other hand, Chen and Qu[79] showed that in polypropylene–graft–maleic anhydride (PE-g-MA) nanocomposite based on a dodecyl sulfate–modified LDH, thermal stability is depressed but charring of the polymer is enhanced. So it seems that the intercalated anion, in particular its acidity, are main factors that influence the degradation pathway by a charring mechanism.

9.5.3.3 Combustion Behavior of Polymeric Nanocomposites Based on Anionic Clays LDH used as a conventional filler contributes to the flame retardancy of polymeric matrix by producing a refractory oxide residue on the surface of the material and releasing aqueous vapor and carbon dioxide during the decomposition. The endothermic nature of these processes and the dilution of combustible gases of pyrolysis increase the ignition time and reduce the heat release during combustion.[76] When LDH is dispersed at a nanolevel, these effects are combined with the typical effects of nanocomposites. In this work the flame retardant effectiveness of nanodispersed LDH in epoxy resins is compared with nanodispersed montmorillonite clay and conventional microfiller such as aluminum trihydroxide and ammonium polyphosphate. The combustion behavior of polymeric nanocomposites is studied by means of the UL-94 burning test and cone calorimetry. The formulations investigated together with their identification names are shown in Table 9.1.

The UL-94 horizontal burning test points out that LDH-based nanocomposites (epoxy–LDH1 and epoxy–LDH2) show higher flame retardant properties than those of conventional nanocomposites based on organically modified montmorillonite (epoxy–MMT1 and epoxy–MMT2) and of microcomposites containing traditional flame retardants such as aluminum hydroxide (epoxy–ATH) and carbonate-form LDH (epoxy–LDH/CO_3). In fact, only LDH nanocomposites showed self-extinguishing behavior in the horizontal UL-94 HB test; LDH

TABLE 9.1 Epoxy Micro-and nanocomposites[a]

Sample	Filler	*d*-Spacing[b] (nm)
Epoxy	—	—
Epoxy + LDH/CO_3	Carbonate LDH (Pural MG61HT)	0.8
Epoxy + LDH1	3-Aminobenzenesulfonate–modified LDH	5.6
Epoxy + LDH2	4-Toluenesulfonate–modified LDH	2.3
Epoxy + MMT1	Bis(2-hydroxyethyl)ammonium montmorillonite (Cloisite 30B)	4.0
Epoxy + MMT2	C14–C18 primary alkylamine montmorillonite (Laviosa)	7.0
Epoxy + ATH	Aluminum hydroxide (Apyral 40D, Nabaltec)	—
Epoxy + APP	Ammonium polyphosphate (Exolit AP422, Clariant)	—

[a] All samples are based on DGEBA cured with polyoxypropylene diamine (Jeffamine D230, Huntsman Corp.) for 5 h at 50°C and 2 h at 110°C. The filler content is 5 wt%.
[b] Values determined by XRD or estimated by TEM when no clear diffraction peak was detected.

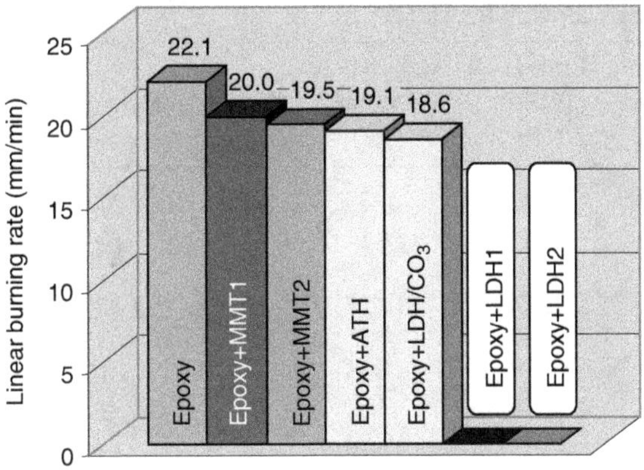

FIGURE 9.26 Linear burning rates of epoxy samples obtained according to the UL-94 horizontal burning test; only LDH-based nanocomposites (i.e., epoxy + LDH1 and epoxy + LDH2) show self-extinguishing properties (the front flame did not reach the first mark at 25 mm or extinguished in the first 30 mm of the specimen).

microcomposites and montmorillonite-based nanocomposite samples burned completely, showing that the unique flame resistance of LDH nanocomposites is related to both the level of dispersion and the intrinsic properties of the modified LDH (Figure 9.26). To our knowledge these are the first examples of self-extinguishing nanocomposites which contain no other flame retardant used for the nanoadditive.

The combustion of the LDH nanocomposites leads to the formation of an intercalated nanostructure of mixed metal oxides (produced by thermal degradation of LDH) and char. This hypothesis is supported by XRD, due to the presence of a peak at 1.28 nm in the epoxy–LDH1 and epoxy–LDH2 after the UL-94 test. It is interesting to note that Gilman et al.[3] reported an intercalated structure of char from montmorillonite nanocomposites, with the same interlayer spacing of the chars, 1.3 nm, independent of the chemical structure of the polymer (thermoplastic or thermosetting) or nanostructure (delaminated or intercalated). The same value observed in an LDH-based nanocomposite (1.28 nm) shows that the interlayer spacing in the char may be independent of the nature of the layered crystal.

Cone calorimetry confirms the higher performance of LDH nanocomposites compared to montmorillonite nanocomposites (Table 9.2). Epoxy–LDH1 and epoxy–LDH2 show a reduction in the peak heat release rate (PHRR) of 51 and 40%, respectively, compared to neat epoxy resin. This result is much better than the one achieved with montmorillonite epoxy–MMT2 (27% reduction in PHRR). The slow heat release of epoxy–LDH1 and epoxy–LDH2 can be justified by the formation of a compact continuous residue with intumescent behavior, whereas the surface of epoxy–MMT2 after a cone calorimeter test is fragmented completely (Figure 9.27). The residue of epoxy–LDH1 after a cone calorimeter test shows a thin shell structure. The thickness of this shielding layer is about 1 mm,

TABLE 9.2 Cone Calorimeter Data[a]

Sample	Residue Yield (%)	Peak HRR [kW/m^2 (Δ%)]	Mean HRR [kW/m^2 (Δ%)]	Mean H_c (MJ/kg)	Mean MLR (g/m^2· s)	Mean SEA (m^2/ kg)	t_{ign} (s)
Epoxy	3.3	1181	533	26	23	750	109
Epoxy–MMT2	8.6	862(27)	477(11)	23	21	773	110
Epoxy–LDH1	8.4	715(40)	382(28)	23	17	724	98
Epoxy–LDH2	9.5	584(51)	347(35)	22	15	743	112
Epoxy–APP	90.1	491(62)	105(80)	17	6	720	78

[a] Heat flux, 35 kW/m^2. H_c, specific heat of combustion; SEA, specific extinction area; t_{ign}, ignition time. Peak heat release rate, mass loss rate, and SEA data are reproducible to within ±10%. The heat of combustion and the time to ignition data are reproducible to within ±15%. The cone data reported are the average of three replicated samples. The samples are square plates 100 mm large and 8 mm thick.

FIGURE 9.27 Comparison of the carbonaceous residues obtained after cone calorimeter tests: (*a*) compact char of epoxy–LDH1 (LDH nanocomposite) with intumescent behavior (char thickness up to 50 mm); (b) fragmented char of Epoxy–MMT2 (montmorillonite nanocomposite). (From Ref. 40, copyright © 2005, Elsevier, with permission.)

and it reaches a maximum height of 5 cm. This structure has good mechanical strength, integrity, coherence, and adherence to the substrate. The protective shell formed on the surface of the samples after burning has a bilayer structure: the internal face, white and porous, is formed by metal oxides, whereas the external face, black and compact, is formed by carbonaceous residues (some of which may be from soot formed in the gas phase).

LDH-based nanocomposites after UL-94 HB presented an intercalated structure of the residue with an interlayer spacing of 1.28 nm (discussed above), whereas after cone calorimeter, no peak could be detected by XRD. Actually, cone calorimetry and UL-94 HB present different testing conditions. In UL-94 the flame on the sample is the only source of heat, so after the flame is extinguished the char is not subjected to oxidation. Instead, in the cone calorimeter the residue is exposed to an external heat flux during burning and after flame extinction until it is removed from the device (≈2 min). After flame extinction, LDH-based nanocomposites show a strong incandescent afterglow, which forces the residual carbon to be at least partially oxidized to CO and CO_2. This is due to the products of decomposition of LDH (i.e., magnesium and aluminum oxides) that catalyze oxidation reactions to produce afterglowing.[80,81] Therefore, we propose that during the combustion of LDH-based nanocomposites, first an intercalated structure of char and metal oxides is formed, and then, during afterglowing, the intercalated carbonaceous residue is oxidized to CO and CO_2 by a reaction that is catalyzed by the metal oxides themselves.

HRR plots are compared in Figure 9.28. It is worth noting that the strong reduction in heat release rate observed in epoxy–LDH1 and epoxy–LDH2 after about 160 s is coincident with the rapid expansion of char. Thus, LDH modified with organic sulfonates act as a nanointumescent system in which the epoxy resin itself is the source of the char, the sulfonate is the charring agent (see

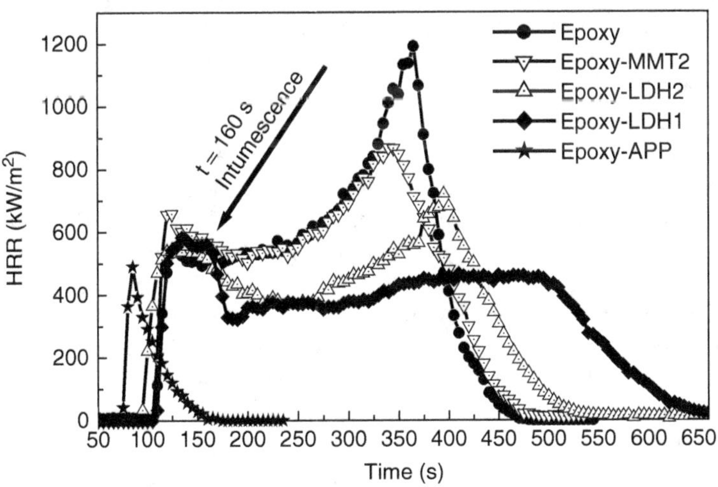

FIGURE 9.28 HRR plots for epoxy control sample (epoxy), epoxy–LDH nanocomposites (epoxy–LDH1 and epoxy–LDH2), epoxy–montmorillonite nanocomposite (epoxy–MMT2) nanocomposite, and epoxy/ammonium polyphosphate (epoxy–APP). The intumescent behavior of LDH-based nanocomposites starts at about $t = 160s$. Reprinted (From Ref. 40, Copyright © 2005, Elsevier, with permission.)

Section 9.5.3.2), and water and CO_2 (evolved during the thermal decomposition of hydroxyl layers) are the blowing agents.

In terms of weight loss and heat release, nanocomposites are less effective than epoxy–APP, but on the other hand, APP reduces the thermal stability and hence the ignition time by 29%. Epoxy–LDH1 and epoxy–LDH2 show ignition times comparable to that of neat resin. Furthermore, APP causes a depletion of mechanical properties of epoxy, whereas LDH-based epoxy nanocomposites show an increase in tensile strength, Young's modulus, and strain at break, and a reduction of thermal expansion and permeability.[70,74]

Most important, even if the effectiveness of LDH alone is lower than that of APP, LDH shows a synergistic effect with APP. In fact, to comply with the UL-94-V-0 standard, for 3-mm-thick samples a 30 wt% loading of APP in epoxy was necessary, whereas using a 4 wt% of 3-aminobenzenesulfonate/LDH, the APP content necessary was 16 to 20 wt%.

The thickness of the samples used for the cone calorimeter test is an important parameter that affects the relative reduction of HRR in nanocomposites compared to pure polymer. We observed, for example, that the relative reduction in the PHRR is 45 and 19% for sample thicknesses of 3 and 8 mm, respectively (5 wt% of 3-aminobenzenesulfonate/LDH in DGEBA cured with triethylene tetraamine). Therefore, the flame retardant effectiveness of LDH is increased significantly for thin samples compared to thick samples. Gilman et al.[82] reported that the trend in the function of the thickness of polyamide-6/montmorillonite (5 wt%) is opposite. The reduction in PHRR compared with neat polymer is about 60% for 8-mm-thick samples and 20% for 1.6-mm-thick samples. This is the usual result seen in charring systems. The superior performance of thin samples of LDH nanocomposites indicates that they may have excellent potential for thin-walled FR products.

For this reason we investigated the use of LDH in epoxy-based FR coatings. 3-Aminobenzenesulfonate/LDH was added to a standard epoxy formulation (DGEBA–polyoxypropylene diamine, Jeffamine D230) containing ammonium polyphosphate. The standard formulation, containing 30 wt% APP (Microcoat), was compared with a nanocomposite formulation containing 3.7 wt% 3-aminobenzenesulfonate/LDH and 26.3 wt% APP (Nanocoat). The total FR content was 30 wt% in both formulations. A glass-reinforced polypropylene laminate (60 wt% glass fiber) was coated with 1 mm of Microcoat or Nanocoat formulation. The combustion behavior of the as-prepared samples was investigated by cone calorimetry at a heat flux of 25 kW/m^2. The cone data obtained are the average of two replicates and are shown in Table 9.3. The use of LDH in the formulations allowed decreasing the PHRR by 45%, the SEA (specific extinction area) by 55%, and the CO/CO_2 ratio by 93%. A stunning effect was also observed in terms of intumescence, as shown in Figure 9.29. The reductions in SEA and CO/CO_2 ratio are probably due to the catalytic action of the nano-dispersed magnesium and aluminum oxides[81,83] (produced by the decomposition of LDH), which promotes the oxidation of CO to CO_2 and reduces the soot formed in the gas phase.

TABLE 9.3 Cone Calorimeter Data at 25 kW/m² for PP–GF Laminates Coated with 1 mm of Nanocomposite Formulation (Nanocoat) or Standard Microformulation (Microcoat)[a]

	t_{ign} (s)	Mass Loss (g)	HRR (kW/m^2)	PHRR (kW/m^2)	H_c (MJ/kg)	SEA (m^2/ kg)	CO/CO_2
Microcoat	95	18.3	37.3	511	17.6	790	0.288
Nanocoat	108	12.7	22.3	280	16.9	355	0.020

[a] t_{ign}, time to ignition; HRR, average heat release rate; PHRR, peak heat release rate; H_c, effective heat of combustion; SEA, specific extinction area.

FIGURE 9.29 Residue after a cone calorimeter test of laminate PP–GF coated 1 mm thick: on the left side a standard epoxy–APP formulation (microcoat), and on the right side an LDH modified formulation (nanocoat). (Courtesy of Consorzio CETMA, Italy.)

9.6 POLYURETHANE NANOCOMPOSITES

Polyurethanes (PUs) represent one of the most useful and widespread families of commercial polymeric materials. They are obtained by reacting polyols with polyisocyanate. PUs are available in a large variety of forms: rigid or flexible foams, adhesives, coatings, elastomers, and rubbers. Their properties can be tuned to match a wide class of applications by changing the types or proportions among the reagents, catalysts, surfactants, or curing conditions. The common reactions involved can be divided into two main groups: reaction of isocyanates with compounds containing reactive hydrogen (e.g., hydroxyl or amine groups, to give urethanes and ureas, respectively) and self-addition reaction of isocyanates.[84]

9.6.1 Preparation Procedures

The preparation of several layered silicate[55,85–90]- or silica[91] -based polyurethane nanocomposites have been reported in the literature. The most common approach to forming polyurethane nanocomposites consists of solvation of a nanofiller by polyols. Wang and Pinnavaia[85] found that montmorillonites exchanged with C12–C18 onium ions are easily solvated by several commercial polyols. The *d*-spacing of the clay after solvation is mainly dependent on the chain length of the onium ion; the influence of clay layer charge or of molecular weight and functionality of the polyol is not significant. Yao et al.[89] showed that by selecting a proper polyol, use of an organic surfactant is not necessary, and intercalated nanocomposites can be prepared directly with inorganic Na^+-montmorillonite.

9.6.2 Characterization of the Composite

The mechanical properties of PU nanocomposites with a subambient glass transition temperature are improved notably over those of a pristine polymer.[85] At a loading of only 10 wt% organoclay, tensile strength, tensile modulus, and strain at break are all increased by more than 100%. Similar results (discussed in Section 9.4.2), are observed in elastomeric epoxies. These improvements in mechanical properties are common to different nanocomposite systems when the testing temperature is above T_g. They can be ascribed to an energy-dissipating mechanism via orientation and alignment of clay layers under an applied stress field.[92] A remarkable decrease in O_2 permeability (about 50%) is observed when the nanosilicate layers (4 wt%) are dispersed in a polyurethane matrix.[90] As measured for two different vapors (water and dichloromethane), the permeability decreases with increasing clay loading up to 20 wt% and then levels off at higher contents.[86]

9.6.3 Thermal Stability and Combustion Behavior

The thermal decomposition, combustion, and fire retardancy of PU materials have been reviewed.[84] Upon heating, the polymer decomposes primarily by regenerating the precursor moieties: polyols, isocyanates, and amines. Complete volatilization of the resulting products is prevented by the formation of thermally stable isocyanurate rings (produced by trimerization of isocyanates) and the formation of substituted ureas (produced by the reaction of hydroxy compounds with the carbodiimides generated by dimerization of isocyanates).[87]

The presence of a flame retardant can modify the pathway of thermal decomposition. In thermoplastic PU, for example, ammonium polyphosphate or melamine polyphosphate generates polyphosphoric acid that catalyzes dehydration and promotes charring by reacting with urethane groups.[84] Organoclays may have a similar effect on the thermal degradation of PUs. The acidic sites generated on the surface of the clay by decomposition of the onium salt may, in fact, catalyze decomposition of the polymer.[55] Charring reactions are further enhanced when inorganic phosphorus-based flame retardants (i.e., ammonium and melamine polyphosphate) are used in combination with layered silicates.

The clay also has a synergistic effect on the LOI values of flame retardant PUs containing melamine polyphosphate (MPP). In fact, the values of LOI observed are 19.0, 20.5, 24.0, and 27.5% for neat polymer, a nanocomposite

(5 wt% hexadecyltrimethylammoniumchloride–modified montmorillonite), a MPP microcomposite (6 wt% MPP), and a nanocomposite containing MPP (6 wt% MPP and 5 wt% organoclay), respectively. The flame retardant performance of polyurethane/layered-silicate nanocomposites is improved significantly over that of pristine polymer in terms of PHRR.[55,87,88] A reduction in the PHRR of 68% is obtained with only 2.5 wt% of organically modified montmorillonite clay (Cloisite 30B, Southern Clay Products San Antonio, Texas), as shown in Figure 9.30. However, in the nanocomposite, the time to ignition is similar or reduced, and the initial rate of heat release is increased. This is consistent with the mechanism of catalyzed polymer degradation in the presence of clay mentioned previously.

As discussed in Section 9.4.3.4, clays can reduce the amount of traditional flame retardants required to comply with the actual standards, but the approach

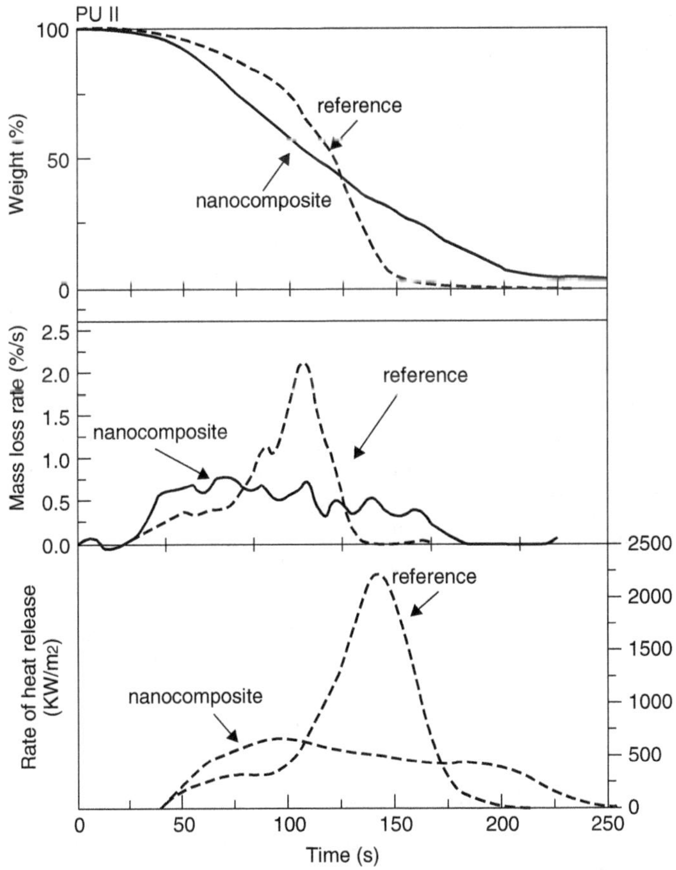

FIGURE 9.30 Cone calorimeter plots for a polyurethane nanocomposite containing 2.5 wt% organoclay (nanocomposite) and neat polymer (control). (From Ref. 87, copyright © 2006, Elsevier, with permission.)

to nanocomposites, by itself is not sufficient for passing commercial tests. As an example, when a polyurethane reference material (V-2 ranking in the UL-94 burning test) is used for the preparation of a 2.5 wt% organoclay nanocomposite, the dripping is strongly suppressed but the specimens burn completely, leading to an unclassified ranking.[87]

9.7 VINYL ESTER NANOCOMPOSITES

Vinyl ester resins (VEs) are used largely for the preparation of high-performance fiber-reinforced composites.[93] They are obtained by reacting methacrylated epoxy with styrene* in the presence of a peroxide initiator and a catalyst. Vinyl ester monomers have several reactive vinyl end groups that provide cross-linking ability. They are solid or highly viscous materials at room temperature, depending on their molecular weight. Styrene acts as a solvent and chain extender for vinyl ester monomers, and is necessary for decreasing the viscosity and increasing processability.

9.7.1 Preparation Procedures

Vinyl ester/layered silicates nanocomposites are prepared by in situ polymerization.[3,94,95] Gilman et al.[3] obtained intercalated nanocomposites (layer spacing between 4.8 and 6.2 nm) by short mixing (5 min) the resin with organoclay (dimethyl ditallow ammonium montmorillonite), the catalyst, and the initiator. No further increase in the *d*-spacing of the cured nanocomposite is observed for longer mixing times.[94] Therefore, organoclays are easily solvated by VEs during a short mixing time at room temperature. Longer swelling times and higher temperatures are necessary for epoxies.[7,36] This is probably due to the high diffusion coefficient of styrene compared to epoxy or vinyl ester monomers.[35] Taking in account the high styrene content in VE resins (40 to 70 wt%) and the difference in the diffusion coefficients, during mixing the styrene will diffuse into the clay gallery much faster than will vinyl ester monomers. Thus, a heterogeneous distribution of styrene is obtained in the formulation. In such a condition, styrene homopolymer is produced in the intragallery region mainly during curing, and cross-link density in the extragallery region is affected by a depletion of styrene. Such a phenomenon has been investigated thoroughly for unsaturated polyester resins–layered-silicate nanocomposites in terms of dispersion of the clay, cross-linking, glass transition temperature, and mechanical properties.[35]

9.7.2 Characterization of the Composite

The mechanical properties and water diffusivity of vinyl ester/clay nanocomposites have been investigated by Shah et al.[95] A vinyl ester resin (bisphenol A

*Methacrylate or acrylate low-molecular-weight monomers (e.g., butyl methacrylate or butyl acrylate) can be used as an alternative to styrene.

epoxy-based, mass fraction of 45% in styrene: Derakane 411-350, Dow Chemical Co.) and two different clays, a benzyl(hydrogenated tallow alkyl)dimethyl quaternary ammonium–modified montmorillonite (Cloisite 10A, Southern Clay Products) and a vinylbenzyltrimethylammonium–modified montmorillonite (VMC), are used. Cloisite 10A gives formation to clay intercalated aggregates with a *d*-spacing between 5.4 and 6.0 nm, which is close to results reported previously.[3] Instead, VMC gives formation to a microcomposite; in fact, no significant increase in *d*-spacing is observed after curing by x-ray diffraction. This is probably due to cross-linking reactions between unsaturation sites on the surfaces of adjacent clay layers.

Diffusion tests conducted by immersing the composites in water show, on the one hand, that a seven-fold decrease in the diffusion coefficient is achieved with 5 wt% organoclay, but on the other, that the equilibrium moisture content has a three-fold increase with the same amount of clay. These values are only slightly affected by the type of clay surface treatment. The increment in equilibrium water content is due to the tendency of clays to adsorb water even after organic surface treatment.

An increase in the glass transition temperature was observed with both clays compared to the temperature of neat polymer (see Figure 9.31). The T_g is affected by the clay concentration for Cloisite 10A, but it is independent of the concentration of VMC. It must be mentioned that when the amount of Cloisite 10A increases, more styrene may intercalate and segregate in the intragallery region,

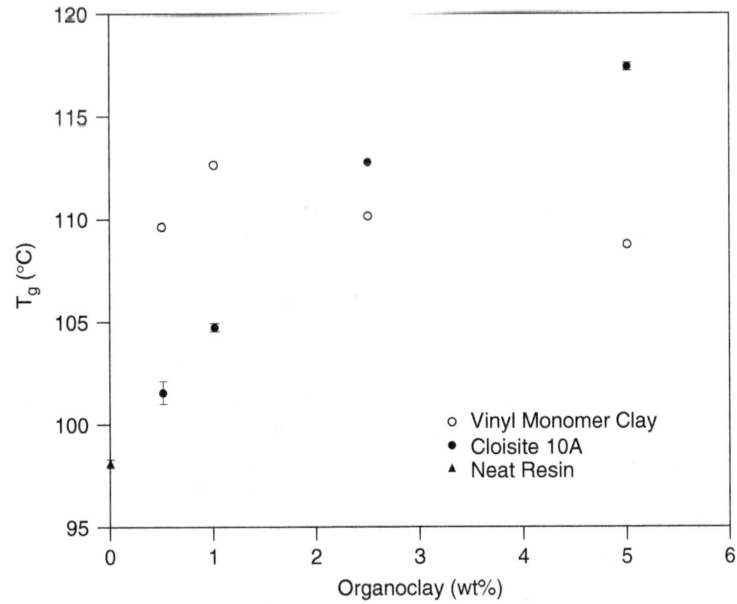

FIGURE 9.31 Glass transition temperature versus clay content for an intercalated vinyl ester nanocomposite (Cloisite 10A), a microcomposite (vinyl monomer clay), and neat resin. (From Ref. 95, copyright © 2002, John Wiley & Sons, Inc., with permission.)

so that the cross-link density of the vinyl ester resin in the extragallery region is increased. The same phenomenon does not occur in the microcomposite.

The mechanism proposed might also explain the trends observed for tensile strength and impact strength. In fact, the microcomposite shows higher values than those of the neat polymer and no significant variation when the clay content varies between 0.5 and 5.0 wt%. On the contrary, in the nanocomposite, both tensile strength and impact strength decrease when the clay content increases.

9.7.3 Thermal Stability and Combustion Behavior

The thermal stability and combustion behavior of vinyl ester nanocomposites have been discussed by Chigwada et al.[94] They investigated the use of different nanofillers: organoclays (i.e., modified montmorillonite and magadiite) and POSS (vinyl-POSS cage mixture, Hybrid Plastics). Thermogravimetric analyses showed that the formation of a nanocomposite does not affect the onset temperature of decomposition in nitrogen. However, an increase from 7 wt% for neat resin to about 13 wt% for 6 wt% nanocomposites was obtained in the char yield at 600°C. No significant difference in the char yields was obtained among montmorillonite, magadiite, and POSS nanocomposites.

Cone calorimetry was used to compare the flame retardant performance of the various formulations. Montmorillonite modified with either dimethyl dehydrogenated tallow ammonium (Cloisite 15A, Southern Clay Products) or an oligomeric unit of styrene and diphenyl vinylphenylphosphate gave a reduction of 31% in the PHRR at a loading of 6 wt%. POSS gave a similar reduction in PHHR (29%) at a loading of 5 wt%. The PHHR did not decrease further when the POSS content was increased up to 10 wt%. The effectiveness of magadiite, modified with the same organomodifiers used for montmorillonite, is lower (14% reduction in PHRR).

Gilman et al.[3] found similar reductions in the PHRR of clay nanocomposites. They prepared 6 wt% intercalated nanocomposites with Cloisite 15A dispersed in a nitrile rubber–modified bisphenol A epoxy-based vinyl ester (mod-bis-A) or a combination of bisphenol A and novolac epoxy–based vinyl ester (bis-A/novolac). The PHRR was reduced by 25 and 39% for mod-bis-A and bis-A/novolac, respectively. The clay promoted charring; in fact, no residue was obtained for the neat resins, while in the nanocomposites the residue yields were 8 wt% (mod-bis-A) or 9 wt% (bis-A/novolac). The heat of combustion, specific extinction area, and carbon monoxide yields were unchanged.

The possible synergy between nanofillers and phosphorus-based fire retardants such as tricesylphosphate (TCP) and resorcinol diphosphate (RDP), has been investigated.[94] RDP, for example, gave a reduction in PHRR between 47 and 61% at a loading of 30 wt%. No improvement was obtained by combining an organoclay (Cloisite 15A) with RDP (30 wt%); the reduction in the PHRR was between 48 and 55%.

A formulation containing phosphorus fire retardant and clay (30 wt% TCP and 6 wt% clay) gave a PHHR of 299 kW/m^2, which is lower than the PHHR achieved with a standard brominated vinyl ester resin. This trend was inverted

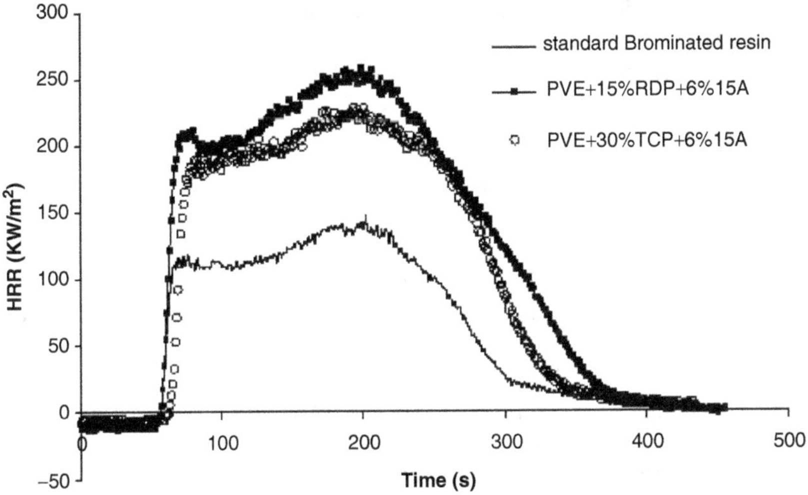

FIGURE 9.32 Comparison of the heat release rate plot for a standard brominated vinyl ester resin and vinyl ester nanocomposites (6 wt% Cloisite 15A) additivated with 15 wt% RDP or 30 wt% TCP. (From Ref. 94, copyright © 2005, Elsevier, with permission.).

when glass fiber–reinforced composites were prepared with these two formulations. The values of PHRR observed in the glass fiber–reinforced composites were 141 and 197 kW/m^2 for the brominated resin and the resin containing TCP and clay, respectively (see Figure 9.32). Similar results were observed for the formulation containing RDP and clay. Furthermore, the high flame retardant content affects the mechanical properties of the composite for two main reasons: (1) reduced mechanical properties of the matrix itself, and (2) reduced wetability of fibers and matrix fiber adhesion. From this point of view, intrinsically flame-retarded systems such as brominated resins, appear to be the best choice for fiber reinforced composites, but halogen-free systems have advantages in terms of lower smoke generation, toxicity, and environmental impact.[2]

9.8 SUMMARY AND CONCLUSIONS

Thermosetting nanocomposites exhibit a reduced rate of heat release compared to neat polymer. However, the approach to nanocomposites itself is not sufficient to comply with the actual fire test standards. For this reason, traditional flame retardants are currently used in combination with nanofillers, and researchers are focusing on the individuation of synergistic systems. As an alternative to the most common cationic clays, anionic clays show improved performance in terms of flame retardancy. Epoxy nanocomposites based on anionic clay exhibit unique self-extinguishing behavior in a UL-94 horizontal burning test never observed before in a pure nanocomposite. The formation of a continuous intumescent ceramic layer on the surface of a polymer during combustion reduces the heat release rate to a higher extent than do montmorillonite nanocomposites.

Furthermore, anionic clays can be combined profitably with phosphorus-based flame retardants to enhance intumescence, reduce the heat release rate, and decrease the production of smoke.

Acknowledgments

This work has been financially supported by CimTecLab Srl, Trieste, Italy. The author thanks NIST for providing travel funds to support a guest research position at NIST. An expression of gratitude to Olaf Torno (Sasol Germany GmbH) for helpful discussions and for providing synthetic LDH carbonate, and to Cinzia Della Porta (Laviosa Chimica Mineraria, Italy) for her skilfull assistance in montmorillonite modifications. The author thanks Ugo Ferri and Massimiliano Franceschi (CimTecLab), Monica Celotto (Electrolux), Sergio Meriani and Loris Sartore (University of Trieste), Piero Talmesio, Alessandra Passaro, and Simona Corigliano (Consorzio CETMA–Brindisi), and Jeffrey W. Gilman (Fire Research Division, NIST) for their fundamental contributions. Part of this work was carried out in the framework of an Italian research program (Progetto MAVET) funded by the Ministry of Education, University and Research (MIUR).

REFERENCES

1. Dumler, R.; Thoma, H.; Lenoir, D.; Hutzinger, O. Thermal formation of polybrominated dibenzodioxins (PBDD) and dibenzofurans (PBDF) from bromine containing flame retardants. *Chemosphere* **1989**, 19, 305–308.
2. Camino, G.; Lomakin, S. Intumescent materials, in: Horrocks, A.R.; Price, D., Eds., *Fire Retardant Materials*. Woodhead Publishing, Cambridge, England, 2001, pp. 318–336.
3. Gilman, J.W.; Kashiwagi, T.; Nyden, M.; Brown, J.E.T.; Jackson, C.L.; Lomakin, S.; Giannelis, E.P,; Manias, E. Flammability studies of polymer–layered silicate nanocomposites: polyolefin, epoxy, and vinyl ester resins, in: S. Ak-Malaika, A. Colovoy, and C.A. Wilkie Eds., *Chemistry and Technology of Polymer Additives*. Blackwell Science, Malden, MA, 1990, pp. 249–265.
4. LeBaron, P.C.; Wang, Z.; Pinnavaia, T.J. Polymer–layered silicate nanocomposites: an overview. *Appl. Clay Sci.* **1999**, 15, 11–29.
5. Okada, A.; Kawasumi, M.; Kuraurchi, T.; Kamigaito, O. Synthesis and characterization of a nylon 6–clay hybrid. *Polym. Prepr.* **1987**, 28, 447–452.
6. Lan, T.; Pinnavaia, T.J. On the nature of polyimide clay hybrid composites. *Chem. Mater.* 1994, 6, 573–575.
7. Messersmith, P.; Giannelis, E.P. Synthesis and characterization of layered silicate–epoxy nanocomposites. *Chem. Mater.* **1994**, 6, 1719–1725.
8. Burnside, S.D.; Giannelis, E.P. Synthesis and properties of new poly(dimethylsiloxane) nanocomposites. *Chem. Mater.* **1995**, 7, 1597–1600.
9. Pinnavaia, T.J. Intercalated clay catalysts. *Science* 1983, 220, 365–371.
10. Giannelis, E.P. Polymer layered silicate nanocomposites. *Adv. Mater.* **1996**, 8, 29–35.
11. Vaia, R.A.; Giannelis, E.P. Polymer nanocomposites: status and opportunities. *MRS Bull.* **2001**, 26, 394–401.

12. Giannelis, E.P.; Krishnamoorti, R.; Manias, E. Polymer–silicate nanocomposites: model systems for confined polymers and polymer brushes, in: Granick, S., Ed., *Polymers in Confined Environments*. Springer-Verlag, Berlin, 1998, pp. 107–147.

13. Kashiwagi, T.; Du, F.; Douglas, J.F.; Winey K.I.; Harris, R.H.; Shields, J.R. Nanoparticle networks reduce the flammability of polymer nanocomposites. *Nat. Mater.* **2005**, 4, 928–933.

14. Hartwig, A.; Putz, D.; Schartel, B.; Bartholmai, M.; Wendschuh-Josties, M. Combustion behaviour of epoxide based nanocomposites with ammonium and phosphonium bentonites. *Macromol. Chem. Phys.* **2003**, 204, 2247–2257.

15. Kashiwagi, T.; Harris, R.H.; Zhang X.; Briber, R.M.; Cipriano, B.H.; Raghavan, S.R.; Awad, W.H.; Shields, J.R. Flame retardant mechanism of polyamide 6–clay nanocomposites. *Polymer* **2004**, 45, 881–891.

16. Zanetti, M.; Camino, G.; Thomann, R.; Mullhaupt, R. Synthesis and thermal behavior of layered silicate–EVA nanocomposites. *Polymer* **2001**, 42, 4501–4507.

17. Zanetti, M.; Kashiwagi, T.; Falqui, L.; Camino, G. Cone calorimeter combustion and gasification studies of polymer layered silicate nanocomposites. *Chem. Mater.* **2002**, 14, 881–887.

18. Tang, T.; Chen, X.; Chen, H.; Meng, X.; Jiang, Z.; Bi, W. Catalyzing carbonization of polypropylene itself by supported nickel catalyst during combustion of polypropylene/clay nanocomposite for improving fire retardancy. *Chem. Mater.* **2005**, 17, 2802.

19. Zhu, J.; Uhl, F.M.; Morgan, A.B.; Wilkie, C.A. Studies on the mechanism by which the formation of nanocomposites enhances thermal stability. *Chem. Mater.* **2001**, 13, 4649–4654.

20. Hsiue, G.H.; Liu, Y.L.; Liao, H.H. Flame-retardant epoxy resins: an approach from organic–inorganic hybrid nanocomposites. *J. Polym. Sci. A Polym. Chem.* **2001**, 39, 986–996.

21. Vaccari, A. Preparation and catalytic properties of cationic and anionic clays. *Catal. Today* **1998**, 41, 53–71.

22. Crepaldi, E.L.; Pavan, P.C.; Valim, J.B. Anion exchange in layered double hydroxides by surfactant salt formation. *J. Mater. Chem.* **2000**, 10, 1337–1343.

23. Meyn, M.; Beneke, K.; Lagaly, G. Anion-exchange reactions of layered double hydroxides. *Inorg. Chem.* **1990**, 29, 5201–5207.

24. Borja, M.; Dutta, P.K. Fatty-acids in layered metal-hydroxides: membrane-like structure and dynamics. *J. Phys. Chem.* **1992**, 96, 5434–5444.

25. Zhao, Y.; Li, F.; Zhang, R.; Evans, D.G.; Duan, X. Preparation of layered double-hydroxide nanomaterials with a uniform crystallite size using a new method involving separate nucleation and aging steps. *Chem. Mater.* **2002**, 14, 4286–4291.

26. Ogawa, M.; Kaiho, H. Homogeneous precipitation of uniform hydrotalcite particles. *Langmuir* **2002**, 18, 4240–4242.

27. Newman, S.P.; Jones, W. Comparative study of some layered hydroxide salts containing exchangeable interlayer anions. *J. Solid State Chem.* **1999**, 148, 26–40.

28. Choy, J.H.; Kwon, Y.M.; Song, S.W.; Chang, S.H. Structural phase transformation of layered hydroxy double salts depending on hydration degree. *Bull. Korean Chem. Soc.* **1997**, 18, 450–453.

29. Alberti, G.; Bartocci, M.; Santarelli, M.; Vivani, R. Zirconium phosphate chloride dimethyl sulfoxide, a reactive precursor of a large family of layered compounds. *Inorg. Chem.* 1997, 36, 3574–3575.
30. Vivani, R.; Costantino, U.; Nocchetti, M. Crystal engineering on layered zirconium phosphonates: crystal structure (from x-ray powder data) and non-covalent interactions on the layered zirconium compound of 4-[bis(phosphonomethyl)amino]butanoic acid. *J. Mater. Chem.* **2002**, 12, 3254–3260.
31. Becker, O.; Cheng, Y.B.; Varley, R.J.; Simon, G.P. Layered silicate nanocomposites based on various high-functionality epoxy resins: the influence of cure temperature on morphology, mechanical properties, and free volume. *Macromolecules* **2003**, 36, 1616–1625.
32. Bharadwaj, R.K.; Mehrabi, A.R.; Hamilton, C.; Trujillo, C.; Murga, M.; Fan, R.; Chavira, A.; Thompson, A.K. Structure–property relationships in cross-linked polyester–clay nanocomposites. *Polymer* **2002**, 43, 3699–3705.
33. Camino, G.; Tartaglione, G.; Frache, A.; Manferti, C.; Costa, G. Thermal and combustion behaviour of layered silicate–epoxy nanocomposites. *Polym. Degrad. Stab.* **2005**, 90, 354–362.
34. Lan, T.; Kaviratna, P.D.; Pinnavaia, T.J. Epoxy self-polymerization in smectite clays. *J. Phys. Chem. Solids* **1996**, 57, 1005–1010.
35. Suh, D.J.; Lim, Y.T.; Park, O.O. The property and formation mechanism of unsaturated polyester–layered silicate nanocomposite depending on the fabrication methods. *Polymer* **2000**, 41, 8557–8563.
36. Lewis, A.F.; Doyle, M.J.; Gillham, J.K. Effect of cure history on dynamic mechanical properties of an epoxy-resin. *Polym. Eng. Sci.* **1979**, 19, 683–686.
37. Xie, W.; Gao, Z.; Pan, W.P.; Hunter, D.; Singh, A.; Vaia, R.A. Thermal degradation chemistry of alkyl quaternary ammonium montmorillonite. *Chem. Mater.* **2001**, 13, 2979–2990.
38. Gilman, J.W.; Awad, W.H.; Davis, R.D.; Shields, J.R.; Harris, R.H.; Davis, C.; Morgan, A.B.; Sutto, T.E.; Callahan, J.; Trulove, P.C.; DeLong, H.C. Polymer/layered silicate nanocomposites from thermally stable trialkylimidazolium-treated montmorillonite. *Chem. Mater.* **2002**, 14, 3776–3785.
39. Lan, T.; Kadiratna, K.D.; Pinnavaia, T.J. Mechanism of clay tactoid exfoliation in epoxy–clay nanocomposites. *Chem. Mater.* **1995**, 7, 2144–2150.
40. Zammarano, M.; Franceschi, M.; Bellayer, S.; Gilman, J.W.; Meriani, S. Preparation and flame resistance properties of revolutionary self-extinguishing epoxy nanocomposites based on layered double hydroxides. *Polymer* **2005**, 46, 9314–9328.
41. Park, J.H., Jana, S.C. Mechanism of exfoliation of nanoclay particles in epoxy–clay nanocomposites. *Macromolecules* **2003**, 36, 2758–2768.
42. Chen, B.; Liu, J.; Chen, H.; Wu, J. Synthesis of disordered and highly exfoliated epoxy/clay nanocomposites using organoclay with catalytic function via acetone–clay slurry method. *Chem. Mater.* **2004**, 16, 4864–4866.
43. Ma, J.; Yu, Z.Z.; Zhang, Q.X.; Xie, X.L.; Mai, Y.W.; Luck, I. A novel method for preparation of disorderly exfoliated epoxy/clay nanocomposite. *Chem. Mater.* **2004**, 16, 757–759.
44. Wang, Z.; Pinnavaia, T.J. Hybrid organic–inorganic nanocomposites: exfoliation of magadiite nanolayers in an elastomeric epoxy polymer. *Chem. Mater.* **1998**, 10, 1820–1826.

45. Zilg, C.; Mulhaupt, R.; Finter, J. Morphology and toughness/stiffness balance of nanocomposites based upon anhydride-cured epoxy resins and layered silicates. *Macromol. Chem. Phys.* **1999**, 200, 661–670.
46. Alexandre, M.; Dubois, P. Polymer-layered silicate nanocomposites: preparation, properties and uses of a new class of materials. *Mater. Sci. Eng. Rep.* **2000**, 28, 1–63.
47. Wang, Z.; Massam, J.; Pinnavaia, T.J. Epoxy–clay nanocomposites, in: T.J. Pinnavaia and G.W. Beall, Eds. *Polymer–Clay Nanocomposites.* Wiley, New York; **2001**, pp. 127–149.
48. Wang, Z.; Lan, T.; Pinnavaia, T.J. Hybrid organic–inorganic nanocomposites formed from an epoxy polymer and a layered silicic acid (magadiite). *Chem. Mater.* **1996**, 8, 2200–2204.
49. Levchik, S.V.; Camino, G.; Costa, L.; Luda, M.P. Mechanistic study of thermal behaviour and combustion performance of carbon fibre–epoxy resin composites fire retarded with a phosphorus-based curing system. *Polym. Degrad. Stab.* **1996**, 54, 317–322.
50. Levchik, S.V.; Camino, G.; Luda, M.P.; Costa, L.; Muller, G.; Costes, B. Epoxy resins cured with aminophenylmethylphosphine oxide—II. Mechanism of thermal decomposition. *Polym. Degrad. Stab.* **1998**, 60, 169–183.
51. Rose, N.; LeBras, M.; Delobel, R.; Costes, B.; Henry, Y. Thermal oxidative-degradation of an epoxy-resin. *Polym. Degrad. Stab.* **1993**, 42, 307–316.
52. Davis, R.D.; Gilman, J.W.; Sutto, T.E.; Callahan, J.H; Trulove, P.C.; DeLong, H.C. Improved thermal stability of organically modified layered silicates. *Clays Clay Miner.* **2004**, 52, 171–179.
53. VanderHart, D.L.; Asano, A.; Gilman, J.W. Solid-state NMR investigation of paramagnetic nylon-6 clay nanocomposites, 2: Measurement of clay dispersion, crystal stratification, and stability of organic modifiers. *Chem. Mater.* **2001**, 13, 3796–3809.
54. Zanetti, M.; Camino, G.; Reichert, P.; Mulhaupt, R. Thermal behaviour of poly(propylene) layered silicate nanocomposites. *Macromol Rapid Commun.* **2001**, 22, 176–180.
55. Song, L.; Hu, Y.; Tang, Y.; Zhang, R.; Chen, Z.; Fan, W. Study on the properties of flame retardant polyurethane/organoclay nanocomposite. *Polym. Degrad. Stab.* **2005**, 87, 111–116.
56. Hussain, M.; Varley, R.J.; Mathys, Z.; Cheng, Y.B.; Simon, G.P. Effect of organophosphorus and nano-clay materials on the thermal and fire performance of epoxy resins. *J. Appl. Polym. Sci.* **2004**, 91, 1233–1253.
57. Triantafillidis, C.S.; LeBaron, P.C.; Pinnavaia, T.J. Thermoset epoxy–clay nanocomposites: the dual role of alpha,omega-diamines as clay surface modifiers and polymer curing agents. *J. Solid State Chem.* **2002**, 167, 354–362.
58. Zanetti, M.; Lomakis, L.S.; Camino, G. Polymer layered silicate nanocomposites. *Macromol. Mater. Eng.* **2000**, 279, 1–9.
59. Gilman, J.W.; Kashiwagi, T.; Morgan, A.B.; Harris, R.H.; Brassell, L.; Landingham, M.V.; Jackson, C.L. Flammability of polymer clay nanocomposites consortium. http://fire.nist.gov/bfrlpubs/fire00/PDF/f00026.pdf. Accessed Mar. 28, 2006.
60. Bourbigot, S.; Le Bras, M.; Dabrowski, F.; Gilman, J.W.; Kashiwagi, T. PA-6 clay nanocomposite hybrid as char forming agent in intumescent formulations. *Fire Mater.* **2000**, 24, 201–208.

61. Jacobson, A.J. Colloid dispersions of compounds with layer and chain structures. *J. Mater. Sci. Forum* **1994**, 152–153, 1–12.
62. Hibino, T.; Jones, W. New approach to the delamination of layered double hydroxides. *J. Mater. Chem.* **2001**, 11, 1321–1323.
63. Hussein, M.Z.B.; Yun-Hin, T.Y.; Tawang, M.M.; Shahadan, R. Thermal degradation of (zinc–aluminium-layered double hydroxide–dioctyl sulphosuccinate) nanocomposite. *Mater. Chem. Phys.* **2002**, 74, 265–271.
64. Oriakhi, C.O.; Farr, I.V.; Lerner, M.M. Thermal characterization of poly(styrene-sulfonate) layered double hydroxide nanocomposites. *Clays Clay Miner.* **1997**, 45, 194–202.
65. Oriakhi, C.O.; Farr, I.V.; Lerner, M.M. Incorporation of poly(acrylic acid), poly(vinylsulfonate) and poly(styrenesulfonate) within layered double hydroxides. *J. Mater. Chem.* **1996**, 6, 103–107.
66. Leroux, F.; Aranda, P.; Besse, J.P.; Ruiz-Hitzky, E. Intercalation of poly(ethylene oxide) derivatives into layered double hydroxides. *Eur. J. Inorg. Chem.* **2003**, 6, 1242–1251.
67. Zammarano, M.; Bellayer, S.; Gilman, J.W.; Franceschi, M.; Beyer, F.L.; Harris, R.H.; Meriani, S. Delamination of organo-modified layered double hydroxides in polyamide 6 by melt processing. *Polymer* **2006**, 47, 652–662.
68. Liang, L.; Ma, R.; Ebina, Y.; Iyi, Y.; Sasaki, T. Positively charged nanosheets derived via total delamination of layered double hydroxides. *Chem. Mater.* **2005**, 17, 4386–4391.
69. Leroux, F.; Adachi-Pagano, M.; Intissar, M.; Chauviere, S.; Forano, C.; Besse, J.P. Delamination and restacking of layered double hydroxides. *J. Mater. Chem.* **2001**, 11, 105–112.
70. Hsueh, H.B.; Chen, C.Y. Preparation and properties of LDHs/epoxy nanocomposites. *Polymer* **2003**, 44, 5275–5283.
71. Chen, J.S.; Poliks, M.D.; Ober, C.K.; Zhang, Y.M.; Wiesner, U.; Giannelis, E.P. Study of the interlayer expansion mechanism and thermal–mechanical properties of surface-initiated epoxy nanocomposites. *Polymer* **2002**, 43, 4895–4904.
72. Hartwig, A.; Schneider, B.; Luhring, A. Influence of moisture on the photochemically induced polymerisation of epoxy groups in different chemical environment. *Polymer* **2002**, 43, 4243–4250.
73. Kornman, X.; Lindbergh, H.; Berglunda, L.A. Synthesis of epoxy–clay nanocomposites: influence of the nature of the curing agent on structure. *Polymer* **2001**, 42, 4493–4499.
74. Gensler, R.; Groppel, P.; Muhrer, V.; Muller, N. Application of nanoparticles in polymers for electronics and electrical engineering. *Part. Part. Syst. Char.* **2002**, 19, 293–299.
75. Bellotto, M.; Rebours, B.; Clause, O.; Lynch, J.; Bazin, D.; Elkaim, E. Hydrotalcite decomposition mechanism: a clue to the structure and reactivity of spinel-like mixed oxides. *J. Phys. Chem.* **1996**, 100, 8535–8542.
76. Camino, G.; Maffezzoli, A.; Braglia, M.; De Lazzaro, M.; Zammarano, M. Effect of hydroxides and hydroxycarbonate structure on fire retardant effectiveness and mechanical properties in ethylene–vinyl acetate copolymer. *Polym. Degrad. Stab.* **2001**, 74, 457–464.

77. Hörold, S. Phosphorus flame retardants in thermoset resins. *Polym. Degrad. Stab.* **1999**, 64, 427–431.

78. Lewin, M.; Brozek, J.; Marvin, M.M. The system polyamide/sulfamate/dipentaerythritol: flame retardancy and chemical reactions. *Polym. Adv. Technol.* **2002**, 13, 1091–1102.

79. Chen, W.; Qu, B.J. Structural characteristics and thermal properties of PE-g-MA/MgAl-LDH exfoliation nanocomposites synthesized by solution intercalation. *Chem. Mater.* **2003**, 15, 3208–3213.

80. Delfosse, L.; Baillet, C.; Brault, A.; Brault, D. Combustion of ethylene–vinyl acetate copolymer filled with aluminium and magnesium hydroxides. *Polym. Degrad. Stab.* **1989**, 23, 337–347.

81. Rychly, J.; Vesely, K.; Gal, E.; Kummer, M.; Jancar, J.; Rychla, L. Use of thermal methods in the characterization of the high-temperature decomposition and ignition of polyolefins and EVA copolymers filled with $Mg(OH)_2$, $Al(OH)_3$ and $CaCO_3$. *Polym. Degrad. Stab.* **1990**, 30, 57–72.

82. Gilman, J.W.; Bourbigot, S.; Shields, J.R.; Nyden, M.; Kashiwagi, T.; Davis, R.D.; VanderHart, D.L.; Demory, W.; Wilkie, C.A.; Morgan, A.B.; Harris, J.; Lyon, R.E. High throughput methods for polymer nanocomposites research: extrusion, NMR characterization and flammability property screening. *J. Mater. Sci.* **2003**, 38, 4451–4460.

83. Delfosse, L.; Baillet, C.; Brault, A.; Brault, D. Combustion of ethylene vinyl-acetate copolymer filled with aluminum and magnesium hydroxides. *Polym. Degrad. Stab.* **1989**, 23, 337–347.

84. Levchik, S.V.; Weil, E.D. Thermal decomposition, combustion and fire-retardancy of polyurethanes: a review of the recent literature. *Polym. Int.* **2004**, 53, 1585–1610.

85. Wang, Z.; Pinnavaia, T.J. Nanolayer reinforcement of elastomeric polyurethane. *Chem. Mater.* **1998**, 10, 3769–3771.

86. Tortora, M.; Gorrasi, G.; Vittoria, V.; Galli, G.; Ritrovati, S.; Chiellini, E. Structural characterization and transport properties of organically modified montmorillonite/polyurethane nanocomposites. *Polymer* **2002**, 43, 6147–6157.

87. Berta, M.; Lindsay, C.; Pans, G.; Camino, G. Effect of chemical structure on combustion and thermal behaviour of polyurethane elastomers layered silicate nanocomposites. *Polym. degrad. Stab.* **2006**, 91, 1179–1191.

88. Devaux, E.; Rouchery, M.; Bourbigot, S. Polyurethane/clay and polyurethane/POSS nanocomposites as flame retarded coating for polyester and cotton fabrics. *Fire Mater.* **2002**, 26, 149–154.

89. Yao, K.J.; Song, M.; Hourston, D.J.; Luo, D.Z. Polymer/layered clay nanocomposites, 2: Polyurethane nanocomposites. *Polymer* 2002, 43, 1017–1020.

90. Chang, J.H.; An, Y.U. Nanocomposites of polyurethane with various organoclays: thermomechanical properties, morphology, and gas permeability. *J. Polym. Sci. B Polym. Phys.* **2002**, 40, 670–677.

91. Petrovic, Z.S.; Javni, I.; Waddan, A.; Banhegyi, G. Structure and properties of polyurethane–silica nanocomposites. *J. Appl. Polym. Sci.* **2000**, 76, 133–151.

92. Shah, D.; Maiti, P.; Jiang, D.D.; Giannelis, E.P. *Adv. Mater.* **2005**, 17, 525.

93. Abadie, M.J.M.; Mekhissi, K.; Burchill, P.J. Effects of processing conditions on the curing of a vinyl ester resin. *J. Appl. Polym. Sci.* **2002**, 84, 1146–1154.

94. Chigwada, G.; Jash, P.; Jiang, D.D.; Wilkie, C.A. Fire retardancy of vinyl ester nanocomposites: synergy with phosphorus-based fire retardants. *Polym. Degrad. Stab.* **2005**, 89, 85–100.

95. Shah, A.P.; Gupta, R.K.; Gangarao, H.V.S.; Powell, C.E. Moisture diffusion through vinyl ester nanocomposites made with montmorillonite clay. *Polym. Eng. Sci.* **2002**, 42, 1852–1863.

10

PROGRESS IN FLAMMABILITY STUDIES OF NANOCOMPOSITES WITH NEW TYPES OF NANOPARTICLES*

TAKASHI KASHIWAGI

Fire Research Division, National Institute of Standards and Technology, Gaithersburg, Maryland

10.1 INTRODUCTION

Although the incorporation of microscale particles as fillers into polymers has been well explored scientifically, the decrease in size of particles to nanometers, and the simultaneous increase in interface area, results in extraordinary new material properties.[1–4] In one such application, the flammability properties of polymers have been improved with the addition of nanoscale particles. These filled nanocomposites provide an attractive alternative to conventional flame retardants. At present, the most common approach to improving flammability is the use of layered silicates such as clays, as described in Chapter 3. However, there are many different shapes and types of nanoparticles. (Here, a *nanoscale particle* is defined as having at least one dimension on the nanometer scale.) When all three dimensions are on the order of nanometers, we are dealing with true nanoparticles, such as spherical silica particles, having an aspect ratio of 1. Another type of nanoparticle has only one dimension on the nanometer scale. Such nanoscale

Flame Retardant Polymer Nanocomposites, edited by Alexander B. Morgan and Charles A. Wilkie

particles are sheets or layers, such as layered silicate or graphite, which are one to a few nanometers thick and hundreds to thousands of nanometers in the other two dimensions. When two dimensions are on the nanometer scale and the third is larger, the particles form elongated structures such as nanotubes, whiskers, or rods with a high aspect ratio.

It is of interest to determine the flame retardant effectiveness of shapes or types of nanoparticles other than layered silicates, to find what shape or type of nanoparticle is most effective for improving the flammability properties of commodity polymers. In this chapter, flammability properties of nanocomposites containing nanoscale oxides such as nanoscale silica particles and metal oxides, polyhedral oligomeric silsesquioxanes (POSSs), and carbon-based nanoparticles such as graphite, single-walled carbon nanotubes (SWNTs), multiwalled carbon nanotubes (MWNTs), and carbon nanofibers (CNFs) are described and a flame retardant mechanism of these nanoparticles is discussed.

10.2 NANOSCALE OXIDE-BASED NANOCOMPOSITES

10.2.1 Nanoscale Silica Particles

Nanoscale silica particles can have a huge interfacial area as long as the diameter of the particles is in the range of nanometers. Although they do not have the narrow gallery structure of a layered clay, an improvement in physical properties[5–8] and an improvement in thermal stability[9, 10] by the addition of nanoscale silica particles to polymer were reported. The latter improvement was attributed to the formation of tightly packed particles in various polymers that significantly reduced both heat-bound and loosely bound polymer chains around the particles.[11] It was also reported that the addition of mesoscale silica to various polymers significantly reduced the heat release rate of the polymers.[12, 13] Flammability properties of poly(methyl methacrylate) (PMMA)–nanoscale silica nanocomposites[14–16] and polyimide–nanoscale silica nanocomposites[17] have been reported. Samples have been prepared by solvent blending,[14, 17] melt blending utilizing single-screw extrusion,[16] or in situ polymerization[15] in order to obtain well-dispersed nanoscale silica particles in the sample.

Dispersion of the particles in a polymer is critical for obtaining better flame retardant performance, as described in earlier chapters. Transmission electron microscopy (TEM) was used to determine the dispersion. The example shown in Figure 10.1 indicates well-dispersed particles having an average diameter of about 12 nm. Roughly a 50% reduction in the peak heat release rate was reported with the addition of a 13% mass fraction of silica particles,[15] as shown in Figure 10.2. Little to no improvement was reported in a limiting oxygen index (LOI) measurement[14] with up to a 10% mass fraction of silica particles of diameter as small as 7 nm (the dispersion of these particles was not shown). Although the LOI values increased from 36 to 44, the addition of a 28% mass fraction of silica particles (diameter 50 to 300 nm) was required. In the heat release rate curves shown in Figure 10.2, the addition of nanoscale silica particles hardly reduced

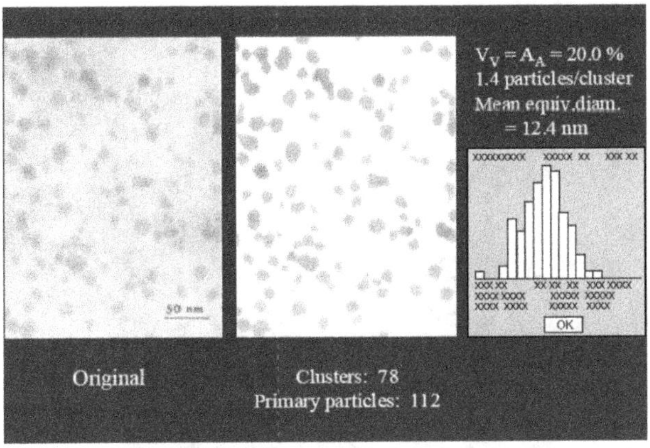

FIGURE 10.1 TEM image of a PMMA–nanosilica nanocomposite (*left*), analyzed image (*middle*), and a histogram distribution of diameter (*right*). (From Ref. 15.) (*See insert for color representation of figure.*)

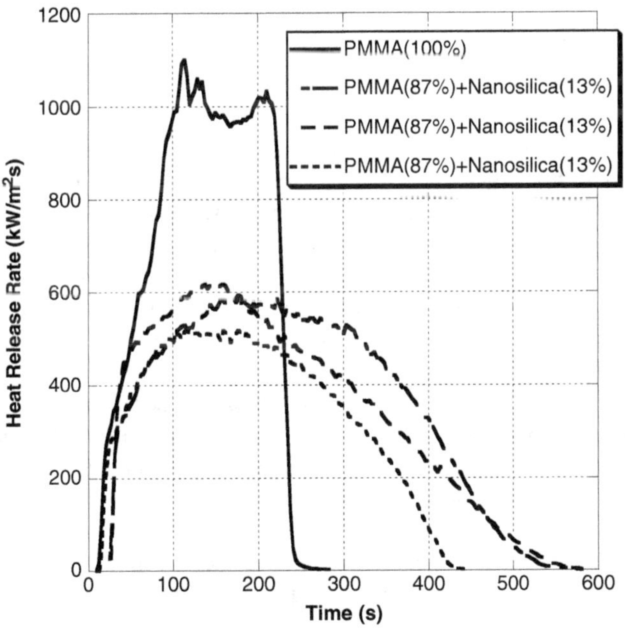

FIGURE 10.2 Effects of the addition of nanosilica on the heat release rate of PMMA at 50 kW/m^2.

the heat release rate at the early stage of burning, and it was demonstrated that the addition of nanoscale silica particles did not significantly modify the UL-94 rating.[16] Therefore, the overall flame retardant effectiveness of nanoscale silica particles appears to be less than that of clay particles, as described in earlier chapters.

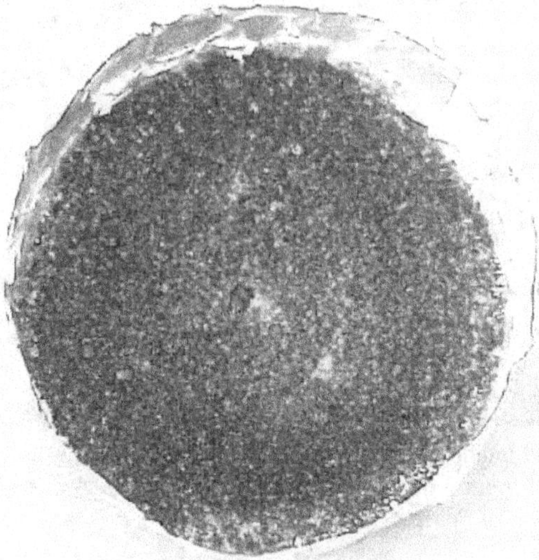

FIGURE 10.3 Residue of a PMMA–nanosilica nanocomposite after a gasification test in nitrogen at 40 kW/m^2. (From Ref. 15.)

Observation of sample behavior during gasification in a nitrogen atmosphere at an external radiant flux of 40 kW/m^2 reported the formation of many small bubbles, followed by the formation of many rigid white islands.[15] Vigorous bursting of small bubbles was observed around the islands. The islands appeared to be made of coarse granular particle clumps. Since the sample surface was covered only partially by these loose granular particles or clumps, part of the sample surface was still exposed to the external heat between the coarse particles, and the barrier performance of the layer to slow the evolution of the degradation products of PMMA was not effective. Similar behavior was observed with polycarbonate containing 15 nm coated silica particles.[18] At the end of the test, a dark, coarse powdery layer was left at the bottom of the sample container (Figure 10.3). No network-structured protective layer covering the entire sample surface was formed. One possible approach to forming such an in situ silica network during gasification would be to enhance the formation of cross-links among the particles by appropriate surface treatments on the surface of nanoscale silica particles, but no work along these lines has been reported.

10.2.2 Metal Oxides

The flammability properties of nanocomposites consisting of nanoscale titanium oxide (TiO_2 with a median diameter of 21 nm) and iron oxide (Fe_2O_3 with a median diameter of 23 nm) in PMMA were measured.[19] The nanocomposites were prepared by melt blending. A morphology study of the nanocomposites showed that the particles were well distributed in the sample but had some

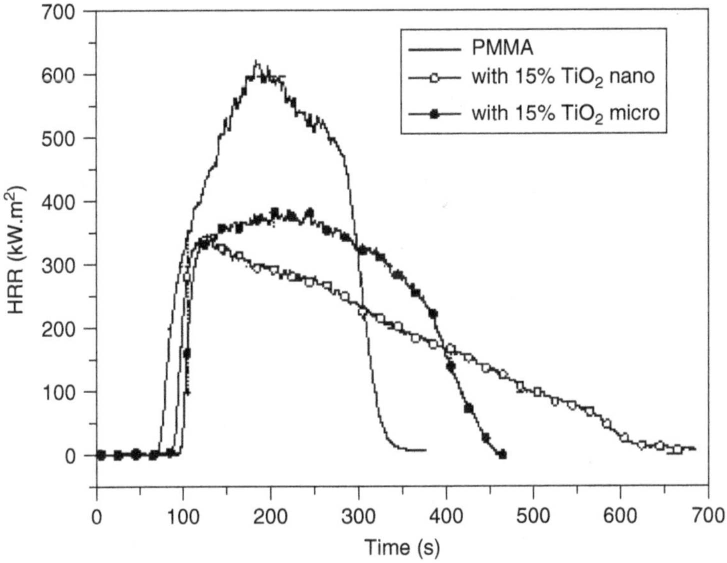

FIGURE 10.4 Effect of particle size on the heat release rate for PMMA–TiO_2 at 35 kW/m^2. (From Ref. 19, copyright © 2005, Elsevier, with permission.)

tendency to aggregate since no surface treatment was done on the oxides. The effect of particle size was studied by comparing the heat release rate of a nanocomposite with nanoscale TiO_2 particles to that of a microcomposite with micrometer scale TiO_2 (0.2 μm); the comparison is shown in Figure 10.4. The peak heat release rate of the nanocomposite was about 10% lower than that of the microcomposite. A similar result was observed for PMMA–Fe_2O_3 samples. An increase in nanoparticle TiO_2 concentration reduced the heat release rate of PMMA, as shown in Figure 10.5, but the amount of the reduction is not as significant as with other types of nanoparticles, such as clay or carbon nanotubes, as described later. The combination of these nanoparticles with organo-modified montmorillonite (OMMT) was used to determine the synergistic effect on a reduction in the heat release rate of PMMA; the results are shown in Figure 10.6. The improvement observed via the incorporation of oxide particles was explained by several effects: (1) TiO_2 to act as a "heat shield," which can limit the thermal conduction into the sample, (2) limitation of evolved gas release due to an increase in melt viscosity, and (3) enhanced wetting of mineral particles by the molten polymer.

10.2.3 Polyhedral Oligomeric Silsequioxanes

With the recent development of nanostructured chemical feedstocks based on polyhedral oligomeric silsequioxane (POSS),[20–22] POSS-based hybrid nanocomposites have received increasing attention because of the unique three-dimensional

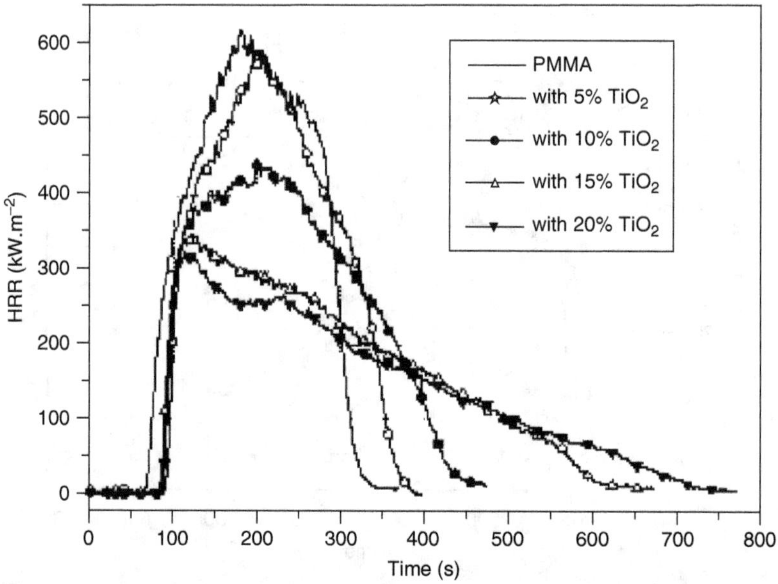

FIGURE 10.5 Effect of TiO_2 nanoparticle concentration on the heat release rate of PMMA at 35 kW/m^2. (From Ref. 19, copyright © 2005, Elsevier, with permission.)

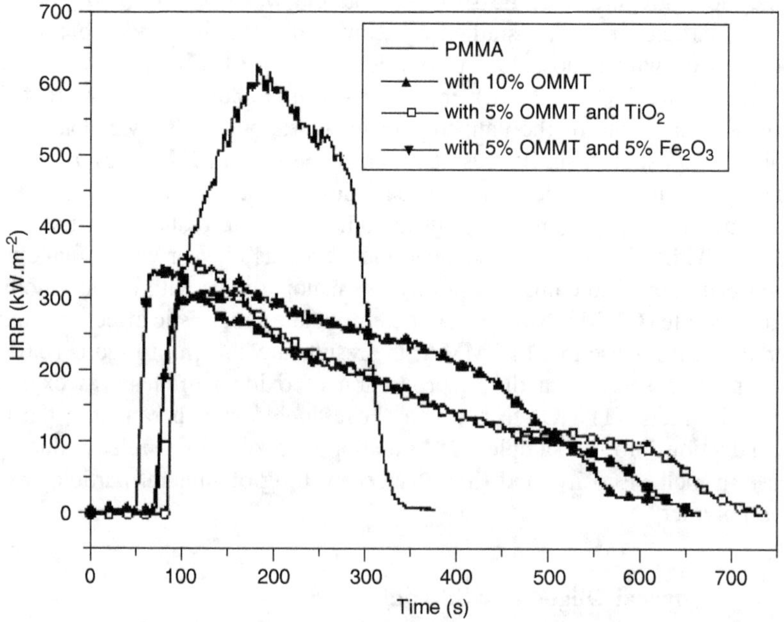

FIGURE 10.6 Comparison of heat release rates of nanocomposites with OMMT, OMMT–TiO_2, and OMMT–Fe_2O_3 at 35 kW/m^2. (From Ref. 19, copyright © 2005, Elsevier, with permission.)

structure of the POSS macromonomer,[23] as shown in Figure 10.7. POSS represents an intermediate structure between that of silicone and that of silica, explaining its excellent oxidation stability and reaction to fire. It consists of an inorganic silicalike core (Si_8O_{12}) surrounded by eight organic groups at the corners to enhance compatibility with organic polymers. Its nanoscale enables the POSS segment to reinforce polymer chain segments effectively and to control polymer chain motion at the molecular level through maximizing the interface area and chemical interactions of reinforcement with polymers. Early examples were presented with siloxanes,[24, 25] followed by numerous applications showing enhancement of thermal stability and improving the flammability properties of polymers.

POSS macromers generally sublime at high temperatures provided that they contain functionalities that do not readily undergo cross-linking reactions. Once incorporated into a polymeric form, POSS macromers do not sublime; rather, they decompose primarily through partial loss of their organic substituents without significantly affecting the degradation of the matrix polymers[26] or with subsequent cross-linking reactions, which incorporates the remaining composition into the SiO_xC_y network (residue) in a POSS–siloxane copolymer.[24] Thermal gravimetric analysis of these nanocomposites shows that the initial decomposition temperatures and residue (ceramic and/or char) yield increased with increasing POSS concentration.[24, 26–28] In terms of the initial decomposition temperature and residue yield, the thermal stability of the nanocomposites was enhanced significantly with increasing inorganic component.

The thermal analysis study above demonstrated the enhanced thermal stability of POSS–polymer nanocomposites and suggested that there is a potential to improve the flammability properties of matrix polymers. However, studies clearly demonstrating such improvement by means of the use of POSS-based

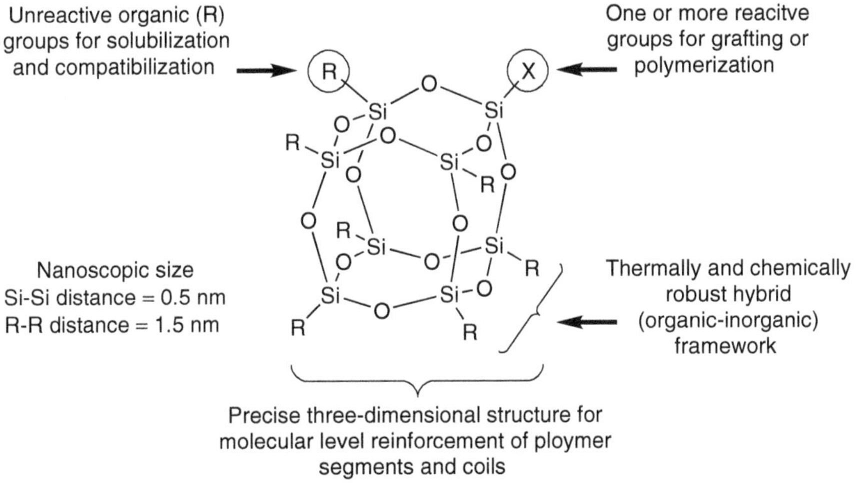

FIGURE 10.7 General structure of POSS. (From Ref. 23.)

nanocomposites are rather limited. One study[29] is of nanocomposites consisting of polytetramethylene ether-glycol-*b*-polyamide-12, 1% polyimide-12 (PTME–PA), polystyrene–polybutadiene–polystyrene (SBS), and polypropylene (PP) prepared with POSS (the structure is described in Figure 10.8) ranging from 10 to 20% via solution blending in tetrahydrofuran (THF). For comparison purposes, composites based on other silicone compounds, such as polycarbosilane (PCS) and polysilastyrene (PSS), were also prepared by solution blending. The flammability properties of these blends were characterized using a cone calorimeter. The results, shown in Figure 10.9 and Table 10.1, reveal that both PCS (although twice as high in concentration as POSS) and POSS are reasonably effective for reducing the heat release rate measured at 35 kW/m^2. However, the total heat release (integrating the heat release rate with respect to time) of the nanocomposites was not significantly reduced from that of the matrix resins. Furthermore, the residue yields are about the same as the theoretical yields shown in parentheses in Table 10.1. This means that the addition of POSS to the nanocomposites does not significantly increase the yield of carbonaceous char. The residue is mainly the inorganic component of the POSS.

Simultaneous significant reduction of heat release rate and total heat release was achieved with polyurethane (PU) POSS nanocomposites (POSS 10% mass fraction) used as a coating on poly(ethylene terephthalate) (PET) knitted fabric,[30] as shown in Figure 10.10. Two different POSSs were used in this study. One was octamethyl POSS (POSS MS) with R = methyl in the structure described in Figure 10.7, and the other was poly(vinylsilsequioxane) (POSS FQ) with R = vinyl in the structure described in Figure 10.8 (all ends are vinyl). For comparison purposes, clay (Closite 30B) was also used as a nanocomposite filler. It was observed that a significant reduction of heat release rate and total heat release of the PET knitted fabric was achieved with POSS FQ2 ("2" indicates incorporation

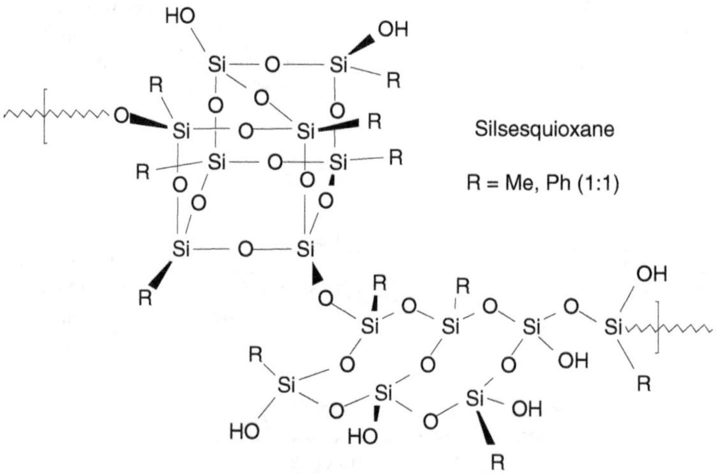

FIGURE 10.8 Structure of POSS.

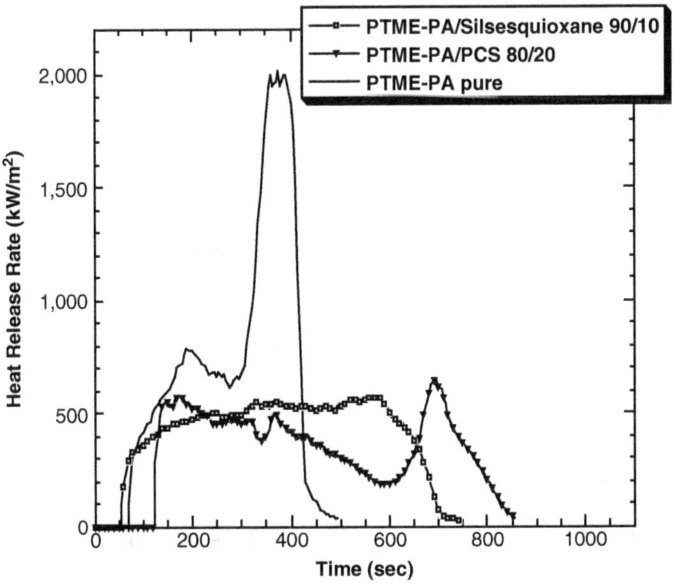

FIGURE 10.9 Heat release rate curves of PTME–PA with siloxane and the POSS at 35 kW/m^2. (From Ref. 29.)

TABLE 10.1 Summary of Cone Calorimeter Data of PP, PTME–PA, and SBS with Siloxanes and POSS at 35 kW/m^{2a}

Sample	Residue Yield[b] (%)	Mean Mass Loss Rate (g/m^2·s)	Peak HRR [kW/m^2 (Δ%)]	Mean HRR [kW/m^2 (Δ%)]	H_c (MJ/kg)	SEA (m^2/kg)	Mean CO Yield (kg/kg)
PP	0	25.4	1466	741	34.7	650	0.03
PP/POSS 80/20	17 (16)	19.1	892 (40%)	432 (42%)	29.8	820	0.03
PTME–PA	0	34.2	2020	780	29.0	190	0.02
PTME–PA/PCS 80/20	15 (15)	14.8	699 (65%)	419 (46%)	28.5	260	0.02
PTME–PA/POSS 90/10	6 (8)	19.8	578 (72%)	437 (44%)	25.2	370	0.02
SBS	1	36.2	1405	976	29.3	1750	0.08
SBS/PCS 80/20	20 (15)	18.5	825 (42%)	362 (63%)	26.4	1550	0.07
SBS/POSS 90/10	6 (8)	31.2	1027 (27%)	755 (23%)	26.9	1490	0.07

Source: After Ref. 29.
[a] H_c = mean heat of combustion; SEA, specific extinction area (smoke measurement). Uncertainties: ±5% of reported value for residue yields for heat release rate (HRR) and H_c data; ±10% for carbon monoxide and SEA data.
[b]Theoretical residue yields in parentheses.

of the nano additives in the second stage during sample preparation), but no reduction was observed with POSS MS2. The thermal stability of coated knitted fabric containing POSS MS2 was lower than that of fabric coated with virgin PU in the TGA measurement. Destabilization of the fabric at 200°C by POSS MS2 and sublimation of POSS MS2 at around 300°C could explain the lack of

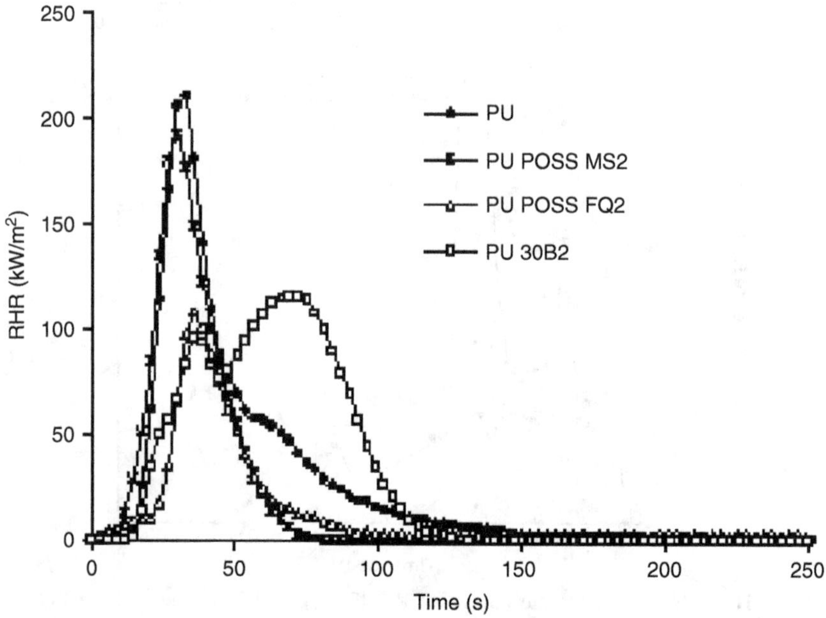

FIGURE 10.10 Heat release rate curves of PU nanocomposites on PET knitted fabrics at 35 kW/m^2; incorporation of nanoadditives in the second stage of sample preparation. (From Ref. 30, copyright © 2002, John Wiley & Sons Ltd., with permission.)

FR performance shown in Figure 10.10. On the other hand, POSS FQ showed a remarkable thermal stability without mass loss up to 380°C, and a mass loss of only 6% occurred at 700°C, due to the formation of cross-links.

In the case of the PU–POSS FQ2 coating, the residue (consisting of carbonaceous char and possible preceramic components) was more uniform, and only small cracks were observed on the surface. This residue was more resistant and could suppress the flame.[30] The formation of a hard uniform barrier over a polycarbonate surface was also reported with polycarbonate–coated POSS nanocomposites.[18] However, PP–POSS multifilament yarns prepared from PP–POSS FQ nanocomposites did not show either any reduction of the heat release rate or total heat release, but the ignition delay time was much longer that that of PP, as shown in Figure 10.11.[31] These results suggest that POSS FQ did not act as a flame retardant for PP, only as a thermal stabilizer. These results reveal that flammability of polymer–POSS nanocomposites depends on the type of polymer matrix, the structure of POSS, and incorporation of POSS into the polymer structure. If a certain POSS structure significantly enhances cross-links with a matrix polymer to form a significant amount of SiO_xC_y network, not only reduction of the heat release rate but also reduction of the total heat release could be achieved. Another important factor for the previous inconsistent flame retardant performance of POSS could be due to difference in the dispersion of POSS in the matrix polymer. The importance of the dispersion of

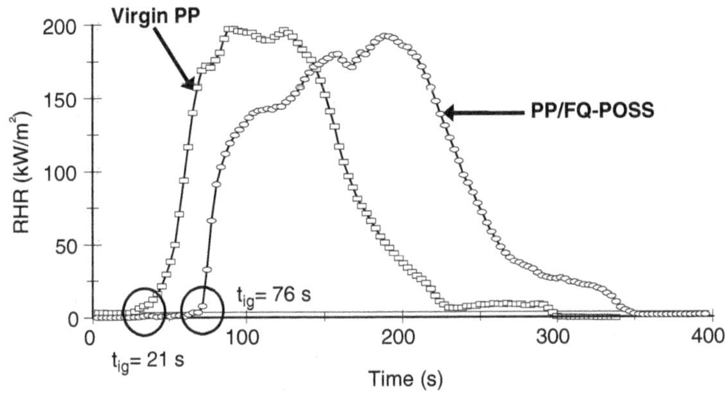

FIGURE 10.11 Heat release rate curves of PP and PP/POSS–FQ knitted fabrics at 35 kW/m^2. (From Ref. 31, with permission from the Royal Society of Chemistry.)

POSS on the flame retardant performance was described for the formation of an oxidatively stable, uniformly covered nonpermeable surface char layer.[32] Some previous studies demonstrated reasonably effective flame retardant performance of POSS, but a recent study with trisilanol phenyl POSS in PMMA did not show any flame retardant performance as measured in a cone calorimeter.[33] It was suggested that POSS has the potential to reduce the heat release rate, but one must be careful in selecting the POSS material to be evaluated.

10.3 CARBON-BASED NANOCOMPOSITES

There are several different types of carbon-based nanoparticles. One is graphite, a layered material having a thickness of a nanometer, similar to that of clay particles, and others are based on a tubular shape having a diameter of nanometers. Since expandable graphite (an intumescent material) is discussed in Chapter 6, it is not discussed in this chapter.

10.3.1 Graphite Oxide

The graphite structure consists of carbon layers in a stacked configuration. The carbon atoms are bonded covalently in a hexagonal arrangement within each layer, and these layers are weakly bonded by van der Waals forces between the layers, which makes intercalation possible. Graphite does not undergo any ion exchange process, but graphite oxide (GO) can add organophilic ammonium cations between the layers. Molecular dynamic simulations of the thermal degradation of a series of PP–graphite nanocomposites at 873 K were performed as a function of the distance of separation between the graphite sheets.[34] The mass loss curves obtained from these simulations indicate that there is a pronounced stabilization of the polymer at a distance of 3.0 nm that results from both PP–PP and PP–graphite interactions. Below a distance of 2.5 nm, the van der Waals repulsions between atoms destabilize the polymer due to high density in the narrow space between the graphite sheets. However, at a larger spacing between the sheets, interactions with the polymer melt do not provide sufficient resistance to prevent the rapid escape of the degradation products from the spaces between the sheets.

Motivated by the foregoing study, Uhl and Wilkie studied the thermal stability and flammability properties of polystyrene (PS)–graphite nanocomposites.[35, 36] Nanocomposites with graphite concentrations of 1, 3, and 5% were prepared by two different methods: in situ polymerization in the presence of graphite oxide and melt blending. Graphite oxide was organically modified using three different surfactants (GO-C14, GO-10A, and GO-VB16), the structures of which are shown in Figure 10.12. The x-ray diffraction (XRD) data showed that no peaks were observed at 1% for all three modified graphite oxides, indicating their exfoliation, and also for two of the three modified graphite oxides at 3% loading; at 5% loading, XRD peaks were clearly seen, with the d-spacing (plate-to-plate spacing) much larger than that in the GO (nonmodified). Intercalation of graphite oxide plates was suggested for those samples having XRD peaks. Similar results were observed for melt-blended samples with narrower d-spacing than for those observed in in situ polymerized samples. The reduction in peak heat release rate ranged from 27 to 54%, as shown in Figure 10.13; the reduction increased as the loading of GO increased. It was also observed that GO as well as modified GOs gave qualitatively similar reduction (from 1 to 27%) in the peak heat release rate. The time to ignition was decreased drastically for in situ polymerized nanocomposites compared to pristine PS. The amount of reduction in time to ignition for the melt-blended samples was less than that for the in situ polymerized nanocomposites. Since there were no significant difference in thermal stability among all samples (actually, there was a slight increase in thermal stability for both in situ polymerized nanocomposites and melt-blended samples), the reduction observed in time to ignition must be due to some other reason. Some other works claimed a more significant increase in thermal stability for epoxy–graphite composites[37]

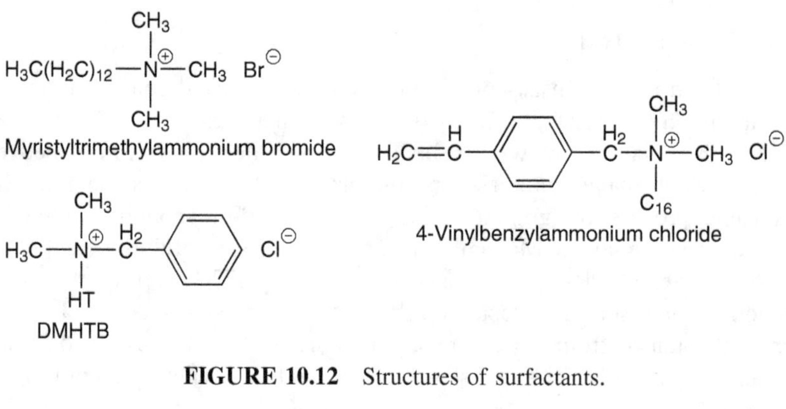

FIGURE 10.12 Structures of surfactants.

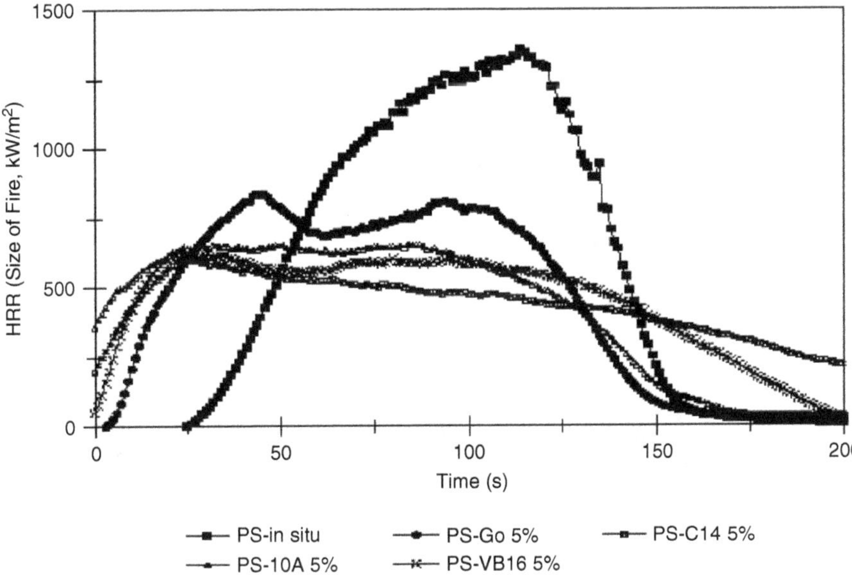

FIGURE 10.13 Heat release rate curves for in situ polymerized PS–GO nanocomposites at 35 kW/m^2. (From Ref. 36, copyright © 2004, Elsevier, with permission.)

and poly(vinyl alcohol)/graphite oxide nanocomposites.[38] One possible reason for the reduction in time to ignition by graphite-based nanocomposites is discussed in Section 10.4.1. The reduction in peak heat release rate observed for in situ polymerized PS–graphite nanocomposites is comparable to the roughly 50% reduction for PS–clay nanocomposites with 3 and 5% clay content.[39]

Mixed flame retardant performance of polymer–graphite samples was reported for phenolic–graphite and epoxy–graphite composites compared to fiberglass and aramid as a filler. Phenolic–graphite had the highest flame resistance, but epoxy–graphite had the lowest flame resistance.[40] Since the dispersion of graphite in the polymers was not discussed, it is not clear whether the samples studied were nanocomposites or microcomposites.

Very effective flame retardant data were obtained with styrene–butyl acrylate copolymer/graphite oxide (St–BA/GO) nanocomposites.[41, 42] The GO was prepared by oxidation of expandable graphite, and the St–BA/GO nanocomposites (GO content of up to 4% mass fraction) were synthesized by exfoliation– adsorption of monomer followed by in situ emulsion polymerization. The distribution of the GO particles was examined by XRD, TEM, and electron diffraction; exfoliated GO layers in crystalline structures were observed. The thermo-gravimetric analysis (TGA) data show a slight increase in thermal stability (up to 15°C with a 3% mass fraction of GO). Significant reduction in heat release rate by increasing GO content has been reported; all nanocomposites reduced about 40% of total heat released compared with that of St–BA, as shown in Figures 10.14 and 10.15.

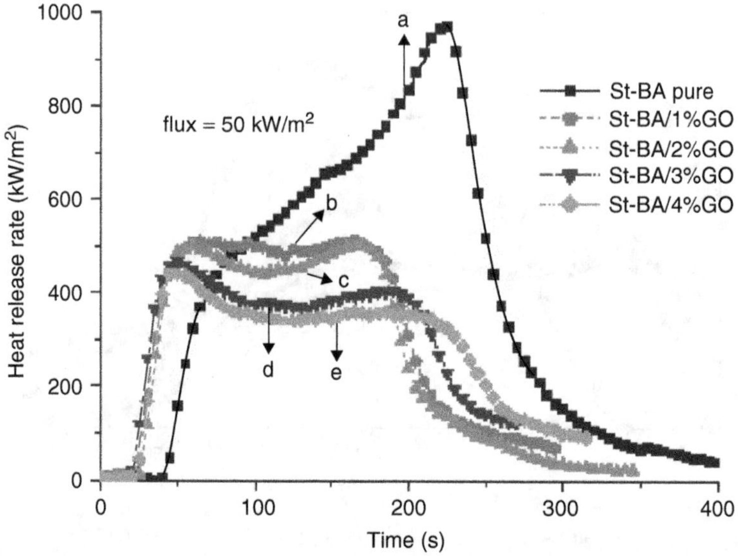

FIGURE 10.14 Heat release rate curves of St–BA and St–BA/GO nanocomposites. (From Ref. 41, copyright © 2004, Elsevier, with permission.) (*See insert for color representation of figure.*)

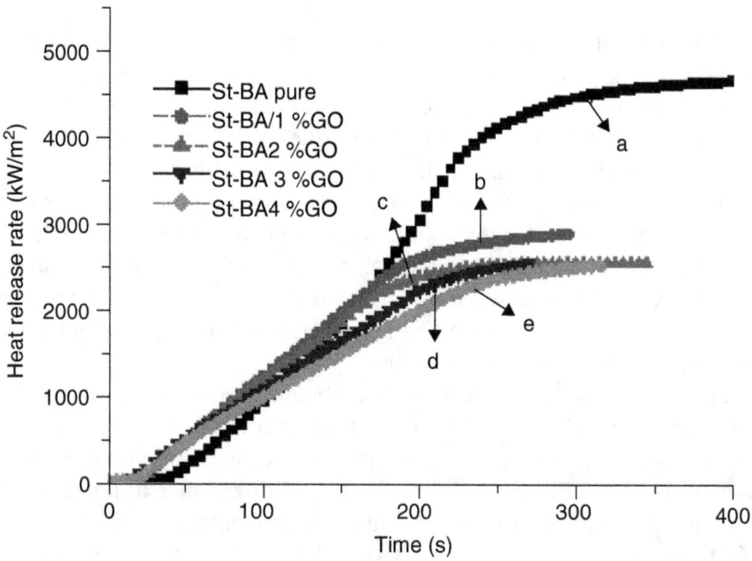

FIGURE 10.15 Total heat released for St–BA and St–BA nanocomposites at 50 kW/m^2. (From Ref. 41, copyright © 2004, Elsevier, with permission.) (*See insert for color representation of figure.*)

However, ignition delay times for the nanocomposites were shorter than that of the pristine sample, and it was suggested that this could be caused by thermal degradation of its organic emulsifier, resulting in the formation of volatile combustibles[43] or a catalytic effect on the initial stages of thermal degradation brought by Lewis or Brønsted acid sites on the GO layer. The proposed flame retardant mechanism of the addition of GO is that the formation of a char layer consisting of GO acts as a thermal insulator and a mass transport barrier, slowing the escape of the volatile products generated from the degradation of St–BA.

10.3.2 Carbon Nanotubes

Since carbon nanotubes were first synthesized in 1991,[44] there have been numerous studies on the preparation of carbon nanotubes and their many different applications which take advantage of their unique physical properties, such as high thermal conductivity (more than 3000 W/mK[45]) and high electrical conductivity. There are two different types of carbon nanotubes: single-walled (SWNT) with small diameters (1 to 2 nm) and multiwalled (MWNT) with larger diameters (10 to 100 nm). The manufacturing processes of the nanotubes include direct-current arc discharge,[46] laser ablation,[47] thermal- and plasma-enhanced chemical vapor-growth deposition,[48, 49] and flame synthesis.[50] After the synthesis of these nanotubes, the tubes contain various impurities, such as residual catalysts, amorphous carbons, and fullerenes. Therefore, these tubes are generally purified by various processes, such as oxidation in concentrated acids,[51] wet air oxidation,[52] or high-temperature treatment.[53] Detailed studies of the effects of cleaning on the characteristics of carbon nanotubes have been reported over the last several years.[51, 54–57] Cleaning of the tubes is critical for obtaining thermal stability and for the preparation of nanocomposites with well-dispersed tubes.

10.3.2.1 SWNT A TEM picture of SWNTs is shown in Figure 10.16. Generally, SWNTs form bundles (or ropes) due to van der Waals forces between the tubes. The black spots in the picture are residual catalyst particles. There have been many studies on the enhancement of physical properties of polymers by polymer–SWNT nanocomposites such as electric conductivity[59, 60] and mechanical strength.[61–63] There are also several papers reporting on the thermal stability of nanocomposites,[60, 63–65] but as far as this author is aware, only two papers reporting on the flammability of polymer–SWNT nanocomposites.[58, 66] Significantly enhanced thermal stability in air was reported for PMMA–SWNT nanocomposites.[60] However, a decrease in thermal stability in nitrogen was reported for epoxy-fluorinated SWNT nanocomposites compared to pristine epoxy, and no significant change in thermal stability in nitrogen was observed for PMMA–SWNT nanocomposites[58, 63] compared to pristine PMMA.

SWNTs for the flammability study of PMMA–SWNT nanocomposites were synthesized by the high-pressure carbon monoxide method (HiPCO)[48] and the coagulation method was used to produce PMMA–SWNT nanocomposites[60] in

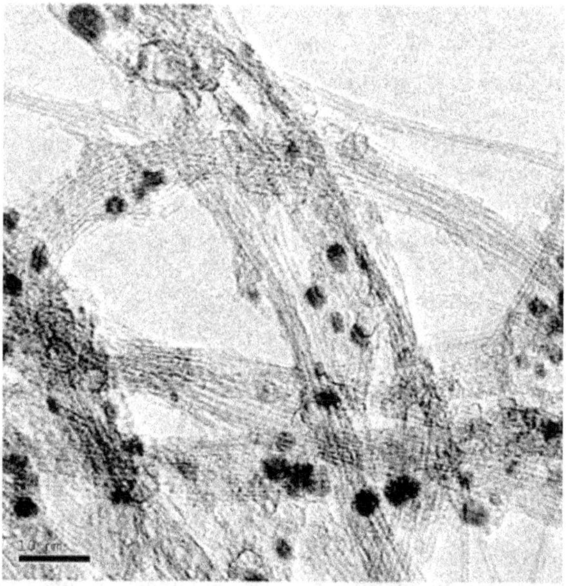

FIGURE 10.16 TEM image of SWNT ropes. (Scale bar, 10 nm.) (From Ref. 58.)

order to control dispersion of the SWNTs in the nanocomposites. In the coagulation method, dimethylformamide (DMF) was chosen to dissolve the PMMA and to permit dispersion of the SWNT by bath sonication. The nanotube dispersion in the nanocomposites was controlled by changing the nanotube concentration in DMF. The effects of nanotube dispersion in the nanocomposites on the flammability properties of the nanocomposites were investigated by comparing the flammability properties of those nanocomposite with poor nanotube dispersion to those with good tube dispersion. The global nanotube dispersion was determined by optical microscopy; images are shown in Figure 10.17. Figure 10.17*a* indicates that the nanotubes are relatively uniformly distributed within the polymer matrix on a micrometer scale. By using a higher concentration of SWNT in the DMF suspension, the sample in Figure 10.17*b* shows regions of nanotube aggregation. The former sample is designated as having good dispersion and the latter sample is designated as having poor dispersion. A TEM image of the purified original SWNT shows many nanotube bundles with a small amount of amorphous carbon and large carbon fullerenes with many iron particles in the nanotubes from the residual catalyst (see Figure 10.16).

Heat release rate histories of three different samples—PMMA, PMMA–SWNT (0.5%, good dispersion), and PMMA–SWNT(0.5%, poor dispersion)—were measured in a cone calorimeter at an external radiant flux of 50 kW/m^2; the results are shown in Figure 10.18. The heat release rate of the sample with good nanotube dispersion is much lower than those of pristine PMMA and of the sample with poor nanotube dispersion. The heat release rate of the sample with poor nanotube dispersion is not reduced appreciably from that of pristine PMMA. However, the total

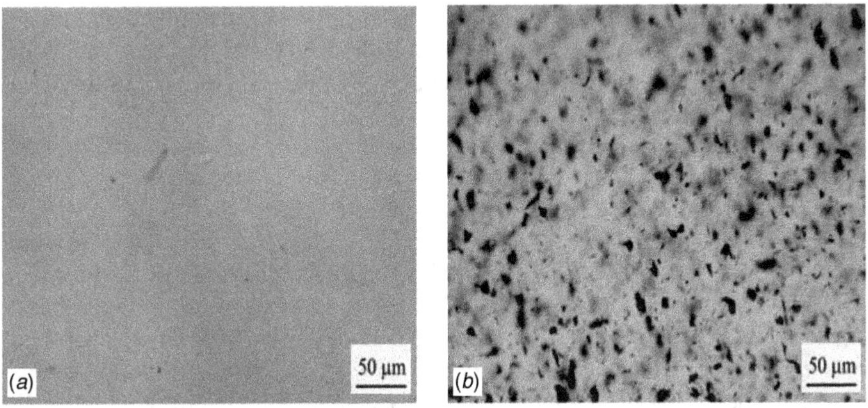

FIGURE 10.17 Optical microscopy images of PMMA–SWNT(0.5%) with two different dispersion of nanotubes with numerous agglomerates: (*a*) good dispersion; (*b*) poor dispersion. (From Ref. 58.) (*See insert for color representation of figure.*)

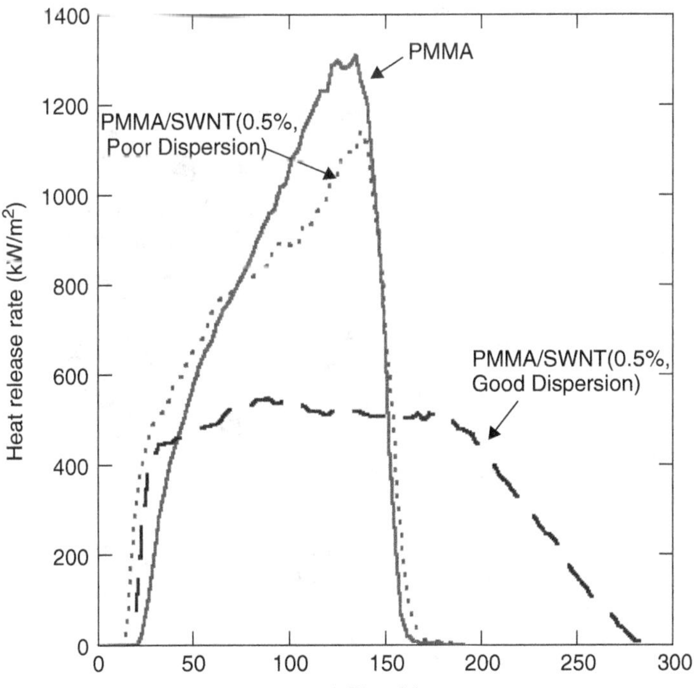

FIGURE 10.18 Effect of SWNT dispersion on heat release rate of PMMA–SWNT (0.5%) nanocomposites at an external radiant flux of 50 kW/m^2. (From Ref. 58.)

heat release values of all samples are comparable. This indicates that the sample with relatively good nanotube dispersion burns much slower than that with poor nanotube dispersion, but both samples eventually burn almost completely at an external radiant flux of 50 kW/m^2.

In order to understand how the difference in dispersion of the nanotubes affects the heat release rate of the nanocomposite, the behavior of the two samples during a gasification test in nitrogen atmosphere at an external radiant flux of 50 kW/m^2 was observed by taking video images. Selected pictures from the video images are shown in Figure 10.19. For the sample with good nanotube dispersion, numerous small bubbles formed initially, and their bursting was observed at the surface. This was followed shortly by formation of a solidlike behavior with no overt fluid motion. The final residue was a continuous dark layer covering the entire sample container. The sample with poor nanotube dispersion initially formed many small bubbles, and their bursting at the surface was followed by the formation of many small black islands. Vigorous bubbling was subsequently observed between the islands. Later, the islands coalesced into a connected structure, and their size gradually increased during the course of the test. The mass loss rate curves of samples with good and poor nanotube dispersion in the gasification tests have very similar trends, as shown by the heat release rate curves in Figure 10.18.

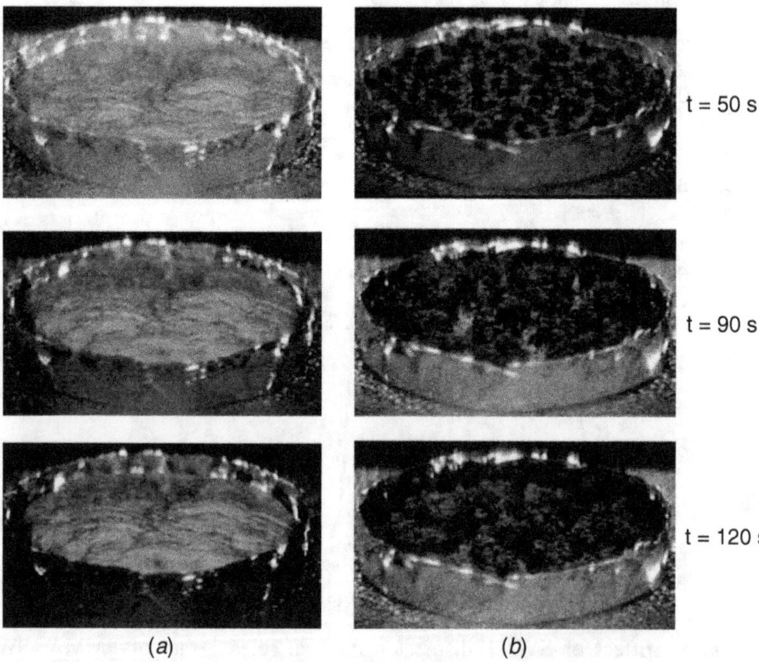

FIGURE 10.19 Selected video images of PMMA–SWNT(0.5%) during gasification tests at 50 kW/m^2 in nitrogen: (*a*) with good nanotube dispersion; (*b*) with poor nanotube dispersion. (From Ref. 58.)

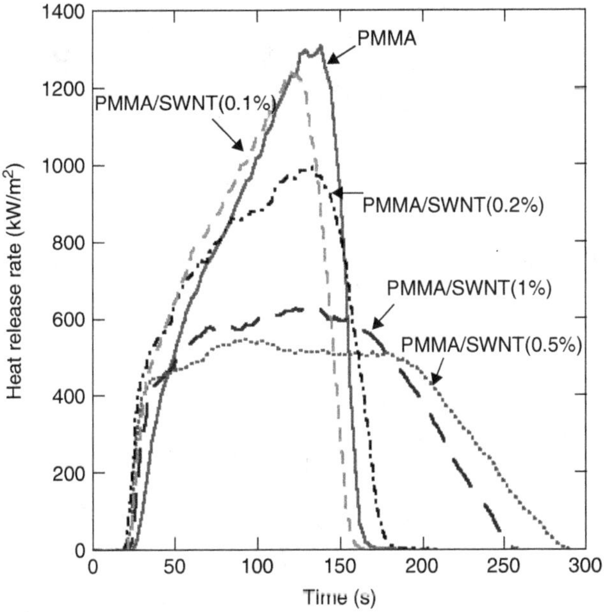

FIGURE 10.20 Effects of SWNT concentration on the heat release rate curve of PMMA–SWNT at 50 kW/m^2.

The effects of SWNT concentration on flammability properties of the nanocomposites were determined by measuring the heat release rate curves of PMMA–SWNT nanocomposites that have good dispersion of the nanotubes at levels from 0.1 to 1% prepared by the coagulation method. The results are shown in Figure 10.20. The addition of a 0.1% mass fraction of SWNT did not significantly reduce the heat release rate of PMMA. The most reduction in heat release rate was achieved by a 0.5% mass fraction. The amount of the reduction with 0.5% SWNT (about 60% reduction) is much larger than that with clay (about 28% reduction) even at 3% loading.[67] The behavior of the nanocomposite sample with 0.2% SWNT during a gasification test in nitrogen atmosphere was similar to that of PMMA–SWNT(0.5%, poor dispersion) (i.e., formation of many small, black discrete islands after initial numerous small bubbles and their bursting at the surface). Bubbling was observed between islands. It appeared that bubbling pushed nanotubes to the islands, the size of islands gradually became larger, and eventually some of the islands were connected to each other. The connected black islands were left behind at the end of the test, as shown in Figure 10.21*b*. Samples with 0.5 and 1% both appeared to be solidlike throughout their gasification; a network structured layer covered the sample surface during the entire test period and was left behind as a residue without any major open cracks, as shown in Figure 10.21*c* and *d*. An SEM image of the residue of PMMA–SWNT(1%) shows a network structure consisting of bundled, interwined carbon nanotubes, as shown in Figure 10.22. The residue was strong enough to be handled readily

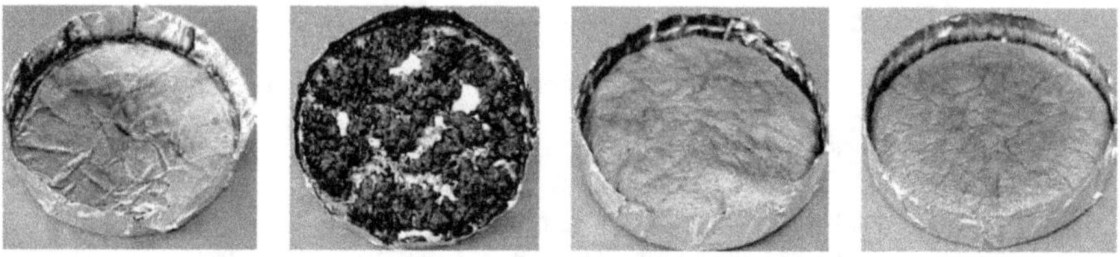

(*a*) (*b*) (*c*) (*d*)

FIGURE 10.21 Residues of PMMA–SWNT after gasification tests in a nitrogen atmosphere at 50 kW/m^2: (*a*) PMMA; (*b*) PMMA–SWNT(0.2%); (*c*) PMMA–SWNT(0.5%); (*d*) PMMA–SWNT(1%). (*See insert for color representation of figure.*)

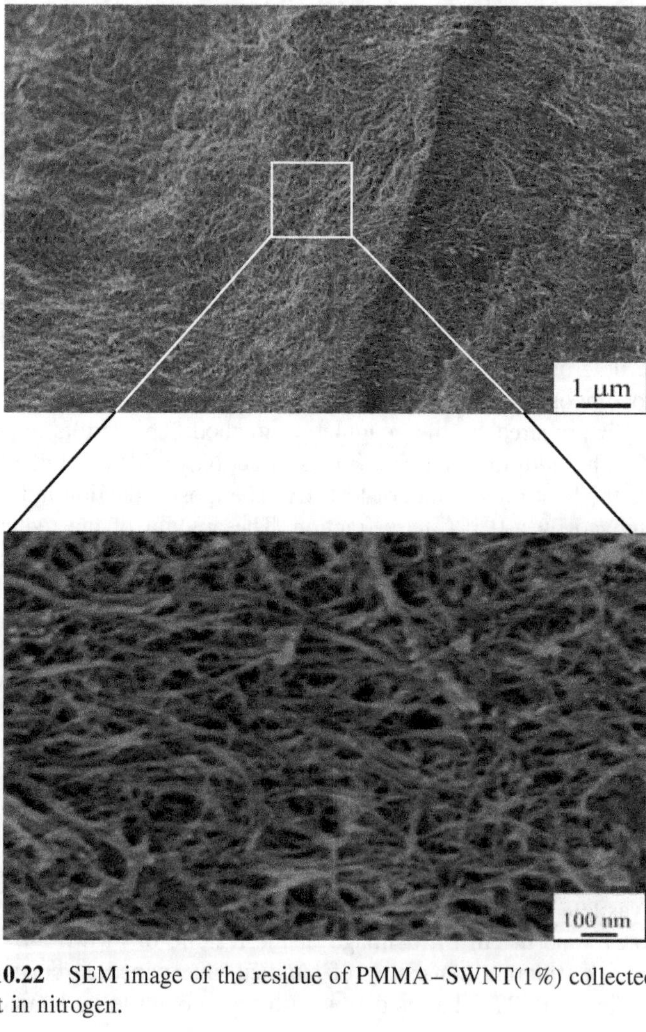

FIGURE 10.22 SEM image of the residue of PMMA–SWNT(1%) collected after gasification test in nitrogen.

without breaking. The amount of each residue collected after the gasification test was measured. The results indicate that the addition of the nanotubes increases only slightly the amount of the residue from PMMA.

Despite an effective flame retardant performance in the previous study,[58] another recent study shows no flame retardant effectiveness for PE–SWNT samples with SWNT concentrations of 5 and 10%.[66] These samples were prepared by melt blending, and dispersion of SWNTs in the sample was not determined. Considering the difficulty of the dispersion of SWNTs in a polymer, the results could be due to poor dispersion of the SWNTs.

10.3.2.2 MWNT TEM images of MWNT are shown in Figure 10.23. The lower maginification picture in the figure shows that the tubes appear to be flexible and have more of an appearance of noodles than of rods. Many studies have been published on the enhancement of electric conductivity[68–71] and of mechanical properties[72–75] of polymers by polymer–MWNT nanocomposite, and on the flammability of polymer–MWNT nanocomposites.[76–81] It was also reported that the oxidation of PS, PP, and poly(vinylidene fluoride) is retarded by the addition of carbon nanotubes.[82]

PP–MWNT nanocomposite samples with MWNT loading of 1, 2, and 4% by mass were melt-blended in a shear mixer. The MWNTs were prepared by chemical vapor deposition using xylene as a carbon source and ferrocene as a catalyst at about 675°C.[83] The distribution of nanotubes in the blended samples was examined by two different methods and magnifications. A scanning electron micrograph (SEM) picture of the MWNT dispersion in PP–MWNT(4%) nanocomposites after solvent removal is shown in Figure 10.24*a*. An optical microscopy image of PMMA–MWNT(1%) is shown in Figure 10.24*b*, which shows globally well-dispersed nanotubes in PP at large scales and a wide range of diameters and lengths of nanotubes, as shown in Figure 10.24*a*. The residual catalyst particles (iron) are encapsulated at various locations inside the nanotubes

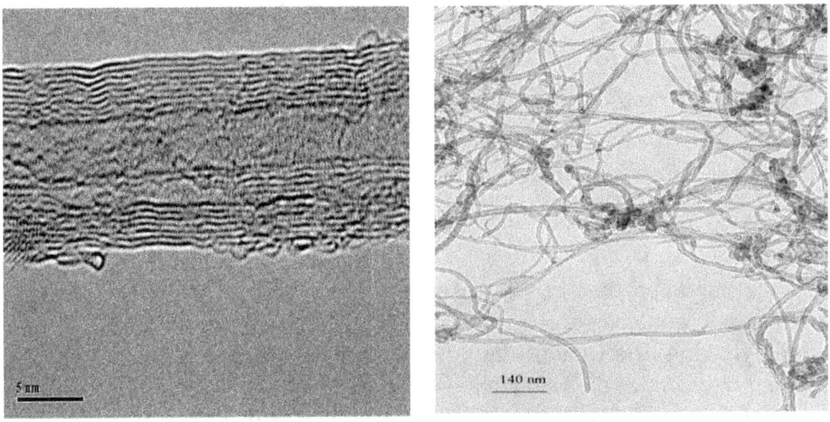

FIGURE 10.23 TEM images of MWNT: (*a*) scale bar 5 nm; (*b*) scale bar 140 nm.

FIGURE 10.24 (*a*) SEM picture of PP–MWNT(4%) after solvent removal of PP; (*b*) optical microscopy image of PP–MWNT(1%) nanocomposite in the melt.

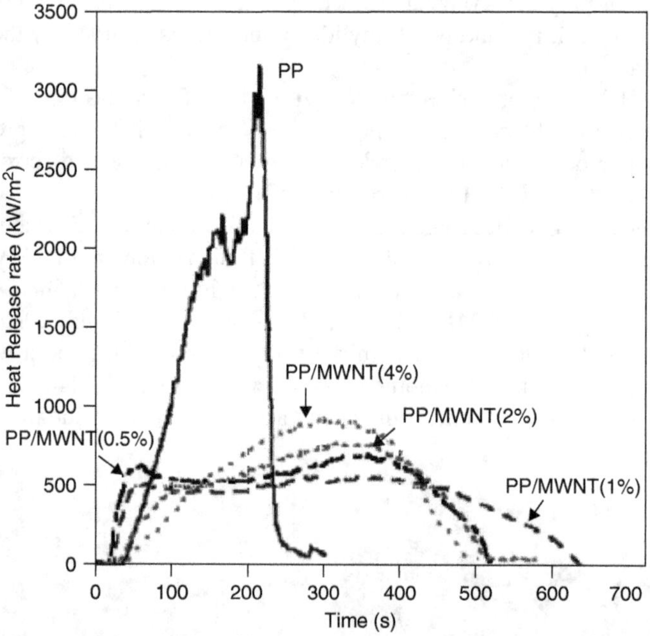

FIGURE 10.25 Effects of concentration of MWNT in PP on the heat release rate of PP–MWNT nanocomposites at 50 kW/m^2.

and at the nanotube tips, as shown in Figure 10.24*a*. Nanoparticulate iron is pyrophoric and could reduce the thermal oxidative stability of MWNT, as well as possibly acting as a catalyst during oxidative degradation of the PP–MWNT nanocomposites. Since the heat release rate curves of the PP–graphitized MWNT nanocomposites (with iron particles removed by high-temperature annealing[53, 84]) were similar to those of PP–MWNT (with iron particles), it was concluded that residual iron particles did not have significant effects on the heat release rate of PP–MWNT nanocomposites during flaming combustion[78] (little oxygen reaches the sample surface because oxygen is consumed by gas-phase oxidation reactions). However, strong glowing combustion (smoldering) of the sample residues (PP–MWNT with iron) was observed after flaming combustion was over (oxygen could then reach the residue surface) during cone calorimeter tests. (Smoldering was not observed with the residue of PP–graphitized MWNT under the same conditions.)

The effects of the concentration of MWNTs in PP on the heat release rate curves of the nanocomposites are shown in Figure 10.25. The results show two distinct characteristics brought on by the addition of MWNTs; first, there is a shortened ignition delay time with the PP–MWNT(0.5%), followed by an increase in ignition delay time with an increase in the concentration of MWNT; second, there is a gradual increase in peak heat release rate above about 1% by mass of MWNT. A similar trend was observed for PMMA–SWNT nanocomposites (less obvious for PMMA–SWNT, due to a lower concentration of SWNT, as shown in Figure 10.20). The lowest heat release rate curve for PP–MWNT is achieved with about 1% by mass of MWNT compared to about 0.5% by mass of SWNT. The increase in peak heat release rate with concentration of MWNT above 1% appears to be due to an increase in thermal conductivity of the nanocomposite.[78]

The physical behavior of PP–MWNT nanocomposites was significantly different from that of PP during a gasification test in a nitrogen atmosphere, as shown in Figure 10.26. The PP sample behaved like a liquid, with a fine frothy top layer generated by the bursting of numerous small bubbles at the sample surface. No char was left at the end of the test. However, all the PP–MWNT samples behaved like a solid without any visible melting except at the very beginning of the test, and the shape of the sample or size of the sample did not change significantly during the test. The residue of each sample was collected. No cracks were observed in any residue of the PP–MWNT nanocomposites. The network-structured layer of the PP–MWNT samples covered the entire sample surface and extended to the bottom of the residue, as shown in Figure 10.27. The residue consisted of tangled and roped carbon nanotubes. The tubes in the residue were more interwined and larger than those in the original sample. The network layer was porous but had physical integrity and did not break when picked lightly using one's fingers. The structure of the residue of PP–MWNT was very similar to that of the residue of PMMA–SWNT nanocomposites. The mass of the network-structured layer was very close to the initial mass of carbon nanotubes in the original nanocomposites. This indicates that the network-structured layer did not enhance char formation from PP. The importance of the formation of a network structure and of melt viscosity on flame retardant effectiveness was reported for PA6–MWNT nanocomposites prepared from a commercially available master batch sample.[80]

Since carbon black (CB) has been used as a filler to enhance the physical properties of rubbers, the flame retardant performance observed for MWNTs and

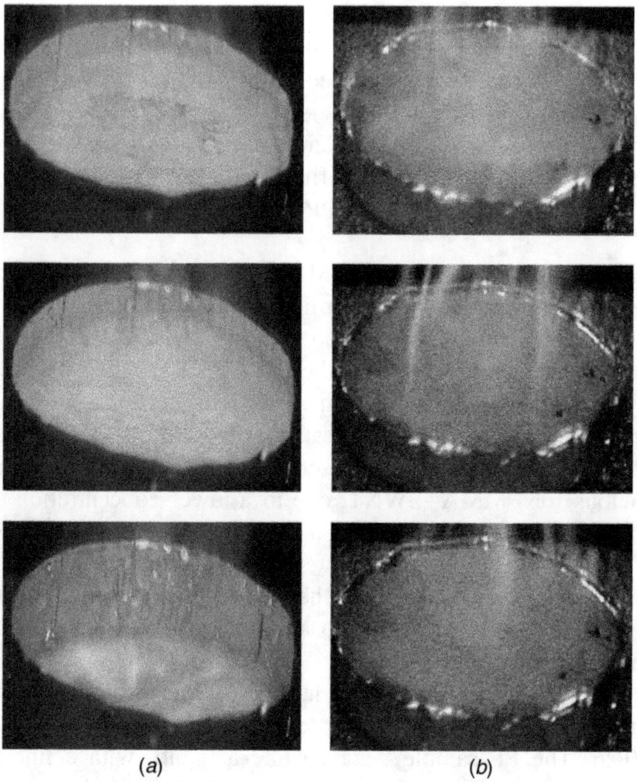

FIGURE 10.26 Sample behavior in a gasification test at 50 kW/m^2 in nitrogen: (*a*) PP; (*b*) PP–MWNT(1%).

FIGURE 10.27 Cross section of the residue of PP–MWNT (1%). (*See insert for color representation of figure.*)

SWNTs could possibly be due to the addition of carbon alone, independent of size or/and shape. To test this hypothesis, two different carbon blacks having different surface areas were compounded with PP at the same level of carbon concentration in PP as those of the PP–MWNT nanocomposites. The surface area of the carbon black designated as N299 was 102 m^2/g, and that designated as N762 was

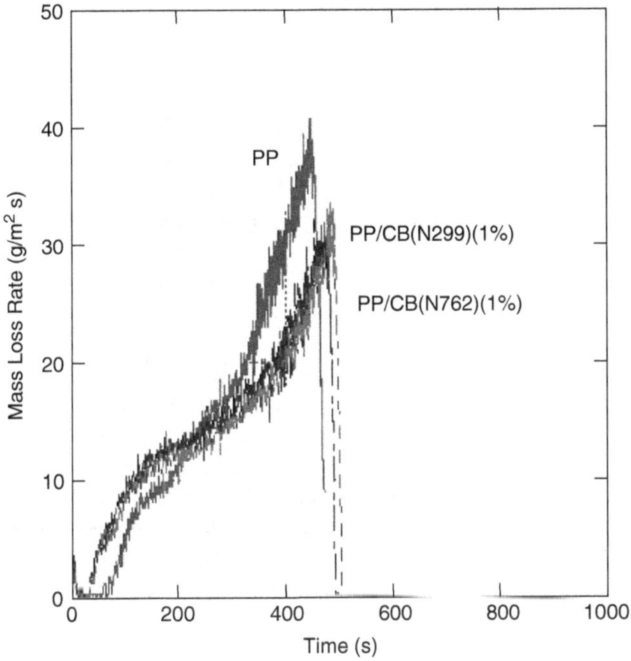

FIGURE 10.28 Effects of addition of carbon black on mass loss rate of PP at 50 kW/m^2 in nitrogen.

27.3 m^2/g. The mass loss rate curves of the PP–CB measured at 50 kW/m^2 in a nitrogen atmosphere were compared with that of PP in Figure 10.28. The addition of either carbon black increased the initial mass loss rate compared to that of PP. This trend is similar to the addition of MWNTs to PP, as shown in Figure 10.25 (the trend of the heat release rate was very similar to the mass loss rate[76]), but the reduction in the peak mass loss rate was much less than that for PP–MWNT(1%).

During the gasification test in a nitrogen atmosphere with the PP–CB samples, the sample behaved like a viscous liquid, with the formation of large bubbles which frequently burst at the sample surface. The residue of the PP–MWNT(1%) was a smooth layer filling the sample container without any cracks (almost the same size as the original sample). However, both residues of the PP–CB samples consisted of dispersed, aggregated granular particles left at the bottom of the sample container, as shown in Figure 10.29.

Three studies[66, 77, 79] investigated the synergistic flame retardant performance of combined organoclay–MWNT/EVA [poly(ethylene-co-vinyl acetate)] nanocomposites prepared by melt blending. The MWNTs were prepared by catalytic decomposition of acetylene with cobalt and iron as catalysts supported on alumina. The synthesized MWNTs were used directly as a crude sample, and purified MWNTs were also made by boiling concentrated sodium hydroxide water solution and removing mainly alumina in concentrated hydrochloric acid

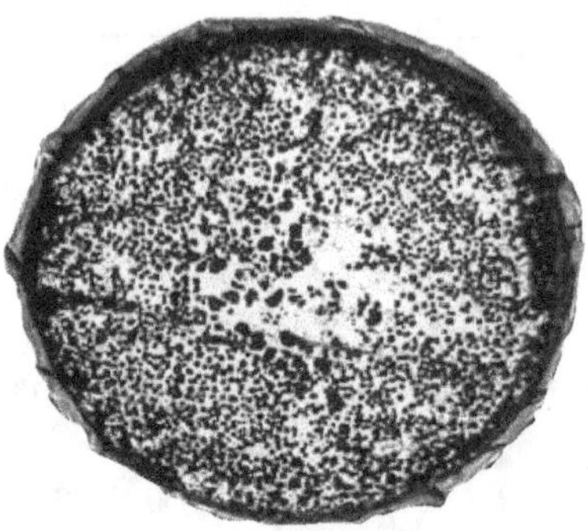

FIGURE 10.29 Residue of PP–CB(N299)(1%) after a gasification test at 50 kW/m^2 in nitrogen.

water solution. For same filler content, either the purified or crude MWNTs act as better flame retardants than organoclays do, with a larger reduction in the peak heat release rate and no significant influence on the time of ignition, as shown in Table 10.2. The crude MWNT was as effective as the purified MWNT in the reduction of the peak heat release rate. The peak heat release rate of the ternary nanocomposites, filled with 2.4% of an organoclay and 2.4% of purified MWNTs, was slightly less than that of nanocomposites with either purified MWNT(4.8%) or crude MWNT(4.8%). A comparison of the heat release rate curves for EVA–clay(4.8%), EVA–MWNT(4.8%), and VA–clay(2.4%)–MWNT (2.4%) is shown in Figure 10.30. It was speculated that the formation of graphitic carbon in char is enhanced when both carbon nanotubes and clay particles are applied, and this may contribute directly to the reduction of the peak heat release rate. The nanotubes also tend to reduce surface cracks of chars, as demonstrated above, leading to an increase in barrier resistance to the evolution of flammable volatiles and oxygen ingress to the condensed phase.[81]

TABLE 10.2 Properties of Samples with MWNT at 35 kW/m^2

	MWNT				
Sample	(% by Purified Mass)	(% by Crude Mass)	Organoclay (% by mass)	Ignition Time (s)	PHRR (kW/m^2)
1	—	—	—	84	580
2	2.4	—	—	85	520
3	4.8	—	—	83	405
4	—	—	2.4	70	530
5	—	—	4.8	67	470
6	2.5	—	2.5	71	370
7	—	4.8	—	83	403

Source: After Ref. 77.

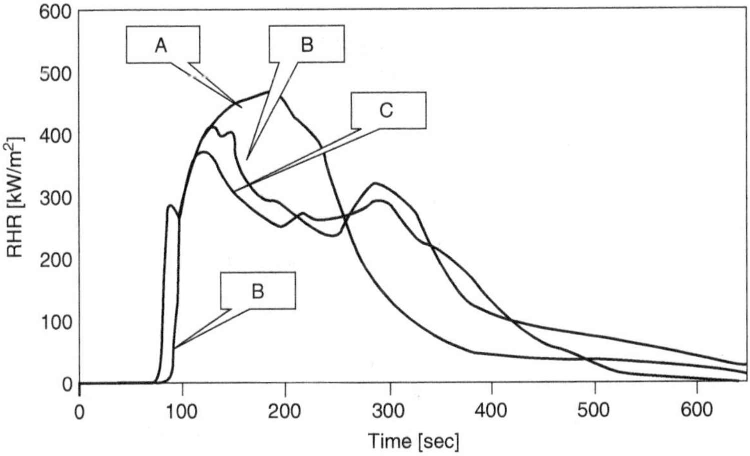

FIGURE 10.30 Heat release rate curves of EVA–clay(4.8%): A, EVA–MWNT(4.8%); B, EVA–clay(2.4%)/MWNT(2.4%); C, at an external flux of 35 kW/m^2. (From Ref. 77, copyright © 2002, John Wiley & Sons, Ltd., with permission.)

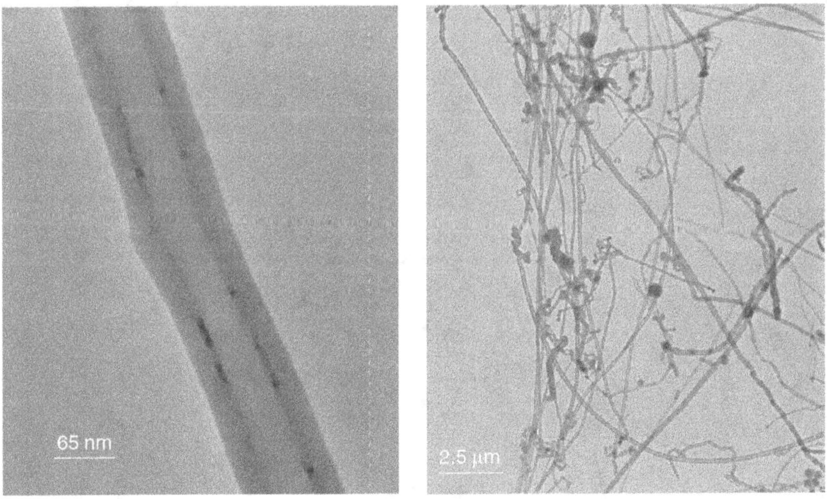

FIGURE 10.31 TEM images of CNF at two different magnifications.

10.3.2.3 Carbon Nanofibers Another type of nanoscale carbon-based particles is vapor-grown carbon nanofibers (VGCNFs) [or carbon nanofibers (CNFs)]. These diameters are in the range 60 to 200 nm and lengths are tens to hundreds of micrometers, which are much larger than those of SWNTs and MWNTs. They are commercially readily available at the kilogram level and with different levels of purified samples. TEM images of these nanofibers are shown in Figure 10.31. Many polymer nanocomposites were prepared with CNFs by melt blending for rheological studies,[85–87] reinforcement of physical properties,[88] and enhancement of electric conductivity.[85, 89] However, published studies on the flammability properties of polymer–CNF nanocomposites are rare. Since enhancement of the physical properties by polymer–CNF has been demonstrated, it was expected that flame retardant performance by the addition of CNF could be as effective as that of SWNTs and MWNTs, probably with a higher loading level of CNF than of SWNT or MWNT. Our recent results of PMMA–CNF nanocomposites and PP–CNF are discussed in this chapter.

The PMMA–CNF nanocomposites were prepared by the coagulation method using DMF as a solvent. The method was the same as that used for the PMMA–SWNT discussed in Section 10.3.2.1. Two different CNFs were used: PR-1 and PR-24LHT. TEM images of PR-24LHT are shown in Figure 10.31. According to the manufacturer of the CNFs (Applied Science Inc.), PR-1 is as-grown material with a diameter of 100 to 200 nm containing amorphous carbons, and PR-24LHT is fiber graphitized by heat treatment, with diameters of 60 to 150 nm without amorphous carbons. The flame retardant effectiveness of these nanofibers was investigated by measuring heat release rates of PMMA–CNF nanocomposites at 50 kW/m^2. The results are shown in Figure 10.32. An increase in loading of the PR-24 reduced the heat release rate of the nanocomposites up to 4% by mass (although the reduction in heat release from a 2% mass fraction to a 4% mass

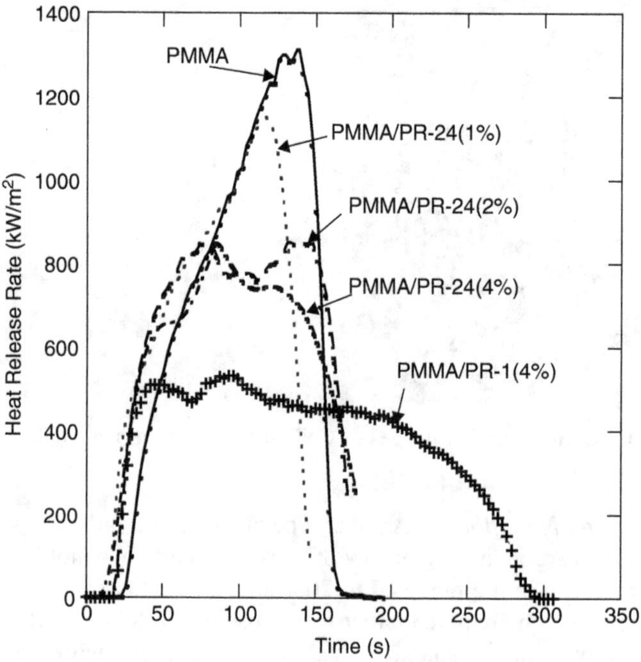

FIGURE 10.32 Mass loss rate curves of PMMA–CNF nanocomposites at 50 kW/m^2 in nitrogen.

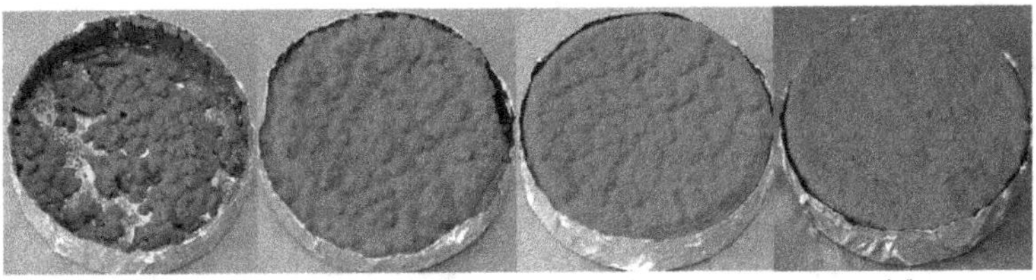

FIGURE 10.33 Residues after the gasification tests at 50 kW/m^2 in nitrogen: (*a*) with PR-24(1%); (*b*) with PR-24(2%); (*c*) with PR-24(4%); (*d*) with PR-1(4%). (*See insert for color representation of figure.*)

fraction becomes less than that from a 1% mass fraction to a 2% mass fraction). The PMMA/PR-24(1%) nanocomposite showed muddy liquidlike behavior followed by the formation of many small black islands during the test. The islands gradually coalesced as the test progressed, and a thin connected mass of black islands was left on the bottom of the sample container at the end of the test, as shown in Figure 10.33*a*. Similar behavior was observed for the PMMA/PR-24(2%) nanocomposite, but it appeared to be more viscous, with the formation of large islands followed by a rugged solidlike appearance accompanied by large bubbles and their bursting. A rugged layer without any cracks was left at the end of test, as shown in Figure 10.33*b*. The PMMA/PR-24(4%) nanocomposite appeared to be solidlike, accompanied by several large bubbles and their bursting, followed by small swelling. A slightly rugged surfaced solid layer without cracks was left at the end of the test, as shown in Figure 10.33*c*. On the other hand, the PMMA/PR-1(4%) nanocomposite remained solidlike with a smooth surface, without forming any significant number of bubbles over the entire duration of the test. The shape of the residue was nearly the same as that of the original sample, as shown in Figure 10.33*d*.

The heat release rate of PMMA/PR-1(4%) is much less than that of PMMA/PR-24, as shown in Figure 10.32. Considering the high-purity nature of PR-24 (without amorphous carbons) compared to PR-1, which contains amorphous carbons, it is surprising to observe better flame retardant effectiveness PR-1 than with PR-24 in PMMA. The heat treatment for PR-24 could remove any defects and −COOH and −OH from the fibers if they existed on the nanofibers.[51] If so, the PR-24 could be less polar than the PR-1. Then PR-24 in polar PMMA may not be dispersed as well as PR-1dispersed in PMMA. Optical microscopy image of the PMMA/PR-24 (2%) shows well-dispersed nanofibers with some agglomerated nanofibers, as shown in Figure 10.34*a*. However, the image of the PMMA/PR-1(2%) shows well-dispersed nanofibers without agglomerates, as seen in Figure 10.34*b*. Another possibility to explain the difference in flame retardant effectiveness of the two nanofibers is the difference in the size of the two nanofibers. The images indicate that the PR-24 might be much smaller in diameter and have much shorter fibers than the PR-1. Therefore, the better flame retardant performance observed for PR-1 in PMMA than for PR-24 could be due to better dispersion of PR-1 in PMMA and to the difference in the size of the nanofibers.

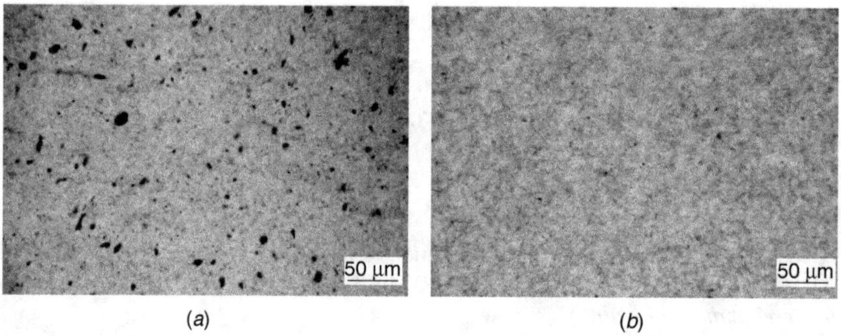

FIGURE 10.34 Optical microscopy images: (*a*) PMMA/PR-24(2%); (*b*) PMMA/PR-1 (2%).

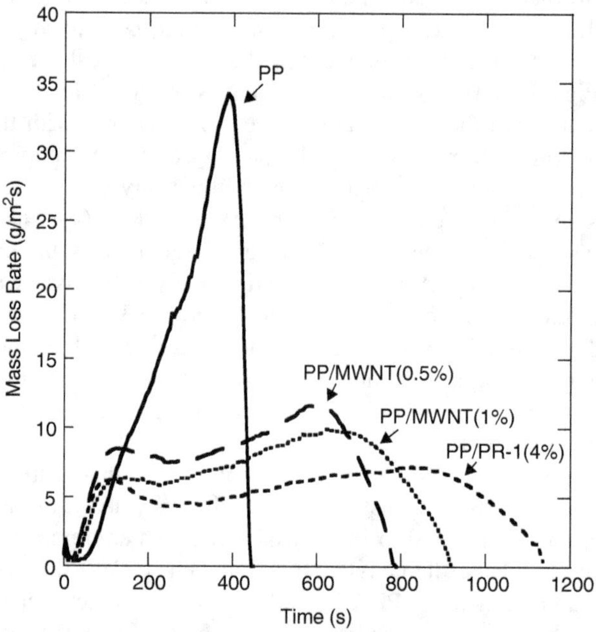

FIGURE 10.35 Comparison of mass loss rate curves for PP, PP–MWNTs, and PP/PR-1 nanocomposites at 50 kW/m^2 in nitrogen.

Another example of the excellent flame retardant performance by PR-1 could be seen with PP. A PP/PR-1(4%) nanocomposite was prepared by melt mixing. The mass loss rate of the PP/PR-1(4%) nanocomposite was measured at 50 kW/m^2 in a nitrogen atmosphere. The nanocomposite appeared to be solid-like with a smooth surface without cracks during the most of the test period except for an initial, brief period of formation of numerous small bubbles and their bursting at the surface. The size of the residue collected at the end of the test was nearly the same as that of the original sample. The measured mass loss rate curve is compared with those of PP–MWNT nanocomposites described in Section 10.3.2.2 and the comparison is shown in Figure 10.35. The mass loss rate of the PP/PR-1(4%) is slightly less than those of PP–MWNT(0.5%) and PP–MWNT(1%). Thus, the heat release rate curve (Figure 10.32) and the mass loss rate curve (Figure 10.35) show effective flame retardant performance by an *appropriate CNF* as long as CNFs are *well dispersed without agglomerates*. The flame retardant effectiveness of such CNFs appears to be as good as those of SWNT and of MWNT except that CNF requires higher loading (roughly four to eight times by mass). This indicates that the use of CNF is much more economical (at least 1/1000) than SWNT to obtain a similar FR performance.

10.4 DISCUSSION OF RESULTS

10.4.1 Flame Retardant Mechanism

It appears that the flame retardant mechanism for the nanoparticles discussed in this chapter is the formation of a continuous protective layer consisting of a *network* of the nanoparticles, and the layer appears to act as a heat shield. All data shown in this chapter show that the peak heat release rate could be reduced significantly using nanoscale tube-shaped particles as long as these particles are well dispersed in a polymer matrix as a filler. Recent study indicates a direct relationship between viscoelastic measurement (storage modulus) and reduction in heat release rate.[90] This suggests that we might be able to screen for promising flame-retarded polymer nanocomposites by performing viscoelastic measurements on the initially fabricated samples. Although heat release rate is the key parameter for fire growth,[91] total heat releases of these nanocomposites are not reduced significantly except in the results shown in Figure 10.15. This means that the burning rate or flame size of these nanocomposites would be low or small, but they would burn slowly for a longer time, and eventually most of the matrix would be thermally decomposed to provide flammable gaseous products. Furthermore, ignition delay times of the nanocomposites based on carbons measured in a cone calorimeter tend to be shorter than those of polymer matrixes, despite little difference in thermal stability between the nanocomposite and the polymer matrix. (In some cases, the thermal stability of the nanocomposite is slightly higher than that of the matrix.) This shorter ignition delay time for carbon-based nanocomposite can be explained using PP–MWNT as an example.

In a cone calorimeter test, ignition is initiated by thermal radiation from an electrically heated element at a temperature of about 750°C. It is expected that the emission spectra from the heater element is that of a gray body covering the visible to the far infrared but peaking at about 2.7 μm. Therefore, there might be a significant difference in the absorption characteristics of the external emission by PP–MWNT compared to that of PP. The measured infrared transmission spectra of a PP sample was compared with that of PP–MWNT(1%) nanocomposite (see Figure 10.36). The PP shows many absorption bands based on various vibrational modes, but there is substantial transmission between these bands. This

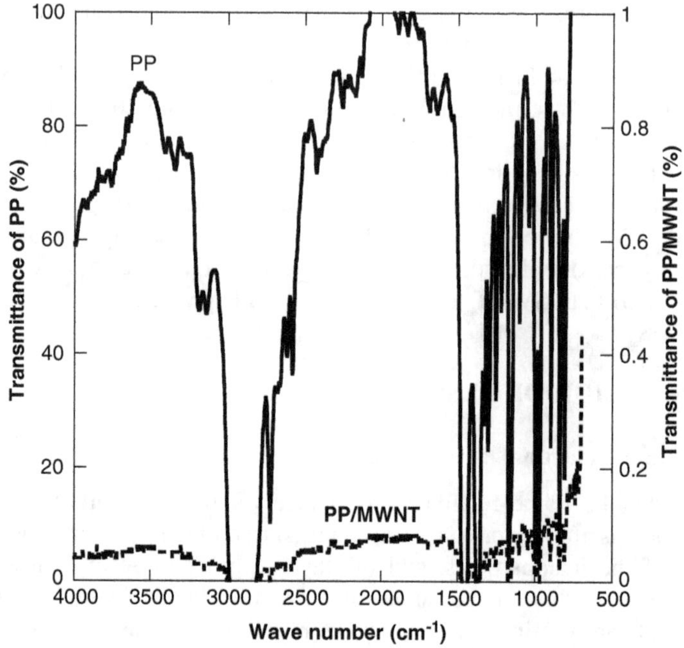

FIGURE 10.36 Comparison of transmission spectra of PP and PP–MWNT (1%) through 200-μm-thick film. (From Ref. 78.).

indicates that the external thermal radiant flux of 50 kW/m^2 is absorbed by the PP sample over some depth. On the other hand, the PP–MWNT nanocomposite shows no significant transmission bands and all of the 50-kW/m^2 flux is absorbed very near the sample surface, within a distance of 200 μm. Therefore, a narrow layer in the vicinity of the PP–MWNT sample surface is heated rapidly and its temperature becomes high enough to initiate thermal degradation of PP and to generate evolved degradation products of monomer dimer, trimer, and oligomers to initiate ignition. On the other hand, the PP sample is heated over a greater depth, and it thereby takes longer to heat the sample to initiate degradation. Thus, the ignition delay time of PP–MWNT, in particular at low concentrations of MWNT, tends to be shorter than that of PP. This explanation applies to any polymer/carbon-based nanocomposites because of absorption of incident radiant energy by *discrete* bands based on the polymer structure.

10.4.2 Morphology

The dispersion of nanoparticles in polymer nanocomposites has a significant effect on their flammability properties, as shown in Figure 10.18. Many studies used TEM and/or SEM images to demonstrate the quality of the dispersion of the nanoparticles in the nanocomposites. However, these images tend to observe an extremely small area of the samples, on the order of 100 × 100 nm. These images show the shape, size, and interaction of the nanoparticles, but they do not show the overall dispersion of the nanoparticles in the samples. Furthermore, there are two aspects that affect the effectiveness of TEM analysis. First is the sample preparation. The sample prepared is extreme small (as described above) and the observed area may not represent overall dispersion characteristics. The people preparing the sample may select the "good" region to cut. The second is the fact that we are all prone to look at what we want. So with the TEM technique, there is some concern that the images selected may not be fully representative of the actual sample. Researchers using TEM are therefore encouraged to collect multiple images of their samples at multiple magnfications and from multiple sections. This is indeed more work, but is essential to ensure uniform and accurate analysis when using TEM.

Looking at the dispersion on a micrometer scale (e.g., on the order of $100 \times 100\ \mu m$) it appears that the formation of agglomerates of the nanoparticles tends to be more appropriate for flame retardant effectiveness. Such measurements could be made with confocal microscopy or optical microscopy at various locations in the sample. An image by confocal microscopy of the polyamide-6 nanocomposite with 2% clay particles is shown in Figure 10.37 as an example. This image was constructed from 300 images taken from the surface of a 200 – μm-thick sheet looking inside the sample in 0.1 μm steps. This image shows a large-scale distribution pattern of clay particles, including several agglomerates, which cannot be determined by TEM or SEM. Ideally, a statistical analysis should be conducted on these images to quantify the dispersion of the nanoparticles instead of the qualitative image observation that has commonly been used.

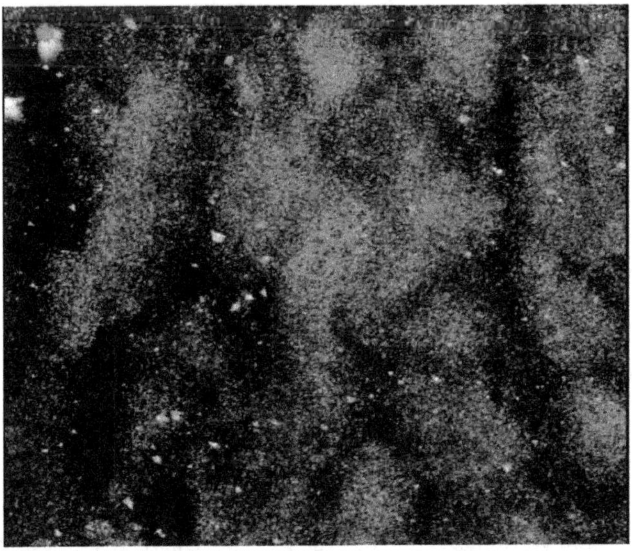

FIGURE 10.37 Confocal microscopy image of PA6/clay(2%). The image dimension is about $100 \times 100\ \mu m$ with a thickness of 30 μm.

10.4.3 Thermal Gravimetric Analysis

Measurement of the thermal stability of the polymer nanocomposites by TGA is useful for understanding their flame retardant mechanism. Since oxygen is consumed primarily by gas-phase oxidation reactions during flaming burning of the nanocomposites, oxygen hardly reaches the thermally degrading sample surface beneath the evolved gaseous products. Therefore, it is recommended that TGA be conducted in an inert atmosphere instead of air. The results of TGA conducted in air would apply to smoldering combustion instead of flaming combustion. Heating rates in TGA are generally at least one to three orders of magnitude slower than heating rates in fire conditions. The composition of the degradation products can be modified significantly by the heating rate of the sample. Furthermore, a TGA sample is generally very small (a few milligrams). Then, secondary reactions of degradation products passing through the sample (the real material is thicker than that for TGA) are not encountered. Therefore, one needs to be cautious in extrapolating the TGA results, in particular, degradation products, to fire conditions.

10.5 SUMMARY AND CONCLUSIONS

It is observed that the heat release rates of the nanocomposites during the *early stage* of burning from the onset of ignition (until the establishment of a protective layer) are not significantly different from those of the matrix. A similar trend was observed with polymer nanocomposites with nanoclay particles as a filler. However, it appears that the flame retardant effectiveness of carbon nanotubes determined by the reduction in heat release rate is better than those with nanoclay particles per unit mass base at low concentration.[92] This might be why these particles might not be considered as an all-around fire retardant since they fail to pass the UL-94 type small ignition test.[80] However, a performance-based fire safety approach instead of a single go/no go type of test is becoming considered increasingly in many countries, looking at material flammability characteristics of ignition, heat release rate, CO production rate, and so on, as inputs. Therefore, these nanocomposites can help to reduce heat release and slow fire growth, but further improvement in their flame retardant effectiveness is needed for a wider application of these nanoparticles. An improvement in flame retardant effectiveness could be achieved by enhancing the formation of char significantly by making numerous cross-links from specifically functionalized nanoparticles (more carbons in the polymer matrix remain in the condensed phase) or by a combined use with conventional flame retardant additives.

Acknowledgments

The author thanks Dr. Severine Bellayer at NIST, Dr. Fangming Du at the University of Pennsylvania, Dr. Ying Yang at the University of Kentucky, and Dr. Kazuya Yamamoto and Dr. Kathryn Butler at NIST for their help with various TEM images, the confocal microscopy image, and the image analysis in this chapter. The author also thanks Dr. Tom Ohlemiller at NIST for his careful review of the text. The author acknowledges funding from NIST with grant 5D1022.

REFERENCES

1. Kojima, Y.; Usuki, A.; Kawasumi, M.; Okada, A.; Fukushima, Y.; Kurauchi, T.; Kamigaito, O. Mechanical properties of nylon 6-clay hybrid. *J. Mater. Res.* **1993**, 8, 1185–1189.
2. Novak, B.M. Hybrid nanocomposite materials: between inorganic glasses and organic polymers. *Adv. Mater.* **1993**, 5, 422–433.
3. Giannelis, E. Polymer layered silicate nanocomposites. *Adv. Mater.* **1996**, 8, 29–35.
4. Alexandre, M.; Dubois, P. Polymer-layered silicate nanocomposites: preparation, properties and uses of a new class of materials. *Mater. Sci. Eng.* **2000**, R28, 1.
5. Landry, C.J.T.; Coltrain, B.K.; Landry, M.R.; Fitzgerald, J.J.; Long, V.K. Poly(vinyl acetate) silica filled materials: material properties of in-situ vs. fumed silica particles. *Macromolecules* **1993**, 26, 3702–3712.
6. Hajji, P.; David, L.; Gerard, J.F.; Pascault, J.P.; Vigier, G. Synthesis, structure, and morphology of polymer–silica hybrid nanocomposites based on hydroxyethyl methacrylate. *J. Polym. Sci. B* **1999**, 37, 3172–3187.
7. Ou, Y.; Yang, F.; Yu, Z.-Z. New conception on the toughness of nylon 6/silica nanocomposite prepared via in situ polymerization. *J. Polym. Sci. B*, **1998**, 36, 789–795.
8. Reynaud, E.; Jouen, T.; Gauthier, C.; Vigier, G.; Varlet, J. Nanofillers in polymeric matrix: a study on silica reinforced PA6. *Polymer* **2001**, 42, 8759–8768.
9. Hsiue, G.-H.; Kuo, W.-J.; Huang, Y.-P.; Jeng, R.-J. Microstructural and morphological characteristics of PS–SiO_2 nanocomposites. *Polymer* **2000**, 41, 2813–2825.
10. Liu, Y.L.; Hsu, C.Y.; Wei, W.L.; Jeng, R.J. Preparation and thermal properties of epoxy–silica nanocomposites from nanoscale colloidal silica. *Polymer* **2003**, 44, 5159–5167.
11. Tsagaropoulos, G.; Eisenberg, A. Direct observation of 2 glass transitions in silica-filled polymers: implications for the morphology of random ionomers. *Macromolecules* **1995**, 28, 396–398.
12. Kashiwagi, T.; Gilman, J.W.; Butler, K.M.; Harris, R.H.; Shields, J.R. Flame retardant mechanism of silica gel/silica. *Fire Mater.* **2000**, 24, 277–289.
13. Kashiwagi, T.; Shields, J.R.; Harris, R.H.; Davis, R.D. Flame-retardant mechanism of silica: effects of resin molecular weight. *J. Appl. Polym. Sci.* **2003**, 87, 1541–1553.
14. Yang, F.; Nelson, G.L. PMMA/silica nanocomposite studies: synthesis and properties. *J. Appl. Polym. Sci.* **2004**, 91, 3844–3850.
15. Kashiwagi, T.; Morgan, A.B.; Antonucci, J.M.; VanLandingham, M.R.; Harris, R.H.; Awad, W.H.; Shields, J.R. Thermal and flammability properties of a silica–poly(methylmethacrylate) nanocomposite. *J. Appl. Polym. Sci.* **2003**, 89, 2072–2078.
16. Yang, F.; Yngard, R.; Nelson, G.L. Flammability of polymer–clay and polymer–silica nanocomposites. *J. Fire Sci.* **2005**, 23, 209–226.
17. Liu, J.; Gao, Y.; Wang, F.; Wu, M. Preparation and characteristics of nonflammable polyimide materials. *J. Appl. Polym. Sci.* **2000**, 75, 384–389.

18. Okoshi, M.; Nishizawa, H. Flame retardancy of nanocomposites. *Fire Mater.* **2004**, 28, 423–429.
19. Laachachi, A.; Leroy, E.; Cochez, M.; Ferriol, M.; Lopez Cuesta, J.M. Use of oxide nanoparticles and organoclays to improve thermal stability and fire retardancy of poly(methyl methacrylate). *Polym. Degrad. Stab.* **2005**, 89, 344–352.
20. Lichtenhan, J.D.; Vu, N.Q.; Carter, J.A.; Gilman, J.W.; Feher, F.J. Silsesquioxane siloxane copolymers from polyhedral silsesquioxanes. *Macromolecules* **1993**, 26, 2141–2142.
21. Lichtenhan, J.D.; Otonari, Y.A.; Carr, M.J. Linear hybrid polymer building-blocks: methacrylate-functionalized polyhedral oligomeric silsesquioxane monomers and polymers. *Macromolecules* **1995**, 28, 8435–8437.
22. Haddad, T.S.; Lichtenhan, J.D. Hybrid organic–inorganic thermoplastics: styryl-based polyhedral oligomeric silsesquioxane polymers. *Macromolecules* **1996**, 29, 7302–7304.
23. http://www.hybridplastics.com.
24. Mantz, R.A.; Jones, P.F.; Chaffee, K.P.; Lichtenhan, J.D.; Gilman, J.W.; Ismail, I.M.K.; Burmeister, M.J. Thermolysis of polyhedral oligomeric silsesquioxane (POSS) macromers and POSS–siloxane copolymers. *Chem. Mater.* **1996**, 8, 1250–1259.
25. Schwab, J.J.; Lichtenhan, J.D. Polyhedral oligomeric silsesquioxane (POSS)-based polymers. *Appl. Organomet. Chem.* **1998**, 12, 707–713.
26. Ni, Y.; Zheng, S. A novel photocrosslinkable polyhedral oligomeric silsesquioxane and its nanocomposites with poly(vinyl cinnamate). *Chem. Mater.* **2004**, 16, 5141–5148.
27. Zheng, L.; Kasi, R.M.; Farris, R.J.; Coughlin, E.B. Synthesis and thermal properties of hybrid copolymers of syndiotactic polystyrene and polyhedral oligomeric silsesquioxane. *J. Polym. Sci. A, Polym. Chem.* **2002**, 40, 885–891.
28. Huang, J.C.; He, C.B.; Xiao, Y.; Mya, K.Y.; Dai, J.; Siow, Y.P. Polyimide/POSS nanocomposites: interfacial, interaction, thermal properties and mechanical properties. *Polymer* **2003**, 44, 4491–4499.
29. Kashiwagi, T.; Gilman, J.W. Silicon-based flame retardants, in: A.F. Grand, and C.A. Wilkie, Eds., *Fire Retardancy of Polymeric Materials.* Marcel Dekker, New York, 2000, pp. 353–389.
30. Devaux, E.; Rochery, M.; Bourbigot, S. Polyurethane/clay and polyurethane/POSS nanocomposites as flame retarded coating for polyester and cotton fabrics. *Fire Mater.* **2002**, 26, 149–154.
31. Bourgiot, S.; Flambard, X.; Rochery, M.; Le Bras, M.; Devaux, E.; Lichtenhan, J.D. Polyhedral oligomeric silsesquioxanes: application to flame retardant textile, in: M. Le Bras, C.A. Wilkie, and S. Bourbigot, Eds., *Fire Retardancy of Polymers.* Royal Society of Chemistry, London, 2005, pp. 189–201.
32. Gupta, S.K.; Schwab, J.J.; Lee, A.; Fu, B.X.; Hsiao, B.S. POSS reinforced fire retarding EVE resins, in: B.M. Rasmussen, L.A. Pilato, and H.S. Kliger, Eds., *Affordable Materials Technology: Platform to Global Value and Performance. SAMPE Pub.* **2002**, 47(2), 1517–1526.
33. Jash, P.; Wilkie, C.A. Effects of surfactants on the thermal and fire properties of poly(methyl methacrylate)/clay nanocomposites. *Polym. Degrad. Stab.* **2005**, 88, 401–406.

34. Nyden, M.R.; Gilman, J.W. Molecular dynamics simulations of the thermal degradation of nano-confined polypropylene. *Compos. Theor. Polym. Sci.*, **1997**, 7, 191–198.

35. Uhl, F.W.; Wilkie, C.A. Polystyrene/graphite nanocomposites: effect on thermal stability. *Polym. Degrad. Stab.* **2002**, 76, 111–122.

36. Uhl, F.M.; Wilkie, C.A. Preparation of nanocomposites from styrene and modified graphite oxides. *Polym. Degrad. Stab.* **2004**, 84, 215–226.

37. Xu, J.; Hu, Y.; Song, L.; Wang, Q.; Fan, W.; Liao, G.; Chen, Z. Thermal analysis of poly(vinyl alcohol)/graphite oxide intercalated composites. *Polym. Degrad. Stab.* **2001**, 73, 29–31.

38. Yasmin, A.; Daniel, I.M. Mechanical and thermal properties of graphite plated/epoxy composites. *Polymer* **2004**, 45, 8211–8219.

39. Zhu, J.; Wilkie, C.A. Thermal and fire studies on polystyrene–clay nanocomposites. *Polym. Int.* **2000**, 49, 1158–1163.

40. Hshieh, F.Y.; Beeson, H.D. Flammability testing of flame-retarded epoxy composites and phenolic composites. *Fire Mater.* **1997**, 21, 41–49.

41. Zhang, R.; Hu, Y.; Xu, J.; Fan, W.; Chen, Z. Flammability and thermal stability studies of styrene–butyl acrylate copolymer/graphite oxide nanocomposite. *Polym Degrad Stab.* **2004**, 85, 583–588.

42. Zhang, R.; Hu, Y.; Xu, J.; Fan, W.; Chen, Z.; Wang, Q. Preparation and combustion properties of flame retardant styrene–butyl acrylate copolymer/graphite oxide nanocomposite. *Macromol. Mater. Eng.* **2004**, 289, 355–359.

43. Zanetti, M.; Kashiwagi, T.; Falqui, L.; Camino, G. Cone calorimeter combustion and gasification studies of polymer layered silicate nanocomposites. *Chem. Mater.* **2002**, 14, 881–887.

44. Iijima, S. Helical microtubules of graphitic carbon. *Nature* **1991**, 354, 56–58.

45. Kim, P.; Shi, L. Majumdar, A. McEuen, P.L. Thermal transport measurements of individual multiwalled nanotubes. *Phys. Rev. Lett.*, **2001**, 87, 215502.

46. Journet, C.; Maser, W.K.; Bernier, P.; Loiseau, A.; Lamy de la Chapelle, M.; Lefrant, A.; Deniard, P.; Lee, R.; Fischer, J.E. Large-scale production of single-walled carbon nanotubes by the electric-arc technique. *Nature* **1997**, 388, 756–758.

47. Rinzler, A.G.; Liu, J.; Dai, H.; Nikolaev, P.; Huffman, C.B.; Todriguez-Macias, F.J.; Boul, P.J.; Lu, A.H.; Heymann, D.; Colbert, D.T.; Lee, R.S.; Fischer, J.E.; Rao, A.M.; Eklund, P.C.; Smalley, R.E. Large-scale purification of single-wall carbon nanotubes: process, product, and characterization. *Appl. Phys. A* **1998**, 67, 29–37.

48. Nikolaev, P.; Bronikowski, M.J.; Bradley, R.K.; Fohmund, F.; Colbert, D.T.; Smith, K.A.; Smalley, R.E. Gas-phase catalytic growth of single-walled carbon nanotubes from carbon monoxide. *Chem. Phys. Lett.* **1999**, 313, 91–97.

49. Hata, K.; Futaba, D.N.; Mizuno, K.; Namai, T.; Yumura, M.; Iijima, S. Water-assisted highly efficient synthesis of impurity-free single-walled carbon nanotubes. *Science* **2004**, 306, 1362–1364.

50. Height, M.J.; Howard, J.B.; Tester, J.W.; Vander Sande, J.B. Flame synthesis of single-walled carbon nanotubes. *Carbon* **2004**, 42, 2295–2307.

51. Furtado, C.A.; Kim, U.J.; Gutierrez, H.R.; Pan, L.; Dickey, E.C.; Eklund, P.C. Debundling and dissolution of single-walled carbon nanotubes in amide solvents. *J. Am. Chem. Soc.*, **2004**, 126, 6095–6105.

52. Chiang, I.W.; Brinson, B.E.; Huang, A.Y.; Willis, P.A.; Bronikowski, M.J.; Margrave, J.L.; Smalley, R.E.; Hauge, R.H. Purification and characterization of single-wall carbon nanotubes (SWNTs) obtained from the gas-phase decomposition of CO (HiPco process). *J. Phys. Chem. B* **2001**, 105, 8297–8301.

53. Andrews, R.; Jacques, D.; Qian, D.; Dickey, E.C. Purification and structural annealing of multiwalled carbon nanotubes at graphitization temperatures. *Carbon* **2001**, 39, 1681–1687.

54. Monthioux, M.; Smith, B.W.; Burteax, B.; Claye, A.; Fischer, J.E.; Luzzi, D.E. Sensitivity of single-wall carbon nanotubes to chemical processing: an electron microscopy investigation. *Carbon* **2001**, 39, 1251–1272.

55. Zhang, M.; Yudasaka, M.; Koshio, A.; Iijima, S. Thermogravimetric analysis of single-wall carbon nanotubes ultrasonicated in monochlorobenzene. *Chem. Phys. Lett.* **2002**, 364, 420–426.

56. Jang, J.; Bae, J.; Yoon, S.H. A study on the effect of surface treatment of carbon nanotubes for liquid crystalline epoxide–carbon nanotube composites. *J. Mater. Chem.* **2003**, 13, 676–681.

57. Ziegler, K.J.; Gu, Z.; Peng, H.; Flor, E.L.; Hauge, R.H.; Smalley, R.E. Controlled oxidative cutting of single-walled carbon nanotubes. *J. Am. Chem. Soc.* **2005**, 127, 1541–1547.

58. Kashiwagi, T.; Du, F.; Winey, K.I.; Groth, K.M.; Shields, J.R.; Bellayer, S.P.; Kim, H.; Douglas, J.F. Flammability properties of polymer nanocomposites with single-walled carbon nanotubes: effects of nanotube dispersion and concentration. *Polymer* **2005**, 46, 471–481.

59. Tchmutin, I.A.; Ponomarenko, A.T.; Krinichnaya, E.P.; Kozub, G.I.; Efimov, O.N. Electrical properties of composites based on conjugated polymers and conductive fillers. *Carbon* **2003**, 41, 1391–1395.

60. Du, F.; Fischer, J.E.; Winey, K.I. Coagulation method for preparing single-walled carbon nanotube/poly(methyl methacrylate) composites and their, modulus, electrical, conductivity, and thermal stability. *J. Polym. Sci. B Polym. Phys.* **2003**, 41, 3333–3338.

61. Ajayan, P.M.; Schadler, L.S.; Giannaris, C.; Rubio, A. Single-walled carbon nanotube–polymer composites: strength and weakness. *Adv. Mater.* **2000**, 12, 750–753.

62. Chang, T.E.; Jensen, L.R.; Kisliuk, A.; Pipes, R.B.; Pyrz, R.; Sokolov, A.P. Microscopic mechanism of reinforcement in single-wall carbon nanotube/polypropylene nanocomposite. *Polymer* **2004**, 46, 439–444.

63. Putz, K.W.; Mitchell, C.A.; Krishnamoorti, R.; Green, P.F. Elastic modulus of single-walled carbon nanotube/poly(methyl methacrylate) nanocomposites. *J. Polym. Sci. B Polym. Phys.* **2004**, 42, 2286–2293.

64. Yang, S.; Castilleja, J.R.; Barrera, E.V.; Lozano, K. Thermal analysis of an acrylonitrile–butadiene–styrene/SWNT composite. *Polym. Degrad. Stab.* **2004**, 83, 383–388.

65. Miyagawa, H.; Drzal, L.T. Thermo-physical and impact properties of epoxy nanocomposites reinforced by single-wall carbon nanotubes. *Polymer* **2004**, 45, 5163–5170.

66. Beyer, G. Filled blend of carbon nanotubes and organoclays with improved char as a new flame retardant system for polymers and cable applications. *Fire Mater.*, **2005**, 29, 61–69.

67. Zhu, J.; Start, P.; Mauritz, K.A.; Wilkie, C.A. Thermal stability and flame retardancy of poly(methyl methacrylate)–clay nanocomposites. *Polym. Degrad. Stab.* **2002**, 77, 253–258.

68. Stephan, C.; Nguyen, T.P.; Lahr, B.; Blau, W.; Lefrant, S.; Chauvet, O. Raman spectroscopy and conductivity measurements on polymer–multiwalled carbon nanotubes composites. *J. Mater. Res.* **2002**, 17, 396–400.

69. Kilbride, B.E.; Coleman, J.N.; Fraysse, J.; Fournet, P.; Cadek, M.; Drury, A.; Hutzler, S.; Roth, S.; Blau, W.J. Experimental observation of scaling laws for alternating current and direct current conductivity in polymer–carbon nanotube composite thin films. *J. Appl. Phys.* **2002**, 92, 4024–4030.

70. Barrau, S.; Demont, P.; Peigney, A.; Laurent, C.; Lacabanne, C Dc and ac conductivity of carbon nanotubes–polyepoxy composites. *Macromolecules* **2003**, 36, 5187–5194.

71. Hsu, W.K.; Koteva, V.; Watts, P.C.P.; Chen, G.Z. Circuit elements in carbon nanotube–polymer composites. *Carbon* **2004**, 42, 1707–1712.

72. Ruan, S.L.; Gao, P.; Yang, X.G.; Yu, T.X. Toughening high performance ultrahigh molecular weight polyethylene using multiwalled carbon nanotubes. *Polymer* **2003**, 44, 5643–5654.

73. Breton, Y.; Desarmot, G.; Salvetat, J.P.; Delpeux, S.; Sinturel, C.; Beguin, F.; Bonnamy, S. Mechanical properties of multiwall carbon nanotubes/epoxy composites: influence of network morphology. *Carbon* **2004**, 42, 1027–1030.

74. Meincke, O.; Kaempfer, D.; Weickmann, H.; Friedrich, C.; Vathauer, M.; Warth, H. Mechanical properties and electrical conductivity of carbon-nanotube filled polyamide-6 and its blends with acrylonitrile/butadiene/styrene. *Polymer* **2004**, 45, 739–748.

75. Liu, T.; Phang, I.Y.; Shen, L.; Chow, S.Y.; Zhang, W.D. Morphology and mechanical properties of multiwalled carbon nanotubes reinforced nylon-6 composites. *Macromolecules* **2004**, 37, 7214–7222.

76. Kashiwagi, T.; Grulke, E.; Hilding, J.; Harris, R.H.; Awad, W.H.; Douglas, J. Thermal degradation and flammability properties of poly(propylene)/carbon nanotube composites. *Macromol. Rapid Commun.* **2002**, 23, 761–765.

77. Beyer, G. Short communication: carbon nanotubes as flame retardants for polymers. *Fire Mater.* **2002**, 26, 291–293.

78. Kashiwagi, T.; Grulke, E.; Hilding, J.; Groth, K.; Harris, R.H.; Butler, K.; Shields, J.; Kharchenko, S.; Douglas, J. Thermal and flammability properties of polypropylene/carbon nanotube nanocomposites. *Polymer* **2004**, 45, 4227–4239.

79. Peeterbroeck, S.; Alexandre, M.; Nagy, J.B.; Pirlot, C.; Fonseca, A.; Morea, N.; Philippin, G.; Delhalle, J.; Mekhalif, Z.; Sporken, R.; Beyer, G.; Dubois, P. Polymer-layered silicate–carbon nanotube nanocomposites: unique nanofiller synergistic effect. *Compos. Sci. Technol.*, **2004**, 64, 2317–2323.

80. Schartel, B.; Pötschke, P.; Knoll, U.; Abdel-Goad, M. Fire behaviour of polyamide 6/multiwall carbon nanotube nanocomposites. *Eur. Polym. J.* **2005**, 41, 1061–1070.

81. Gao, F.; Beyer, G.; Yuan, Q. A mechanistic study of fire retardancy of carbon nanotube/ethylene vinyl acetate copolymers and their clay composites. *Polym. Degrad. Stab.* **2005**, 89, 559–564.

82. Watts, P.C.P.; Fearon, P.K.; Hsu, W.K.; Billingham, N.C.; Kroto, H.W.; Walton, D.R.M. Carbon nanotubes as polymer antioxidants. *J. Mater. Chem.* **2003**, 13, 491–495.

83. Andrews, R.; Jacques, D.; Rao, A.M.; Derbyshire, F.; Qian, D.; Fan, X.; Dickey, E.C.; Chen, J. Continuous production of aligned carbon nanotubes: a step closer to commercial realization. *Chem. Phys. Lett.* **1999**, 303, 467–474.

84. Bom, D.; Andrews, R.; Jacques, D.; Anthony, J.; Chen, B.; Meier, M.S.; Selegue, J.P. Thermogravimetric analysis of the oxidation of multiwalled carbon nanotubes: evidence for the role of defect sites in carbon nanotube chemistry. *Nano Lett.* **2002**, 2(6), 615–619.

85. Lozano, K.; Bonilla-Rios, J.; Barrera, E.V. A study on nanofiber-reinforced thermoplastic composites, II: Investigation of the mixing rheology and conduction properties. *J. Appl. Polym. Sci.* **2001**, 80, 1162–1172.

86. Zeng, J.; Saltysiak, B.; Johnson, W.S.; Schiraldi, D.A.; Kumar, S. Processing and properties of poly(methyl methacrylate)/carbon nano fiber composites. *Composites B*, **2004**, 35, 173–178.

87. Lozano, K.; Yang, S.; Zeng, Q. Rheological analysis of vapor-grown carbon nanofiber-reinforced polyethylene composites. *J. Appl. Polym. Sci.* **2004**, 93, 155–162.

88. Gauthier, C.; Chazeau, L.; Prasse, T.; Cavaille, J.Y. Reinforcement effects of vapour grown carbon nanofibres as fillers in rubbery matrices. *Compos. Sci. Technol.*, **2005**, 65, 335–343.

89. Xu, Y.J.; Higgins, B.; Brittain, J. Bottom-up synthesis of PS-CNF nanocomposites. *Polymer* **2005**, 46, 799–810.

90. Kashiwagi, T.; Du, F.; Douglas, J.F.; Winey, K.I.; Harris, R.H.; Shields, J.R. Nanoparticles networks reduce the flammability of polymer, nanocomposites. *Nat. Mater.*, **2005**, 928–933.

91. Quintiere, J.Q. Surface flame spread, in: *SFPE Handbook of Fire Protection, Engineering*, 3rd ed. Society of Fire Protection Engineers, Bethesda, MD, 2002, Chap. 2–12.

92. Kashiwagi, T. Flammability of nanocomposites: effects of the shape of nanoparticles, in: M. Le Bras, C.A. Wilkie, S. Bourbigot, S. Duquesne, and C. Jama, Eds., *Fire Retardancy of Polymers*. Royal Society of Chemistry, London, 2005, pp. 81–99.

11

POTENTIAL APPLICATIONS OF NANOCOMPOSITES FOR FLAME RETARDANCY

A. Richard Horrocks and Baljinder K. Kandola

Fire Materials Laboratory, Centre for Materials Research and Innovation, University of Bolton, Bolton, UK

11.1 INTRODUCTION

Demonstration of the beneficial effects of including dispersed nanoclay particles as reinforcing species originated with a disclosure by the Toyota research group in the 1980s that increases in tensile strength and modulus were possible.[1,2] In addition to superior mechanical properties, these structures have also been shown to exhibit improved barrier behavior, ablation performance, and thermal and fire stability.[3,4] However, this improved fire performance is usually concerned with reduction in heat release properties only, and ease of ignition and times to self-extinguish are usually affected adversely. In fact, use of simple techniques such as limiting oxygen index show that the addition of nanoclays and other nanoparticles such as fumed silica alone do not significantly increase limiting oxygen index (LOI) values[5,6] unless their presence modifies the burning behavior and melt-dripping character of a polymer as observed with various montmorillonite clays dispersed in polyamide-6 and polyamide-66 polymers.[7] Consequently, as shown in earlier chapters, the flame retardant potential of nanodispersed particles in polymeric matrices lies in their ability to function additively and even synergistically with other flame retardants or the matrix itself, if inherently flame retardant. In this way, nanodispersed flame retardant formulations may

Flame Retardant Polymer Nanocomposites, edited by Alexander B. Morgan and Charles A. Wilkie

be more effective than conventional formulations or as effective but at significantly reduced overall additive concentrations. This is particularly important when conventional flame retardant concentrations may be required in concentrations as high as 60% w/w, yet as Chapters 7 and 8 have shown, and discussed further in this one, the presence of between 1 and 5% nanoparticles may enable significant reductions to be achieved. This has real consequences for fire retardant polymer applications in terms of possible reduced additive costs, improved physical and mechanical properties, and environmental sustainability as well as improved overall fire performance. Such advantages will obviously have possible applications across the entire spectrum of polymer product types but will have special relevance to those applications where flame retardant minimization and polymer physical and mechanical property optimization are of crucial importance. Prime examples will be fibers (and textiles), films, foams, and composites where not only are high specific surfaces often evident but product physical and mechanical properties are also of major importance to their success in a variety of applications.

11.2 REQUIREMENTS FOR NANOCOMPOSITE SYSTEM APPLICATIONS

Since it is evident from discussions in previous chapters that nanodispersed, functionalized, largely inert particles such as clays and synthetic alternatives cannot promote sufficient flame retardant activity alone but only in the presence of more conventional flame retardants, their potential usefulness will be determined by their ease of processing and the manner in which they influence both process and end product. Essential issues to be considered and resolved are nanoparticle compatibility with the polymer matrix and other additives present, the ability to maintain a nanodispersion during all processing stages, their influence on rheology, and the possible compromise between effective concentrations levels and optimization of these.

Currently, there is a number of methods of major significance for the preparation of nanocomposites in polymeric matrices, including sol-gel formation,[8] in situ polymerization,[9] intercalative polymerization,[1] solution blending, and melt intercalation.[2] In all of these, the process-dependent factors above will be crucial in enabling successful small-scale systems to be scaled up to full commercial levels. Not least of these, compatibility is largely determined by the nature of the functionalizing groups present, and so in the case of clays, for example, hydrophobic long-chain aliphatic substitutents within the quaternized functionalizing complex will encourage nanoclay intercalation and exfoliation within a nonpolar polymer matrix typified by the polyolefins and polystyrene. Substituents with variously polar side groups, such as –OH, –NH–, or NH_2, will encourage nanodispersion in polar [e.g., poly(methyl methacrylate)] and hydrogen-bonded polymers [e.g., polyamide-6(PA6) and polyamide-6,6(PA6,6) and poly(vinyl alcohol)]. Typical examples of functionalizing groups used by Southern Clay Products

TABLE 11.1 Typical Characteristics of Commercial (Southern Clay Products) Clays

Clay	Treatment[a]	Modifier Concentration (meq/100 g clay)	d-Spacing (Å)	Density (g/cm^3)	Compatible Polymer Examples[b]
Cloisite Na^+	None	93	11.7	2.86	PVOH, PA6, and PA6,6
Cloisite 10A	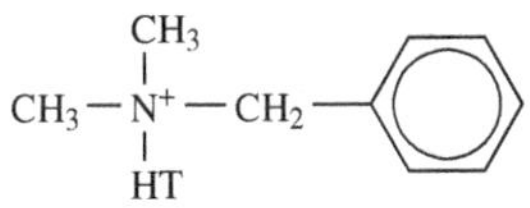	125	19.2	1.90	PET, PBT, PS
Cloisite 15A	CH_3 CH_3-N^+-HT HT	125	31.5	1.66	PLA, EVA, PS
Cloisite 25A	CH_3 $CH_3-N^+-CH_2{\cdot}CH\,(C_2H_5){\cdot}(CH_2)_3{\cdot}CH_3$ HT	95	18.6	1.87	PLA, PMMA, PS
Cloisite 30B	CH_2CH_2OH CH_3-N^+-T CH_2CH_2OH	90	18.5	1.98	EVA, epoxy, PC, PBT

[a]HT, hydrogenated (~65% C_{18}; ~30% C_{16}; ~5% C_{14}), anion: sulfate; T, tallow (~65% C_{18}; ~30% C_{16}; ~5% C_{14}), anion: chloride in 10A, 15A, and 30A, sulfate in 25A.

[b]PVOH, poly(vinyl alcohol); PBT, poly(butylene terephthalate); PLA, poly(lactic acid); EVA, poly(ethylene-co-vinyl acetate); PMMA, poly(methyl methacrylate); PC, polycarbonate.

are shown in Table 11.1 along with polymers with which they are respectively compatible. The respective densities listed show how the presence of functionalizing species, especially those with bulky side groups, reduce clay density by increasing gallery spacings. Similar ranges of functionalized clays are commercially available from companies such as Süd-Chemie in Germany and Nanocor in the United States. Surprisingly, even unfunctionalized clays may nanodisperse in very polar polymers, and recent work in our own laboratories suggests that this may be possible with polyamide-6 and polyamide-6,6.[7,10]

However, as noted by Gilman and co-workers,[11,12] quaternized ammonium salts with aliphatic side chains tend to decompose at temperatures in the range 200 to 250°C and so will degrade during the compounding and processing of most conventional melt-processed polymers such as PA6, PA6,6, poly(ethylene terephthalate) (PET) and polystyrene (PS). From our own recent studies, Figure 11.1*a* and *b* show the respective thermogravimetric analysis (TGA) and differential thermal analysis (DTA) responses in air for the unfunctionalized Cloisite Na^+ and functionalized Cloisite 10A, 15A, 25A, and 30B clays having functionalizing group structures defined in Table 11.1. As expected, the unfunctionalized sodium montmorillonite shows very little weight loss until above 600°C, when some dehydroxylation occurs, as shown in Figure 11.1*a*. However, there is still a

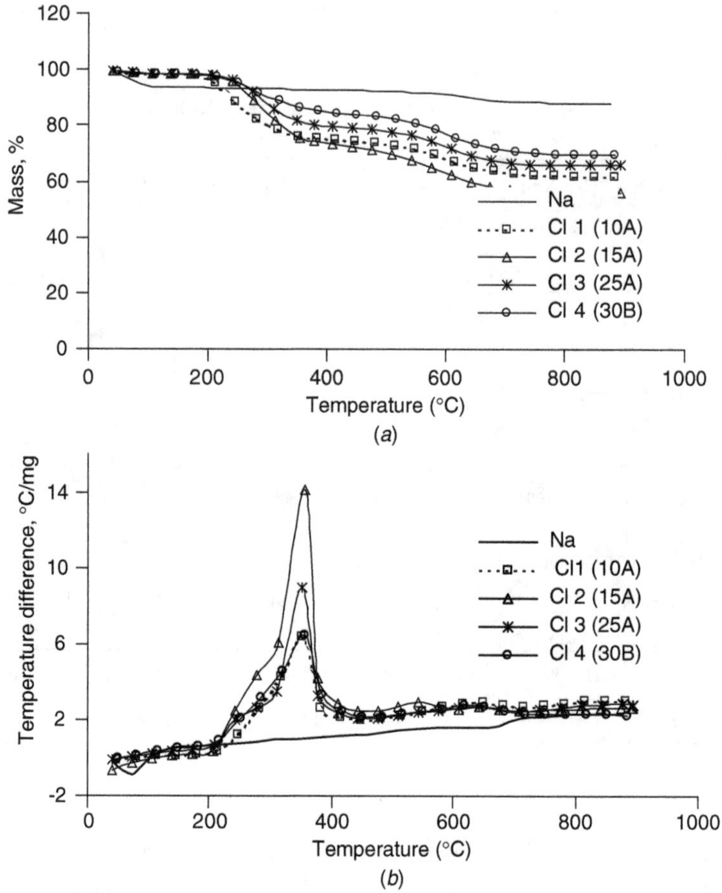

FIGURE 11.1 (*a*) TGA and (*b*) DTA responses for selected Cloisite clays in air. (From Ref. 13, with permission from the Royal Society of Chemistry.)

high residual mass at 800°C, and the related DTA response is featureless (see Figure 11.2*b*). All the organically modified clays, however, show two stages of weight loss, the first represented by double-peaked DTA maxima (in the temperature range 235 to 293 and 307 to 348°C) and the second by single-peaked DTG maxima (575 to 605°C) reflected by the respective DTA exotherms in Figure 11.1*b*. The first stages are probably due to decomposition and oxidation in air of respective organic components of the clays, and the second, single stage is due to dehydroxylation of the clay layers as noted above.[14] Although the TGA data shown are in air, and the behavior under nitrogen might be expected to indicate some improved thermal stability, it is clear that functionalizing groups of these clays would most likely be degraded during normal melt polymer processing, as suggested previously.

That this is indeed the case has prompted the need to generate more stable functionalities. As a consequence, Gilman and his co-workers[11,12] subsequently showed that layered-silicate nanoparticles functionalized with higher-temperature stable groups, such as imidazolium derivatives and crown ethers, can increase stability to temperatures in the range 262 to 343°C under nitrogen compared with a typical alkylammonium–based salt such as dimethyldioctadecylammonium bromide, which starts to degrade at 225°C. Figure 11.2 reproduces the TGA responses in nitrogen of dimethylhexadecylimidazolium (DMHDIM) salts with different anions (Cl^-, Br^-, BF_4^-, PF_6^-) and montmorillonite (MMT) clay ion-exchanged with dimethylhexadecylimidazolium (DMHDIM) salt.[12] Here a dramatic increase in the thermal stability of dimethylhexadecylimidazolium–intercalated montmorillonite clays is evident, as is a dependence of salt stability on the type of anion present. Halide ions obviously destabilize salts, so it is important to remove all halide residue that may contaminate the intercalated product after ion exchange. On the other hand, thermal stabilities of the intercalated tetrafluoroborate and hexafluorophosphate salts are similar. However, for polymers that may be compounded and processed below 200°C, such as EVA,[15] the more simple quaternary ammonium functionalizing groups may be used (e.g., using dimethyldistearylammonium salts) to generate nanocomposite polymers.

Generally, the addition of a nanodispersed phase will increase the viscosity of a polymer melt under a given shear stress and temperature, although

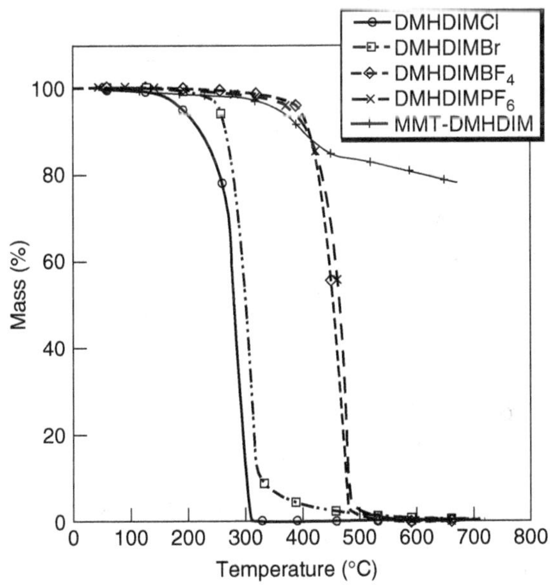

FIGURE 11.2 TGA under nitrogen of dimethylhexadecylimidazolium (DMHDIM) salts with different anions (Cl^-, Br^-, BF_4^-, PF_6^-) and montmorillonite (MMT) clay treated with dimethylhexadecylimidazolium (DMHDIM) salt [12] (From Ref. 12, copyright © 2004, Elsevier, with permission.)

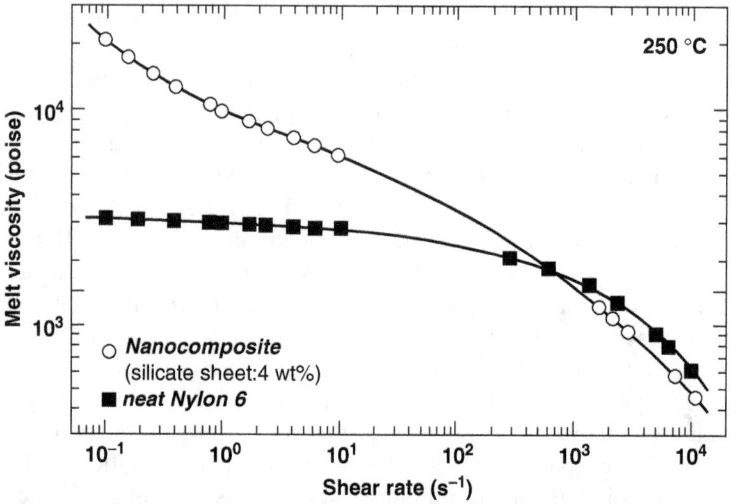

FIGURE 11.3 Shear rate dependence of melt viscosity of polyamide-6 in the presence and absence of a nanoclay. (From Ref. 16, copyright © 2000, John Wiley & Sons Ltd., with permission.)

shear stress sensitivity may be increased as noted for polyamide-6 and shown in Figure 11.3.[16] Similar behavior has been reported by Sihna Ray and Okamoto[17] for molten polylactide (PLA)/layered-silicate nanocomposites at 175°C. These changes have implications on processability rates for high-throughput processes such as melt extrusion of filaments; thus, if nanodispersed clays and similar particles are present, there may have to be an upper limiting concentration determined by the need to compromise between added property and reduced extrusion rate. Furthermore, the increase in melt viscosity will reduce the ease of melt blending, although the increased sensitivity of melt viscosity to increased shear stresses at higher extrusion rates may offset this factor as well as partially restoring process efficiency.[18] However, while the presence of nanoclays has a physical effect, in some polymers [e.g., poly(ethylene terephthalate)], sensitized degradation may occur. Matayabas et al.[19] have shown that during melt compounding the inherent viscosity of a high-molecular-weight PET decreased from 0.98 dL/g to 0.48 dL/g as clay increased from 0.36% to 6.7% (expressed as ash %). Davies et al.[20] have also noted that for in situ polymerized polyamide-6 in the presence of a montmorillonite clay, significant thermal degradation occurs when the compounded melt is subjected to injection molding processes. Clearly, the effects of adding nanoparticulates to a polymer melt or indeed a solution are quite complex and so will have an impact on the processability of nanocomposites designed to have desired improvements in property, including fire performance.

Dispersion at the nano level, is of course, essential if fire performance properties are to be optimized, so the challenge of introducing functionalized clays in a suitable manner to a particular polymer process may be considerable. For instance, the products Nanomer I.30P for film grade and I.44 PA for engineering-grade applications are Nanocor's commercial products and are quaternized ammonium ion modified montmorillonite, designed for maximum compatibility and dispersion in a polyolefin matrix. They are available as free-flowing powders with mean dry particle sizes of 15 to 25 μm or as master batches containing 40 and 50% w/w clay contents, and they are capable of dispersion to the nanoscale in conventional twin-screw compounders.

Rheological effects have also been observed in our own laboratories during the formulation of aqueous copolymeric emulsions for use in textile back-coating formulations.[21] In this work the addition of either a nanoclay [5% w/w Cloisite 15A (Southern Clay Products; see Table 11.1) with respect to coating solids] or fumed silica (up to a maximum of 17% w/w with respect to coating solids) modifies the paste rheology, with the latter especially producing significant viscosity changes and hence difficulty of maintaining uniform and reproducible coating applications. High concentrations of silica were used because of the need to introduce assumed high levels of flame retardancy, which in fact proved to be false (see Section 11.3.3). However, this does illustrate one adverse effect of nanoparticle concentration, as has the work of Matayabas and Turner[18] on PET degradation.

11.3 POTENTIAL APPLICATION AREAS

11.3.1 Bulk Polymeric Components

Although the patent literature is populated with claims of the effectiveness of nanofilled polymeric formulations offering improved fire performance, at present few commercial products are available. The best known example uses nanocomposite technology specifically to improve the fire performance of electrical cable sheathings. Kabelwerk Eupen AG now markets cables that incorporate nanoclays within ethylene-co-vinyl acetate (EVA)–based sheathings, which have the advantage of requiring reduced levels of conventional flame retardants, such as alumina trihydrate.[15,22] To achieve normal flame retardancy requirements, up to 65% w/w alumina trihydrate (ATH) is required, which means that the overall cable sheathing physical properties may be reduced significantly. By incorporation of up to 5% of a functionalized nanoclay (with dimethyldistearylammonium salts), the ATH content may be reduced to about 45%, with a consequent improvement in sheath properties and saving on flame retardant costs. Furthermore, the char formed by the nanofilled ATH-containing formulation is far more rigid and less prone to crack development than is the conventional analog. Later work by Beyer[23] has shown that introduction of multiwalled carbon nanotubes on a weight-for-weight basis (at 5% w/w) into EVA reduces cone calorimetrically determined peak heat release values to slightly greater levels than when a functionalized nanoclay is present. Extending the inclusion of nanoclays into other polymers with a potential for use in cables, such as thermoplastic polyurethanes (TPUs), poly (vinyl chloride) (PVC), and blends of the two, shows mixed results, however. For instance, while inclusion of 5-phr organoclay with TPU and TPU containing a phosphate ester flame retardant again showed reductions in respective peak heat release rate (PHRR) values, in each case the time-to-ignition values decreased and burnout times increased.[24] However, whereas the time to ignition of TPU under a 35-kW/m^2 heat flux increased from about 70 s to 85 s in the presence of phosphate flame retardant, it decreased to about 70 s following subsequent addition of organoclay. On the other hand, PVC–EVA and PVC–TPU polymer blends showed heat release curves little affected by the introduction of organoclays, apart from decreases in times to ignition. Clearly, whether or not a useful level of flame retardancy is achieved depends on the polymer type, the conventional flame retardant present, and the type of nanospecies present.

A second example is the group of nanocomposites for improved flame retardancy based on Nanocor products (Gitto/Global Corp., Lunenburg, Massachusetts).[25] One of these is for heavy-duty electrical enclosure applications. These are typically made of injection-molded polypropylene and vary in size up to 1 m^3. By using a nanocomposite, the level of the flame retardant additive present may be reduced significantly, yielding an overall weight saving of 18%. Although their original UL-94 V-0 ratings are maintained, both flexural and tensile moduli are also claimed to increase by about 25% without loss of impact resistance.

These commercial examples provide evidence that inclusion of nanoparticle species can indeed reduce the overall flame retardant additive levels required to achieve a desired fire performance level. To corroborate this, work in our laboratories[10,26–28] with PA6 and PA6,6 films and selected phosphorus-containing flame retardants suggests this to be the case when selected functionalized nanoclays are also present, and this work is reviewed below.

The situation with other polymers, such as polystyrene, is less clear. Wang et al.[29] have recently shown that introducing nanoclays and brominated species into polystyrene (PS) produces significant reductions in PHRR values determined by cone calorimetry. Although times to ignition are slightly reduced, thereby making the compounded nanocomposites more prone to ignite, their behavior in a practical burning test such as UL-94 is less straightforward. For example, the introduction of 3% Cloisite 30B (Southern Clay Products; see Table 11.1) enables a previously failing compounded copolymeric PS containing 10% dibromostyrene (DBS) to achieve a V-2 pass rating, raising the DBS level to 20% alone yields a pass of V-2, and introducing Cloisite 30B or a quaternized dimethyl-*n*-hexadecyl-4-vinylbenzylammonium (VB16)–functionalized clay produces a fail or "not classified (NC)" result. These results are listed in Table 11.2. However, for copolymers containing 40% DBS, which alone maintains a V-2 rating, addition of Cloisite 30B raises this to the highest pass rating, V-0. Earlier work by the same group[30] investigated the effects of adding a variety of organophosphate additives and the clay Cloisite 10A (see Table 11.1) to polystyrene and studied thermal degradative, burning, and cone calorimetric properties. Table 11.2 also lists some of their results, which show that only when phosphate flame retardants are present at higher concentrations are respective PHRR values reduced

TABLE 11.2 UL-94 Test Results for Polystyrene Nanocomposites Containing Either Dibromostyrene Comonomer or Phosphate Additives

Flame Retardant Comonomer or Additive	Clay	PHRR at 35 kW/m^2	UL-94 Result
Pure polystyrene	Nil	1419	NC
	3% Cloisite 10A	310	—
10% Dibromostyrene	3% Cloisite 30B	—	V-2
20% Dibromostyrene	Nil	—	V-2
	3% Cloisite 30B	—	NC
	3% VB16	—	NC
40% Dibromostyrene	Nil	—	V-2
	3% Cloisite 30B	—	V-0
15% Tricresyl phosphate	Nil	1122	—
30% Tricresyl phosphate	3% Cloisite 10A	378	V-2
	5% Cloisite 10A	342	V-1
	10% Cloisite 10A	324	V-1/V-0?
30% Resorcinol diphosphate	Nil	499	—
	5% Cloisite 10A	110	V-2
	10% Cloisite 10A	307	V-0/V-1?
30% Trixylyl phosphate	Nil	864	—
	5% Cloisite 10A	313	V-2

Source: Data from Refs. 29 and 30.

significantly, although the reduction is not as great as that caused by the presence of 3% nanoclay alone with respect to virgin PS. Furthermore, the presence of both a clay and a phosphate causes not only a reduction in PHRR values but also a very significant reduction in total heat released compared to both virgin polymer and a styrene nanocomposite. In practical UL-94 tests, selected phosphate-containing samples that showed the best flame retardant effect in terms of their respective ability not to sustain a flame after ignition are also listed in Table 11.2. Here it can be seen clearly that not only does the presence of nanoclay together with a retardant raise the fire performance in terms of the UL-94 V-rating, but also, increasing the concentration from 3% to 10% promotes a similar increase.

From these studies, although it is clear that nanoclay-flame retardant interactions are not simple and may be dependent on respective concentrations present, inclusion of nanodispered phases offers the opportunity to reduce overall additive loadings or to improve fire performance at currently acceptable loadings.

In Chapter 7 the combination of nanocomposites with metal hydroxide flame retardants has generally been discussed. Since the use of metal hydroxide usually requires very high concentrations within the polymer matrix (often higher than 50% w/w), to achieve desired levels of flame retardancy as noted above regarding the work of Beyer,[15,22,23] the influence on rheology and hence processability can be significant. Hornsby and Rothon[31] have discussed this issue and they report that compounded polymer melt viscosities and shear sensitivities, for example, are determined by hydroxide type, particle size, surface chemistry, and concentration. Any means of reducing the microparticulate flame retardant content will have a beneficial effect on both processibility and final polymer properties. It is in these areas that the inclusion of nanodispersed particles may provide application opportunities. Lomakin et al.[32] have recently described the effects of introducing nanoparticulate aluminosilicates (i.e., Cloisite 15A, Southern Clay Products, San Antonio, Texas) into magnesium hydroxide (MgH)–filled polypropylene and show that a 60% w/w MgH-filled polymer has thermal stability similar to that of a polymer containing 50% MgH and 10% w/w Cloisite 15A; no effects of improved processibility were reported. However, work reported by Song et al.[33] reports that the mechanical properties of PA6,6 are improved if a nanoclay is added in the presence of MgH and red phosphorus as flame retardants. Furthermore, the two flame retardants and nanoclay act synergistically, thereby enabling lower concentrations to be used. A similar flame retardant synergism was reported by Fu and Qu,[34] who noted that addition of fumed silica to MgH-filled EVA not only enabled low levels of MgH to be used but increased elongation-at-break values.

The combination of intumescents and nanoparticles (also discussed in Chapter 6) was reviewed initially by Gilman and Kashiwagi[35] and more recently by Duquesne et al.[36] The possible synergies between micro-dispersed intumescent flame retardants and nano-dispersed species offer obvious opportunities to reduce overall filler contents, with beneficial effects on polymer processing and properties, although no such examples worthy of application have been reported to date. One of the first reported instances of this was by Bourbigot et al.,[37] who showed that inclusion of a polyamide-6/montmorillonite nanocomposite within an EVA matrix with APP as a conventional flame retardant still enabled a UL-94 test rating of V-0 to be achieved with only two-thirds of the concentration of the latter with respect to an equivalent intumescent formulation without nanoclay present. Subsequently, Vyver-Berg and Chapman[38] reported that combining functionalized clays at 1 to 3% w/w in the presence of intumescent flame retardants such as melamine phosphate, ammonium polyphosphate, pentaerythritol phosphate, and zinc borate in appropriate combinations enables lower concentrations than normal to be used to achieve UL-94 V-0 ratings in polypropylene at total concentrations just below 20% w/w.

11.3.2 Films, Fibers, and Textiles

The need to minimize flame retardant concentrations is especially important in synthetic fibers, where additive levels in excess of 10% w/w usually reduce their ease of extrusion and subsequent processing as well as affecting adversely their normally desirable textile properties. The major difference between fibers and bulk polymers, including films and composites, is the small thickness of individual fibers, typically being 15 to 30 μm in diameter, yielding yarns 50 to 100 μm in diameter, and fabrics having thicknesses varying from as low as 100 μm to several millimeters. As noted previously, although fire performance, often using cone calorimetry, of bulk polymers[11,35,39] typically shows that the presence of nanoclays reduces peak heat release rates, they also most often reduce times to ignition and extend total burning periods while little affecting the overall heat release of the polymeric substrate. However, while slowing down the burning process and encouraging more rapid ignition, they also encourage increased char formation. In fact, in some cases where polymers are not char-formers, some char development has been reported,[35,39] and this is of special importance to extremely thermoplastic and negligible char-forming fiber-forming polymers such as PET and polypropylene.

The first fire performance studies of nanocomposite polyamide-6 filaments reported were by Bourbigot et al.,[40,41] and these were converted into fabric having an area density of 1020 g/m^2 and thickness 2.5 mm. These fabrics were exposed to 35 kW/m^2 heat flux in a cone calorimeter and ignition times of 70 and 20 s and PHRR values of 375 and 250 kW/m^2, respectively, for the normal and nanocomposite polyamide-6 fabrics were recorded. Although the latter represents a significant 33.3% reduction in PHRR, ignition resistance was reduced significantly and total heat release was affected little, if any. Thus, flame retardancy in terms of increased ignition resistance was not observed; in fact, the converse held true. However, thermogravimetric analysis suggested that the presence of nanoclay had little effect up to 400°C, but above 450°C there appeared higher char formation. One problem with fibers and fabrics with respect to bulk polymers is their high-specific-surface areas and their thermally thin character, and it is interesting to note that recent work by Kashiwagi et al.[42] suggested that the effectiveness of polyamide-6 nanocomposites in having reduced PHRR values and related fire performance may be a function of sample or composite thickness. Results tabulated from this research for 8- and 1.6-mm polyamide-6/clay nanocomposites exposed in a cone calorimetry at 50 kW/m^2 are shown in Table 11.3, and the effect reported for nanoclay presence may be expressed as a percentage retention of PHRR; these are included in parentheses.

The poor performance of the thinner composite, which is corroborated by data for mass loss rate for composites having intermediate thicknesses (3.2 and 4 mm), may be explained in terms of competition between the formation of a

TABLE 11.3 Tabulated and Calculated PHRR Results for Polyamide-6/Clay Nanocomposites

Nanoclay	PHRR [kW/m^2 (% retention)] for a Thickness of:		
Level (%)	8 mm	1.6 mm	220 μm[a]
0	1950	1690	*1634*
2	1025 (53)	1615 (95.5)	*1742 (~100)*
5	690 (36)	1360 (80)	*1505 (92)*

Source: Data from Refs. 27 and 42.
[a]Result predicted for 250-g/m^2 fabric.

surface carbonaceous–silica shield and the volatilization to fuel of surrounding polymer. In thicker composites, the competition favors ceramic barrier formation, while for thin composites, volatilization dominates.[43] This can be considered as the difference between thick and thin thermal behavior,[44] so in similarly "thin" textile fabrics it is possible that the "shield-forming" mechanism observed for bulk polymer nanocomposites may be too slow for effective improvement in fire performance. It is likely, however, that the thickness effect observed by Kashiwagi and co-workers will be influenced by the heat flux since both competing mechanisms are thermally driven, but to different extents.

Assuming Kashiwagi's results to be reasonably valid and that a simple negative linear relationship exists between composite thickness and heat release rate, in a recent paper[27] we have shown that these former results for 8- and 1.6-mm polyamide-6/clay nanocomposites exposed in a cone calorimeter at 50 kW/m^2 suggest a minimal reduction in PHRR reduction for a typical polyamide-6 textile fabric having an area density of about 250 g/m^2, which equates to a film with a thickness of 220 μm (see Table 11.3). Furthermore, we have proposed that the 33.3% reduction in PHRR observed by Bourbigot et al.[40,41] for 2.5-mm-thick 1020-g/m^2 polyamide-6 fabrics, equivalent to films about 0.9 mm in thickness, is a consequence of the lower heat flux, 35 kW/m^2. This is suggested because the thickness effect observed by Kashiwagi and co-workers[42] will be influenced by the heat flux since both competing mechanisms are thermally driven but to different extents, and a lower heat flux will favor the diffusion of clay particles to the surface and formation of a surface clay layer.

There is a possibility that selection of char-promoting nanoclay or nanoparticle functionalizing groups may further encourage char formation, but since these are present at low concentrations within the particle substrate and the functionalized nanoparticles themselves are introduced only at 2 to 5% w/w loadings, the effectiveness of char-promoting or even vapor-phase-active functional groups might be questioned when present at very low ($\ll$1%) levels in the polymer. However, since the thermal stability of the functionalizing species during processing significantly affects the resulting nanoclay behavior as discussed previously,[11,12] possible char-promoting or vapor-phase effects at such low levels should not be ruled out.

Work in our own laboratories has shown, however, that in the presence of conventional flame retardants, nanoclays can promote additive and synergistic effects in PA6, PA6,6 films that have been used as models for respective fibers.[7,10,13,26,27] This work has provided evidence that significant reductions in flame retardant additive concentrations may be achievable, as has been noted for other polymers in Section 11.3.1. Normally, minimal flame retardant additive contents of about 15 to 20% w/w are required, which are too high for inclusion in conventional synthetic fibers. This is because for fusible fiber-forming polymers such as PA6, PA6,6, PET, and polypropylene, flame retardant property trends versus concentration are not linear but follow an S-shaped curve.[45–47] This phenomenon is believed to be a consequence of the need to generate a threshold char level having an extended coherence throughout the polymer. It follows that this will relate to a critically minimal concentration of a given flame retardant that must be present in a given polymer in order to achieve acceptable flame retardancy.

We have reported both additive or synergistic effects of adding selected flame retardants based on ammonium polyphosphate (Antiblaze MCM, Rhodia), melamine phosphate (Antiblaze NH, Rhodia), pentaerythritol (PER), pentaerythritol phosphate [Chemtura (formerly, Great Lakes) NH 1197], cyclic phosphate (Antiblaze CU, Rhodia), intumescent mixtures of APP, PER, and melamine (Amgard MPC, Rhodia), ammonia cross-linked polymer of a tetrakis(hydroxylphosphonium salt)–urea condensate (Proban CC polymer, Rhodia), and related formulations into PA6 and PA6,6 polymer films (~80 μm thick) in the presence of nanoclay (polyamides supplied by RTP Plastics).[10,13,26,27] Analysis of the flame retardant performance of each FR–nanoclay–polyamide film compared with respective FR–polyamide films, for polyamide-6,6 films shows that only for APP, Proban CC, and the intumescent Amgard MPC does the presence of nanoclay significantly increase LOI values, and that using Lewin's *synergistic effectivity* measure,[48] these demonstrate possible synergy, as reported elsewhere.[10,27] The effect of the addition of nanoclay is seen clearly in the previously published LOI versus FR concentration plots presented in Figure 11.4.[27] Why only a few of the selected flame retardants show positive effects in the presence of nanoclay is not understood. Suffice it to say that ammonium polyphosphate is not only the most synergistic but also has a decomposition temperature in the range 250 to 300°C, and this overlaps the melting point of polyamide-6,6 (~265°C). It is considered that this will encourage flame retardant mechanisms to start alongside polymer fusion. Surprisingly, however, the APP–PER combination does not show synergy and, indeed, shows

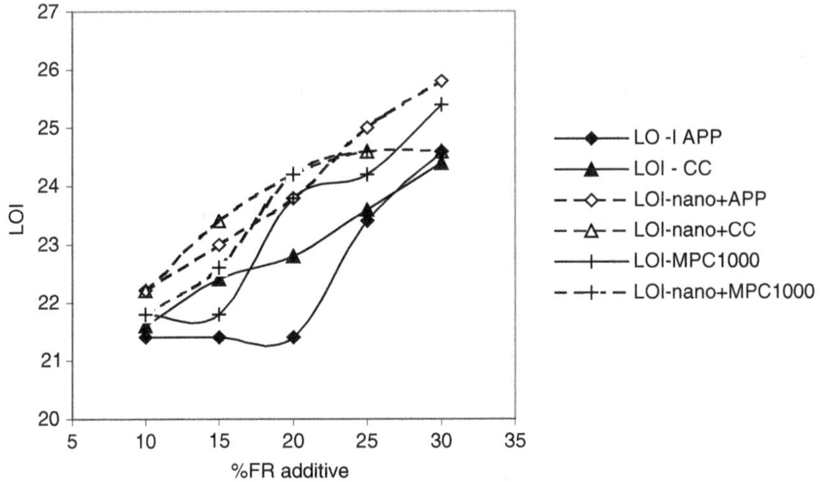

FIGURE 11.4 LOI values for flame retardant/PA6,6 and flame retardant PA6.6 nanocomposite films for flame retardants showing positive behavior in the presence of nanoclay. (From Ref. 27, with permission from Interscience Communications.)

TABLE 11.4 Flame Retardant Concentrations Required to Achieve Defined LOI Values in Polyamide-6,6 Films

	LOI = 23		LOI = 24		LOI = 25	
Flame Retardant	PA6,6	PA6,6 nano	PA6,6	PA6,6 nano	PA6,6	PA6,6 nano
APP	23.8	15	28.5	20.1	33.3[a]	25
MPC	16.3	14.5	20.5	18	30	>30
CC	20.5	10.5	28.5	17.5	36.3[a]	25

Source: After Ref. 26.
[a]Extrapolated from Figure 11.4.

slightly less than additive flame retardant effects when both flame retardant and nanoclay are present. However, the intumescent Amgard MPC does show evidence of synergy. The pentaerythritol phosphate–derived species [Chemtura (formerly, Great Lakes) NH 1197] have higher decomposition temperatures than APP, as does melamine phosphate (Antiblaze NH).[49] The curves in Figure 11.4 are variants of the typical S-shaped curve,[45–47] and from them the reductions in flame retardant concentration to achieve a specific level of flame retardancy may be assessed. These concentrations are shown in Table 11.4.

Thus, to achieve LOI values up to 24, the addition of nanoclay (at a 2% assumed level) enables flame retardant levels to be reduced significantly. At LOI = 25, similar reductions in concentration are observed for APP and Proban CC polymer–containing films, but a converse situation was observed for the Amgard MPC system. Unfortunately, a similar analysis could not be undertaken for the polyamide-6 film set because the presence of nanoclay significantly changed the burning manner of cast polyamide-6 films in a manner that rendered simple LOI value comparison difficult, although evidence of synergism was generated, as indicated above.[7,10]

Possible effects of a clay functionalizing group have also been investigated using clays selected from those in Table 11.1, and these were introduced with selected flame retardants into cast PA6 and PA6,6 films. Clay selection was based on observing the greatest increase in LOI that each clay alone could promote in a cast polyamide-6,6 film in the first instance. The Cloisite 30B clay fit this criterion, yielding a cast film LOI value of 28.0, compared with a clear polyamide,-6,6 film value of 21.0. This clay was selected along with the nonfunctionalized Cloisite Na^+ which was found to disperse quite easily in polar polyamide films, yielding an LOI value of 25.2. The variations in LOI for 2% clay-containing films were noted to be mainly a consequence of the effect that each clay had on the burning character of the film in an LOI test vertical geometry. A similar effect was noted for polyamide-6 films in which previously reported flame retardant–nanoclay interactions for polyamide-6 formulations appeared to be less efficient.[7,10] It was considered that this lower activity could be a consequence of the lower melting point of polyamide-6, about 215°C, a temperature too low for most flame retardants to start functioning efficiently.

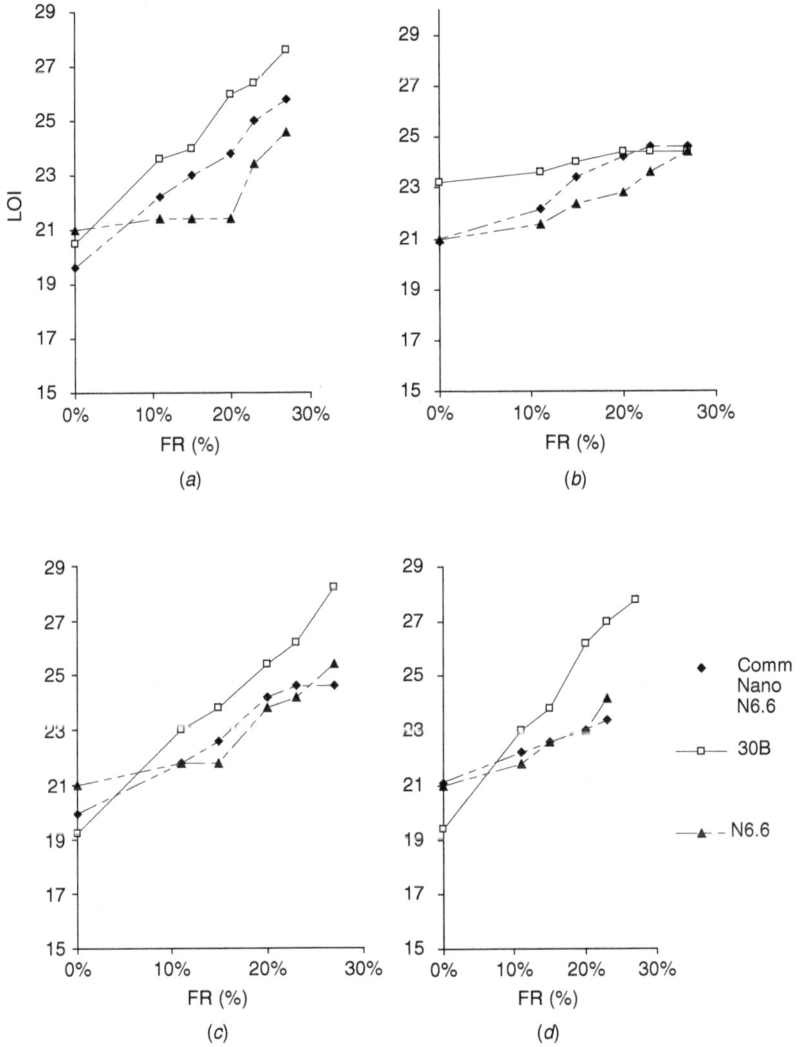

FIGURE 11.5 LOI values for polyamide-6,6 films in the absence and presence of various flame retardant/commercial and Cloisite 30B nanoclay combinations; (*a*) ammonium polyphosphate; (*b*) Proban CC polymer; (*c*) intumescent MPC1000; (*d*) ammonium polyphosphate/pentaerythritol; 0% flame retardant values for flame retardant and nanoclay-containing films are extrapolated. (From Ref. 26, with permission from the Textile Institute.)

The LOI results for polyamide-6,6 films in the absence and presence of nanoclays are shown in Figure 11.5 for Cloisite 30B compared with clay-free and previously reported commercial clay-containing polyamide-6,6 films.[7,26,28] Because the clay modifies burning behavior in a way that appeared to be canceled when flame retardant was added, LOI values for 0% flame retardant in nanoclay-containing films were obtained by extrapolating LOI versus flame retardant concentrations for each flame retardant. These values are included in Figure 11.5, which shows that all the polyamide-6,6 film samples comprising each individual flame retardant additive, in the absence of clay, provide lower LOI values than those of their clay-containing analogs. Furthermore, these curves exhibit the expected S-shaped curve behavior, with LOI values increasing significantly only when respective flame retardant concentrations exceed 20% w/w. Introduction of nanoclays appears to linearize respective LOI versus flame retardant percentage trends, and generally the nanoclays selected have a positive effect on the burning behavior of polymer when flame retardant is present.

Similar results are shown in Table 11.5 for polyamide-6 films in which either Cloisite Na^+ or 30B clays were included together with a flame retardant. Again, as seen for polyamide-6,6 films, the presence of nanoclay alone changed the burning behavior of polyamide-6 significantly by enhancing char formation and reducing melt dripping,[7] thus giving rise to increased LOI values in the absence of flame retardant. Hence, 0% flame retardant LOI values were again obtained by extrapolation and included in Table 11.5. The results obtained for the polyamide-6 films comprising the cyclic organophosphate (Antiblaze CU, Rhodia) and the control clay are interesting, inasmuch as they yield the highest LOI values at each additive level, in comparison with the APP and APP–PER flame retardant systems. However, during testing, these films continued to melt and drip when exposed to a flame, as opposed to forming a carbonaceous char, irrespective of

TABLE 11.5 LOI Values for Flame Retardant Polyamide-6 in Combination with Cloisite Na^+ and 30B Nanoclays and Selected Flame Retardants

	Polyamide-6	Polyamide-6 + Na^+	Polyamide-6 + Cloisite 30B
Standard film	22.6	23.0	23.4
0% (extrapolation)	*20.9*	*18.5*	*17.5*
11% APP	23.4	22.6	21.8
15% APP	23.4	23.4	24.2
20% APP	24.2	25.6	26.4
23% APP	26.0	26.8	27.2
27% APP	26.0	28.0	28.8
0% (extrapolation)	*23.1*	*21.7*	*24.1*
11% APP–PER	23.8	23.8	25.0
15% APP–PER	24.2	23.8	25.0
20% APP–PER	24.6	23.8	25.0
23% APP–PER	24.6	25.0	25.0
27% APP–PER	—	26.2	26.2
0% (extrapolation)	—	*19.3*	—
11% CU	—	24.2	—
15% CU	—	26.8	—
23% CU	—	30.0	—

Source: After Ref. 26.

additive loading. Thus, it is proposed that higher LOI values were obtained due to the difficulty in sample ignition as a consequence of enhanced melt dripping. It is interesting to note that as with the polyamide-6,6 films, the presence here of Cloisite Na^+ clay had an effect similar to that of the functionalized analog, Cloisite 30B, which contradicts the expectations that the 30B clay would be more fully nanodispersed and so should be expected to yield higher LOI values.

For APP $< 11\%$, the order of decreasing LOI at the same FR concentration appears to be

pure polyamide-6 > Cloisite Na^+ > Cloisite 30B > commercial nanoclay

while at higher APP concentrations of 20% and above, the order is

Cloisite 30B > Cloisite Na^+ > commercial nanoclay > pure polyamide-6

For an APP–PER system at most concentrations, the decreasing LOI order is

Cloisite 30B > pure polyamide-6 > Cloisite Na^+ > commercial nanoclay

This general observation is opposite to that seen for polyamide-6,6 film analogs, where the LOI order at a given retardant concentration tended to be[26,28]

Cloisite Na^+ > Cloisite 30B > commercial nanoclay > pure polyamide-6,6

The influence of the actual physical morphology of the clay dispersions and the associated polyamide within each set of films could have a significant influence on the orders noted above and on whether or not there is a true nanocomposite effect at all. Cursory work in this respect suggests that films reported previously, formed from commercially available nanocomposite PA6 and PA6,6 retain their nanodispersed clay formations.[7] However, films containing both Cloisite Na^+ and 30B, while promoting spherulitic crystal formation of both PA6 and PA6,6 polymers during film formation, have varying degrees of nano- and micro-dispersed clay particles present. The formation of spherulitic structures has, of course, serious implications for resulting fiber-forming characteristics.

Clearly, the presence of each clay significantly reduces the concentration required of each flame retardant required to achieve a specific LOI value, as shown in Table 11.3 for commercial nanocomposite polyamide-6,6 films and in Figure 11.6, where for each flame retardant the percentage reduction is plotted for each clay type in polyamide-6,6 films. There is an indication that the two clays used in this study show significant advantages over the commercial nanocomposite polyamide-6,6 studied previously, and that for LOI values of about 23, levels of retardant approaching only 10% w/w may be applicable, and to achieve LOI = 24, levels of about 15% may suffice. To explain these results, our previous research has also developed a model[26–28] that assumes a network of interlinked domains for an imaginary micro-dispersed flame retardant having an average particle (nano-, micro-, or a mixture of both) diameter of σ μm and a separation or

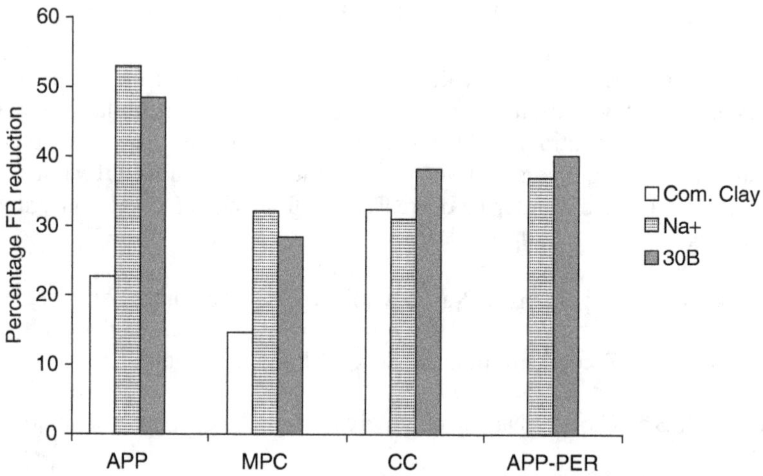

FIGURE 11.6 Percentage reduction of each flame retardant concentration required to produce an LOI value of 24 when present with an unspecified commercial clay the specified clays Cloisite Na^+ and 30B. (From Ref. 26, with permission from the Textile Institute.)

"reaction zone" of l μm. Any micro- and nano-dispersion would become active when heated above the temperature at which the flame retardant particles decomposed, and in the case of ammonium polyphosphate, this would be in the region 250 to 300°C where char-forming reactions would also begin.[50]

This simple model may be used to visualize the mechanisms taking place in the examples above. For example, Table 11.4 shows that for APP present without nanoclay, to achieve an LOI value of 24, 26.0% w/w is required in polyamide-6,6. The model predicts an average APP particle separation l or reaction zone = 8 μm, assuming an average diameter of 25 μm. However, in the presence of 2% w/w Na^+ clay, only 12.3% APP is required, which corresponds to an increased value of $l = 19$ μm, alongside which the inter-nanoclay reaction zone separation is on the order of 0.5 μm. This means, in effect, that the larger interparticle reaction zone distances between the suspended APP particles are bridged by a large number of nanoparticles which themselves promote char formation. Clearly, the reduced retardant concentrations illustrated in Figure 11.6 support this hypothesis, although much more detailed work is necessary to substantiate it more fully. Unfortunately, because of the different burning behavior of the polyamide-6 films in the presence and absence of clays, a similar analysis could not be undertaken.

Although there have been other attempts to produce nanoclays in the presence of flame retardants in other fiber-forming polymers, such as polypropylene[51] and polyester,[52] their combined effects in fibers have not been directly observed. In the case of polypropylene, the addition of nanoclay to a flame retardant formulation based on a hindered amine stabilizer and a char-promoting ammonium polyphosphate at concentrations on the order of only about 5% w/w does enhance char formation, although insufficiently to increase the LOI value above 22. A similar char-enhancing effect of added functionalized montmorillonite clay was observed by Wang et al.[52] in a copolymer of PET and a phosphorus-containing monomer, in that higher residues above 450°C were recorded.

Very recent work in our laboratories as yet unpublished[53] has shown that poly(acrylonitrile) copolymer having properties suitable for fibers, when polymerized in the presence of a functionalized nanoclay, may absorb ammonium polyphosphate during filament extrusion and yield fibers with LOI > 35. In these fibers, a clear synergy between nanoclay and flame retardant is observed, and filament properties are little changed from those acceptable for normal textile applications.

11.3.3 Coatings

Examples of the potential for the use of nanoparticulate fillers to enhance the fire performance of polymer coatings have largely been restricted to coatings for textile substrates, including back-coatings. Bourbigot et al.[40,41,54,55] have shown that addition of nanoclays and poly(silsosesquioxanes) can reduce the peak heat release rates in polyurethane-coated knitted polyester fabrics as shown in Figure 11.7. However, the presence of these nanoparticles alone reduces the time to ignition and prolongs the time of burning—exactly the opposite of what is required for flame-retarded coated textiles.

More recently, Horrocks et al.[21] have shown that if a back-coating is to be effective, it must have a transferable flame retardant activity from the coating on

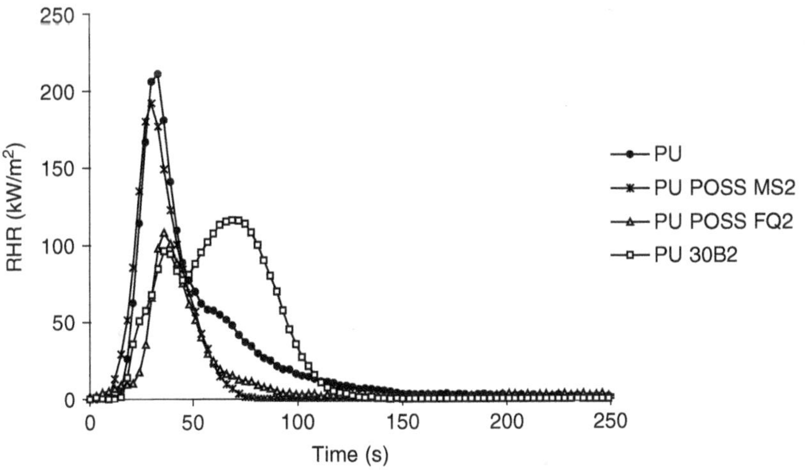

FIGURE 11.7 Rate of heat release curves of PU-nanocomposite coatings on PET knitted fabric samples at 35 kW/m^2 treated with an octamethyl POSS (POSS MS2), a poly(vinylsilsosesquioxane) (POSS FQ2), or a nanoclay (Cloisite 30B) at 10% loadings with respect to PU. (From Ref. 41, copyright © 2002, John Wiley & Sons, Ltd., with permission.)

the reverse face of the textile when ignited from the front face in tests such as BS 5852 Part 1 1979 and 1990 and EN 8191 Parts 1 and 2.[56] The use of purely char-promoting flame retardants within a coating does not allow this to occur unless the retardant species becomes mobile and can diffuse through the fabric to the front face.[49] The addition of a nanoclay alone to a back-coating polymeric film has been shown to have no beneficial effect,[21] but when fumed (nanoparticulate) silica is added with ammonium polyphosphate to the back-coating formulation, not only was an adverse effect noted with respect to formulation rheology, but the flame retardant character as determined by the LOI was reduced with increasing silica content. Using a dispersion of the silica equivalent to a maximum 17% w/w within a solid coating, a series of Vycar PVC (Noveon, Cleveland) copolymeric dispersions were prepared comprising a constant total retardant concentration (250 parts by mass with respect to 100 parts dry resin mass) with silica/APP molar ratios varying from 1 : 0 to 0 : 1 respectively, in 0.1 increments. Coated fabric samples at a nominal 30% dry add-on were prepared, but with increasing silica content and associated degradation in rheology, add-ons decreased to below 10% for SiO_2/APP > 0.6 : 1. However, subjecting samples to LOI testing enabled the results in Figure 11.8 to demonstrate the negative effect that silica has on the overall flame retardancy of such a coated material. Clearly, the potential applications of nanocomposites within the coating area, especially with respect to coated textiles, must be questioned based on the data presently available.

11.3.4 Composites

Within the area of rigid, reinforced composites, the major requirements for acceptable fire performance are resistance to ignition, minimal flame spread and heat release rate, and low smoke generation. The bromine-based flame retardants used

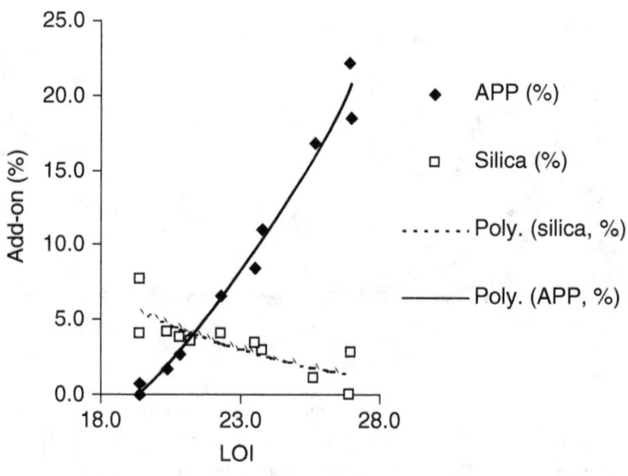

FIGURE 11.8 Effect of ammonium polyphosphate and fumed silica concentrations present in coatings on cotton fabric on the fabric LOI values. (From Ref. 21.)

currently, whether as additives or resin comonomeric modifications, present major challenges posed by their increased smoke-generating propensity, coupled with questions regarding environmental concerns related to the use of halogenated flame retardants in general. The search for nonbrominated alternatives has led to the potential offered by intumescent-based systems on the one hand and by nanocomposite formation on the other. This entire area has been reviewed,[57,58] and Sorathia[59] has reviewed the requirements of end users such as the U.S. Navy. He cites published attempts by Wilkie[60] to use nanocomposites in combination with nanoclays and phosphorus-based flame retardants such as tricresyl phosphate and resorcinol diphosphate within vinyl ester resin matrices. Interestingly, the latter work indicated that although the presence of clay did not increase times to ignition during cone calorimetry, clay introduction at the 6% w/w level reduced the peak heat release rate relative to the pure resin, and that subsequent decreases were proportional to the amounts of phosphate added.

Work in our own laboratories has investigated the thermal degradation effects[13] of introducing functionalized nanoclays along with phosphorus-containing flame retardants in vinyl ester resins. Subsequent work has reported the effect of nanoclays and flame retardants on cone calorimetric properties.[61] In this work, a typical polyester resin [Crystic 471 PALV (Scott Bader)] was investigated by DTA–TGA in the presence of a range of clays [Cloisite Na^+, 10A, 15A, 25A, and 30B (Southern Clay Products, San Antonio, Texas); see Table 11.1)] and phosphorus-containing (ammonium polyphosphate (Antiblaze MCM, Rhodia), melamine phosphate (Antiblaze NH, Rhodia), dipentaerythritol–melamine phosphate intumescent mixture (Antiblaze NW, Rhodia), and alumina trihydrate. Initial results[13] reported that in the derived unsaturated resin nanocomposites, nanoclays reduce thermal stability and the char formation tendency of the resin up to 600°C, with no change at higher temperatures. Whereas introduction of different condensed-phase active flame retardants increased char formation of the resin above 400°C, when nanoclays were added, char formation was not greatly affected, and in fact for ammonium polyphosphate–containing resins, the char was reduced. Figure 11.9 shows this quantitatively as percentage residual mass differences between resin–flame retardant–clay and respective resin–flame retardant TGA responses. Thus, the introduction of Cloisite 25A clay appears to have a minimal effect (except for APP) on the thermal degradation of a vinyl ester resin containing the flame retardants above, suggesting that fire performance may be little influenced by its addition. Furthermore, in a later publication,[61] while only x-ray diffraction was used as a means of understanding whether or not each resin–clay composition had a nanocomposite structure, a major conclusion was that inclusion of flame retardants neither influences the level of clay dispersion present nor facilitates nanocomposite formation. The fire performances were derived by cone calorimetry at 50 kW/m^2 incident flux and expressed in terms of peak heat release rate (PHRR), total heat release (THR), fire growth index (FIGRA), and smoke evolution. The differences in these parameters with respect to pure resin behavior as a consequence of adding clays at 5% levels alone to vinyl ester are shown by revisiting our data[61] and plotting as in Figure 11.10*a*, and

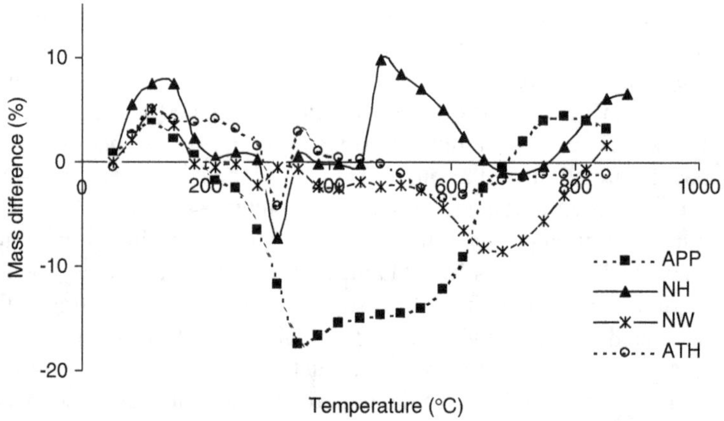

FIGURE 11.9 Percentage residual mass differences between resin-flame-retardant clay and respective resin-flame-retardant TGA responses to show the effect of Cloisite 25A clay on the thermal degradation of a vinyl ester resin containing the flame retardants ammonium polyphosphate (APP), melamine phosphate (NH), melamine phosphate and dipentaerythritol (NW), and alumina trihydrate (ATH). (From Ref. 13, with permission from the Royal Society of Chemistry.)

these may be compared with the effects on cone calorimetric behavior following addition of each flame retardant at 20% w/w in Figure 11.10*b*. To a first approximation, each clay has a similar PHRR- and FIGRA-suppressing effect, although smoke generation is increased generally. The presence of each flame retardant had a similar beneficial effect in terms of reducing heat release rate parameters, although again, smoke is generally increased. The effects of adding clays at 5% w/w and different flame retardants at 20% w/w on cone calorimetric parameters with respect to respective flame-retarded resin formulations are shown in Figure 11.11*a* for Cloisite 25A with each flame retardant and in Figure 11.11*b* for each clay with a single flame retardant, APP. Figure 11.11*a* shows that whether a given clay improves the fire performance of an already flame-retarded resin depends on the latter type. Thus, Antiblaze NW (melamine phosphate and dipentaerythritol), and ATH in particular, show further reductions in PHRR and FIGRA while having little effect on smoke formation, whereas melamine phosphate alone (Antiblaze NH) shows a reduced fire performance with increases in PHRR and FIGRA, and particularly with smoke evident. However, all clays behave similarly in the presence of APP (Figure 11.11*b*) in promoting reductions in PHRR, FIGRA, and apart from Cloisite 25A, smoke generation.

In conclusion, although their appears to be no general improvement in fire performance when nanoclays are added to conventionally flame-retarded resins, there is evidence that in certain formulations, such as those containing APP and ATH, some benefits are observed, and this opens opportunities for favorable introduction of nanoclays and other nanoparticles in flame retardant resin formulations for use in reinforced composites that have improved fire properties.

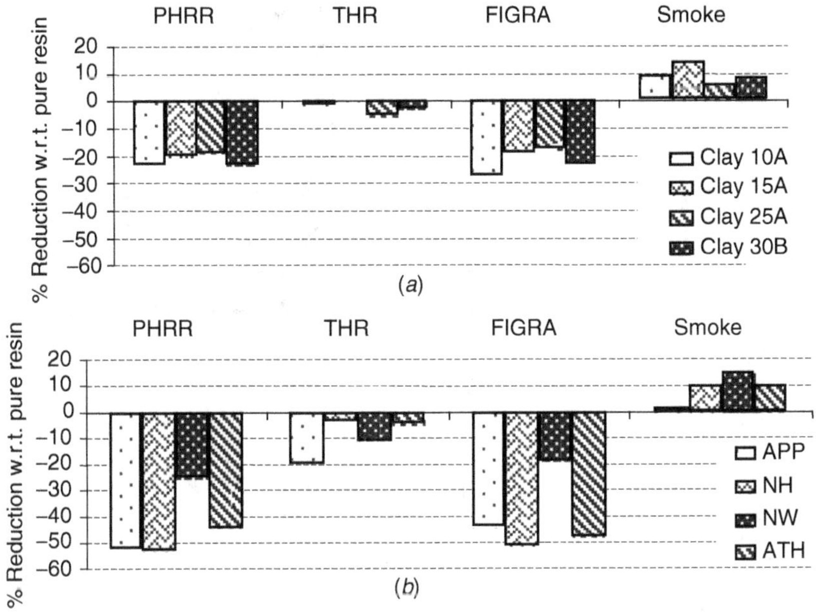

FIGURE 11.10 Cone calorimetric behavior with respect to that of pure vinyl ester resin of formulations containing (*a*) different functionalized clays at 5% (w/w) only and (*b*) various flame retardants, at 20% (w/w) only. (Plotted from data presented in Ref. 61.)

11.3.5 Foams

The challenge of developing novel flame retardant foams containing nanoparticulates will lie not only in selecting and optimizing a synergistic formulation itself with respect to improved fire performance but also in accommodating the increased rigidity that will be expected to accompany the formation of a nanocomposite structure. This could render any potential suitable exploitation relevant only to rigid-foam applications, although flexible foams for use in less critical areas such as packing, as opposed to fillings for upholstered products, might be feasible. Furthermore, since many foams are based on polyurethane chemistry produced by in situ polymerizations, including concurrent inflation, the generation of a truly nano-dispersed phase will be a significant challenge in itself. There is evidence, however, that nanodispersions can be achieved if the functionalizing substituents are chosen carefully and contain a significantly high number of hydroxyl groups.[62] Notwithstanding the dispersion issue, recent research has also shown that in the case of polylactide foams, where high levels of nanodispersion can be achieved by melt processing and physical inflation with an inert gas, layered silicates influence cell dimensions and density by nucleating cell formation.[17] This gives rise to foams having a finer and more uniform cell structure. A similar phenomenon has been observed in clay-filled PS foams,[17] where improved fire retardancy has also been reported.

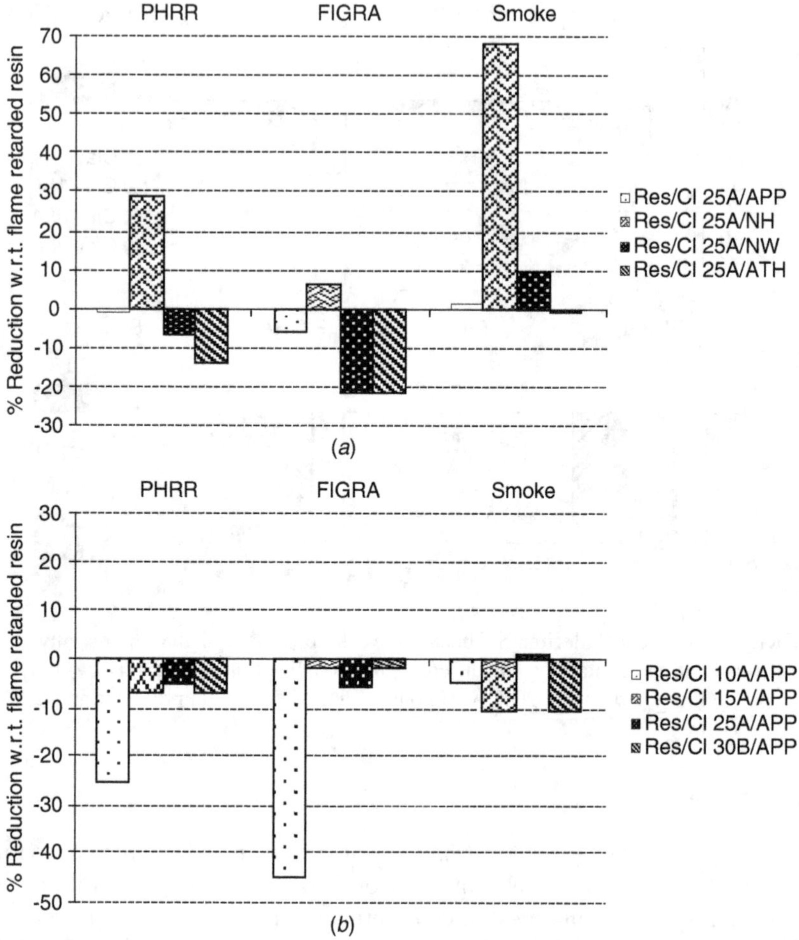

FIGURE 11.11 Differences in burning behavior of resin formulations containing clays (5% w/w) and flame retardants (20% w/w) with respect to resins containing flame retardant only: (*a*) Cloisite 25A with each flame retardant; (*b*) each clay with APP. (Plotted from data presented in Ref. 61.)

11.4 FUTURE OUTLOOK

It is evident that future successful commercial application of nanocomposites in developing improved fire performance will take the lead from the work of Beyer[15,22–24] and Kabelwerk Eupen AG, where the combination of appropriate nanoparticulates and conventional flame retardants (e.g., ATH) enables a reduction in the concentration of the latter to be made in EVA-based cable sheathings. Furthermore, because of synergy between the nano- and micro-dispersed phases, the total concentration of both components is less than the previous higher concentration of the single flame retardant (ATH) necessary to create the desired levels of fire performance, and this gives rise to improved mechanical properties. Evidence from the literature shows that under experimental conditions, similar nanoparticle conventional flame retardant synergies are observed in other polymer matrices. Therefore, although this is by no means a general observation, opportunities exist for significant reductions in additive concentrations being made on the one hand to achieve given fire retardancy levels, while on the other, gaining mechanical performance and cost advantages. These may, of course, be realized only if inclusion of nanophases can be achieved with minimal or controllable effects on polymer rheology during processing. This will be especially valuable in the film and fiber sectors, where minimal flame retardant concentrations are mandatory if the tensile properties expected are to be maintained.

REFERENCES

1. Kojima, Y.; Usuki, A.; Kawasumi, M.; Okada, A.; Kurauchi, T.; Kamigaito, O. Synthesis of polyamide 6–clay hybrid by montmorillonite intercalated with ε-caprolactam. *J. Polym. Sci. Polym. Chem.* **1993**, 34(4), 983–986.
2. Usuki, A.; Kojima, Y.; Kawasumi, M.; Okada, A.; Fukushima, Y.; Kurauchi, T.; Kamigaito, O. Synthesis of nylon 6–clay hybrid. *J. Mater. Res.*, **1993**, 8, 1179.
3. Pinnavia, T.J.; Beall, G.W., Eds. *Polymer–Clay Nanocomposites*. Wiley Series in Polymer Science. Wiley, New York, 2000.
4. Vaia, R.A. Structure characterisation of polymer–layered silicate nanocomposites, in: T.J. Pinnavia and G.W. Beall, Eds., *Polymer–Clay Nanocomposites*. Wiley Series in Polymer Science. Wiley, New York, 2000, pp. 229–266.
5. Yngard, R.; Yang, F.; Nelson, G.L. Flame retardant or not: fire performance of polystyrene/silica nanocomposites prepared via extrusion, in: *Proceedings of the 14th Conference on Advances in Flame Retardant Polymers*, Stamford, CT, 2003.
6. Yang, F.; Nelson, G.L. PETg/PMMA/silica nanocomposites prepared via extrusion, in: *Proceedings of the 5th Conference on Advances in Flame Retardant Polymers*, Stamford, CT, 2004.
7. Padbury, S.A. Possible interactions between char-promoting flame retardants and nanoclays in polyamide films. Ph.D. dissertation. University of Bolton, Bolton, Lancashire, England, 2004.
8. Carrado, K.A.; Xu, L.; Seifert, S.; Csencsits, R.; Bloomquist, C.A.A. Polymer–clay nanocomposites derived from polymer–silicate gels, in: T.J. Pinnavia and G.W. Beall, Eds., *Polymer–Clay Nanocomposites*. Wiley Series in Polymer Science. Wiley, New York, 2000, pp. 47–63.
9. Lan, T.; Pinnavaia, T.J. Clay-reinforced epoxy nanocomposites. *Chem. Mater.* **1994**, 6, 2216.
10. Padbury, S.A.; Horrocks, A.R.; Kandola, B.K. The effect of phosphorus containing flame retardants and nanoclay on the burning behaviour of polyamides 6 and 6.6, in: *Proceedings of the 14th Conference on Advances in Flame* Retardant Polymers, Stamford, CT, 2003.
11. Gilman, J.W.; Awad, W.; Davis, R.; Morgan A.B.; Trulove, P.C.; DeLong, H.C.; Sutto, T.E.; Mathias, L.; Davies, C.; Chiraldi, D. Improved thermal stability of

crown ether and imidazolium treatments for flame retardant polymer-layered silicate nanocomposites, in: *Flame Retardants 2002*. Interscience Publications, London, 2002, pp. 139–146.

12. Awad, W.H.; Gilman, J.W.; Nyden, M.; Harris, R.H., Jr.; Sutto, T.E.; Callahan, J.; Trulove, P.C.; DeLong, H.C.; Fox, M. Thermal degradation studies of alkyl-imidazolium salts and their application in nanocomposites. *Thermochim. Acta* **2004**, 409(1), 3–11.
13. Kandola, B.K.; Nazaré, S.; Horrocks, A.R. Thermal degradation behaviour of flame retardant unsaturated polyester resins incorporating functionalised nanoclays, in: M. Le Bras, C.A. Wilkie, S. Bourbigot, S. Duquesne, and C. Jama, Eds., *Fire Retardancy of Polymers: New Applications of Mineral Fillers*. Royal Society of Chemistry, London, 2005, pp. 147–160.
14. Pramoda, K.P.; Liu, T.; Liu, Z.; He, C.; Sue, H.-J. Thermal degradation behaviour of polyamide 6/clay nanocomposites. *Polym. Degrad. Stab.* **2003**, 81, 47–56.
15. Beyer, G. Flame retardant properties of EVA-nanocomposites and improvements by combination of nanofillers with aluminium trihydrate. *Fire Mater.* **2001**, 25, 193–197.
16. Yasue, K.; Katahira, S.; Yoshikawa, M.; Fujimoto, K. In situ polymerisation route to nylon 6–clay nanocomposites, in: T.J. Pinnavia and G.W. Beall Eds., *Polymer–clay Nanocomposites*. Wiley Series in Polymer Science. Wiley, New York, 2000, pp. 111–126.
17. Sinha Ray, S.; Okamoto, M. New polylactide/layered silicate nanocomposites, 6: Melt rheology and foam processing. *Macromol. Mater. Eng.* **2003**, 288, 936–944.
18. Matayabas, J.C.; Turner, S.R. Nanocomposite technology for enhancing the gas barrier of polyethylene terephthalate, in: T.J. Pinnavia and G.W. Beall Eds., *Polymer–Clay Nanocomposites*. Wiley Series in Polymer Science. Wiley, New York, 2000, pp. 207–226.
19. Matayabas, J.C.; Turner, S.R.; Sublett, B.J.; Connell G.W.; Barbee, R.B. (Eastmann Chemical Co.). PCT Int. Patent Appl. WO 98/29499, Aug. 9, 1998.
20. Davies, R.D.; Gilman, J.W.; VanderHart, D.L. Processing degradation of polyamide 6–montmorillonite nanocomposites in: *Proceedings of the 13th Conference on Advances in Flame Retardant Polymers*, Stamford, CT, 2002.
21. Horrocks, A.R.; Davies, P.J.; Alderson, A.; Kandola, B.K. The challenge of replacing halogen flame retardants in textile applications: phosphorus mobility in back-coating formulations, in *Proceedings of the 10th European Conference on Flame Retardant Polymeric Materials* (FRPM05), Berlin, 6–9Sept. 2005.
22. Beyer, G. Flame retardancy of nanocomposites: from research to technical products. *J. Fire Sci.* **2005**, 23, 75–87.
23. Beyer, G. Carbon nanotubes as flame retardants in polymers. *Fire Mater.* **2002**, 26, 291–294.
24. Beyer, G. Progress with nanocomposites and new nanostructures, in: *Flame Retardants 2006*. Interscience Publications, London, 2006, pp. 123–133.
25. Lan, T.; Qian, G.; Liang, Y.; Cho, J.W. FR application of plastic nanocomposites. Technical paper. http://www.nanocor.com/tech_papers/FRAppsPlastic.asp.
26. Horrocks, A.R.; Kandola, B.K.; Padbury, S.A. The effect of functional nanoclays in enhancing the fire performance of fiber-forming polymers. *J. Text. Inst.* **2003** (published 2005), 94(3), 46–66.

27. Horrocks, A.R.; Kandola, B.K.; Padbury, S.A. Effectiveness of nanoclays as flame retardants for fibers, in: *Flame Retardants 2004*. Interscience Publications, London, 2004, pp. 97–108.

28. Horrocks, A.R.; Kandola, B.K.; Padbury, S.A. Interaction between nanoclays and flame retardant additives in polyamide 6 and polyamide 6.6 films, in: M. Le Bras, C.A. Wilkie, S. Bourbigot, S. Duquesne, and C. Jama, Eds., *Fire Retardancy of Polymers: New Applications of Mineral Fillers*. Royal Society of Chemistry, London, 2005, pp. 223–238.

29. Wang, D.; Echols, K.; Wilkie, C.A. Cone calorimetric and thermogravimetric analysis evaluation of halogen-containing polymer nanocomposites. *Fire Mater.* **2005**, 29, 283–294.

30. Chigwada, G.; Wilkie, C.A. Synergy between conventional phosphorus fire retardants and organically-modified clays can lead to fire retardancy of styrenics. *Polym. Degrad. Stab.* **2003**, 81, 551–557.

31. Hornsby, P.R.; Rothon, R.N. Fire retardant fillers for polymers, in: M. Le Bras, C.A. Wilkie, S. Bourbigot, S. Duquesne, and C. Jama, Eds., *Fire Retardancy of Polymers: New Applications of Mineral Fillers*. Royal Society of Chemistry, London, 2005, pp. 19–41.

32. Lomakin, S.; Zaikov, G.E.; Koverzanova, E.V. Thermal degradation and combustibility of polypropylene filled with magnesium hydroxide micro-filler and polypropylene nano-filled aluminosilicate composites, in: M. Le Bras, C.A. Wilkie, S. Bourbigot, S. Duquesne, and C. Jama, Eds., *Fire Retardancy of Polymers: New Applications of Mineral Fillers*. Royal Society of Chemistry, London, 2005, pp. 100–113.

33. Song, L.; Hu, Y.; Lin, Z.; Xuan, S.; Wang, S.; Chen, Z.; Fan, W. Preparation and properties of halogen-free flame-retarded polyamide 6/organoclay nanocomposite. *Polym. Degrad. Stab.* **2004**, 86, 535–540.

34. Fu, M.; Qu, B. Synergistic flame retardant mechanism of fumed silica in ethylene–vinyl acetate/magnesium hydroxide blends. *Polym. Degrad. Stab.* **2004**, 85, 633–639.

35. Gilman, J.W.; Kashiwagi, T. Polymer-layered silicate nanocomposites with conventional flame retardants, in: T.J. Pinnavia and G.W. Beall, Eds., *Polymer–Clay Nanocomposites*. Wiley Series in Polymer Science. Wiley, New York, 2000, pp. 193–206.

36. Duquesne, S.; Bourbigot, S.; Le Bras, M.; Jama, C.; Delobel, R. Use of clay–nanocomposite matrixes, in: M. Le Bras, C.A. Wilkie, S. Bourbigot, S. Duquesne, and C. Jama, Eds., *Fire Retardancy of Polymers: New Applications of Mineral Fillers*. Royal Society of Chemistry, London, 2005 pp. 239–247.

37. Bourbigot, S.; Le Bras, M.; Dabrowski, F.; Gilman, J.W.; Kashiwagi, T. PA-6 clay nanocomposite hybrid as char-forming agent in intumescent formulations. *Fire Mater.* **2000**, 24, 201–208.

38. Vyver-Berg, F.J.; Chapman, R.W. (Great Lakes Chemical Corporation). World Patent WO 01/10944, Feb. 15, 2001.

39. Gilman, J.W. Flammability and thermal stability studies of polymer layered-silicate (clay) nanocomposites. *Appl. Clay Sci.* **1997**, 15(1–2), 31–49.

40. Bourbigot, S.; Devaux, E.; Rochery, M.; Flambard, X. Nanocomposite textiles: new routes for flame retardancy, in: *Proceedings of the 47th International SAMPE Symposium*, May 12–16, 2000. **2002**, 47, 1108–1118.

41. Bourbigot, S.; Devaux, E.; Flambard, X. Flammability of polyamide-6/clay hybrid nanocomposite textiles. *Polym. Degrad. Stab.* **2002**, 75, 397–402.

42. Kashiwagi, T.; Shields, J.R.; Harris, R.H., Jr. Awad, W.A. Flame retardant mechanism of a polymer clay nanocomposite, in: *Proceedings of the 14th Conference on Advances in Flame Retardant Polymers*, Stamford, CT, 2003.

43. Kashiwagi, T.; Harris, R.H., Jr.; Zhang, X.; Briber, R.H.; Cipriano, B.H.; Raghavan, S.R.; Awad, W.H.; Shields, J.R. Flame retardant mechanism of polyamide 6–clay nanocomposites. *Polymer* **2004**, 45(23), 881–891.

44. Drysdale, D. *An Introduction to Fire Dynamics*, 2nd ed. Wiley, Chichester, West Sussex, England, 1999, pp. 212–222.

45. Levchik, S.V.; Weil, E.D. Combustion and fire retardancy of aliphatic nylons. *Polym. Int.* **2000**, 49(10), 1033–1073.

46. Horrocks, A.R.; Price, D.; Tankard, C. Unpublished results, 1995. See also Tankard, C. Flame retardant systems for polypropylene. M.Phil. dissertation. University of Manchester, Manchester, Lancashire, England, 1995.

47. Zhang, S.; Horrocks, A.R. A review of flame retardant polypropylene fibres. *Prog. Polym. Sci.* **2003**, 28(11), 1517–1538.

48. Lewin, M.; Weil, E.D. Mechanisms and modes of action in flame retardant polymers, in: A.R. Horrocks and D. Price, Eds., *Fire Retardant Materials*. Woodhead Publishing Cambridge, England, 2001, p. 39.

49. Horrocks, A.R.; Wang, M.Y.; Hall, M.E.; Sunmomu, F.; Pearson, J.S. Flame retardant textile back-coatings, 2: Effectiveness of phosphorus-containing retardants in textile back-coating formulations. *Polym. Int.* **2000**, 49, 1079–1091.

50. Horrocks, A.R. Developments in flame retardants for heat and fire resistant textiles: the role of char formation and intumescence. *Polym. Degrad. Stab.* **1996**, 54, 143–154.

51. Zhang, S.; Horrocks, A.R.; Hull, T.R.; Kandola, B.K. Flammability, degradation and structural characterization of fiber-forming polypropylene containing nanoclay–flame retardant combinations. *Polym. Degrad. Stab.* **2006**, 91(4), 719–725.

52. Wang, D.-Y.; Wang, Y.-Z.; Wang, J.-S.; Chen, D.-Q.; Zhou, Q.; Yang, B.; Li, W.-Y. Thermal oxidative degradation behaviours of flame-retardant copolyesters containing phosphorous linked pendent group/montmorillonite nanocomposites. *Polym. Degrad. Stab.* **2005**, 87, 171–176.

53. Hicks, J. Flame retardant investigations in acrylic fibre-forming copolymers. Ph.D. dissertation. University of Bolton, Bolton, Lancashire, England, 2005.

54. Bourbigot, S.; Devaux, E.; Rochery, M. Polyurethane/clay and polyurethane/POSS nanocomposites as flame retarded coating for polyester and cotton fabrics. *Fire Mater.* **2002**, 26, 149–154.

55. Bourbigot, S.; Le Bras, M.; Flambard, X.; Rochery, M.; Devaux, E.; Lichtenhan, J.D. Polyhedral oligomeric silsesquioxanes: applications to flame retardant textiles, in: M. Le Bras, C.A. Wilkie, S. Bourbigot, S. Duquesne, and C. Jama, Eds., *Fire Retardancy of Polymers: New Applications of Mineral Fillers*. Royal Society of Chemistry, London, 2005, pp. 189–201.

56. Horrocks, A.R. Textiles, in: A.R. Horrocks and D. Price, Eds., *Fire Retardant Materials*. Woodhead Publishing, Cambridge, England, 2001, pp. 128–181.

57. Kandola, B.K.; Horrocks, A.R. Composites, in: A.R. Horrocks and D. Price, Eds., *Fire Retardant Materials*. Woodhead Publishing, Cambridge, England, 2001, pp. 182–203.
58. Horrocks, A.R.; Kandola, B.K. Flammability and fire resistance of composites, in: A.C. Long, Ed., *Design and Manufacture of Textile Composites*. Woodhead Publishing, Cambridge, England, 2005, pp. 330–363.
59. Sorathia, U. Improving the fire performance characteristics of composite materials for naval applications, in: *Proceedings of the Conference on Fire and Materials, 2005*. Interscience Communications, London, 2005, pp. 415–424.
60. Wilkie, C.A. Fire retardancy of vinyl ester nanocomposites: synergy with phosphorus-based fire retardants. *Polym. Degrad Stab.* **2005**, 89, 85–100.
61. Nazaré, S.; Kandola, B.K.; Horrocks, A.R. Flame-retardant unsaturated polyester resin incorporating nanoclays, in: B. Schartel C.A. Wilkie, Eds., special edition, *Polymers for Advanced Technologies*, **2006**, 17, 294–303.
62. Tien, Y.I.; Wei, K.H. High-tensile-property layered silicates/polyurethane nanocomposites by using reactive silicates as pseudo chain extenders. *Macromolecules* **2001**, 34(26), 9045.
63. Han, X.; Zeng, C.; Lee, L.J.; Koelling, K.W.; Tomasko, D.L. Extrusion of polystyrene nanocomposite foams with supercritical CO_2. *Polym. Eng. Sci.* **2003**, 43(6), 1261.

12

PRACTICAL ISSUES AND FUTURE TRENDS IN POLYMER NANOCOMPOSITE FLAMMABILITY RESEARCH

ALEXANDER B. MORGAN
Nonmetallic Materials Division, University of Dayton Research Institute, Dayton, Ohio
CHARLES A. WILKIE
Marquette University, Milwaukee, Wisconsin

12.1 INTRODUCTION

The current state of polymer nanocomposite flammability research should by now be obvious, but some summary and review is needed to put this information into perspective. To date, we understand the following to be fundamentally important to polymer nanocomposite flammability:

1. The type of nanocomposite structure or dispersion is controlled by many factors, all of which must be considered to produce a polymer nanocomposite successfully (Chapters 2 and 4); the extent of flammability reduction observed in a nanocomposite is due to the nanofiller dispersion and polymer degradation chemistry (Chapters 3, 5, and 10).
2. Polymer–clay nanocomposites reduce flammability by slowing the mass loss rate of fuel to the flame, thus keeping the heat release rate (HRR) low (Chapter 3). However, the material will eventually burn completely, leaving only a small amount of noncombusted carbon, with very little reduction in

Flame Retardant Polymer Nanocomposites, edited by Alexander B. Morgan and Charles A. Wilkie

total heat release, which will not be sufficient to meet existing regulatory tests (Chapter 5).

3. The flame retardancy observed with polymer–clay nanocomposites also seems to occur with other nanofillers, such as carbon nanotubes and nanofibers, and to some extent with colloidal particles, and the mechanism for flame retardancy is very similar (Chapter 10).
4. Since polymer nanocomposites have been unable to meet regulatory fire tests by themselves (Chapters 5 and 11), additional flame retardants (Chapter 1) have been necessary (Chapters 6, to 9) to allow them into commercial use and end-use applications (Chapters 6 and 11).

Although nanocomposites have already leapt into commercial use, there are remaining challenges, and in this chapter we discuss these and explain how they relate to future trends in the field. It should be noted, though, that all new technologies have had challenges when moving from the laboratory to commercialization, and in this respect nanocomposites are no different than many other fields of research.

12.2 POLYMER NANOCOMPOSITE STRUCTURE AND DISPERSION

12.2.1 Synthesis Procedures

With all of the techniques discussed in this synthesis section, the primary goal is to break up the primary nanoparticle agglomerates, which is one of the most important steps in producing a polymer nanocomposite. To achieve this, the experimental conditions must be tailored in such a way that they work with the polymer chemistry, preferred and accepted processes and engineering practices, and the nanoparticle chemistry. As detailed in Chapter 2, there are three main techniques for polymer nanocomposite synthesis: in situ polymerization, solvent-based blending, and melt compounding. Although it may seem that newer techniques are being published, in reality all of these newer techniques can fit into one of these three categories.

12.2.1.1 In Situ Polymerization In situ polymerization involves polymerizing monomer in the presence of well-dispersed nanoparticles. In the case of clays, this means exfoliated or well-intercalated clay particles. The in situ polymerization process usually yields a polymer nanocomposite in which the clay is well dispersed, either because the clay was exfoliated in the monomer (and/or solvent) prior to polymerization or because the polymerization process causes expansion of the clay galleries, pushing the clay apart to yield the final polymer nanocomposite. In-situ polymerization usually yields nanocomposites that are well dispersed at both the micro- and nanoscale, but depending on the final interface between polymer and nanoparticle, these may not be the thermodynamically stable forms, and during processing they may revert to a different morphology.[1] Since there have been so many examples of this technique, the reader is encouraged to read some of the key reviews in this area.[1–4]

One of the techniques frequently used to prepare a polymer–clay nanocomposite is emulsion or suspension polymerization. This is similar to in situ polymerization in that the clay is exfoliated prior to polymerization, but it is exfoliated in a solvent (or aqueous phase) rather than in a monomer. In some cases the surfactant used for the polymerization process also becomes part of the final nanocomposite, either yielding an organoclay as part of the polymerization process or serving as a compatibilizer between organoclay and polymer.[5–8] Emulsion or suspension processes are being used increasingly as a tool for breaking the clay particles apart before polymerization. Part of the reason for the increase in the use of this technique is that bulk polymerization does not always lend itself to industrial scale-up, whereas emulsion and suspension polymerizations may be more suitable for this purpose. Ultimately, the choice between those methods depends on the polymer system in question.

For industrial applications, the commercial polyamide-6 nanocomposite from Unitika and Ube/Toyota seems to be the only example of an in situ process that has been scaled up. Since most polymerization reactors that could take advantage of an in situ process are used to produce base resin constantly, there is concern about taking plants (pilot or full scale) off-line temporarily for modification to handle new processes; steps like these are normally not taken, due to profit and operating expense concerns. More important, if one were to take advantage of existing capital equipment, the expenses associated with developing a new process, and even engineering changes to handle the addition of new nanofillers or nanocomposite polymerization approaches, may present a large hurdle to commercial application. It might be easier to start the process and reactor from scratch rather than retrofit an existing polymerization line. The business case for creating more in situ polymerized nanocomposites, at least at the multiton thermoplastic scale, is not attracting many new customers at this time. With the right business model and a targeted market, it may make sense to revisit the in situ polymerization process and build a plant to make nanocomposites with polymers other than polyamide-6. Whoever can get to the economy of scale the fastest will probably capture the market for these polymer nanocomposites and eventually make enough profit to pay back the cost of capital—but someone must be willing to take the risk, and aversion to risk seems to be the key stopping point for most potential industrial suppliers.

In the preceding paragraph a point was made about the capital concerns associated with multiton amounts of thermoplastic polymer nanocomposites. The point about scale needs to be elaborated again here, as the capital issue does not really come into play with thermosetting resins. For thermoset nanocomposites, the synthesis process is almost always an in situ polymerization where nanoparticle and monomer are mixed together. Certainly, there are processing tricks that can be played (e.g., solvent addition for better mixing) to help this along, but in situ remains the only feasible technique for thermoset polymer nanocomposites.

In this case, one does not have to worry about modifying capital equipment to make specialized monomer or polymerization conditions to make nanocomposite. A ready supply of monomer and nanoparticles that can be well mixed and then processed easily to produce the final thermoset part are all that is required. The processing of thermoset nanocomposites is an issue that continues to plague the commercialization of these materials, an issue that differs from the commercial concerns seen with thermoplastics. Many nanoparticles, and clays especially, can greatly modify the rheological properties of a fluid[9] such that a manufacturer who routinely works with a particular thermoset formulation can be seriously impeded by characteristics that are quite different from those of virgin polymer. Studies have looked at ways of solving this with solvent techniques,[10] or at a minimum, have sought a better understanding of the rheology of the clay and nanofiller.[11] It may be that this processing problem will be solved through clever engineering and processing steps rather than through chemical changes. Although this viscosity increase can be a problem, it can also be used to arrange the viscosity to deliberately set up an ordered, aligned polymer nanocomposite structure. For example, a magnetic field was used to align clay particles in an epoxy matrix before final polymerization and cure with interesting results.[12]

With both thermoset and thermoplastic in situ polymerization techniques, there are some fundamentals of polymer nanocomposite synthesis that are well understood. One fundamental to consider is the interface between polymer and nanoparticle in the final application; without a well-designed interface between nanoparticle and polymer, no synthetic technique will yield a good polymer nanocomposite. For in situ polymerization, the focus is on clay nanocomposites, as this field has a wealth of information on structure–property relationships between clay and polymer. Some factors that must be considered include:

- *Organic treatment interface with polymer*: Are the polymer and the treatment miscible?
- *Organic treatment functionality*: Is the treatment reactive or passive during polymerization?
- *Organic treatment thermal stability*: Can the treatment handle postpolymerization processing?
- *Organic treatment loading on the nanoparticle*: Does the loading level balance dispersion needs without degrading the final nanocomposite performance?

For the purposes of this chapter, the term *organic treatment* is used to mean a chemical interface between nanoparticle and polymer. The organic treatment could be an onium ion which would be ion-exchanged onto a clay surface to yield an organoclay. It could also be a reactive molecule which coats the outer surface of a nanoparticle through a covalent reaction, yielding a new surface for the nanoparticle that allows it to interface with a polymer to create a polymer nanocomposite. *Organic treatment* will not refer to the use of processing aids or compatibilizers that help in dispersing a nanoparticle into a polymer matrix.

Instead, those materials that help set up a polymer nanocomposite structure but may not necessarily be part of the interface between polymer and nanoparticle are described as compatibilizers or processing aids, depending on their mode of action.

Miscibility with the polymer matrix has typically been addressed by having a chemical structure similar to that of the polymer matrix, such as the use of a benzyl ammonium treatment for aromatic polymers. There have been a few attempts to use the solubility parameter to predict the compatibility between a clay and a polymer.[13, 14] However, such factors as the ability of polymer chains to entangle with the organic treatment or to undergo hydrogen bonding, or the alignment of the organic treatment on the clay surface are also important.[15–18] Additional factors of entropy and enthalpy of mixing between polymeric and organic treatment alkyl chains are also important to consider. For clays, the length of the alkyl chains plays a role, as longer chains seem to be able to expand the gallery space far enough to allow for intercalation of polymeric material. Chain lengths of 12 to 16 carbons seem optimal for this purpose.[19–21] A final consideration is the dispersion of the organoclay in the monomer or polymerization solvent. Usually, an organoclay is miscible with the polymer matrix and will disperse in the same types of solvents that will dissolve the polymer. There are techniques for analyzing the degree to which an organoclay disperses in solvent,[22–25] and these should be used as a screening method before beginning in situ polymerization. A simpler but cruder method is to disperse the clay into the solvent with simple mixing and see if the clay settles out over time.

A closely related consideration in the organic treatment of nanofillers is the presence or absence of functional groups on the nanofiller surface for the in situ polymerization process. The presence of functional groups can allow the polymerization to occur on the nanoparticle surface, so that some amount of polymer is bound to the nanoparticle. The presence of the functional groups can also lead to potential cross-linking of the final polymer, with the multifunctional nanoparticle serving as the cross-link site. Using a treatment without functional groups will eliminate possible cross-linking, but the interface between nanoparticle and polymer may not be as strong as it would be with the functional group. The decision as to which route to choose is still open for debate and in some cases seems to be polymer dependent. For polymer–clay nanocomposites, the presence of functional groups seems to yield higher amounts of exfoliation in the final nanocomposite, and functional groups that initiate polymerization on the particle surface seem to be even better for this purpose.[26–30] Certainly, this was the technique used in the original polyamide-6 nanocomposite produced by Toyota,[31, 32] and with this system, the organic treatment used on the clay works very well and is commercially available. For other polymers, however, potential clay organic treatments that can act in this way are not commercially available, so for thermoplastics this approach is currently limited to polyamides. Thermosetting resins, especially polyurethanes and epoxies, can exhibit some reactivity problems with functionalized organoclays[9, 33, 34] such that it becomes difficult to control polymerization conditions.

One must also consider the thermal stability of the organic treatment. This concern over thermal stability has been shown mostly for organoclay nanocomposites, where the alkylammonium treatment degrades via a Hofmann elimination reaction between 180 and 200°C.[35–38] While in situ polymerization processes do not usually rise to these temperatures, due to polymerization temperature or exothermic reaction, some might exceed this value, and this can result in degradation of the organic treatment.[39] Postpolymerization processing and its effect on the polymer nanocomposite must also be considered, as there are not many polymers that can be processed below 200°C. This concern is of key importance for thermoplastics, which may undergo additional heating during injection or compression molding processes or postpolymerization extruder processing. Therefore, organic treatments are needed that are thermally stable and miscible with the final polymer. There have been some breakthroughs: namely, with the use of functionalized imidazolium treatments for clays,[40] but more work is needed in this area.

Finally, the effects of organic treatment loading on the clay must be considered. The exchange capacity can be quite variable from one clay to the next; for example, that of montmorillonite is close to 100 meq-per 100 g, where as it is more than twice this value for a layered double hydroxide. This can have a great effect on the exchange of the inorganic ion for an organophilic ion. Too much organic treatment may enable better dispersion of nanoparticle in the polymer matrix but could yield losses in mechanical and thermal properties. Not enough organic treatment would prevent nanocomposite dispersion; the nanoparticles would probably not disperse and break up from their primary particles, as their interfacial area with the polymer would be limited. The amount of clay that is used also plays a role; the greater the amount of clay, the less likely that good dispersion will be obtained. For flammability applications, 5 wt% inorganic content seems to be optimal for flammability reduction, but other balance of property issues (thermal, mechanical) may dictate a different loading in the final polymer nanocomposite. For more details on how the total loading of nanoparticle can affect nanocomposite properties, see Chapter 2.

12.2.1.2 Solvent Blending Solvent blending for production of polymer–clay nanocomposites has been perceived as a process to be used only on a research scale, but in reality it may be more industrially friendly than the in situ process, at least for primary polymer producers rather than the downstream users. Although aspects of solvent blending are used in processing thermoset nanocomposites, this technique is really limited to thermoplastics or to polymers that can swell extensively in solvent, allowing polymer chains and clay to mix freely.

In industry, many polymerization processes use solvent in one form or another, either during the polymerization itself (such as free-radical-initiated processes) or postpolymerization (such as metallocene-based polyolefin processes). The solvent is then removed through large-scale evaporation distillation and devolatilization techniques, and the resulting polymer is processed into its final pellet form. There may be a need for significant engineering changes to handle feeding nanoparticles into the solvent during the polymerization process, and if a polymerization reactor is used to produce general-purpose polymer as well as solvent-blended nanocomposite, there could be concerns about potential nanoparticle contamination from nanocomposite runs in the general-purpose reactor runs. Although capital equipment concerns are likely to be less with a solvent-blending approach than with an in situ polymerization approach, they still exist and would have to be overcome for a product to be commercially feasible.

With solvent blending one must consider three parameters in picking an interface, or organic treatment, between polymer and nanoparticle. These are, again, the organic treatment structure, its thermal stability, and its miscibility with the polymer. Thermal stability is important only for post–solvent blending processing such as molding, or for end-use conditions that may have high thermal exposure. The organic treatment and its miscibility are more important factors in solvent blending. Not only must the organic treatment be miscible with the final polymer, it must also be miscible with the solvent chosen to dissolve the polymer. Some polymers can be blended only under specific solvent conditions which may attack the nanoparticle treatment (or even the nanoparticle itself) and change the chemistry of the final nanocomposite. For example, polystyrene can be used with a variety of solvents, so it is easy to pick a solvent that would be compatible with both the nanoparticle and the polymer. Polyolefins, on the other hand, often require high-boiling solvents that could thermally degrade the surfactant, and polyamides may require acidic solvents, which could react with some organic treatments.

As mentioned with the in situ approach, breaking apart the primary nanoparticle agglomerates is important for obtaining good polymer nanocomposites. With the solvent blending approach, the intensity and uniformity of mixing is of key importance, as is dilution–solids content. Several high–intensity mixing methods have been employed to break apart the primary nanoparticle agglomerates, with varying degrees of success.[41–43] One approach that appears to provide enough energy to break agglomerates apart is sonication, which has shown success in polystyrene for clay nanocomposite applications.[44, 45] With any mixing technique, the ability to break up the agglomerates completely (or in the case of clays, provide nano-dispersion) is limited by the total loading of nanoparticles in the matrix, which is, in turn, dictated by particle geometry. In the case of clays, which have been well studied, about 1 to 3 wt% clay (assuming an aspect ratio similar to montmorillonite) is the maximum amount that can be put into solution and still have an exfoliated clay structure; this amount is also true for the final polymer nanocomposite. Therefore, any solvent-blending experiments must be run at high dilution. To revisit the issue of the commercial feasibility of this technique: Certainly, high amounts of solvents can be used, although they are usually fully recovered during the devolatilization step postpolymerization. However, such a high level of solvent may limit throughput and production rates of a nanocomposite, and there could be considerable engineering challenges to mix high quantities of solvent evenly so that the clay is exfoliated. Clearly, this would be a challenge, but it seems like one that can be overcome with good chemical and mechanical engineering research, perhaps including continuous rather than batch processes.

12.2.1.3 Melt Compounding Of all the approaches used for nanocomposite synthesis, melt compounding is, to date, the most widely studied. Widespread use of this technique is due to the ready availability of thermoplastic processing equipment, which is easily amenable to scale-up, and, relatively speaking, is low in capital equipment costs. The technique is usually limited to thermoplastics, but concepts from the melt-compounding process can be used for thermoset processes, especially if one considers reactive injection molding (RIM) or resin transfer molding (RTM) processes for final nanocomposite parts.

Melt compounding is an all-encompassing term that includes extrusion (twin and single screw), roll mixing (heated rollers, also known as two- or three-roll milling), batch mixing (heated mixing bowls/heads), and static mixing (melting together nanoparticle and polymer). With any of these techniques, one must again consider the nanoparticle interface. As with other techniques, the thermal stability of the organic treatment and its miscibility with polymer are the most important considerations for melt-blending operations. The need for functional groups that may covalently link polymer and nanoparticle becomes less important with this technique, but it can be a factor in synthetic design if there is a desire for in situ grafting or reaction between polymer and nanoparticle. Along with the interface design, the intensity of mixing is another consideration; the types of mixing provided by melt compounding can bring interesting effects to the final nanocomposite.

Attention will again be focused on polymer–clay nanocomposites as we consider the thermal stability of the organic treatment. As mentioned earlier, the currently used commercial organoclays, or any organoclay based on alkylammonium organic treatments, begin to decompose between 180 and 200°C[35–38] (Figure 12.1). It is important to note what will happen if the organic treatment decomposes, as there can be more than one effect; normally, the clay will reaggregate as the clay surface becomes increasingly polar. Therefore, the attempt to form a nanocomposite structure will fail and the result will be a traditional filled composite, or microcomposite. The degradation of the organic treatment can also lead to polymer molecular-weight degradation, as the acid sites formed on the clay surface are known to be able to break C–C bonds at elevated temperature. In effect, the acid site can "crack" the polymer into smaller fragments, which will have an impact on mechanical and flammability properties. The presence

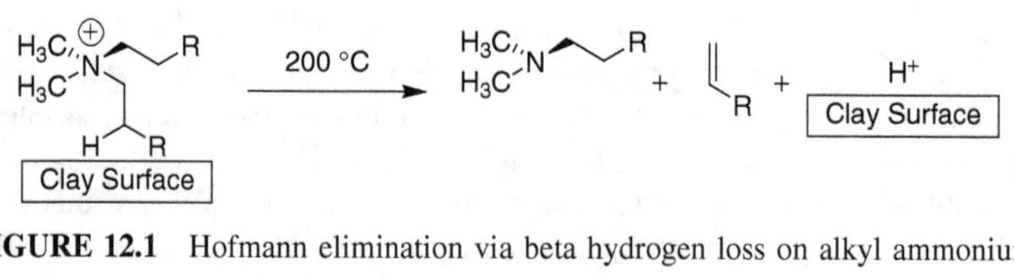

FIGURE 12.1 Hofmann elimination via beta hydrogen loss on alkyl ammonium.

of oxygen during extrusion can also have an effect. It was found that when a nanocomposite was extruded successively, each extrusion brought about a reduction in the shear viscosity. This has been explained as due to the consumption of oxygen by the double bond produced in the degradation, leading to polymer chain degradation.[46] This loss of organic treatment has been shown to lead to losses in flammability improvement,[47] and discussion continues about whether early degradation of the organic treatment may be responsible for the early ignition seen with polymer nanocomposites[48, 49] (see Chapters 3 and 5 for more details).

Some improvements in thermal stability can be obtained from the use of phosphonium treatments[50] or substoichometric amounts of organic treatment on montmorillonite.[36] Extraction of excess surfactant using various solvent washes[51] or exhaustive Soxhlet extraction[52] can also improve stability for alkyl ammoniums, sometimes with improvements up to 220°C. However, the C−N bond is the weak link in the ammonium structure, and it will eventually break with prolonged heating. Since most thermoplastics are processed above 200°C, especially under injection-molding conditions, the organic treatment must be stable under higher processing temperatures. Further, melt processing temperatures can exceed their set point by 20 to 30°C in polymer processing equipment due to shear, which may accelerate decomposition of the organic treatment. There are significant improvements in thermal stability with the use of imidazolium organic treatments[53] (Figure 12.2), allowing successful melt compounding in high-temperature plastics such as syndiotactic polystyrene[54] and polyamide-6 (PA6).[55] The drawback to these imidazolium surfactants is that they are not commercially available at this time. However, several imidazolium materials are beginning to appear in larger commercial quantities, most commonly as tonic liquids, and in time these materials may be scaled up and commercialized.

A recent area which has been explored is the use of oligomerically modified clays where the organic treatment is an oligomeric ion with more than one cationic group. These, in general, have enhanced thermal stability relative to the typical organically modified clays. A variety of new oligomeric polycations have been produced with thermal stability that is, in some cases, sufficient for them to be melt-blended with poly(ethylene terephthalate), which requires a temperature of about 280°C; an example of these oligomeric cations is shown in Figure 12.3. These systems have been based on styrenic, methacrylate, butadiene, lauryl acrylate oligomers, and other systems.[56–62]

Assuming that thermal stability has been addressed, the next factor in nanocomposite experimental design using melt-compounding techniques is the choice of processing equipment. Two roll mill and mixing bowl experiments are batch-type processes, so can produce only limited amounts of nanocomposite material. These batch processes are fine for laboratory or prototype work, and they continue to be workhorse techniques for smaller-scale experiments. However, for commercialization purposes, or even to make enough material for full-scale materials testing, twin-screw extrusion is the preferred method. An alternative approach to the use of a twin-screw extruder is to use a single-screw extruder together with an extensional flow mixer. It has been reported that such a system leads to better

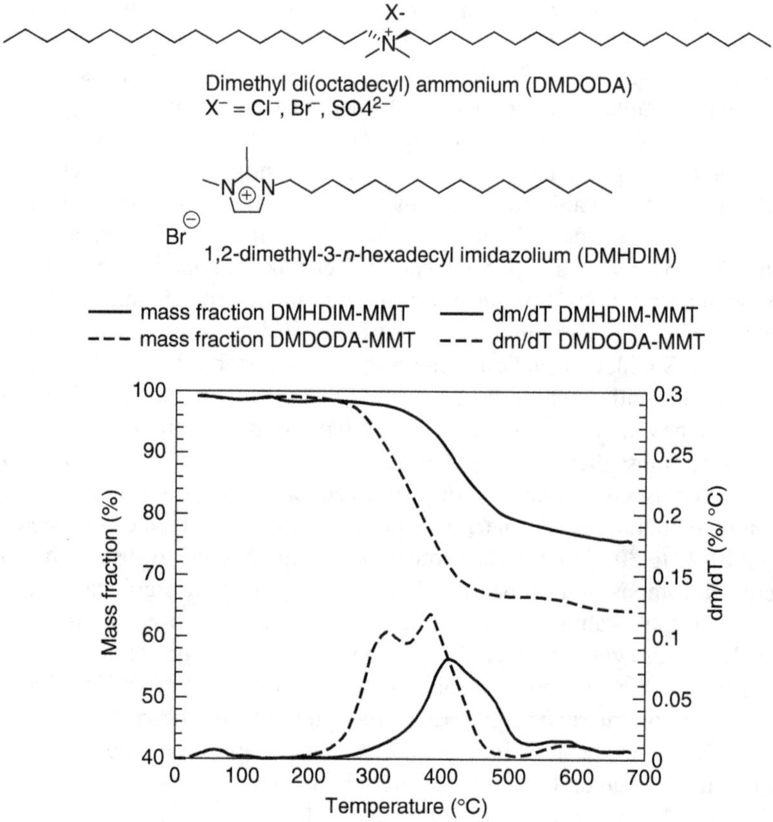

FIGURE 12.2 Imidazolium structure and TGA data comparison for imidazolium versus alkylammonium-treated clays. (From Ref. [53].)

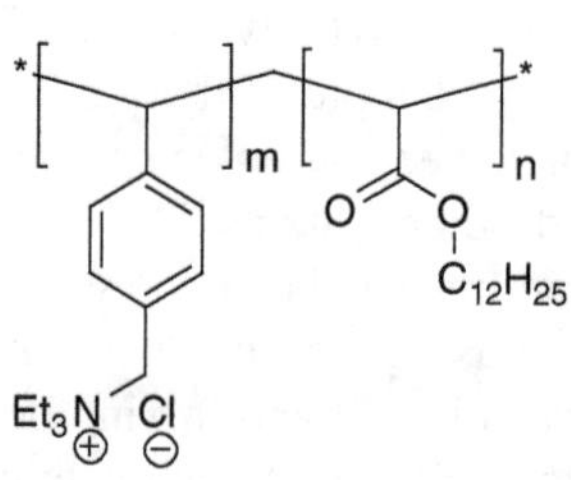

FIGURE 12.3 Example of oligomeric polycation treatments for clays.

dispersion of both PA6 and polypropylene (PP) in clay than does a twin-screw extruder.[63] The twin-screw extruder may cause intercalant and matrix degradation and reaggregation of platelets, and there may be local high-stress zones where shear heating can lead to a temperature increase of up to 50°C. Obviously, the temperature of extrusion has an effect on nanocomposite design, as the temperature must be kept below that at which the organic treatment degrades. Screw design also plays a role, and several papers have looked at these effects.[64–67] The residence time in the extruder can affect degradation of the organic modification of the clay. Researchers have observed that as the residence time increases, the intensity of the X-may diffraction (XRD) peak due to the nanocomposite decreases and a new peak, due to degradation, appears.[68] There have been some interesting new results where significant engineering changes can be implemented allowing one to feed exfoliated clay directly into the polymer matrix. Specifically, there has been recent work which demonstrated that a sodium clay in a water slurry can be fed into an extruder along with PA6 or PP, and exfoliated clay polymer nanocomposites can be produced once the water has been devolatilized using a custom-made twin-screw extruder with an L/D ratio of 77.[69,70] Presumably this could be done with other solvents or liquids as well if the design controls are met. One startup company has begun to use a similar process by exfoliating and dispersing the nanoparticles in solvent, and then extruding this material into an epoxy master batch from which the nanoparticles can then be easily handled.[71,72]

12.3 POLYMER NANOCOMPOSITE ANALYSIS

After the nanocomposite is prepared, it must be analyzed to confirm nanocomposite structure and dispersion. Polymer–clay nanocomposites have been described using the terms *exfoliated*, *mixed intercalated–exfoliated*, *intercalated*, and *immiscible* (or *microcomposite*); in all of these categories there are both ordered and disordered versions.

For other nanoparticles, such as nanotubes, nanofibers, and colloidal (spherical particles), terms for the degree of nanoparticle dispersion have not been clearly defined. The term *exfoliated* has been used for nanotubes and nanofibers to describe unbundling of the primary nanofiber aggregates as well as breakup of colloidal nanoparticle primary aggregates. Since there is no layered structure in nanotubes, fibers, or colloidal particles, the *intercalated* description cannot be used for these systems. *Immiscible* and *microcomposite* descriptions are still applicable to these types of nanoparticles, and are used to describe examples where the nanoparticles never disperse from their primary particles into the polymer matrix.

These descriptive terms are used to classify the degree of nanoscale dispersion as well as global micro- and macroscale dispersion of the nanoparticles in the polymer matrix. Since no numerical standards exist for rating the degree of nanoparticle dispersion in the polymer matrix, use of these terms is strictly qualitative and continues to be area of some controversy, as the classification of dispersion is mostly the opinion of the user. Unfortunately, not all researchers in the nanocomposite area use these terms in the same ways. The definition of *intercalated* comes from an

XRD measurement in which the clay layers are seen to be still in registry; *exfoliated* is defined as the situation when this registry is lost. Some researchers use the term *exfoliated* simply to mean well dispersed at the nanometer level, either intercalated or exfoliated, others mean as stated above. Another problem is that some use the terms *exfoliated* and *delaminated* interchangeably where as others will argue that delamination is a better state of dispersion than exfoliation. It is unlikely that there is any near-term solution to this problem, so one must critically evaluate the type of system that is being described.

Some papers have attempted to provide standard descriptions for levels of clay dispersion,[73,74] and others have used image analysis[74–76] to try to quantify the degrees of nanoparticle dispersion, but more work is needed in these areas, as well as standards for comparison. At the time of this writing, however, there are some commonly accepted techniques for nanocomposite analysis that are widely used and accepted and do help in understanding the degree of nanoparticle dispersion. There are two broad categories, nanoscale and micro/macroscale, that can be used for nanocomposite analysis, each having its own advantages and disadvantages. It should become clear from the discussion below that no one technique provides all the information that a nanocomposite researcher requires. Multiple techniques are needed to understand the system properly and to develop relationships between nanoparticle dispersion and the observed (or desired) material properties.

12.3.1 Nanoscale Analysis Techniques

12.3.1.1 X-ray Diffraction (XRD) Two types of XRD or scattering are typically used: wide-angle x-ray scattering (WAXS) and small-angle x-ray scattering (SAXS). Each covers a different domain size and degree of nano-dispersion, as described in key references.[25,77] It is important to note about any x-ray technique that only materials ordered enough to scatter or diffract the x-rays can be detected; disordered materials (see Figure 12.4) will show no pattern with x-ray techniques. X-ray scattering and diffraction have been used extensively in the characterization of polymer–clay nanocomposites, as clays diffract readily and changes in spacing between clay layers due to polymer intercalation can easily be detected with this technique. In fact, the terms *exfoliation* and *intercalation* were derived from XRD analysis of clays and are used to describe the degree of clay dispersion in the polymer matrix. XRD is best suited for polymer–clay nanocomposites rather than for nanotube or colloidal particle nanocomposites, because clays and other layered materials (such as layered double hydroxides; see Chapter 9) show changes in the XRD trace depending on the nanocomposite structure formed, but with nanotubes and colloidal nanoparticles, these materials maintain a static XRD pattern that does not change, regardless of interactions with the polymer.

There are some practical experimental details to be considered with XRD techniques, some of which are listed here:

- Scattering/diffraction mode (transmission or reflection)
- Sample form (powder or solid)

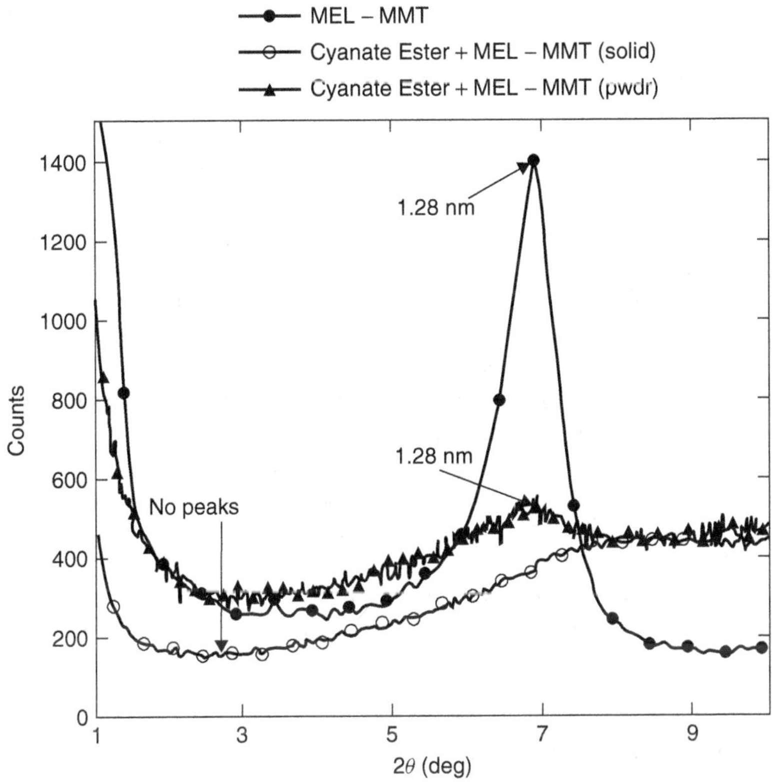

FIGURE 12.4 XRD data for solid and powder samples of a cyanate ester clay nanocomposite. (From Ref. [73].)

- Data collection parameters (2θ step size and count time)
- Nanofiller diffraction peaks and signals
- Equipment parameters (e.g., beam intensity, slit size, detection modes)

Details on why these parameters matter can be derived from key references[77] as well as from some example books and manuscripts on XRD.[78–80] To elaborate on the points above does somewhat distill the references listed in this paragraph, but it does not properly substitute for them. The reader is strongly advised to learn more about the technique before definitively stating what XRD data indicate about a nanocomposite material.

(a) For scattering or diffraction mode, the use of transmission or reflection can be of great importance. Thick materials will block x-rays from fully transmitting, and a signal that is present may not be detected. Each polymeric material can block x-rays with varying depth, but a thickness of less than 2 mm seem to be optimal for transmission through a solid polymeric nanocomposite. Reflection mode is most often used for powder XRD and solid samples, but it can lead to some misleading results with solid samples. Since the x-rays only penetrate a certain depth into a sample in reflection mode, only ordered particles on the surface of the sample will yield XRD signals. Several references have shown that clay particles align along flow fields during injection molding, especially at the surfaces of the mold.[81–83] This type of processing yields very aligned and ordered clay particles at the surface of sample, but disorder deeper in the polymer sample.[2,65,84] Therefore, reflection mode results may not always be representative of the whole sample, but it may provide a snapshot of the surface of the sample. Transmission and powder reflection modes eliminate this by capturing a global image of the material.

(b) Sample form (solid or powder) relates closely to the diffraction modes mentioned previously. Powder mode forces the sample to be randomized with no preferred order, so XRD data collected on powder samples do provide a true global overview of scattering materials throughout the sample; but not all polymer nanocomposites can easily be converted to powders, including soft materials, elastomers, or materials whose glass transition temperatures (T_g) are below room temperature. Even cryogrinding these materials rarely produces good powders for XRD analysis from polymer–clay nanocomposites, which is why solid samples are most often tested by XRD. When solid samples are used, a rough surface can affect XRD data quality, so the experimenter should try to ensure a smooth surface, if possible.

(c) Data collection parameters are important in regard to data quality, especially when small amounts of clay are present in the sample. At 1 to 3 wt% inorganic clay content, XRD signals can be quite weak; longer count times to pick up weak signals, as well as smaller step sizes, may be needed to collect quality data. Sample surface smoothness, powder versus solid, and other parameters can also play a role in the data collected.[73] For example, a cyanate ester clay nanocomposite containing 10% organoclay showed no signal with a rough surface solid sample in reflection mode, but when powdered, a signal was observed that matched the morphology determined by other techniques (Figure 12.4).

(d) Nanofiller diffraction peaks and signals are another area of consideration. Clays have their own distinctive patterns that should be observed, but in the case of crystalline polymers, the signals and peaks from the polymer may overlap and hide the clay signals. Collecting XRD data on the base crystalline polymer and base clay is always recommended before attempting to analyze XRD data from a polymer–clay nanocomposite.

It is important to note that the absence of signal with any scattering technique is simply an absence of signal and is not proof of exfoliation of the nanoparticle.[73] Additional analysis is needed to confirm the morphology of a material and the meaning of the XRD signal.

On a final note, neutron scattering has been used successfully to investigate the features of organoclays by themselves in various solvents,[22–24] which does help illuminate some basic features of organoclay dispersion in organic matrices. No scattering on polymeric materials has been collected at this time, but the technique may prove useful in polymer–clay nanocomposite analysis.

12.3.1.2 Transmission Electron Microscopy (TEM) Next to XRD analysis, TEM is the most commonly used technique for the determination of morphology. Unlike XRD, which is used predominantly for clays, TEM can be used for all nanofiller materials and can yield information about the nanocomposite morphology and the level of nano-dispersion. By observing the nanoparticle dispersion through TEM, the researcher can make a well-informed qualitative decision about what type of nanocomposite has been made and how well dispersed the nano-dimensional material is in the polymer. This last point requires elaboration, as each TEM image is a very small snapshot (length scales of micrometers to nanometers) of the entire material, so multiple images may be needed to make an informed decision about the nanoparticle dispersion and morphology. Further, this decision is only qualitative, not quantitative, since at this time there are no well-defined standards for exfoliation, intercalation, and other structures, although TEM examples of these structures do exist in the open literature[2,3,39,73] to which researchers can compare their materials. Some attempts at image analysis have been used to quantify what has been observed by TEM,[74–76] but at this time, this still appears to be a labor-intensive technique and one that requires "training" the software to quantify the results, especially when unusually shaped nanofillers (clays, layered structures, tubes/fibers) are present.

Since TEM is only qualitative, it is essential to combine this technique with others to get a better overall picture of the nanocomposite dispersion and to confirm nanoscale structures observed with the TEM result. It is recommended that researchers using TEM collect several images at high and low magnifications, and several sections from different parts of the nanocomposite sample. The low-magnification images are particularly important for determining overall nanoparticle distribution and dispersion in the polymer matrix. Although this increases the amount of sample analysis time and cost, it does give the researcher far more confidence in describing the morphology, especially when combined with other data, such as XRD or material property measurements, as described later in this section.

12.3.1.3 Nuclear Magnetic Resonance (NMR) Spectroscopy This technique[85–89] utilizes solid-state NMR to analyze nanoscale dispersion for the overall sample. The iron in the montmorillonite structure facilitates the relaxation of nearby protons, which provides information on the dispersion of the clay in the polymer matrix. In the cases reported, a ^{1}H signal in the polymer is identified and its relaxation time (T_1) is measured; the relaxation time depends on how close the proton is to a paramagnetic iron atom. On average, the protons of the polymers will be closer to the iron in the clay in a well-exfoliated system and this will have the smallest relaxation time; in a microcomposite, the protons will be farthest from the iron and show the largest relaxation time. This information can be correlated with TEM and XRD information and can also be used as a stand-alone technique to ascertain morphology. Since the terminology (exfoliated, intercalated, immiscible) was first described from XRD analysis, it is not certain if the NMR description is the same as the XRD–TEM description. Only styrenics and PA6 have been studied, but the technique offers a great capability as a stand-alone technique for the determination of morphology, and unlike TEM, it is a rapid technique in which many determinations can be completed in one day. It also provides a number, so the subjectivity of TEM is removed.[90]

At this time the technique seems to be limited by a lack of widespread availability of solid-state NMR equipment and the need to use the same batch of clay each time. If a different clay is used, it may contain a different fraction of paramagnetic iron, which will influence the relaxation time. Thus, one cannot compare samples from different suppliers or even from the same supplier obtained at a different time. The technique is also limited to those materials that contain a paramagnetic species, which can facilitate proton relaxation; montmorillonite and other natural clays that contain iron would be acceptable, but double-layered hydroxides and synthetic clays (or even layered silicas, such as magadiite or kenyite) could not be analyzed by this technique.

12.3.1.4 Other Nanoscale Analysis Techniques There are other techniques that have also been used to better understand polymer nanocomposite structure: namely, atomic force microscopy (AFM), fluorescence, and dielectric constant changes.

AFM has been reported by a few researchers,[91–95] but usually in combination with other analysis techniques. It has also been used to look at the morphology of the nanoparticles themselves,[96,97] sometimes yielding great insights on nanoparticle structure that might be missed by other techniques.

Fluorescence is a recently reported technique[98] that measures nanoscale morphology by showing a change in material properties. When a fluorescent molecule is on the nanoparticle surface, the way that molecule interfaces with the polymer network changes its fluorescent spectra or emission lifetimes, which can be measured. In the one reported example, a fluorescent tag (Nile Blue A) was placed on a clay surface, and this clay was melt-compounded in PA6. As the material exfoliated, the emission wavelength changed, as did the color of the polymer nanocomposite, from violet (for an intercalated system) to red, indicating an exfoliated structure. As with AFM, additional techniques were used to verify the structures deduced by the fluorescence technique. Photographs of the samples are shown in Figure 12.5; the TEM images that correspond are provided in Figure 12.6.

The dielectric constant is another measurable property that appears to change with nanoparticle dispersion and morphology. Changes in dielectric constant were observed when clay concentration and dispersion changed in a molten polymer matrix.[99–101] The original intent was to use the technique to provide nanocomposite analysis during melt compounding by monitoring changes at the die of a

FIGURE 12.5 Polyamide-6 nanocomposite after 1 min of processing (purple, intercalated) and 7 min of processing (red, exfoliated). (From Ref. [98].) (*See insert for color represent of figure.*)

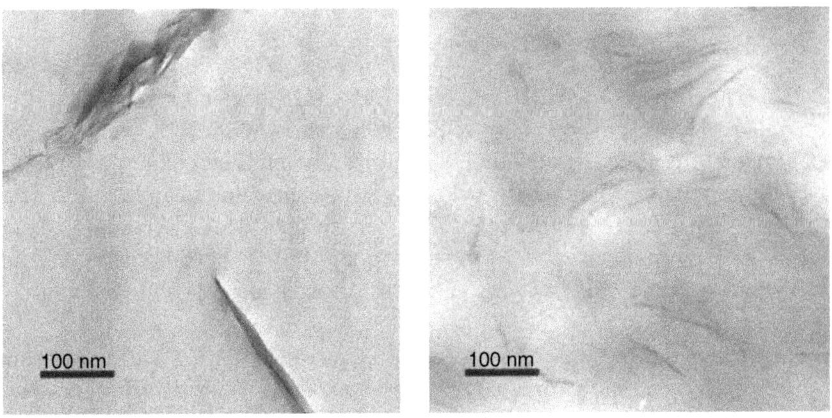

(*a*) (*b*)

FIGURE 12.6 Polyamide-6 nanocomposite TEM after (*a*) 1 min and (*b*) 7 min of processing. (From Ref. [98].)

twin-screw extruder, but there may be opportunities for nanocomposite analysis off-line as well.

12.3.2 Microscale Analysis Techniques

12.3.2.1 Scanning Electron Microscopy (SEM) and Optical Microscopy In addition to TEM, other types of microscopy can provide details of polymer nanocomposite morphology. With SEM and optical microscopy, the analysis is at much larger length scales: micrometers and millimeters. This can give some global insight into dispersion in the polymer nanocomposite, provided that the nanoparticles can be seen at these magnifications. As their name implies, the nanometer size of most nanoparticles prevents them from being imaged easily by optical microscopy, and they can just barely be imaged by SEM. SEM seems to be a good technique for carbon nanotubes and nanofibers but a poor choice for colloidal particles and a borderline choice for clays. Optical microscopy and SEM become very useful when one is rapidly screening for dispersion in a nanocomposite.[102] Since the agglomerates for most nanoparticles are in the micrometer size range, they can easily be detected with optical microscopy and SEM, which may indicate that poor nanoparticle dispersion has been achieved and that changes are needed in the synthetic procedure to obtain the desired nanocomposite. More specifically, SEM has been used mainly for detecting large clay particles, or tactoids, that were either not fully dispersed or formed due to degradation of nanoparticle organic treatment. Optical microscopy is useful in analyzing crystalline polymer nanocomposites because the effects the nanoparticles bring to the crystallite structure can be seen with high-magnification optical microscopy, yielding useful information to the researcher about how the nanocomposite is changing the properties of the material.[103–105]

12.3.3 Macroscale Analysis Techniques

12.3.3.1 Thermogravimetric Analysis (TGA) TGA does not appear useful for ascertaining whether a nanocomposite has been formed. With some polymers, the onset temperature of the degradation is significantly enhanced (e.g., polystyrene) while with many other polymers, including PA6, there is no change in the TGA parameters. It has been observed that the TGA residue does maintain its shape, which is probably related to the lack of dripping of a nanocomposite, and this may be used in some crude way as an indicator of nanocomposite formation.

12.3.3.2 Cone Calorimeter The original observation during early nanocomposite research was that a microcomposite showed no change in the peak heat release rate, whereas a nanocomposite, regardless of whether it was intercalated or exfoliated, did show a reduction.[106] Based on this general effect, cone calorimeter has been used as a tool to identify nanocomposite formation. If the peak heat release rate is less than the best value that has been observed for a particular polymer, this can serve as a rough indicator of the presence of a significant amount of immiscible material in the system.

12.3.3.3 Material Property Tests One way to confirm a polymer nanocomposite morphology is to collect the various properties of interest in the final material (mechanical, thermal, conductivity, gas barrier, etc.) in relation to a traditional filled or composite material, or the base polymer, and see how much these properties have been improved. It can be a successful argument that when normally $X\%$ of an additive is added to a polymer, it yields property improvement Y, but when nanofiller A is added at the same or lower amount, it yields greatly improved property improvement $Y+$. There is the danger that the improvement observed could be an artifact of the experimental procedure, but if the finding has already been seen in similar systems, or is known for that particular nanofiller when properly dispersed, chances are good that a nanocomposite has been formed. Although no single test can provide everything the researcher would like to know, using a material property test along with one of the nanoscale or microscale techniques can help prove a nanocomposite structure, as well as develop an improved understanding of the structure–property relationships. Somewhat related to the material property tests are rheological studies which help us to understand how the polymeric material flows under heat and shear. Since nanocomposites do affect rheological properties, this type of analysis has been used not just to understand the behavior, but also to quantify levels of clay dispersion in regard to the rheological performance.[107–111] By studying changes in the frequency sweeps of lower-frequency regions, one can study how the nanofiller changes the rheological behavior of the plastic (or thermoset monomer before polymerization) when molten. From some studies it appears that organoclay nanocomposites with a slope in the lower-frequency region nearer to 0° are more exfoliated or delaminated than those with higher slopes. However, it requires TEM or XRD analysis up front before conducting this type of analysis to make this determination properly. Once that correlation has been established, rheology can stand on its own as a nanocomposite analysis technique.

12.4 CHANGING FIRE AND ENVIRONMENTAL REGULATIONS

Unique to flame retardant and fire-safe materials research is that fire safety codes and regulations drive the field more than any other phenomena. In addition, there are new trends, codes, and regulations which appear on the surface to have no relation to the flammability performance of materials, but are currently driving the field of flame retardant research. These codes, typically environmental in nature, can adversely affect fire safety for polymeric materials and, in turn, limit the existing flame retardant solutions that could yield acceptable fire safety.

There continue to be perception issues over some species of flame retardants, specifically relating to persistence, bioaccumulation, and toxicity (PBT). Most of these PBT issues are legislatively mandated in the European Union (EU), but these issues are beginning to show up at the state level in the United States and at the national level in Japan and a few other Pacific Rim nations. In addition to PBT issues, there is an increased requirement for recycling plastics used in commercial products, which is most prevalent in the area of information technology equipment (ITE). ITE sale and use in the EU is covered by the Waste Electronic and Electrical Equipment (WEEE) protocol,[112] which dictates that plastics used for ITE need to be either recycled or incinerated after use. Due to the perception issues behind halogenated flame retardants, some have been deselected for ITE applications in favor of nonhalogenated flame retardants. So with the emphasis on recycling, many non–flame retarded materials are used, which raises the fire risk.[113–116] The removal of flame retardants to address recycling results in more fires, which, in turn, causes higher fire risk and more environmental issues than would using a flame retardant additive in the same plastic to prevent the fire.[91] On the one hand, it seems to be a good idea just to continue to use halogenated flame retardants, which are very effective, but this is not necessarily a practical solution, especially in light of the fact that halogenated flame retardants do release small amounts of very corrosive gases (hydrogen halides). These gases are not the most toxic gas in a fire (carbon monoxide is always the primary toxic gas in a fire),[117] but after the fire the hydrogen halides, carried away from the fire in the smoke, can cause damage to sensitive electronics in areas nearby that were not completely destroyed by the original fire.[118,119] Also, halogenated flame retardants do generate large amounts of smoke,[120] which can be an impediment to firefighters trying to work their way through buildings and to those trying to work their way out of a building.[121] The movement to nonhalogenated flame retardant technology may prove a boon to the nanocomposite field, since these lower the total amount of flame retardant required, which is not only good from a cost and properties perspective, but could also be quite beneficial from an environmental and life-cycle assessment perspective.

Along with environmental regulatory changes, there have been changes in fire codes and regulations that are driving advances in flame retardant science. Of more importance is the increase in use of polymeric materials in many different areas, combined with new technological advances, which has resulted in significant increases in fire risk. One of these areas is, again, ITE, where external ignition may be a problem.[122–124] Most flame retardants used in ITE today were designed to handle an internal ignition scenario, where a power source shorts out and arcs to the plastic casing. The external fire risk would be an open flame (such as a candle) that falls against the plastic and ignites it. Materials that use flame retardants and have at least a UL-94 V-1 or better rating seem to handle both internal and external risk very well[125,126] but the external fire risk scenario has risen again with new types of ITE. One way to address the internal ignition scenario and meet electronic interference regulations is to use a metal shell between the outer plastic casing and internal powered components. This means that any plastic (especially one easy to recycle) can be used for the outer casing. Not surprisingly, often an inexpensive one without flame retardant is chosen, which will fail to resist external ignition when exposed to a fire source. This external scenario appears with flat screen and liquid-crystal display (LCD) monitors and televisions, where most of the power supplies are lower in energy, so the internal ignition scenario is no longer relevant; potentially, these types of equipment may not require flame-retardants under existing regulations. It appears that some sort of flame retarded plastic will always be needed for ITE, but what the final scenario and regulatory test will be is unknown. Nanocomposites could play a role here, especially if multifunctional nanocomposite materials are used that provide both electromagnetic shielding and flame retardancy.

Another area of change in fire risk and possible regulations is automotive fire safety. At the moment, plastics used in automobiles in the United States are covered by Federal Motor Vehicle Safety Standard 302 (FMVSS 302), which was originally designed to address cigarette-type ignition scenarios. When this standard originated in the 1970s, a car contained about 10 kg of total plastic, where as a modern car contains about 150 kg of total plastic;[127] recent full-scale fire testing has shown that a modern car can rapidly go to flashover once ignition occurs.[128,129] It is, therefore, likely that there will be a mandated increase in fire safety for automotive materials sometime soon, as there is a NFPA working document on the subject that has been published and may be read by regulators as guidance.[130] Along with a need for improved fire safety with existing automotive plastics, there is a technological change in automobiles which further changes the fire risk scenario: A projected increase in automotive electrical systems from 12 V to 42 V to handle hybrid car technology, as well as increased electrical and electronic components in modern cars. This higher-voltage system has been shown to rapidly ignite FMVSS 302-rated plastic,[127,131] and therefore additional flame retardants will be needed.[127] Since there is a drive to utilize polymer nanocomposites for automobile applications to save weight and improve fuel economy (mileage),[132] nanocomposites could again provide multifunctionality by bringing lighter-weight components to cars and improving the fire safety of plastics used in this application.

Similar to automotive applications, there is an increase in flammability requirements for mass transport, especially as polymer composites are used more and more for interior and structural components to save on weight and improve durability in these applications. For example, the newest large commercial aircraft from Airbus (the A-380) and Boeing (787 Dreamliner) both rely heavily on composites for significant weight and fuel savings. Presumably, the materials identified by the Federal Aviation Administration as fire safe[133] will be used in construction of these composite aircrafts, but given their high potential cost, nanocomposites may be a more attractive alternative for these polymer systems. Multifunctionality also comes into play for these aircraft composites, as they will need to resist lightning strikes and ignition due to electrically induced thermal degradation. This may be addressed through the use of conductive nanofillers (such as carbon nanofibers or nanotubes) that allow the material to conduct the lightning strike through the composite rather than absorbing it. In the event that this is not sufficient, the nanocomposite structure would burn slowly. Such a flame could be extinguished, or would eventually self-extinguish with the addition of other flame retardant systems. There are stringent requirements in the EU for smoke and flame spread in subway and rail applications, and polymer nanocomposites could help reach that goal.

12.5 CURRENT ENVIRONMENTAL HEALTH AND SAFETY STATUS FOR NANOPARTICLES

The regulatory topic of the safety of nanoparticles is one that should be addressed. There have been a few published studies on the health and safety aspects of nanoparticles[134–136] but no comprehensive studies for all possible nanoparticles in all possible exposure scenarios, especially those relating to human health. There is a strong perception in both scientific and nonscientific circles that nanotechnology presents health and safety unknowns, and care should be exercised with these materials, but there are some known factors associated with nanoparticle safety that can be addressed here.

From most studies it appears that the primary hazard with nanoparticles is inhalation, and in this case, there are some commonly accepted practices that seem well suited to handle this potential hazard. It is recommended that any nanoparticle be treated as if it is an organic vapor and to use appropriate engineering and personal protective equipment accordingly. In their dry form, nanoparticles should be handled in fume hoods with sufficient ventilation enclosures. In wet form, where dust clouds cannot easily form, the only ventilation controls needed would seem to be those that would capture any fumes from the wet form of the nanoparticle (such as solvent in the case of nanoparticle dispersions). For personal protective equipment when handling nanoparticles outside fume hood enclosures, respirators with HEPA dust-type filters have been recommended. Obviously, safety glasses, goggles or face shields should be utilized along with laboratory coats and appropriate gloves when handling nanoparticles inside or outside a fume hood to minimize other possible areas of nanoparticle contamination or unforeseen absorption or skin irritation issues.

It needs to be stated here that these are recommended practices, not standards from any official regulatory agency. All researchers are strongly advised to look at their own process involving nanocomposite synthesis and decide what is appropriate for their operations. In the meantime, safety organizations are looking into handling and the limits of exposure for these materials, and additional guidance may be forthcoming. For additional resources, consult the National Institute for Occupation Safety and Health (NIOSH) for more details at http://www.cdc.gov/niosh/topics/nanotech/

There are also concerns about the safe handling of final parts and components that contain nanoparticles. For a polymer nanocomposite, it seems that the nanoparticle is thoroughly encapsulated within the polymer matrix, so that it would not appear to be available for later exposure. However, environmental degradation such as sanding or abrasion may permit nanoparticles to escape from the polymer matrix. Abrasion or sanding of polymeric components has yielded studies in the past regarding flame retardant exposure to workers, and the same argument could be made for exposure to nanoparticles from sanding or abrasion on a polymeric nanocomposite. It is likely that the nanoparticles would still be encased in the polymer particles that are removed from the polymeric part, but this cannot be said with certainty, and so again, caution is recommended.

12.6 COMMERCIALIZATION HURDLES

Throughout previous sections of this chapter, commercialization hurdles have been discussed where appropriate. The issues associated with polymerization techniques or flammability performance will not be readdressed here; rather, in this section we address regulatory concerns and cost issues that must be overcome to yield commercial nanocomposite products.

Some of the environmental concerns mentioned previously fall into the realm of regulatory concerns, and certainly these must be addressed before a polymer nanocomposite can enter the market. Given that there are commercial polymer nanocomposites currently available, it appears that these hurdles can be overcome. At the moment there do not appear to be any specific regulatory requirements that mandate whether nanoparticles are present in a product, but some of the existing broad requirements on chemical substances may apply. For example, the reduction of hazardous substances (RoHS) act in the EU mandates that particular products be free of various heavy metals or particular elements. This could be a problem for natural clay nanocomposites, as they may contain trace amounts of these heavy metals, depending on where they were obtained and how they were purified after mining. It is important at this point to discuss the differences between natural and synthetic clays, as these differences may hinder commercialization.

Natural clays differ from one location to another as local geological forces shaped them and determined their composition millennia ago. Therefore, natural clays somewhat prevent the creation of a global polymer nanocomposite product, as each local source of clay used in nanocomposite production will be different from batch to batch and mineral vein to mineral vein. This could be overcome by fine-tuning local formulations to account for the difference, but this adds greatly to product costs. Color is an additional concern with natural clays. Most natural clays, especially montmorillonite, have some color that will be imparted to the nanocomposite; the manufacturer cannot always provide a neutral color. In some cases, this may not be a problem, but in others, additional pigment may be needed, which, of course, increases cost and may lead to other problems. Also, carbon-based nanofillers such as carbon nanotubes and nanofibers allow only one color in the final nanocomposite: black. Parts can be painted, but this may present a problem for applications where color and aesthetics are required. It would seem that using a synthetic rather than a natural clay could solve the color and batch consistency issue. However, synthetic clays bring along a cost issue, and this higher cost is a barrier to commercialization.

A final barrier to commercialization with nanocomposite technology relates to processing changes brought by nanofiller effects on polymer rheology. This is more of an effect for thermosetting resins than for thermoplastics, as described earlier. Almost all nanofillers change the rheology of polymeric fluids, especially with thermoset monomers prior to polymerization. The large increases in viscosity make it difficult to handle the thermosetting resin in normal ways, and also make it difficult to make nano-reinforced traditional fiber composites. Besides traditional fiber-reinforced composites, another class of thermoset materials affected by these processing conditions are thermoset foams, such as a polyurethane or polyisocyanurate foams, which may also be adversely affected by processing conditions. Although the clays may facilitate foaming due to an increase of elongational viscosity, this increase in viscosity may cause problems for existing processes, such as an inability to fill molds or handle liquid monomers easily. Processes may need to be modified through clever engineering to address this viscosity change while still producing the desired density and foam structure. Since nanofillers cause nucleation during a foaming process,[137–141] they will create high-density foams, which may not be the preferred structure in the final foam product. This type of high density is preferred for structural foams, which can be useful materials in applications desiring light weight but some structural strength, but high foam density may not be preferred for an insulation or flexible foam product.

There have been reports that polymer nanocomposites, while having some greatly improved properties, can actually result in a major imbalance of overall properties for the material or can cause some unintended effect on the final material. For example, a few papers highlight the fact that the UV degradation of polymer–clay nanocomposites is greater than that for virgin polymer.[142–144] This is apparently due to the absorption of the UV stabilizers on the clay surface, but other mechanisms may be involved. If additional additives have to be included to retain the UV stability, this adds cost, which may overwhelm the benefits brought about by nanocomposite formation. The balance of properties has been an issue for all polymeric materials in all applications, and it appears to be a theme again with polymer nanocomposite technology.

There have been some commercial polymer nanocomposite successes in the areas of automotive, wire and cable, and niche sporting goods.[145] Two flame retardant polymer nanocomposites are reported in the open literature and in Internet searches at this time. The first is a poly(ethylene-co-vinyl acetate)–alumina trihydrate system from Kabelwerk Eupen which uses clay nanocomposite technology to obtain better than expected flame spread performance and mechanical properties,[146] described in Chapter 6. The second example is a series of polyolefin (PE and PP) systems from PolyOne Corporation, but the exact details of the systems are not revealed. According to their Web site, halogenated and nonhalogenated flame retardant nanocomposites with a variety of UL-94 ratings (HB, V-2, V-0, 5VA) are available, but the type of nanocomposite technology used and flame retardant system are not identified. It is expected that additional polymer nanocomposites for flame retardancy applications will be commercialized in the coming years, but at this time, neither the source nor the type is known.

Since only two examples of flame retardant systems are currently available, it seems that the technology is still in its commercial infancy. However, even with the significant improvements in properties brought by polymer nanocomposite technology, it is not always a drop-in replacement for existing materials. Further, there are times when it does not make sense to use a polymer nanocomposite for these applications, especially when the existing material is far less expensive for the potential application in question. Flammability is only one of many requirements for a commercial material, and it may be that inadvertently, enhanced flammability performance materials come to the market as polymer nanocomposite technology becomes more important for other applications. Since it is known that almost all polymer nanocomposites will lower the peak heat release rate, one could argue that even if they do not pass a regulatory fire test, all polymer nanocomposites have flame retardant properties. They do provide a significantly reduced peak heat release rate and slower flame spread when used alone; therefore, a polymer nanocomposite brought to market because of improved mechanical properties in an automotive application would probably offer enhanced fire safety compared to the polymeric material it replaced. As the fire codes change, polymer nanocomposites currently used in those fields for non-fire-related applications will already have improved flammability performance, and thus more commercial polymer flame retardant nanocomposites will automatically be ushered into commercial use.

12.7 FUNDAMENTALS OF POLYMER NANOCOMPOSITE FLAMMABILITY

Before discussing the future directions of polymer nanocomposite flammability, it is important to revisit what is known so that what is still unknown or undecided can be better defined. The flammability effects brought by polymer nanocomposites have been best studied for layered materials, or clays, followed by carbon nanotubes and colloidal particles. The flammability reduction mechanism for clays is the best understood and has been elaborated extensively throughout the book. As the polymer nanocomposite is exposed to fire and heat, the polymer is pyrolyzed and either through collapse of the clay layers or through ablation of polymeric material, the surface of the polymer nanocomposite becomes carbon- and clay-rich, forming an ever-increasing char.[47,147–150] Eventually, the heat from the fire fully penetrates the sample, forming a uniform clay-rich char which usually has the shape of the starting polymer nanocomposite. This process is illustrated in a simplified manner in Figure 12.7.

Because the release rate of flammable fuel is reduced, the heat release rate is reduced as well. However, the char–clay barrier only slows the release of fuel—it does not fully prevent it—so a polymer nanocomposite will slowly burn until almost all the carbon mass has been pyrolyzed and combusted, which means that the total heat release for a polymer–clay nanocomposite is unchanged from that of the base polymer, but the peak heat release and average heat release rates are lowered.

The mechanism of flammability reduction for polymer nanotube and nanofiber nanocomposites (Chapter 10) is similar to that for clays: a nanofiller-rich surface or barrier forms, which slows the rate of mass loss and therefore the rate of heat release.[151–153] There is little reduction in the total heat release, indicating that the carbon nanofibers and nanotubes only reduce the flammability of the

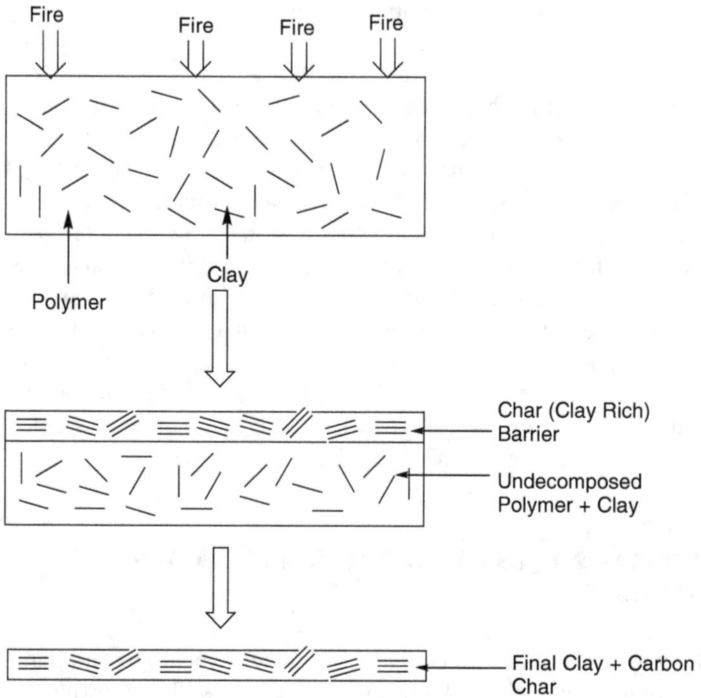

FIGURE 12.7 Idealized mechanism of polymer–clay flame retardancy.

polymer nanocomposite; they do not eliminate it. For colloidal nanoparticles the mechanism of flammability reduction is less well defined and appears to be more polymer- and nanoparticle-specific than the effects seen with clays and nanotubes. Work is ongoing in this field, and much more information is needed for the fundamentals of these polymer nanocomposites and their flammability effects to be understood.

Since polymer–clay nanocomposites are the most heavily studied, most of this section focuses on those materials, but the other two nanofiller types are discussed where appropriate. Although it is well understood that for clay nanocomposites, the clay-rich barrier slows the rate of mass loss, which in turn lowers heat release rate, *why* the mass loss is reduced in the first place is not understood. There have been several hypotheses as to why the polymer–clay nanocomposite reduces flammability, including strictly physical effects[154–156] and chemical effects,[157,158] but most likely it is a hybrid phenomena of physical effects enabled by chemical effects. Clays are known to serve as catalysts for carbon–carbon bond breakage (such as cracking catalysts in refining operations) and for carbon–carbon bond formation (aromatization or char formation, also in refining operations as an undesirable process). Also, most clays have edge hydroxyls, which allows the possibility for hydrogen bonding and rheology modifications in molten organic liquids. It may be that as the polymer decomposes, smaller fragments have a longer residence time in the condensed phase as the clay plates thicken the flow of the molten polymer, a phenomenon that has been seen in gasification studies early in the fire exposure of the polymer nanocomposite.[150,159–161] These smaller fragments have more time to recombine with other fragments and form stable bonds since the clay holds them in the condensed phase longer. These fragments are in turn catalyzed to pregraphitic chars by the clays themselves. Not all of the fragments have enough time to form char; otherwise, the polymer nanocomposites would self-extinguish and show large amounts of carbon char formation. As the nanocomposite is exposed to the flame, polymer deintercalates from between the clay sheets, and the clay organic treatment pyrolyzes as well.[150] Polymer nanocomposite underneath the flame-exposed surface survives for some period of time before eventually decomposing. Analysis of final polymer nanocomposite chars by XRD shows that the clays collapse to d-spacings similar to those seen with dehydrated clays,[150] but sometimes to d-spacings indicating that small amounts of carbon are intercalated in the clay galleries.[162,163] However, since the carbon is not seen with all polymer nanocomposite chars, this additional carbon is probably an effect due to polymer decomposition, modified by the presence of the clay, which agrees with the belief that a hybrid physical (prolonged condensed-phase residency for polymer degradation products) and chemical (clays catalyzing reactions that normally would not have occurred) mechanism is responsible for the mass loss rate reduction. The chemical products of degradation are changed by the presence of the clay, and this has been explained as indicating that the clay either physically or chemically retains the degrading radicals long enough to permit recombination reactions. A correlation between the stabilization energy of the radicals produced during the degradation and the possibility of radical recombination reactions fits this hypothesis.[164–169]

Another topic that must be addressed is the role of morphology in flammability assessment, in particular the time to ignition and the peak heat release rate in a cone calorimeter experiment. It is known that microcomposites do not provide any reduction in the peak heat release rate, while both intercalated and exfoliated structures provide a significant reduction. The amount of peak HRR reduction has been shown to be polymer-dependent, but the degree of flammability reduction due to exfoliated or intercalated structures remains an area where additional insight is needed. For many purposes, notably enhanced mechanical and barrier properties, it is felt that exfoliation is required, and there is a temptation to assume that it is required to obtain enhanced performance for any parameter. The effect of morphology and clay dispersion on material flammability continues to be difficult to quantify and, in some cases, to qualify. It does appear that the more uniformly dispersed the clay is throughout the polymer nanocomposite, the better the reduction in peak HRR and the more reproducible the flammability performance. It is also evident that the greatest relative amount of flammability reduction is obtained around 5 wt% inorganic clay content, regardless of polymer type,[47,147,162] but the effects of other factors, such as the polymer degradation pathway, clay organic treatment, and interface between polymer and clay on flammability, are not evident.

Can specific polymer nanostructures overcome certain drawbacks, such as the early ignition of clay nanocomposites under cone calorimeter conditions, or does the entire intercalation–exfoliation issue really matter as long as the clay dispersion level is uniform? If an ordered aligned exfoliated structure is achieved, such as in the case of injection molded versus compression molded materials,[81–83] does this play a role in flammability performance? It may be that exfoliation is more important for the balance of properties in the final polymer nanocomposite than for flammability performance. The issue with the early time to ignition in a polymer nanocomposite may be related to the early decomposition of the clay alkyl ammonium treatment, as more thermally stable organic treatments appear to solve this problem, but many flame retardant materials also show earlier times to ignition than do the base polymers, so the early ignition behavior may be due to the presence of an additive and unrelated to the organic treatment.[170] However, the Toyota PA6 nanocomposite does not show early ignition in the cone calorimeter,[147,162,163] and perhaps this is because when the organic treatment decomposes on the clay in this system, small molecular-weight fragments are not volatilized but simply detach from the clay surface and remain as a part of the polymer backbone (Figure 12.8). Perhaps tethering of clay and polymer will prevent this early ignition problem, or maybe this is simply a perceived problem unique to certain polymer–clay combinations and is not universal. It is not known if 5 wt% inorganic clay in an exfoliated form will reduce flammability more, less, or the same as 5 wt% inorganic clay in a mixed intercalated–exfoliated form. The effect may be different for each polymer type, since substantial differences in flammability reduction have been observed by cone calorimeter, depending on the chemical decomposition pathway under fire conditions. Papers on poly(ethylene-co-vinyl acetate) (EVA) nanocomposites have shown that the clay helps catalyze char formation[158,159] and yields substantial reductions in flammability that are not observed with other polymers, even when the level of dispersion is similar. The issues are complicated and may never be universally understood, but imply that multiple factors do play a role in flammability reduction.

The final need is to understand the degree to which a nanocomposite reduces flammability and its relevance to actual fire risk scenarios. This subject is covered extensively in Chapter 5, but it is worth revisiting here. To some extent it can be argued successfully that cone calorimeter results for a polymer nanocomposite are not relevant to a final fire risk scenario at all, yet cone calorimeter does yield fundamental flammability performance data for a material. It is necessary to put

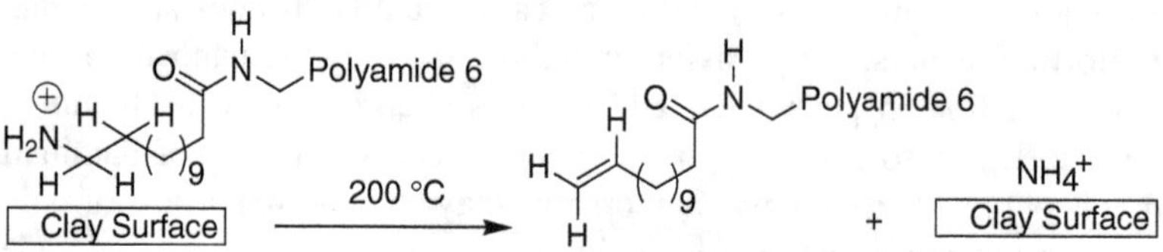

FIGURE 12.8 Polyamide-6 nanocomposite (tethered) thermal degradation via Hofmann elimination.

the existing data into context and then decide if that performance is acceptable, or useful, for the particular fire test at hand. This leads to a question about how the reduction in peak HRR relates to the use of additional flame retardants and the occasional flame retardant antagonism.

Observations from cone calorimetry on polymer–clay nanocomposites are that the peak heat release rate is lowered, with an earlier time to ignition and time to peak HRR, while the total heat released is unchanged. One might reason that the lowered peak HRR would yield improved flame spread properties, or at least lessened rates of fire growth. Indeed, if one calculates the FIGRA[171] (fire growth rate: peak HRR/time to peak HRR; units of $kW/m^2 \cdot s$) for most polymer–clay nanocomposites, the FIGRA is lowered, even with the earlier time to peak HRR. However, the significance of this lowered FIGRA in relation to full-scale testing is not clear, and there have been no tests to date looking at the fire growth rates and rates of flame spread for a polymer–clay nanocomposite. The combination of polymer nanocomposites with additional flame retardants provides quite encouraging results: synergistic reductions in the peak and total heat release rates in the cone calorimeter, but this synergistic performance has not always been observed with other flammability tests. Polymer–clay nanocomposites do little to improve the limiting oxygen index (LOI, ASTM D2863) when used alone, and only sometimes improve LOI upon combination with other flame retardants. The same is true with UL-94 V testing, where great improvements in flammability are sometimes observed with polymer–clay nanocomposites, and in other cases no improvement is seen. For UL-94 V, there does seem to be one important observation of flame retardant antagonism between clay nanocomposites and flame retardants which promote polymer dripping as an flame retardant mechanism. Since clay nanocomposites inhibit polymer dripping,[48] materials that pass flammability tests by dripping are unlikely to work when combined with nanocomposite technology; this has been observed with a polyamide–melamine cyanurate example.[172] Despite this antagonism example and the lack of consistent synergistic performance across many different fire tests, clay nanocomposites seem to work well with both vapor- and condensed-phase flame retardants (Chapters 6 to 9). It is unknown if clays will be antagonistic with other systems, or if there are particular fire scenarios where a clay would be well suited in combination with other flame retardants.

12.8 FUTURE OUTLOOK

Enough progress has been made in this field to consider the next advances in technology for polymer nanocomposite flammability. We begin this section with trends that may be expected in the next 10 years. The section consists of opinion based on the facts at hand rather than a de facto statement of what will certainly occur. It is meant to guide researchers as to what appears to be important for further development of the field.

We think it unlikely that nanocomposites will ever be useful as a stand-alone fire retardant system, but they may be useful as a component of such systems.

The combinations of nanocomposite formation with a variety of conventional fire retardants, including halogen, phosphorus, mineral fillers, and other systems, have been examined and are reported in other places in this book. We feel that this combination approach must continue and will involve other nano-dimensional materials, including other clays, such as the layered double hydroxides, polyhedral silsesquioxanes (POSSs), carbon nanotubes and spherical nanoparticles, and other putative flame retardants.

POSS additives should be explained here in a little more detail. POSS consists of an inorganic siloxanelike core, Si_8O_{12}, with organic substituents at each of the silicones. There continues to be some interest in this material for flammability applications, due to reported results. One patent reports significant reduction in the peak heat release rate for several polymers that contain POSS.[173] A recent paper examines the possibility of using POSS materials as fire retardants for textiles and concludes that although the PHRR is not lowered, the time to ignition is increased.[174] There have also been some recent studies on the thermal degradation of POSS materials, which may be a precursor to utilization as fire retardants.[175–177] What seems to limit these materials from broader use is their higher cost and the apparently higher loadings of additives necessary to obtain good flame retardant performance. It may be that POSS best fits solely as a nanofiller to be used in addition with other flame retardant additives or nanofillers, as described in the paragraph above.

In light of the issues with natural clays, one likely trend is an increase in the use of synthetic clays, such as fluorinated synthetic mica, magadiite, and layered double hydroxides (LDH). This last clay, since it has the potential to release water under fire conditions [much like $Mg(OH)_2$ or $Al(OH)_3$], may find even more use in flame retardant applications. Cost issues and limited sources for synthetic clays will slow the adaptation of these materials, so most of the work will probably be seen in the patent or open literature. More work will be seen for nanocomposites containing nanofillers, such as carbon nanotubes and nanofibers, and these will probably also be combined with additional flame retardants.

A trend that has already begun to arise is the use of multiple types of nanofillers in the same polymer to yield a multicomponent nanocomposite. Some workers have found that some types of nanofillers cannot bring all of the desired properties to the final material, so clays have been combined with multiwall carbon nanotubes to bring enhanced properties.[178] The observation for most polymer additives is that they cannot be used for all applications in all polymers, and the same observation will surely be made about nanocomposites. A clay may be used to enhance the flammability performance, but it could also be combined with a conductive nanofiller to impart antistatic aspects or electrical conductivity in the final system. One potential way to look at the use of multiple nanoparticles is that each nanoparticle plays a complementary role in flammability reduction. For example, one could choose a clay for mass loss rate or fuel release reduction, but then use a colloidal particle to fill in the gaps between clay plates as the nanocomposite thermally decomposes. Perhaps even more useful, the colloidal particle could have catalytic or flame retardant properties that encourage char formation, thus protecting the underlying polymer and building a protective char faster. There have been some papers on nanosized mineral flame retardant fillers [$Al(OH)_3$ and $Mg(OH)_2$],[179–181] and perhaps these, combined with clays or nanotubes, would accomplish the same goal.

Related to the multiple nanofiller type of approach is the use of other additives, which may not be nanosized, to assist in flame retardancy at the nanoscale level. One could envision an approach similar to that of colloidal particles mentioned in the preceding paragraph in which another material can seal the cracks in the char to set up a glassy barrier that enables the polymer nanocomposite to self-extinguish. This concept has been elaborated by one group of researchers,[182–184] where a siloxane or borosiloxane oligomer was used to form an "interlayer" between polymer and nanofiller, thus forming reinforced chars under fire conditions. One may be able to use a polymeric additive, such as a siloxane or some other low-melting inorganic oxides, that will rapidly fuse with the clay particles to form a glassy protective barrier. The approach of forming char faster follows the understanding of how the polymer–clay nanocomposite provides flame retardancy. In theory, if one can slow the rate of fuel release sufficiently, it becomes far more likely that the nanocomposite can self-extinguish. The approach of setting up char rapidly works only with additional flame retardants, but perhaps there will be breakthroughs that allow the polymer nanocomposite char structure to form faster and prevent fuel release as well as offer additional thermal degradation protection to the underlying material.

The use of nanocomposites to reinforce traditional composites has also been increasing and will continue to be a near-term trend. The emphasis for these applications, however, has been on additional mechanical reinforcement rather than flammability reduction.[185,186] Since more traditional fiber-filled composites are exposed to fire risk scenarios, it makes sense to use a nanocomposite with the traditional composite to improve both mechanical and flammability performance. Of course, this does create an additional level of complexity, especially in handling the large increases in viscosity seen with nanocomposites used with thermoset composites. At this time, most nanocomposite–fiberglass/carbon fiber composites are used for military and aerospace applications, but the benefit of lightweight materials may also move these materials into automotive and mass transportation (e.g., bus, rail), where flammability performance is strongly needed.

There are two areas of focus for future trends: specific nanocomposite structural design and true multifunctional materials. Specific nanocomposite structural design is not just exfoliation or perfect dispersion of the nanoparticle, but also deliberate nanoscale ordering and alignment in such a way as to bring very specific properties to the final nanocomposite. For example, fiberglass mats used in traditional composites are woven in particular ways to bring certain properties (e.g., stiffness, impact resistance) to the final system. If one could order and align nanoparticles to mimic this macroscale structure on a nanoscale, greatly improved properties could result. It is not at all clear how to set up these ordered structures or what specific changes in performance would be brought by these structures. This type of work is currently under way at the National Research

Council of Canada.[187] While clays and other nanofillers do align in flow and shear fields during injection molding,[81–83] this effect is not deliberate. For example, the ordered nanocomposite structure desired could be set up by utilizing a deliberate injection molding scheme that takes advantage of these flow and shear fields to align the nanoparticles into the desired orientation. Such a system has already been described.[188] The only example of deliberate alignment of nanoparticles was due to magnetic field alignment in an epoxy nanocomposite.[12] In this case, the researchers took advantage of the magnetic effects in the specific type of clay and used electrical fields to align the clays prior to polymerization. This may be achievable in thermoplastic melts as well, but due to the viscosity of those molten fluids, it could be difficult to obtain. The solution to achieving ordered nanocomposite materials may be a combination of chemistry and engineering, and it is very likely that more of these proposed solutions will be available in the future. There is a rich literature on ordering of polymer matrices or nanostructured polymeric materials and blends that may be utilized for this purpose. It is unknown if this ordering will enhance material performance in regard to flammability. For example, it is reasonable to assume that aligning all of the nanoparticles in one direction could permit the char to set up faster as the clay plates collapse in an orderly fashion. However, since these are one–dimensional (in terms of nanoscale dimensions, i.e., thickness) materials, flammability performance may be enhanced in only one direction. If the flame impinges on the sample such that the clay plates are perpendicular to the flame, the char will probably set up quickly. On the other hand, if the clay plates are parallel to the flame, the fuel should pyrolyze more quickly (depending on the rate at which fuel can flow through the slits in the clay), making the sample more flammable. For carbon nanofibers and nanotubes, a crosshatch pattern would be desired, since a network structure is preferred for flammability reduction,[153] but a structure aligned only in one direction may result in poorer flammability, or show little, if any, improvement.

True multifunctional materials continue to be one of the ultimate goals for materials science. However, this is something that has continued to be elusive for polymer nanocomposites, despite the great improvements in the properties observed. Today, a polymer nanocomposite can obtain one or more improved properties in a general area (such as mechanical or thermal), but not in more than one area of interest (not mechanical, thermal, and electrical, for example). It can be argued that these extra properties are not needed in the final application, but having the ability to impart true multifunctionality is something that will be desired and therefore should be an area of focus for research. To some extent this is the natural process for materials science development in that single-component systems are initially discovered and then applied to a variety of solutions. When it becomes clear that a single-component system does not have the correct balance of properties, multicomponent systems are developed. Eventually, the final commercial product is designed through fundamentals rather than discovered through trial and error. In the polymer nanocomposite field, design through fundamental understanding is now under way.

This book has shown multiple examples in which polymer–clay nanocomposites replace some of the flame retardant to reduce total loading of flame retardant in the final system while still maintaining (or exceeding) desired fire performance and regaining or exceeding some physical properties. An important research goal is to design a polymer nanocomposite system that replaces the flame retardant completely while maintaining fire performance, improving thermal and mechanical properties, and perhaps, contributing electromagnetic properties such as conductivity or shielding. A multifunctional material with this range of properties would be of interest to the aerospace and military markets and probably also to the area of flame retardant polymer enclosures for electrical and electronic devices. One way to accomplish this goal would be to use more than one type of nanofiller additive, but another way would be to engineer all of these properties into the same nanofiller. This could be achieved by placing flame retardant at the interface between nanofiller and polymer or by creating a better interface for a nanosized flame retardant additive. To a certain extent this is already achieved with nanofiller-sized mineral hydroxides,[179–181] but these nanofillers do not bring along additional improved properties (such as mechanical or thermal). For these types of additives, one would choose an interface that gives good mechanical property reinforcement, and perhaps the same nanoparticle coating or interface would include conductive nanoparticles (even smaller than the primary mineral filler) that have the desired electromagnetic property (such as EMI shielding for consumer electronics). This is essentially designing a new interface for an existing nanofiller that has the desired functionality (i.e., good polymer interface, flame retardant structure, conductive coating) and synthesizing the polymer nanocomposite accordingly. This could be achieved through polymer blends rather than through a polymer nanocomposite. One can envision polymer blends and alloys where the domains of each polymer type are dispersed and ordered on the nanoscale to achieve the desired multifunctionality. For example, a polymer blend could be prepared using one polymer for mechanical, another for flame retardant, and another for conductivity properties. Using a nano-ordered polymer, perhaps one can propose a polymer nanocomposite base formulation with two other types of polymers present, yielding a hybrid material where the nanoparticle could be present in more than one phase of the nanostructured polymer, or even located at the interphase between polymer domains, leading to a variety of possible structures and materials properties. If this is possible, it may be interesting to use polymer nanocomposite technology to bring multifunctionality, especially flame retardancy, with another polymer rather than several different nanosized particles.

In conclusion, polymer nanocomposite flammability has made great strides and is a well-developed field. We should expect to see more commercial products utilizing this technology in the coming decade, as well as new advances. There are areas of research that need to be developed further, especially in nanoscale ordering and design of nanoparticle dispersion, but some preliminary results in those areas have already been obtained. We believe that this technology will greatly advance the flame retardant field as more and more researchers understand the advantage of using nanoparticles as a synergist with other flame retardants. Most important, polymer nanocomposite technology will contribute to improved fire safety for society in the years and decades to come. Fire safety is only one small aspect of the effect that the presence of nano-dispersed materials can have in polymers. As noted above, in some cases a nano-dimensional material may be incorporated to improve some other property, with the enhanced fire safety being only one of many potential advantages brought by a polymer nanocomposite. This is certainly the nano-age, and as research in this area progresses, more and more materials will be discovered to enhance the various properties of the polymers.

REFERENCES

1. Utracki, L.A. *Clay-Containing Polymeric Nanocomposites*. Rapra Technology Ltd., Shawbury, Shrewsbury, Shropshire, England, **2004**, p. 496.
2. Ray, S.S.; Okamoto, M. Polymer/layered silicate nanocomposites: a review from preparation to processing. *Prog. Polym. Sci.* **2003**, 28, 1539–1641.
3. Alexandre, M.; Dubois, P. Polymer-layered silicate nanocomposites: preparation, properties, and uses of a new class of materials. *Mater. Sci. Eng. Rep.* **2000**, 28, 1–63.
4. Pinnavaia, T.J.; Beall, G.W., Eds. *Polymer–Clay Nanocomposites*. Wiley, Chichester, West Sussex, England, 2000.
5. Hou, S.-S.; Schmidt-Rohr, K. Polymer–clay nanocomposites from directly micellized polymer/toluene in water and their characterization by WAXD and solid-state NMR spectroscopy. *Chem. Mater.* **2003**, 15, 1938–1940.
6. Xu, M.; Choi, Y.S.; Kim, Y.K.; Wang, K.H.; Chung, I.J. Synthesis and characterization of exfoliated poly(styrene-co-methyl methacrylate)/clay nanocomposites via emulsion polymerization with AMPS. *Polymer* **2003**, 44, 6387–6395.
7. Yei, D.-R.; Kuo, S.-W.; Fu, H.-K.; Chang, F.-C. Enhanced thermal properties of PS nanocomposites formed from montmorillonite treated with a surfactant/cyclodextrin inclusion complex. *Polymer* **2005**, 46, 741–750.
8. Choi, Y.S.; Xu, M.; Chung, I.J. Synthesis of exfoliated poly(styrene-co-acrylonitrile) copolymer/silicate nanocomposite by emulsion polymerization; monomer composition effect on morphology. *Polymer* **2003**, 44, 6989–6994.
9. Dean, D.; Walker, R.; Theodore, M.; Hampton, E.; Nyairo, E. Chemorheology and properties of epoxy/layered silicate nanocomposites. *Polymer* **2004**, 46, 3014–3021.
10. Chen, B.; Liu, J.; Chen, H.; Wu, J. Synthesis of disordered and highly exfoliated epoxy/clay nanocomposites using organoclay with catalytic function via acetone–clay slurry method. *Chem. Mater.* **2004**, 16, 4864–4866.
11. Park, J.H.; Jana, S.C. Mechanism of exfoliation of nanoclay particles in epoxy–clay nanocomposites. *Macromolecules* **2003**, 36, 2758–2768.
12. Koerner, H.; Hampton, E.; Dean, D.; Turgut, Z.; Drummy, L.; Mirau, P.; Vaia, R.A. Generating triaxial reinforced epoxy/montmorillonite nanocomposites with uniaxial magnetic fields. *Chem. Mater.* **2005**, 17, 1990–1996.
13. Ishida, H.; Campbell, S.; Blackwell, J. General approach to nanocomposite preparation. *Chem. Mater.* **2000**, 12, 1260–1267.

14. Jang, B.N.; Wang, D.; Wilkie, C.A. The relationship between the solubility parameter of polymers and the clay-dispersion in polymer–clay nanocomposites and the role of the surfactant. *Macromolecules* **2005**, 38, 6533–6543.

15. Fornes, T.D.; Yoon, P.J.; Hunter, D.L.; Keskkula, H.; Paul, D.R. Effect of organoclay structure on nylon 6 nanocomposite morphology and properties. *Polymer* **2002**, 43, 5915–5933.

16. McAlpine, M.; Hudson, N.E.; Liggat, J.J.; Pethrick, R.A.; Pugh, D.; Rhoney, I. Study of the factors influencing the exfoliation of an organically modified montmorillonite in methyl methacrylate/poly(methyl methacrylate) mixtures. *J. Appl. Polym. Sci.* **2006**, 99, 2614–2626.

17. Zeng, Q.H.; Yu, A.B.; Lu, G.Q.; Standish, R.K. Molecular dynamics simulation of organic–inorganic nanocomposites: layering behavior and interlayer structure of organoclays. *Chem. Mater.* **2003**, 15, 4732–4738.

18. Heinz, H.; Koerner, H.; Anderson, K.L.; Vaia, R.A.; Farmer, B.L. Force field for mica-type silicates and dynamics of octadecylammonium chains grafted to montmorillonite. *Chem. Mater.* **2005**, 17, 5658–5669.

19. Reichert, P.; Nitz, H.; Klinke, S.; Brandsch, R.; Thomann, R.; Mulhaupt, R. Poly(propylene)/organoclay nanocomposite formation: influence of compatibilizer functionality and organoclay modification. *Macromol. Mater. Eng.* **2000**, 275, 8–17.

20. Usuki, A.; Kawasumi, M.; Kojima, Y.; Okada, A.; Kurauchi, T.; Kamigaito, O. Swelling behavior of montmorillonite cation exchanged for ω-amino acids by ε-caprolactam. *J. Mater. Res.* **1993**, 8, 1174–1178.

21. Maiti, P.; Yamada, K.; Okamoto, M.; Ueda, K.; Okamoto, K. New polylactide/layered silicate nanocomposites: role of organoclays. *Chem. Mater.* **2002**, 14, 4654–4661.

22. Ho, D.L.; Briber, R.M.; Glinka, C.J. Characterization of organically modified clays using scattering and microscopy techniques. *Chem. Mater.* **2001**, 13, 1923–1931.

23. Ho, D.L.; Glinka, C.J. Effects of solvent solubility parameters on organoclay dispersions. *Chem. Mater.* **2003**, 15, 1309–1312.

24. Hanley, H.J.M.; Muzny, C.D.; Ho, D.L.; Glinka, C.J. A small-angle neutron scattering study of a commercial organoclay dispersion. *Langmuir* **2003**, 19, 5575–5580.

25. Vaia, R.A.; Liu, W.; Koerner, H. Analysis of small-angle scattering of suspensions of organically modified montmorillonite: implications to phase behavior of polymer nanocomposites. *J. Polym. Sci. B Polym. Phys.* **2003**, 41, 3214–3236.

26. Weimer, M.W.; Chen, H.; Giannelis, E.P.; Sogah, D.Y. Direct synthesis of dispersed nanocomposites by in-situ living free radical polymerization using a silicate-anchored initiator. *J. Am. Chem. Soc.* **1999**, 121, 1615–1616.

27. Fan, X.; Zhou, Q.; Xia, C.; Cristofoli, W.; Mays, J.; Advincula, R. Living anionic surface-initiated polymerization (LASIP) of styrene from clay nanoparticles using surface bound 1,1-diphenylethylene (DPE) initiators. *Langmuir* **2002**, 18, 4511–4518.

28. Zhu, J.; Morgan, A.B.; Lamelas, F.J.; Wilkie, C.A. Fire properties of polystyrene–clay nanocomposites. *Chem. Mater.* **2001**, 13, 3774–3780.

29. Imai, Y.; Nishimura, S.; Abe, E.; Tateyama, H.; Abiko, A.; Yamaguchi, A.; Aoyama, T.; Taguchi, H. High modulus poly(ethylene terephthalate)/expandable fluorine mica nanocomposites with a novel reactive compatiblizer. *Chem. Mater.* **2002**, 14, 477–479.

30. Paul, M.-A.; Alexandre, M.; Degee, P.; Calberg, C.; Jerome, R.; Dubois, P. Exfoliated polylactide/clay nanocomposites by in-situ coordination-insertion polymerization. *Macromol. Rapid Commun.* **2003**, 24, 561–566.
31. Usuki, A.; Kojima, Y.; Kawasumi, M.; Okada, A.; Fukushima, Y.; Kurauchi, T.; Kamigaito, O. Synthesis of nylon 6–clay hybrid. *J. Mater. Res.* **1993**, 8, 1179–1184.
32. Okada, A.; Usuki, A. The chemistry of polymer–clay hybrids. *Mater. Sci. Eng. C* **1995**, 3, 109–115.
33. Xu, W.-B.; Bao, S.-P.; Shen, S.-J.; Hang, G.-P.; He, P.-S. Curing kinetics of epoxy resin–imidazole–organic montmorillonite nanocomposites determined by differential scanning calorimetry. *J. Appl. Polym. Sci.* **2003**, 88, 2932–2941.
34. Ton-That, M.-T.; Ngo, T.-D.; Ding, P.; Fang, G.; Cole, K.C.; Hoa, S.V. Epoxy nanocomposites: analysis and kinetics of cure. *Polym. Eng. Sci.* **2004**, 44, 1132–1141.
35. Xie, W.; Gao, Z.; Pan, W.-P.; Hunter, D.; Singh, A.; Vaia, R. Thermal degradation chemistry of alkyl quaternary ammonium montmorillonite. *Chem. Mater.* **2001**, 13, 2979–2990.
36. He, H.; Ding, Z.; Zhu, J.; Yuan, P.; Xi, Y.; Yang, D.; Frost, R.L. Thermal characterization of surfactant-modified montmorillonites. *Clays Clay Miner.* **2005**, 53, 287–293.
37. Dharaiya, D.; Jana, S.C. Thermal decomposition of alkyl ammonium ions and its effect on surface polarity of organically treated nanoclay. *Polymer* **2005**, 46, 10139–10147.
38. Gelfer, M.; Burger, C.; Fadeev, A.; Sics, I.; Chu, B.; Hsiao, B.S.; Heintz, A.; Kojo, K.; Hsu, S.-L.; Si, M.; Rafailovich, M. Thermally induced phase transitions and morphological changes in organoclays. *Langmuir* **2004**, 20, 3746–3758.
39. Morgan, A.B.; Gilman, J.W.; Jackson, C.L. Characterization of the dispersion of clay in a polyetherimide nanocomposite. *Macromolecules* **2001**, 34, 2735–2738.
40. Bottino, F.A.; Fabbri, E.; Fragala, I.L.; Malandrino, G.; Orestano, A.; Pilati, F.; Pollicino, A. Polystyrene–clay nanocomposites prepared with polymerizable imidazolium surfactants. *Macromol. Rapid Commun.* **2003**, 24, 1079–1084.
41. Koerner, H.; Liu, W.; Alexander, M.; Mirau, P.; Dowty, H.; Vaia, R.A. Deformation–morphology correlations in electrically conductive carbon nanotube–thermoplastic polyurethane nanocomposites. *Polymer* **2005**, 46, 4405–4420.
42. Gelves, G.A.; Sundararaj, U.; Haber, J.A. Electrostatically dissipative polystyrene nanocomposites containing copper nanowires. *Macromol. Rapid Commun.* **2005**, 26, 1677–1681.
43. Mitchell, C.A.; Bahr, J.L.; Arepalli, S.; Tour, J.M.; Krishnamoorti, R. Dispersion of functionalized carbon nanotubes in polystyrene. *Macromolecules* **2002**, 35, 8825–8830.
44. Ryu, J.G.; Kim, H.; Lee, J.W. Characteristics of polystyrene/polyethylene/clay nanocomposites prepared by ultrasound-assisted mixing process. *Polym. Eng. Sci.* **2004**, 44, 1198–1204.
45. Morgan, A.B.; Harris, J.D. Exfoliated polystyrene–clay nanocomposites synthesized by solvent blending with sonication. *Polymer* **2004**, 45, 8695–8703.
46. Nassar, N.; Utracki, L.A.; Kamal, M.R. Melt intercalation in montmorillonite/polystyrene nanocomposites. *Int. Polym. Proc.* **2005**, XX, 423–431.

47. Gilman, J.W.; Jackson, C.L.; Morgan, A.B.; Harris, R.; Manias, E.; Giannelis, E.P.; Wuthenow, M.; Hilton, D.; Phillips, S.H. Flammability properties of polymer–layered silicate nanocomposites: polypropylene and polystyrene nanocomposites. *Chem. Mater.* **2000**, 12, 1866–1873.

48. Bartholmai, M.; Schartel, B. Layered silicate polymer nanocomposites: new approach or illusion for fire retardancy? Investigations of the potentials and the tasks using a model system. *Poly. Adv. Technol.* **2004**, 15, 355–364.

49. Morgan, A.B.; Chu, L.-L.; Harris, J.D. A flammability performance comparison between synthetic and natural clays in polystyrene nanocomposites. *Fire Mater.* **2005**, 29, 213–229.

50. Xie, W.; Xie, R.; Pan, W.-P.; Hunter, D.; Koene, B.; Tan, L.-S.; Vaia, R. Thermal stability of quaternary phosphonium modified montmorillonites. *Chem. Mater.* **2002**, 14, 4837–4845.

51. Davis, R.D.; Gilman, J.W.; Sutto, T.E.; Callahan, J.H.; Trulove, P.C.; De Long, H.C. Improved thermal stability of organically modified layered silicates. *Clays Clay Miner.* **2004**, 52, 171–179.

52. Morgan, A.B.; Harris, J.D. Effects of organoclay Soxhlet extraction on mechanical properties, flammability properties and organoclay dispersion of polypropylene nanocomposites. *Polymer* **2003**, 44, 2313–2320.

53. Awad, W.H.; Gilman, J.W.; Nyden, M.; Harris, R.H.; Sutto, T.E.; Callahan, J.; Trulove, P.C.; DeLong, H.C.; Fox, D.M. Thermal degradation studies of alkyl-imidazolium salts and their application in nanocomposites. *Thermochim. Acta* **2003**, 409, 3–11.

54. Wang, Z.M.; Chung, T.C.; Gilman, J.W.; Manias, E. Melt processable syndiotactic polystyrene/montmorillonite nanocomposites. *J. Polym. Sci. B* **2003**, 41, 3173–3187.

55. Gilman, J.W.; Awad, W.H.; Davis, R.D.; Shields, J.; Harris, R.H., Jr.; Davis, C.; Morgan, A.B.; Sutto, T.E.; Callahan, J.; Trulove, P.C.; DeLong, H.C. Polymer/layered silicate nanocomposites from thermally stable trialkylimidazolium-treated montmorillonite. *Chem. Mater.* **2002**, 14, 3776.

56. Su, S.; Jiang, D.D.; Wilkie, C.A. Novel polymerically-modified clays permit the preparation of intercalated and exfoliated nanocomposites of styrene and its copolymers by melt blending. *Polym. Degrad. Stab.* **2004**, 83, 333–346.

57. Su, S.; Jiang, D.D.; Wilkie, C.A. Poly(methyl methacrylate), polypropylene and polyethylene nanocomposite formation by melt blending using novel polymerically-modified clays. *Polym. Degrad. Stab.* **2004**, 83, 321–331.

58. Su, S.; Jiang, D.D.; Wilkie, C.A. Polybutadiene modified clay and its nanocomposites. *Polym. Degrad. Stab.* **2004**, 84, 279–288.

59. Zhang, J.; Jiang, D.D.; Wilkie, C.A. Polyethylene and polypropylene nanocomposites based upon an oligomerically modified clay. *Thermochim. Acta*, **2005**, 430, 107–113.

60. Zhang, J.; Jiang, D.D.; Wang, D.; Wilkie, C.A. Mechanical and fire properties of styrenic polymer nanocomposites based on an oligomerically modified clay. *Polym. Adv. Technol.*, **2005**, 16, 800–806.

61. Costache, M.C.; Heidecker, M.J.; Manias, E.; Wilkie, C.A. Preparation and characterization of poly(ethylene terephthalate)/clay nanocomposites by melt blending using thermally stable surfactants. *Polym. Adv. Technol.* **2006**, 17, 764–771.

62. Zhang, J.; Jiang, D.D.; Wilkie, C.A. Fire properties of styrenics polymer–clay nanocomposites based on an oligomerically modified clay. *Polym. Degrad. Stab.* **2006**, 91, 358–366.

63. Utracki, L.A.; Sepehr, M.; Li, J. Mely compounding of polymeric nanocomposites. *Int. Polym. Proc.* **2006**, XXI, 3–16.

64. Hotta, S.; Paul, D.R. Nanocomposites formed from linear low density polyethylene and organoclays. *Polymer* **2004**, 45, 7639–7654.

65. Lee, H.-S.; Fasulo, P.D.; Rodgers, W.R.; Paul, D.R. TPO based nanocomposites, 1: Morphology and mechanical properties. *Polymer* **2005**, 46, 11673–11689–.

66. Ton-That, M.-T.; Perrin-Sarazin, F.; Cole, K.C.; Bureau, M.N.; Denault, J. Polyolefin nanocomposites: formulation and development. *Polym. Eng. Sci.* **2004**, 44, 1212–1219.

67. Dennis, H.R.; Hunter, D.L.; Chang, D.; Kim, S.; White, J.L.; Cho, J.W.; Paul, D.R. Effect of melt processing conditions on the extent of exfoliation in organoclay-based nanocomposites. *Polymer* **2001**, 42, 9513–9522.

68. Tanoue, S.; Utracki, L.A.; Garcia-Rejon, A.; Tatibouët, J.; Kamal, M.R. Melt compounding of different grades polystyrene with organoclay, 3: Mechanical properties. *Polym. Eng. Sci.* **2005**, 45, 827–837.

69. Hasegawa, N.; Okamoto, H.; Kato, M.; Usuki, A.; Sato, N. Nylon-6/Na–montmorillonite nanocomposites prepared by compounding nylon-6 with na–montmorillonite slurry. *Polymer* **2003**, 44, 2933–2937.

70. Kato, M.; Matsushita, M.; Fukumori, K. Development of a new production method for a polypropylene–clay nanocomposite. *Polym. Eng. Sci.* **2004**, 44, 1205–1211.

71. Nanosperse LLC. www.nanosperse.com.

72. Klosterman, D., Fritts, A.; Galaska, M.; Gagliardi, N. Commercially scaleable and robust nanocomposite concentrate technology, in: *Nanocomposites 2004 Proceedings*, 4th World Congress, San Francisco, CA, Sept. 1–3, 2004.

73. Morgan, A.B.; Gilman, J.W. Characterization of polymer–layered silicate (clay) nanocomposites by transmission electron microscopy and x-ray diffraction: a comparative study. *J. Appl. Polym. Sci.* **2003**, 87, 1329–1338.

74. Vermogen, A.; Masenelli-Varlot, K.; Seguela, R.; Duchet-Rumeau, J.; Boucard, S.; Prele, P. Evaluation of the structure and dispersion in polymer–layered silicate nanocomposites. *Macromolecules* **2005**, 38, 9661–9669.

75. Perrin-Sarazin, F.; Ton-That, M.-T.; Bureau, M.N.; Denault, J. Micro- and nanostructure in polypropylene/clay nanocomposites. *Polymer* **2005**, 46, 11624–11634.

76. Causin, V.; Marega, C.; Mariog, A.; Ferrara, G. Assessing organo-clay dispersion in polymer layered silicate nanocomposites: a SAXS approach. *Polymer* **2005**, 46, 9533–9537.

77. Vaia, R.A.; Liu, W. X-ray powder diffraction of polymer/layered silicate nanocomposites: model and practice. *J. Polym. Sci. B Polym. Phys.* **2002**, 40, 1590–1600.

78. Drits, V.A.; Tchoubar, C. *X-ray Diffraction by Disordered Lamellar Structures: Theory and Application to Microdivided Silicates and Carbons*. Springer-Verlag; New York, 1990.

79. Bish, D.L.; Post, J.E.; Eds. *Modern Powder Diffraction*. Reviews in Mineralogy, vol. 20. Mineralogical Society of America; Washington, DC, 1989.

80. Roe, R.-J. *Methods of X-ray and Neutron Scattering in Polymer Science*. Oxford University Press; New York, 2000; pp. 155–210.

81. Okamoto, M.; Nam, P.H.; Maiti, P.; Kotaka, T.; Hasegawa, N.; Usuki, A. A house of cards structure in polypropylene/clay nanocomposites under elongational flow. *Nano Lett.* **2001**, 1, 295–298.

82. Koo, C.M.; Kim, S.O.; Chung, I.J. Study on morphology evaluation, orientational behavior, and anisotropic phase formation of highly filled polymer–layered silicate nanocomposites. *Macromolecules* **2003**, 36, 2748–2757.

83. Kim, G.-M.; Lee, D.-H.; Hoffmann, B.; Kressler, J.; Stoppelmann, G. Influence of nanofillers on deformation process in layered silicate/polyamide-12 nanocomposites. *Polymer* **2001**, 42, 1095–1100.

84. Wang, K.; Liang, S.; Du, R.; Zhang, Q.; Fu, Q. The interplay of thermodynamics and shear on the dispersion of polymer nanocomposite. *Polymer* **2004**, 45, 7953–7960.

85. VanderHart, D.L.; Asano, A.; Gilman, J.W. NMR measurements related to clay-dispersion quality and organic-modifier stability in nylon-6/clay nanocomposites. *Macromolecules* **2001**, 34, 3819–3822.

86. VanderHart, D.L.; Asano, A.; Gilman, J.W. Solid-state NMR investigation of paramagnetic nylon-6 clay nanocomposites, 1: Crystallinity, morphology, and the direct influence of $Fe3^{+}$ on nuclear spins. *Chem. Mater.* **2001**, 13, 3781–3795.

87. VanderHart, D.L.; Asano, A.; Gilman, J.W. Solid-state NMR investigation of paramagnetic nylon-6 clay nanocomposites, 2: Measurement of clay dispersion, crystal stratification, and stability of organic modifiers. *Chem. Mater.* **2001**, 13, 3796–3809.

88. Bourbigot, S.; VanderHart, D.L.; Gilman, J.W.; Bellayer, S.; Stretz, H.; Paul, D.R. Solid state NMR characterization and flammability of styrene–acrylonitrile copolymer montmorillonite nanocomposite. *Polymer* **2004**, 45, 7627–7638.

89. Bourbigot, S.; Gilman, J.W.; Vanderhart, D.L.; Awad, W.H.; Davis, R.D.; Morgan, A.B.; Wilkie, C.A. Investigation of nanodispersion in polystyrene–montmorillonite nanocomposites by solid state NMR. *J. Polym. Sci. B Polym. Phys.* **2003**, 41, 3188–3213.

90. Gilman, J.W.; Bourbigot, S.; Shields, J.R.; Nyden, M.; Kashiwagi, T.; Davis, R.D.; Vanderhart, D.L.; Demory, W.; Wilkie, C.A.; Morgan, A.B.; Harris, J.; Lyon, R.E. High throughput methods for polymer nanocomposites research: extrusion, NMR characterization and flammability property screening. *J. Mater. Chem.* **2003**, 38, 4451–4460.

91. Strawhecker, K.E.; Manias, E. AFM of poly(vinyl alcohol) crystals next to an inorganic surface. *Macromolecules* **2001**, 34, 8475–8482.

92. Viville, P.; Lazzaroni, R.; Pollet, E.; Alexandre, M.; Dubois, P.; Borcia, G.; Pireaux, J.-J. Surface characterization of poly(ε-caprolactone)-based nanocomposites. *Langmuir* **2003**, 19, 9425–9433.

93. Yalcin, B.; Cakmak, M. The role of plasticizer on the exfoliation and dispersion and fracture behavior of clay particles in PVC matrix: a comprehensive morphological study. *Polymer* **2004**, 45, 6623–6638.

94. Fan, X.; Xia, C.; Advincula, R.C. Grafting of polymers from clay nanoparticles via in-situ free radical surface-initiated polymerization: monocationic versus bicationic initiators. *Langmuir* **2003**, 19, 4381–4389.

95. Zhu, J.; Start, P.; Mauritz, K.A.; Wilkie, C.A. Silicon-methoxide-modified clays and their polystyrene nanocomposites. *J. Polym. Sci. A Polym. Chem.* **2002**, 40, 1498–1503.

96. Schulz, J.C.; Warr, G.G. Adsorbed layer structure of cationic and anionic surfactants on mineral oxide surfaces. *Langmuir* **2002**, 18, 3191–3197.

97. Sakai, H.; Nakamura, H.; Kozawa, K.; Abe, M. Atomic force microscopy observation of the nanostructure of tetradecyltrimethylammonium bromide films adsorbed at the mica/solution interface. *Langmuir* **2001**, 17, 1817–1820.

98. Maupin, P.H.; Gilman, J.W.; Harris, R.H.; Bellayer, S.; Bur, A.J.; Roth, S.C.; Murariu, M.; Morgan, A.B.; Harris, J.D. Optical probes for monitoring intercalation and exfoliation in melt-processed polymer nanocomposites. *Macromol. Rapid Commun.* **2004**, 25, 788–792.

99. Davis, R.D.; Bur, A.J.; McBrearty, M.; Lee, Y.-H.; Gilman, J.W.; Start, P.R. Dielectric spectroscopy during extrusion processing of polymer nanocomposites: a high throughput processing/characterization method to measure layered silicate content and exfoliation. *Polymer* **2004**, 45, 6487–6493.

100. Noda, N.; Lee, Y.-S.; Bur, A.J.; Prabhu, V.M.; Snyder, C.R.; Roth, S.C.; McBrearty, M. Dielectric properties of nylon 6/clay nanocomposites from on-line process monitoring and off-line measurements. *Polymer* **2005**, 46, 7201–7217.

101. Bur, A.J.; Lee, Y.-S.; Roth, S.C.; Start, P.R. Measuring the extent of exfoliation in polymer–clay nanocomposites using real-time process monitoring methods. *Polymer* **2005**, 46, 10908–10918.

102. Fasulo, P.D.; Rodgers, W.R.; Ottaviani, A.; Hunter, D.L. Extrusion processing of TPO nanocomposites. *Polym. Eng. Sci.* **2004**, 44, 1036–1045.

103. Berta, M.; Lindsay, C.; Pans, G.; Camino, G. Effect of chemical structure on combustion and thermal behaviour of polyurethane elastomer layered silicate nanocomposites. *Polym. Degrad. Stab.* **2006**, 91, 1179–1191.

104. Kodgire, P.; Kalgaonkar, R.; Hambir, S.; Bulakh, N.; Jog, J.P. PP/clay nanocomposites: effect of clay treatment on morphology and dynamic mechanical properties. *J. Appl. Polym. Sci.* **2001**, 81, 1786–1792.

105. Hambir, S.; Bulakh, N.; Kodgire, P.; Kalgaonkar, R.; Jog, J.P. PP/clay nanocomposites: a study of crystallization and dynamic mechanical behavior. *J. Polym. Sci. B.* **2001**, 39, 446–450.

106. Gilman, J.W.; Kashiwagi, T.; Nyden, M.; Brown, J.E.T.; Jackson, C.L.; Lomakin, S.; Giannelis, E.P.; Manias, E. Flammability studies of polymer layered silicate nanocomposites: polyolefin, epoxy and vinyl ester resins, in: S.; Al-Malaika, A.; Golovoy, and C.A. Wilkie, Eds., *Chemistry and Technology of Polymer Additives.* Blackwell Science, Oxford, England, 1999, pp. 249–265.

107. Wagener, R.; Reisinger, T.J.G. A rheological method to compare the degree of exfoliation of nanocomposites. *Polymer* **2003**, 44, 7513–7518.

108. Zhao, J.; Morgan, A.B.; Harris, J.D. Rheological characterization of polystyrene–clay nanocomposites to compare the degree of exfoliation and dispersion. *Polymer* **2005**, 46, 8641–8660.

109. Lee, K.M.; Han, C.D. Effect of hydrogen bonding on the rheology of polycarbonate/organoclay nanocomposites. *Polymer* **2003**, 44, 4573–4588.

110. Hsieh, A.J.; Moy, P.; Beyer, F.L.; Madison, P.; Napadensky, E.; Ren, J.; Krishnamoorti, R. Mechanical response and rheological properties of polycarbonate layered-silicate nanocomposites. *Polym. Eng. Sci.* **2004**, 44, 825–837.

111. Ren, J.; Casanueva, B.F.; Mitchell, C.A.; Krishnamoorti, R. Disorientation kinetics of aligned polymer layered silicate nanocomposites. *Macromolecules* **2003**, 36, 4188–4194.

112. Waste electrical and electronic equipment directive (WEEE), http://europa.eu.int/eur-lex/pri/en/oj/dat/2003/l_037/l_03720030213en00240038.pdf.

113. Simonson, M.; Blomqvist, P.; Boldizar, A.; Möller, K.; Rosell, L.; Tullin, C.; Stripple, H.; Sundqvist, J.O. *Fire-LCA model: TV case study*. SP Report 2000: 13. ISBN 91-7848-811-7. Printed in 2000.

114. Troitzsch, J.H. The globalization of fire testing and its impact on polymers and flame retardants. *Polym. Degrad. Stab.* **2005**, 88, 146–149.

115. Wadehra, I. Designing information technology products with flame retarded plastics with special emphasis on current flammability safety and environmental concerns. *Fire Mater.* **2005**, 29, 121–126.

116. Simonson, M.; Andersson, P.; Bliss, D. Fire performance of selected IT-equipment. *Fire Technol.* **2004**, 40, 27–37.

117. Butler, K.M.; Mullholland, G.W. Generation and transport of smoke components. *Fire Technol.* **2004**, 40, 149–176.

118. Tewarson, A. Flammability, in: J.E. Mark, Ed., *Physical Properties of Polymers Handbook*. AIP Press, New York, 1996, pp. 577–604.

119. Tewarson, A. Non-thermal fire damage. *J. Fire Sci.* **1992**, 10, 188–242.

120. Purser, D. The performance of fire retardants in relation to toxicity, toxic hazard, and risk in fires, in: A.F.; Grand, and C.A. Wilkie, Eds., *Fire Retardancy of Polymeric Materials*. Marcel Dekker New York, 2000, pp. 449–500.

121. Fire at Seven Dials: Video CD. BRE Video/Publishing, London, 2002. http://www.brebookshop.com/details.jsp?id=139812 and http://www.read-eurowire.com/london.cfm.

122. Hall, J.R. Fires involving appliance housings—Is there a clear and present danger? *Fire Technol.* **2002**, 38, 179–198.

123. Hoffmann, J.M.; Hoffmann, D.J.; Kroll, E.C.; Kroll, M.J. Full scale burn tests of television sets and electronic appliances. *Fire Technol.* **2003**, 39, 207–224.

124. Hong, S.; Yang, J.; Anh, S.; Mun, Y.; Lee, G. Flame retardancy performance of various UL94 classified materials exposed to external ignition sources. *Fire Mater.* **2004**, 28, 25–31.

125. Bundy, M.; Ohlemiller, T. *Bench-scale flammability measures for electronic equipment*. NISTIR (National Institute of Standards and Technology Internal Report) 7031. U.S. Department of Commerce, Washington, DC, July 2003.

126. Bundy, M.; Ohlemiller, T. *Full Scale Flammability Measures for Electronic Equipment*. NIST Technical Note 1461. U.S. Department of Commerce, Washington, DC, Aug. 2004.

127. Geran, T.; Ben-Zvi, A.; Schneider, J.; Reznik, G.; Finberg, I.; Georlette, P. Reinforcement of fire safety in the car industry, in: *Proceedings of the 16th Annual Business Communication Conference on Fire Retardancy*, May 2005.

128. Carpenter, K.; Janssens, M.; Sauceda, A. Another look at cone calorimeter predictions of FMVSS 302 performance, in: *Proceedings of Fire and Materials 2005*, San Francisco, CA, Jan. 31–Feb. 1, 2005. Interscience Communications, London, pp. 469–476.

129. Hirschler, M.M. New NFPA proposed guide for identification and development of mitigation strategies for fire hazard to occupants of road vehicles, in: *Proceedings of Fire and Materials 2005*, San Francisco, CA Jan. 31–Feb. 1, 2005. Interscience Communications, London, pp. 457–468.

130. *Guide for Identification and Development of Mitigation Strategies for Fire Hazard to Occupants of Passenger Road Vehicles.* NFPA 556. http://www.nfpa.org/aboutthecodes/AboutTheCodes.asp?DocNum=556&cookie%5Ftest=1.

131. Stimitz, J.S. Properties of plastic materials for use in 42 Volt automotive applications, in: *Proceedings of the 14th Annual Conference on Fire Retardancy*, Stanford, CT, May 2003.

132. Garces, J.M.; Moll, D.J.; Bicerano, J.; Fibiger, R.; McLeod, D.G. Polymeric nanocomposites for automotive applications. *Adv. Mater.* **2000**, 12, 1835–1839.

133. http://www.fire.tc.faa.gov/research/targtare.stm.

134. Lam, C.W.; James, J.T.; McCluskey, R.; Hunter, R.L. Pulmonary toxicity of single-wall carbon nanotubes in mice 7 and 90 days after intratracheal instillation. *Toxicol. Sci.* **2004**, 77, 126–134.

135. Oberdorster, G. Pulmonary effects of inhaled ultrafine particles. *Int. Arch. Occup. Environ. Health* **2001**, 74, 1–8.

136. Oberdorster, G.; Sharp, Z.; Atudorei, V.; Elder, A.; Gelein, R.; Kreyling, W.; Cox, C. Translocation of inhaled ultrafine particles to the brain. *Inhal. Toxicol.* **2004**, 16, 437–445.

137. Taki, K.; Yanagimoto, T.; Funami, E.; Okamoto, M.; Ohshima, M. Visual observation of CO_2 foaming of polypropylene–clay nanocomposites. *Polym. Eng. Sci.* **2004**, 44, 1004–1011.

138. Han, X.; Zeng, C.; Lee, J.; Koelling, K.W.; Tomasko, D.L. Extrusion of polystyrene nanocomposite foams with supercritical CO_2. *Polym. Eng. Sci.* **2003**, 43, 1261–1275.

139. Zeng, C.; Han, X.; Lee, L.J.; Koelling, K.W.; Tomasko, D.L. Polymer–clay nanocomposite foams prepared using carbon dioxide. *Adv. Mater.* **2003**, 15, 1743–1747.

140. Fujimoto, Y.; Ray, S.S.; Okamoto, M.; Ogami, A.; Yamada, K.; Ueda, K. Well-controlled biodegradable nanocomposite foams: from microcellular to nanocellular. *Macromol. Rapid Commun.* **2003**, 24. 457–461.

141. Lee, L.J.; Zeng, C.; Cao, X.; Han, X.; Shen, J.; Xu, G. Polymer nanocomposite foams. *Composites Science and Technology* **2005**, 65, 2344–2364.

142. Morlat, S.; Mailhot, B.; Gonzalez, D.; Gardette, J.-L. Photo-oxidation of polypropylene/montmorillonite nanocomposites, 1: Influence of nanoclay and compatibilizing agent. *Chem. Mater.* **2004**, 16, 377–383.

143. Diagne, M.; Gueye, M.; Vidal, L.; Tidjani, A. Thermal stability and fire retardant performance of photo-oxidized nanocomposites of polypropylene–graft–maleic anhydride/clay. *Polym. Degrad. Stab.* **2005**, 89, 418–426.

144. Morlat-Therias, S.; Mailhot, B.; Gardette, J.-L.; Da Silva, C.; Haidar, B.; Vidal, A. Photooxidation of ethylene–propylene–diene/montmorillonite nanocomposites. *Polym. Degrad. Stab.* **2005**, 90, 78–85.

145. (a) Polymer-clay nanocomposites have been used by Toyota for barrier and under-hood engine parts since the 1990s. See Okada, A.; Fukushima, Y.; Kawasumi, M.; Inagaki, S.; Usuki, A.; Sugiyama, S.; Kurauchi, T.; Kamigaito, O. U.S. Patent 4739007 1988. (b) GM has been using polyolefin nanocomposites for automotive applications since 2001. Recent usage has been in 2004 Hummer H2 and Chevrolet Impala. http://www.schwab-kolb.com/hummer/en/hummer01.htm.

146. Beyer, G. Flame retardant properties of EVA-nanocomposites and improvements by combination of nanofillers with aluminum trihydrate. *Fire Mater.* **2001**, 25, 193–197.

147. Gilman, J.W. Flammability and thermal stability studies of polymer layered-silicate (clay) nanocomposites. *Appl. Clay Sci.* **1999**, 15, 31–49.

148. Du, J.; Wang, D.; Wilkie, C.A.; Wang, J. An XPS investigation of thermal degradation and charring on poly(vinyl chloride)–clay nanocomposites. *Polym. Degrad. Stab.* **2003**, 79, 319–324.

149. Wang, J.; Du, J.; Zhu, J.; Wilkie, C.A. An XPS study of the thermal degradation and flame retardant mechanism of polystyrene–clay nanocomposites. *Polym. Degrad. Stab.* **2002**, 77, 249–252.

150. Gilman, J.W.; Harris, R.H.; Shields, J.R.; Kashiwagi, T.; Morgan, A.B. A study of the flammability reduction mechanism of polystyrene–layered silicate nanocomposite: layered silicate reinforced carbonaceous char. *Polym. Adv. Technol.* In press.

151. Kashiwagi, T.; Du, F.; Winey, K.I.; Groth, K.M.; Shields, J.R.; Bellayer, S.P.; Kim, H.; Douglas, J.F. Flammability properties of polymer nanocomposites with single-walled carbon nanotubes: effects of nanotube dispersion and concentration. *Polymer* **2005**, 46, 471–481.

152. Kashiwagi, T.; Grulke, E.; Hilding, J.; Groth, K.; Harris, R.; Butler, K.; Shields, J.; Kharchenko, S.; Douglas, J. Thermal and flammability properties of polypropylene/carbon nanotube nanocomposites. *Polymer* **2004**, 45, 4227–4239.

153. Kashiwagi, T.; Du, F.; Douglas, J.F.; Winey, K.I.; Harris, R.H.; Shields, J.R. Nano particle networks reduce the flammability of polymer nanocomposites. *Nat. Mater.* **2005**, 4, 928–933.

154. Lewin, M. Unsolved problems and unanswered questions in flame retardance of polymers. *Polym. Degrad. Stab.* **2005**, 88, 13–19.

155. Gilman, J.W.; Kashiwagi, T.; Nyden, M.; Harris, R.H. Flame retardant mechanism of silica, in: *New Flame Retardants Consortium: Final Report.* NISTIR (National Institute of Standards and Technology Internal Report) 6357. U.S. Department of Commerce, Washington, DC, June 1999.

156. Kashiwagi, T.; Gilman, J.W.; Butler, K.M.; Harris, R.H.; Shields, J.R.; Asano, A. Flame retardant mechanism of silica gel/silica. *Fire Mater.* **2000**, 24, 277–289.

157. Zhu, J.; Uhl, F.M.; Morgan, A.B.; Wilkie, C.A. Studies on the mechanism by which the formation of nanocomposites enhances thermal stability. *Chem. Mater.* **2001**, 13, 4649–4654.

158. Zanetti, M.; Camino, G.; Thomann, R.; Mulhaupt, R. Synthesis and thermal behaviour of layered silicate–EVA nanocomposites. *Polymer* **2001**, 42, 4501–4507.

159. Zanetti, M.; Kashiwagi, T.; Falqui, L.; Camino, G. Cone calorimeter combustion and gasification studies of polymer layered silicate nanocomposites. *Chem. Mater.* **2002**, 14, 881–887.

160. Kashiwagi, T.; Harris, R.H.; Zhang, X.; Briber, R.M.; Cipriano, B.H.; Raghavan, S.R.; Awad, W.H.; Shields, J.R. Flame retardant mechanism of polyamide-6 nanocomposites. *Polymer* **2004**, 45, 881–891.

161. Morgan, A.B.; Harris, R.H.; Kashiwagi, T.; Chyall, L.J.; Gilman, J.W. Flammability of polystyrene layered silicate (clay) nanocomposites: carbonaceous char formation. *Fire Mater.* **2002**, 26, 247–253.

162. Gilman, J.W.; Kashiwagi, T.; Morgan, A.B.; Harris, R.H.; Brassell, L.; VanLandingham, M.; Jackson, C.L. *Flammability of Polymer–clay Nanocomposites Consortium: Year One Annual Report*. NISTIR (National Institute of Standards and Technology Internal Report) 6531. U.S. Department of Commerce, Washington, DC, 2000.

163. Morgan, A.B.; Kashiwagi, T.; Harris, R.H.; Campbell, J.R.; Shibayama, K.; Iwasa, K.; Gilman, J.W. Flammability properties of polymer–clay nanocomposites: nylon-6 and polypropylene clay nanocomposites, in: G.L. Nelson and C.A. Wilkie, Eds., *Fire and Polymers: Materials and Solutions for Hazard Prevention.* ACS Symposium Series, Vol. 797. American Chemical Society, Washington, DC, 2001, pp. 9–23.

164. Jang, B.N.; Wilkie, C.A. The thermal degradation of polystyrene nanocomposites. *Polymer* **2005**, 46, 2933–2942.

165. Jang, B.N.; Wilkie, C.A. The effect of clay on the thermal degradation of polyamide 6 in polyamide 6/clay nanocomposites. *Polymer* **2005**, 46, 3264–3274.

166. Costache, M.C.; Jiang, D.D.; Wilkie, C.A. Thermal degradation of ethylene–vinyl acetate copolymer nanocomposites. *Polymer* **2005**, 46, 6947–6958.

167. Jang, B.N.; Jiang, D.D.; Wilkie, C.A. The effects of clay on the thermal degradation behavior of poly(styrene-co-acrylonitirile). *Polymer* **2005**, 46, 9702–9713.

168. Jang, B.N.; Costache, M.; Wilkie, C.A. The relationship between thermal degradation behavior of polymer and the fire retardancy of polymer–clay nanocomposites. *Polymer* **2005**, 46, 10678–10687–.

169. Costache, M.C.; Wang, D.; Heidecker, M.J.; Manias, E.; Wilkie, C.A. The thermal degradation of poly(methyl methacrylate) nanocomposites with montmorillonite, layered double hydroxides and carbon nanotubes. *Polym. Adv. Technol.* **2006**, 17, 272–280.

170. Morgan, A.B.; Bundy, M. Cone calorimeter analysis of UL-94 V rated plastics: qualitative correlations and heat release rate understanding. *Fire and Materials*. In review.

171. Sundstrom, B. Fire hazards and upholstery: fire-growth. http://www.sp.se/fire/Eng/Reaction/furniture.pdf.

172. Hu, Y.; Wang, S.; Ling, Z.; Zhuang, Y.; Chen, Z.; Fan, W. Preparation and combustion properties of flame retardant nylon-6/montmorillonite nanocomposite. *Macromol. Mater. Eng.* **2003**, 288, 272–276.

173. Lichtenham, J.D.; Gilman, J.W. Preceramic additives as fire retardants for plastics U.S. Patent 6362279, Mar. 26, 2002.

174. Bourbigot, S.; Le Bras, M.; Flambard, X.; Rochery, M.; Devaux, E.; Lichtenham, J.D. Polyhedral oligomeric silsesquioxanes: application to flame retardant textiles, in: M. Le Bras, C.A. Wilkie, S. Bourbigot, S. Duquesne, and C. Jama, Eds., *Fire Retardancy of Polymers: New Applications of Mineral Fillers*. Royal Society of Chemistry, London, **2005**, pp. 189–201.

175. Fina, A.; Tabuani, D.; Frache, A.; Boccaleri, E.; Camino, G. Octaisobutyl POSS thermal degradation, in: M. Le Bras, C.A. Wilkie, S. Bourbigot, S. Duquesne, and C. Jama, Eds., *Fire Retardancy of Polymers: New Applications of Mineral Fillers*. Royal Society of Chemistry, London, 2005, pp. 202–220.

176. Fina, A.; Tabuani, D.; Carnaito, F.; Frache, A.; Boccaleri, E.; Camino, G. Polyhedral oligomeric silsesquioxanes (POSS) thermal degradation. *Thermochim. Acta* **2006**, 440, 36–42,

177. Fina, A.; Abbenhuis, H.C.L.; Tabuani, D.; Frache, A.; Camino, G. Polypropylene metal functionalized POSS nanocomposites: a study by thermogravimetric analysis. *Polym. Degrad. Stab.* **2006**, 91, 1064–1070.

178. Kotaki, M.; Wang, K.; Toh, M.L.; Chen, L.; Wong, S.Y.; He, C. Electrically conductive epoxy/clay/vapor grown carbon fiber hybrids. *Macromolecules* **2006**, 39, 908–911.

179. Mishra, S.; Sonawane, S.H.; Singh, R.P.; Bendale, A.; Patil, K. Effect of nano-$Mg(OH)_2$ on the mechanical and flame-retarding properties of polypropylene composites. *J. Appl. Polym. Sci.* **2004**, 94, 116–122.

180. Zhang, Q.; Tian, M.; Wu, Y.; Lin, G.; Zhang, L. Effect of particle size on the properties of $Mg(OH)_2$-filled rubber composites. *J. Appl. Polym. Sci.* **2004**, 94, 2341–2346.

181. Okoshi, M.; Nishizawa, H. Flame retardancy of nanocomposites. *Fire Mater.* **2004**, 28, 423–429.

182. Marosi, G.; Marton, A.; Szep, A.; Csontos, I.; Keszei, S.; Zimonyi, E.; Toth, A.; Almeras, X.; Le Bras, M. Fire retardancy effect of migration in polypropylene nanocomposites induced by modified interlayer. *Polym. Degrad. Stab.* **2003**, 82, 379–385.

183. Szep, A.; Szabo, A.; Toth, N.; Anna, P.; Marosi, G. Role of montmorillonite in flame retardancy of ethylene–vinyl acetate copolymer. *Polym. Degrad. Stab.* **2006**, 91, 593–599.

184. Marosi, G.; Anna, P.; Marton, A.; Bertalan, G.; Bota, A.; Toth, A.; Mohai, M.; Racz, I. Flame-retarded polyolefin systems of controlled interphase. *Polym. Adv. Technol.* **2002**, 13, 1–9.

185. Rice, B.P.; Gibson, T.; Lafdi, K. Development of multifunctional advanced composites using a VGNF enhanced matrix, *International SAMPE Symposium and Exhibition Proceedings*, Long Beach, CA, May 2004.

186. Lafdi, K.; Matzek, M. Carbon nanofibers as a nano-reinforcement for polymeric nanocomposites, presented at the 35th International SAMPE Technical Conference, Dayton, OH, Sept.–Oct. 2003.

187. Utracki, L.A.; Sepehr, M. Private communication.

188. Allan, P.S.; Bevis, M.J.; Zadhoush, A. The development and application of shear controlled orientation technology. *Iranian J. Polym. Sci. Technol.*, **1995**, 4, 50–55.

INDEX

Flame Retardant Polymer Nanocomposites, edited by Alexander B. Morgan and Charles A. Wilkie

www.ingramcontent.com/pod-product-compliance
Lightning Source LLC
LaVergne TN
LVHW011559200926
848484LV00001B/332